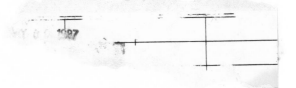

THE MECHANICAL BEHAVIOR OF ELECTROMAGNETIC SOLID CONTINUA

INTERNATIONAL UNION OF THEORETICAL
AND APPLIED MECHANICS

INTERNATIONAL UNION OF PURE
AND APPLIED PHYSICS

NORTH-HOLLAND
AMSTERDAM • NEW YORK • OXFORD

THE MECHANICAL BEHAVIOR OF ELECTROMAGNETIC SOLID CONTINUA

Proceedings of the IUTAM-IUPAP Symposium held in Paris, France 4-7 July, 1983

Edited by
Gérard A. MAUGIN
Laboratoire de Mécanique Théorique Associé au C.N.R.S.
Université Pierre-et-Marie Curie
Paris
France

1984

NORTH-HOLLAND
AMSTERDAM • NEW YORK • OXFORD

ISBN: 0 444 86818 6

Published by:
Elsevier Science Publishers B.V.
P.O. Box 1991
1000 BZ Amsterdam
The Netherlands

Sole distributors for the U.S.A. and Canada:
Elsevier Science Publishing Company, Inc.
52 Vanderbilt Avenue
New York, N.Y. 10017
U.S.A.

PRINTED IN THE NETHERLANDS

EDITOR'S PREFACE

The rapid development of different branches of technology and applied physical sciences in the last decade has revealed the importance of the subject matter of electromagnetic interactions in deformable solids. It is with the aim of providing a forum to stimulate the exchange of ideas among research workers of various backgrounds (mechanical sciences, applied mathematics, solid state physics, electrical engineering) as well as to make informations available to potential users that the Editor proposed in 1980 to the Bureau of the International Union of Theoretical and Applied Mechanics that an IUTAM symposium should be held in Paris in a near future; This symposium should be devoted to the discussion of the various mechanical behaviors of electromagnetic deformable solids and electro-magneto-mechanical interactions, their modelling and their applications to statical and dynamical problems and to engineering devices . Although the International Union of Theoretical and Applied Mechanics faced such a proposal for the first time, a meeting pursuing a rather similar object , but at a more theoretical and speculative level, had been previously organized in Warsaw in 1975 through the initiative of specialists of theoretical mechanics of the Society of Engineering Science (U.S.A.) and the Polish Academy of Sciences . While the present editor enjoyed this first meeting , it was clear to him that any real communication with physicists and electrical engineers was lacking, since these latter were practically absent from the meeting. Therefore, the decision of IUTAM to ask the International Union of Pure and Applied Physics , IUPAP, to join in the sponsorship of the Paris symposium may be considered as a definite step forward to a real dialogue between "mechanicians" and their privileged partners. In such a field of research and development much good will is needed and various approaches, methodologies and techniques can be put together to the benefit of all, on the obvious condition that a minimum of understanding be reached through a common language or an efficient "translation". It is the conviction of the Chairman that this has been achieved at the Paris Symposium , although positive results and cross-disciplinary collaboration will probably take some time to show up and develop to a marked degree. Indeed, out of the eighty five active participants[+] in the symposium from nineteen countries , forty five came from University departments or laboratories of theoretical and applied mechanics, applied mathematics and mathematical physics, twenty nine came from University departments or laboratories of applied physics, solid state physics and electrical engineering and eleven came from laboratories of private companies or government agencies devoted to research and development in various fields of physics, electronics and electrical engineering. For speakers the proportions are thirty five, twenty one and three, respectively, for these three categories. The slight over-representation of the first category is justified both by the fact that the idea of this interdisciplinary symposium was initiated by people from the school of mechanics and by the fact that it is the mechanical behavior of the materials under study which provided the driving force in the discussion. Nonetheless, the concensus may have been large enough since, independently of their origin and initial scientific inte-

---

+ This figure was fixed in agreement with the recommendations for IUTAM symposia. Originally, 250 scientists were contacted through a first circular in 1982.

rests, many participants stayed for a large portion of the programme in spite of the variety of problems and technical jargons, the extremely busy time table and the exceptionally warm weather which should have attracted more participants in the cafés of Paris if it were not for the interest in the lectures.

We hope that this marked interest and conviviality are reflected in the general lectures and invited contributions printed in this volume. With a few exceptions due to occasional changes in the programme, they are here printed in the order in which they were delivered at the symposium with titles of parts added for the publication with a view to helping the reader. These titles reflect essentially the various states of dielectricity, electric conduction and magnetism of the studied materials. Three parts are distinctly important, namely, the acoustoelectricity of dielectric crystals, the magnetoelasticity of magnetically ordered bodies and the magnetoelasticity of electrically conducting structures. This is no mere chance. Nor does it result from an arbitrary choice from the Scientific Committee. Rather, this emphasis reflects the common interest of both theoreticians and engineers for problems which directly materialize in devices and large-scale techniques, the well-known electroacoustic devices in the first case, the intriguing problem of order-disorder transition in the second one and the elaboration of controlled--fusion reactors and magnetically levitated vehicles in the third one. Evidently then, there could be found no better suited local to house the symposium than the ECOLE SUPERIEURE DE PHYSIQUE ET DE CHIMIE INDUSTRIELLES of the City of Paris, with the shadows of Pierre Curie and Paul Langevin following the participants and reminding them constantly of the problems of symmetry and the efforts to apply basic science to useful devices. Marie Slodowska-Curie, Irène, her daughter, and Frédéric Joliot-Curie would have been very pleased to hear about mechanical problems applied to the structure of controlled-fusion reactors, while Marie Curie would have enjoyed the visit of so many Polish scientists. Professor Pierre-Gilles de Gennes, Head of the School, and Professor J.Badoz, its Scientific Director, must be heartfully thanked for the hospitality offered to the symposium. We want to believe that this is more than a mere symbol of cooperation between physicists and "mechanicians", neologism notwithstanding.

The symposium was arranged by members of the Laboratory of Theoretical Mechanics and the Laboratory of Acoustoelectricity of the Pierre and Marie Curie University in Paris. Special thanks go to Professor E.Dieulesaint and his group who were instrumental in the local organization. The financial and technical support from I.U.T.A.M. and I.U.P.A.P., the Centre National de la Recherche Scientifique (Département des Sciences Physiques pour l'Ingénieur), the Commissariat à l'Energie Atomique, Electricité de France , I.B.M.-France, Thomson-C.S.F., Rhône-Poulenc, the Association Universitaire de Mécanique and the United States Army Research, Development and Standardization Group, U.K., is gratefully acknowledged. Thanks are also extended to the speakers, to the members of the International Scientific Committee, to the Chairmen of the sessions, to the members of the Local Organizing Committee and , last but not least, to Mrs. Frémont and Mrs Lazaro who have had the kindness to help us in this enterprise.

Our special thanks belong to North-Holland Publishing Company for the contribution in preparing these proceedings for publication.

Paris, July 17, 1983 Gérard A. Maugin

INTERNATIONAL SCIENTIFIC COMMITTEE APPOINTED BY I.U.T.A.M. AND I.U.P.A.P

G.A. MAUGIN , France, Chairman

J.B. ALBLAS , The Netherlands

S.A. AMBARTSUMIAN, USSR

A.A. MARADUDIN, USA

F.C. MOON , USA

D.F. NELSON, USA

T. TOKUOKA , Japan

H. ZORSKI , Poland

LOCAL ORGANIZING COMMITTEE

E. DIEULESAINT, Université Pierre-et-Marie Curie, Paris

P. GERMAIN , Ecole Polytechnique, Palaiseau

M. KLEMAN, Université de Paris-Sud , Orsay

G.A. MAUGIN, CNRS, Université Pierre-et-Marie Curie, Paris

CHAIRMEN OF THE SESSIONS

S.A. AMBARTSUMIAN

A. ASKAR

J. BAZER

M.A. BREAZEALE

P.J. CHEN

E. DIEULESAINT

A.C. ERINGEN

G.A. MAUGIN

M.F. McCARTHY

F.C. MOON

D.F. NELSON

H.F. TIERSTEN

H. ZORSKI

CONTENTS

+ General lecture.
++ Text not available at the time of printing.

PART 5
PIEZOELECTRIC POWDERS

PART 6
MODELS AND PROBLEMS OF ELECTROELASTICITY

PART 7
MAGNETOELASTICITY OF PARAMAGNETS

PART 8
MAGNETOELASTICITY OF MAGNETICALLY ORDERED BODIES

LIST OF PARTICIPANTS

ALBLAS J.B., (The Netherlands), Department of Mathematics and Computing Science, Eindhoven University of Technology, P.O.Box 513, 5600 MB Eindhoven.

AMBARTSUMIAN S.A., (U.S.S.R), Yerevan State University, Mravian 1, Yerevan, 375049 S.S.R.of Armenia.

ASKAR A., (Turkey), Bogazici Universitesi, Department of Mathematics, Bebek 2, Istanbul.

BAZER J., (U.S.A), Courant Institute of Mathematical Sciences, New York University , 251 Mercer Street, New York, N.Y., 10002.

BILLMANN A., (France), Laboratoire d'Acoustoélectricité, Université Pierre et Marie Curie, 10 rue Vauquelin, 75231 Paris Cedex 05.

BOBROV E.S., (U.S.A), NW14-3209, Francis Bitter National Magnet Laboratory, M.I.T., Cambridge, Mass., 02139.

BONNET J.M., (France), T.R.T., 5 avenue Réaumur, 92350 Le Plessis-Robinson.

BOSSAVIT A., (France), Electricité de France, Etudes et Recherches, 1 avenue du Général de Gaulle, 92141 Clamart.

BOULANGER Ph., (Belgium), Université Libre de Bruxelles, Département de Mathématiques, Campus Plaine, CP 218, 1050 Bruxelles.

BREAZEALE M.A., (U.S.A), Department of Physics, The University of Tennessee , Knoxville, TN 37996-1200.

CAMLEY R., (U.S.A), Department of Physics, University of Colorado, Colorado Spring, CO 80907.

CARABATOS C., (France), Laboratoire de Génie Physique, Université de Metz , Ile du Saulcy, 57045 Metz Cedex.

CARRU M., (France), Centre National d'Etude des Télecommunications, 196 rue de Paris , 92220 Bagneux.

CHEN P.J., (U.S.A), Sandia National Laboratory, Division 1131, Albuquerque , New Mexico , 87185.

COLLET B., (France), Laboratoire de Mécanique Théorique associé au C.N.R.S., Université Pierre et Marie Curie, Tour 66, 4 Place Jussieu, 75230 Paris Cedex 05.

CROWLEY J.M., (U.S.A), Electrical Engineering Department, University of Illinois, 1406 W.Green Street, Urbana, Illinois 61801.

DAHER N., (France), Laboratoire de Mécanique Théorique associé au C.N.R.S., Université Pierre et Marie Curie, Tour 66, 4 Place Jussieu, 75230 Paris Cedex 05.

DEGAUQUE J., (France), I.N.S.A., Département de Physique, avenue de Rangueil, 31077 Toulouse.

DETAINT M., (France), Centre National d'Etude des Télécommunications, 196 rue de Paris, 92220 Bagneux.

DIEULESAINT E., (France), Laboratoire d'Acoustoéléctricité, Université Pierre et Marie Curie, 10 rue Vauquelin, 75231 Paris Cedex 05.

DOST S., (Canada) , Department of Mechanical Engineering, The University of Calgary, Calgary , Alberta ,T2N 1N4.

DUCLOS J., (France), Laboratoire Ultrasons, L.E.A.H., U.E.R. Sciences et Techniques, B.P. 4006, 76077 Le Havre Cedex.

DROUOT R., (France) Laboratoire de Mécanique Théorique associé au C.N.R.S., Université Pierre et Marie Curie, Tour 66, 4 Place Jussieu, 75230 Paris Cedex 05.

ERINGEN A.C., (U.S.A), Princeton University, Department of Civil Engineering, Engineering Quadrangle, Princeton, New Jersey 08544.

ERSOY Y., (Turkey) , Present address: Planetenlaan 1, 5632 DG Eindhoven, The Netherlands.

EYRAUD C., (France) , Laboratoire d'Acoustoélectricité, Université Pierre et Marie Curie, 10 rue Vauquelin, 75231 Paris Cedex 05.

EYRAUD L., (France) , Laboratoire de Génie Electrique et de Ferroélectricité , I.N.S.A., Bât.504, 20 avenue Albert Einstein, 69621 Villeurbanne.

GAGNEPAIN J.J., (France) , L.P.M.O. associé à l'Université de Besançon, 32 avenue de l'Observatoire, 25000 Besançon.

GENG R.S., (United Kingdom), University of London, Chelsea College, Physics Department, Pulton Place, London SW6 5PR.

GERMAIN P., (France), Ecole Polytechnique, Département de Mécanique , 91128 Palaiseau Cedex.

GHALEB A.F., (Egypt), Department of Mathematics, Faculty of Science, Cairo University, Giza, Cairo.

GOUDJO C., (France) , 146 avenue Léon Blum, 92160 Antony.

HAKMI A., (France), Laboratoire de Mécanique Théorique associé au C.N.R.S., Université Pierre et Marie Curie, Tour 66, 4 Place Jussieu, 75230 Paris Cedex 05.

HERPIN S., (France), Laboratoire d'Acoustoélectricité, Université Pierre et Marie Curie, 10 rue Vauquelin, 75231 Paris Cedex 05.

HSIEH R.K.T., (Sweden), Department of Mechanics, Royal Institute of Technology, S-10044 Stockholm.

KACZKOWSKI Z., (Poland), Intitute of Physics, P.A.N., al.Lotnikow 32/46,02-668 Warsaw.

KLEMAN M., (France), Laboratoire de Physique des Solides, Université de Paris-Sud, Orsay, Bât.510, 91405 Orsay Cedex.

KUNIN I.A., (U.S.A), Department of Mechanical Engineering, University of Houston, Houston, TX 77004.

LABBE G., (France), I.R.S.I.D., 185 rue F.D.Roosevelt, 78105 Saint Germain en Laye.

LEDUC M., (France), U.E.R. Sciences et Techniques, place Robert Schuman, B.P. 4006, 76077 Le Havre Cedex.

LENZ J., (West Germany), Institut für Theoretische Mechanik, Universität Karlsruhe, Kaiserstrasse 12, D-7500 Karlsruhe 1.

MARUSZEWSKI B., (Poland), Technical University of Poznan, Institute of Technical Mechanics, ul.Piotrowo 3, 60-965 Poznan.

MATSUMOTO E., (Japan), Department of Aeronautical Engineering, Kyoto University, Kyoto 606.

MAUGIN G.A., (France), Laboratoire de Mécanique Théorique associé au C.N.R.S., Université Pierre et Marie Curie, Tour 66, 4 Place Jussieu, 75230 Paris Cedex 05.

MAYNE G., (Belgium), Université Libre de Bruxelles, Département de Mathématiques, Campus Plaine, CP 218, Boulevard du Triomphe, B-1050 Bruxelles.

McCARTHY M.F., (Ireland), National University of Ireland, University College , Mathematical Physics, Galway.

MEUNIER M., (France), Laboratoire Central, Thomson-CSF, Domaine de Corbeville, B.P. 10, 91401 Orsay Cedex.

MILTAT J., (France), Laboratoire de Physique des Solides, Université de Paris-Sud Orsay, Bât.510, 91405 Orsay Cedex.

MIYA K., (Japan), Nuclear Engineering Research Laboratory, The Faculty of Engineering, University of Tokyo, Tokai-Mura, Ibaraki 319-11.

MOON F.C., (U.S.A), Theoretical and Applied Mechanics, 206 Thurston Hall , Cornell University , Ithaca, New York 14853.

MORGENTHALER F.R., (U.S.A), Massachussetts Institute of Technology, Department of Electrical Engineering and Computing Science, Room 13-3102, Cambridge, Massachussetts 02139.

MORRO A., (Italy), Istituto di Matematica, Via L.Alberti 4, 16132 Genova .

MOTOGI S., (France, Japan), Laboratoire de Mécanique Théorique associé au C.N.R.S., Université Pierre et Marie Curie, Tour 66, 4 Place Jussieu, 75230 Paris Cedex 05 [on leave from : Osaka Municipal Technical Research Institute, 1-6-50, Morinomiya, Joto-ku, Osaka, Japan].

MULLER P., (France) ,Laboratoire de Mécanique Théorique associé au C.N.R.S., Université Pierre et Marie Curie,Tour 66, 4 Place Jussieu, 75230 Paris Cedex 05.

NAGHDI P.M., (U.S.A), Department of Mechanical Engineering, University of California, Berkeley, California 94720.

NELSON D.F., (U.S.A), Bell Laboratories, IC-332, 600 Mountain Avenue , Murray Hill, New Jersey 07974.

NOWACKI J.P., (Poland), I.P.P.T.-P.A.N., Swietokrzyska 21, 00-049 Warsaw.

OTWINOWSKI M., (Poland), Institute of Molecular Physics, P.A.N., ul.Smoluchowskiego 17/19, 60-179 Poznan.

PARTON V.Z., (U.S.S.R), Institute of Chemical Machine Engineering, Department of Mathematics, K.Marx 21/4, 107884 Moscow.

PESQUE P., (France), Laboratoire d'Electronique et de Physique, 3 avenue Descartes, 94450 Limeil-Brévannes.

PLANAT M., (France), L.P.M.O. associé à l'Université de Besançon, 32 avenue de l'Observatoire, 25000 Besançon.

POIREE B., (France), D.R.E.T./SDR/G 63, 26 boulevard Victor,75996 Paris Armées.

POUGET J., (France), Laboratoire de Mécanique Théorique associé au C.N.R.S., Université Pierre et Marie Curie, Tour 66, 4 Place Jussieu, 75230 Paris Cedex 05.

PRECHTL A., (Austria), Techn.Universität Wien, Inst. für Elekt.Maschinen , Gusshausstrasse 25, A-1040 Wien.

ROGULA D., (Poland), I.P.P.T.-P.A.N., Swietokrzyska 21, 00-049 Warsaw.

ROSEAU M., (France), Laboratoire de Mécanique Théorique associé au C.N.R.S., Université Pierre et Marie Curie, Tour 66, 4 Place Jussieu,75230 Paris Cedex 05.

ROYER D., (France), Laboratoire d'Acoustoélectricité, Université Pierre et Marie Curie, 10 rue Vauquelin, 75231 Paris Cedex 05.

SEGALINI S., (France), I.R.S.I.D., 185 rue F.D.Roosevelt, 78105 Saint Germain en Laye.

SINHA K.P., (India), Department of Physics, Indian Institute of Science,Bangalore, 560012.

SIOKE-RAINALDY J., (République Centrafricaine), Faculté des Sciences, B.P.809, Bangui.

SZPUNAR J., (United Kingdom), University of Durham, Department of Physics, Science Laboratories, South Road, Durham DH1 3LE.

TANI J., (Japan), Institute of High Speed Mechanics, Tohoku University, Katahira 2-1-1 , Sendai 980.

TIERSTEN H.F., (U.S.A), Department of Mechanical Engineering, Rensselaer Polytechnic Institute, Troy, New York 12181.

TOURATIER M., (France), Ecole Nationale Supérieure des Arts et Métiers, Laboratoire de Structures, 151 boulevard de l'Hôpital,75640 Paris Cedex 13.

TSYPKIN A.G.,(U.S.S.R), Steklov Institute of Mathematics, Vavilova Avenue,Moscow.

TUROV A.E., (U.S.S.R), Institute of Metal Physics, GSP-170, S.Kovalevskoi 18, Sverdlovsk 620219.

VAN DE VEN A.A.F., (The Netherlands), Department of Mathematics and Computing Science, Technological University of Eindhoven, P.O.Box 513,MB 5600 Eindhoven.

YUSHIN N.K., (U.S.S.R), A.F.Ioffe Physical Technical Institute of the Academy of Science of the U.S.S.R, Politkhnicheskaya 26 , 194021 Leningrad K-21.

ZHAO C.S., (People's Republic of China), Present address: Ecole Nationale Supérieure des Arts et Métiers, Laboratoire de Structures, 151 boulevard de l'Hôpital, 75640 Paris Cedex 05.

ZORSKI H., (Poland) , I.P.P.T.- P.A.N., Swietokrzyska 21, 00-049 Warsaw.

In addition, the following scientists had their contribution accepted for presentation at the Symposium but they were unable to attend:

BAGDASARIAN G.E., (U.S.S.R),Institute of Mechanics, Armenian Academy of Sciences, Magnetoelasticity Department, Barekamutian 24B, 375019 Yerevan, ZSR of Armenia.

BELUBEKIAN M.V., (U.S.S.R), Institute of Mechanics, Armenian Academy of Sciences, Barekamutian 24B, 375019 Yerevan, SSR of Armenia.

KUNIGELIS V., (U.S.S.R),University of Vilnius, Suderves 7-126, 232044 Vilnius, Lithuanian SSR.

OVAKIMIAN R.N.,(U.S.S.R), Mechanics Institute of the Armenian Academy of Sciences, Barekamutian 24B, 375019 Yerevan , SSR of Armenia.

SELEZOV. I.T., (U.S.S.R), Wave Process Hydrodynamics Department, Institute of Hydromechanics of the Academy of Science of the Ukrainian S.S.R., Geliabov 8/4, 252057 Kiev.

Co-authors of contributions not present at the Symposium

ASTIE B., (France), I.N.S.A., Département de Physique, avenue de Rangueil, 3I077 Toulouse.

BASSIOUNY E., (Egypt), Department of Mathematics, Faculty of Education, Cairo University, Fayum, Cairo.

EPSTEIN M., (Canada), Department of Mechanical Engineering, The University of Calgary, Calgary, Alberta T2N 1N4.

FONTANA M.D., (France), Laboratoire de Génie Physique, Université de Metz , Ile du Saulcy, 57045 Metz Cedex.

FULDE P., (West Germany), Max-Plank-Institüt für Festkörperforschung, Heisenbergstrasse 1, Postfach 80-06-65, 7000 Stuttgart 80.

GODZE S., (Canada), Department of Mechanical Engineering, The University of Calgary, Calgary, Alberta T2N 1N4.

GREEN A.E., (United Kingdom), Mathematical Institute, University of Oxford , Oxford OX1 3PH.

KAFHAGY A.H., (United Kingdom), University of London, Chelsea College, Physics Department, Pulton Place, London SW6 5PR.

KUGEL G.E., (France), Laboratoire de Génie Physique, Université de Metz , Ile du Saulcy , 57045 Metz Cedex.

LATIMER P.J., (U.S.A), Department of Physics, The University of Tennessee, Knoxville, TN 37996-1200.

OTOMO K., (Japan), Institute of High Speed Mechanics, Tohoku University ,Katahira 2-1-1 , Sendai 980.

RAJAB A.B., (United Kingdom), University of London, Chelsea College, Physics Department, Pulton Place, London SW6 5PR.

SMOLENSKY G.A., (U.S.S.R), A.F.Ioffe Physical Technical Institute of the Academy of Science of the U.S.S.R., Politekhnicheskaya 26, 194021 Leningrad K-21.

STEPHENS R.W.B., (United Kingdom), University of London, Chelsea College, Physics Department, Pulton Place, London, SW6 5PR.

Part 1

ACOUSTOELECTRICITY OF PIEZOELECTRIC DIELECTRIC CRYSTALS

The Mechanical Behavior of Electromagnetic Solid Continua
G.A. Maugin (editor)
Elsevier Science Publishers B.V. (North-Holland)

RECENT DEVELOPMENTS IN ACOUSTOELECTRICITY

Eugène DIEULESAINT and Daniel ROYER

Laboratoire d'Acoustoélectricité
Université Pierre et Marie Curie
10, rue Vauquelin, 75231 Paris Cedex 05
France

About twelve years ago, the elastic waves brought an elegant solution to the problem of Radar pulse compression. Since then, their applications did not cease to develop. They gave rise to components implementing various functions : delay, generation of analog and numerical codes, matched filtering, bandpass filtering, spectral analysis, convolution.

Technological improvements extend regularly the characteristics of these compact and reliable components and the use of new principles enlarges their domain of applications. The center frequencies exploitable exceed 1000 MHz : The wavelength is then of the order of a micron as the dimensions of the parts determined by this wavelength. They are employed not only in professional electronics (telecommunications, instrumentation) but also in consumer electronics (television). They also play a role in optoelectronics : for instance, acoustooptic modulators are used to restore pictures at the end of transmission lines.

The object of this paper is, after recalling the main principles, to describe the recent developments and the studies in progress of bulk and surface acoustic wave devices especially in the evoked domain of signal processing.

HISTORICAL SUMMARY.

Before talking about recent developments, it seems useful to us to recall the previous history. Table 1 gives a summary of the evolution of acoustoelectricity from its beginnings. Without going into details we can formulate the following remarks.

. There are one hundred years since piezoelectricity was discovered by Pierre and Jacques Curie. About thirty years separate this discovery from its first prominent application : SONAR by Paul Langevin and his coworkers.

. The contribution of electronics, still in its infancy, to the development of acoustics (this term includes the use of all elastic waves whatever their frequency) was essential : without amplifiers, P. Langevin would not have succeeded, to generate acoustic waves in water and to detect a target from their reflected echoes. Acoustics was soon after grateful to electronics : the introduction of quartz resonators into electronic circuits, about 1920, conferred a wonderful stability to oscillators and filters.

. The diffraction of light by acoustic waves was predicted in 1922 by L. Brillouin and experimentally performed in 1932. It gave rise to applications (acoustooptic modulator) only about forty years later, after the laser had become a common optical source.

. As regards the generation and detection of acoustic waves, the piezoelectric effect has supplanted the magnetostrictive effect in most applications after the appearance of piezoelectric ceramics in 1950.

. Eighty years have passed since a convenient and simple technological method (comb-shaped electrodes photoetched on a piezoelectric substrate) was found for exciting Rayleigh waves. The swift implementation of this method brought an elegant solution to the problem of extracting Radar signals from noise (matched filters) and opened the way to the realization of other functions for signal processing. The main ones, implemented since about 1968, are mentioned at the end of table 1.

Table 1 - Main facts about acoustoelectricity.

Year	Facts	Applications
1880	Discovery of piezoelectricity (P. and J. Curie)	Quartz balance
1885	Surface Acoustic Waves (SAW) (Rayleigh)	Seismology
1916	Generation of ultrasounds in water by piezoelectricity (C. Chilowsky and P. Langevin)	SONAR
1920	Electric equivalent circuit of an electromechanical resonator (A. M. Nicholson, W. G. Cady)	Oscillators, filters
1932	Light diffraction by elastic waves (predicted in 1922 by L. Brillouin) Exper. by Lucas-Biquard, Debye-Sears.	Velocity measurements
1940	Low frequency delay lines (water, mercury,silica)	Storage of radar signals.
1950	Piezoelectric ceramics (PZT)	L.F. transducers, SONAR
1960	Piezoelectric thin layers (ZnO)	H.F. transducers and delay lines.
1965	SAW Comb-shaped transducer (Mortley, R.M. White)	
1968	SAW matched filters	Signal Processing (Radar)
.	Multiple output delay lines	Telecommunications
.	Progressive SAW filters	(Television)
.	Bulk wave monolithic filters	Instrumentation
.	Bulk wave acoustooptic modulator	
.	Multistrip coupler (F.G. Marshall, E.G. Paige)	
.	SAW resonator	
.	SAW convolver	
.	Acoustic microscope (C.F. Quate, R.A. Lemon)	
.	Spectrum analysers	

After these short comments on the history of acoustoelectricity, we propose to mention the reasons for the interest of electronics specialists in acoustic waves, to recall the piezoelectric techniques for converting an electrical signal into an acoustic one and their limits, to explain the principles of operation of the devices evoked and to give examples of their implementation.

INTEREST OF ACOUSTIC WAVES FOR ELECTRONICS SPECIALISTS.

Acoustic waves are matter waves propagating in the bulk or on the surface of materials. Why are electronics specialists interested in theses waves since they usually care about electromagnetic waves which can propagate without any support or substrate ? The answer is that elastic waves propagating in solids lend themselves very well to the processing of electronic signals which need to be coded, delayed, filtered, time-compressed ... for the following reasons:

a - The velocity of acoustic waves is very much lower (about 10^5 times lower) than the velocity of electromagnetic waves. Of course, it depends on the material ; however, roughly speaking, this velocity is in the range 1000-10 000 m/s. This property may be expressed by the formulae :

$$\tau(\mu s)/l(cm) \simeq 1 \text{ to } 10 \quad (1)$$
$$f(GHz) \times \lambda(\mu m) \simeq 1 \text{ to } 10 \quad (2)$$

τ : delay, l : path length, f : frequency, λ wavelength

An information of duration 1μs which expands along 300 meters in the atmosphere in the form of electromagnetic waves, once converted into elastic waves, is reduced to a few millimeters. It can be more easily processed. For instance, it is delayed (formula 1) by a few μs after propagating the length of a 1 cm path.

b - Wavelengths of the order of light wavelengths are obtained with elastic waves of comparatively low frequencies (formula 2 : $\lambda \simeq$ a few μm for f = 1 GHz). Consequently, on the one hand, microstructures (for example, integrated circuits and their different layers) can be examined with acoustic waves (acoustic microscope). On the other hand, acoustic gratings able to diffract light can be implemented. Thus, acoustic waves can modify significantly the parameters of light beams, especially laser beams also used by the electronics specialists to carry their signals.

c - The attenuation of acoustic waves in solids, particularly in crystals, is weak. Bulk-wave resonators (plates or bars) or surface wave resonators have Q factors much greater than those of purely electric circuits. Thus, a vibrating crystal constitutes an excellent filter element.

d - Acoustic Rayleigh waves, naturally guided by a free surface, are accessible. This property makes possible a direct processing of the information that they convey. Since they propagate in a material thickness of about one wavelength (i.e a few tens of μm for a frequency of 100 MHz) they lend themselves to the design of planar devices compatible with microelectronics.

e - It is comparatively easy to produce, with bulk or surface guided waves, the dispersion effects necessary to the time compression of a signal.

Nevertheless the exploitation of these seductive properties presumes that the problem of converting electrical signals into acoustic wave trains has been solved.

CONVERSION OF AN ELECTRIC SIGNAL INTO AN ACOUSTIC SIGNAL BY PIEZOELECTRIC EFFECT.

In most applications, electric signals are converted into bulk waves or Rayleigh surface waves.

- Bulk wave transducer

The conversion of an electric voltage into a longitudinal or shear wave is simple in principle. It suffices to cut a plate in a piezoelectric material so that the electric field of the signal applied perpendicularly to the largest faces causes the desired strain : thickness variation or shear displacement, and to bond this plate to the solid where the acoustic wave has to propagate. The cut of the plate can be chosen from the material slowness surface as shown in figure 1.a for lithium niobate.

As a matter of fact, difficulties arise as soon as the frequency of the signal reaches a few hundreds of megahertz. Indeed, the thickness e of the plate is a fraction of the wavelength - $e \simeq \lambda/2$ or $e \simeq \lambda/4$ according as the acoustic impedance of the solid loading the plate is small or large as compared to the acoustic impedance of the piezoelectric material - For Y + 36° cut lithium niobate suitable to launch longitudinal waves : $\lambda \simeq 7$ μm for $f \simeq 1000$ MHz. As a result, the plate has to be designed in the form of a layer.

The deposition of thin piezoelectric layers has been, and still is, the object of numerous studies. Among the materials tested, let us note cadmium sulfide (CdS), aluminium nitride (AlN), selenium (Se), lithium niobate ($LiNbO_3$), zinc oxyde (ZnO). The last material now is the only one used. The thin layer is deposited by diode or triode sputtering. A major step was accomplished a few years ago when Japanese scientists [1] succeeded in increasing the rate of deposition by about 10 times by creating a magnetic field near the target.

However, this technique gives only rise to the growth of layers with the six fold axis of symmetry perpendicular to the substrate (ZnO belongs to the symmetry class 6mm). Consequently, only longitudinal acoustic waves can be launched. The attempts for obtaining layers with the Z axis in the plane of the layer have not yet been successful. Moreover, since this technique is regularly the subject of papers in different symposia, it cannot be considered to be fully mastered. All the same, thousands of delay lines provided with ZnO transducers are made every year. Their frequency range is 500-5000 MHz. The time delay depends on the length of the solid which is usually a single crystal of Al_2O_3 (sapphire, V_L= 11 100m/s, low loss material: 1 dB/cm for f = 2 GHz). Figure 1.b represents a multiple reflection structure suitable for a delay of 100 μs.

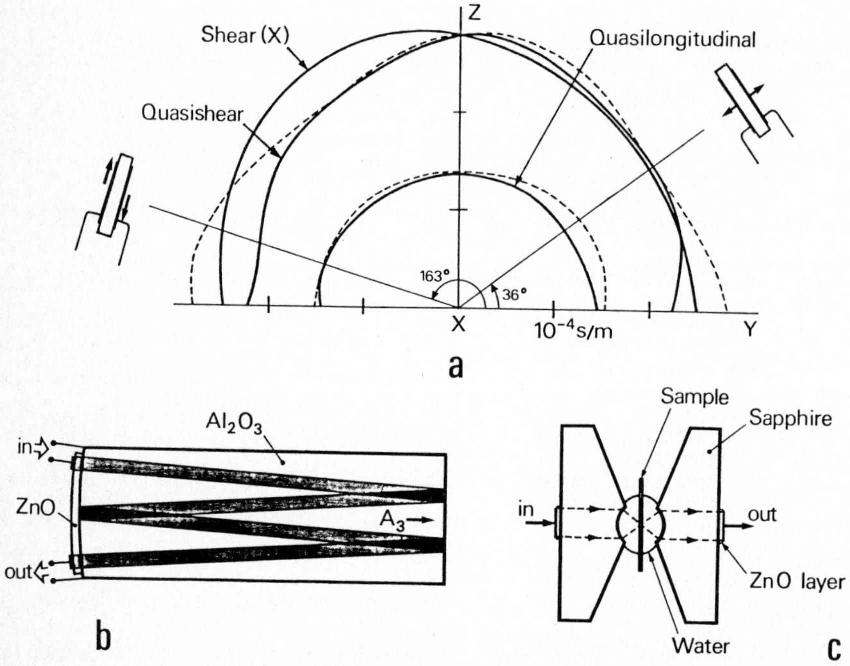

Figure 1 - a) YZ cross-section of $LiNbO_3$ slowness surface (dotted curves ignore piezoelectricity). 36° and 163° rotated Y cut plates are used to excite respectively longitudinal and shear waves. b) Multiple reflection delay line (τ>20μs). c) Schematic of acoustic microscope.

Let us mention here that the ZnO layer transducer is also the source of the waves in the acoustic microscope [2]. This new tool, the principle of which is recalled in figure 1.c, has the advantage over the optical microscope, as far as electronics specialists are concerned, that it allows the observation of the opaque sublayers of integrated circuits.

Surface wave transducer.

The term surface acoustic waves (SAW) here denotes only Rayleigh waves since we do not intend to talk about shear surface waves such as Bleustein-Gulyaev or Love waves which are not much used. The Rayleigh waves propagating in a piezoelectric substrate give rise to many applications. Their success stems from the transducer consisting of a pair of comb-shaped electrodes. In their simplest form these electrodes have identical fingers of same length, width and period (Figure 2).

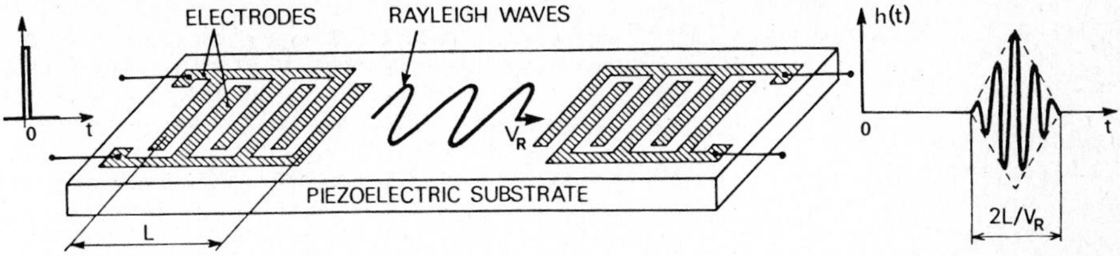

Figure 2 - Generation and detection of Rayleigh waves by interdigital electrodes. Impulse response h(t) of a delay line having two identical transducers.

The application of a short electric pulse between the electrodes generates an acoustic wave train having a rectangular envelop and a duration L/V_R. If this wave train is detected by two electrodes identical to the previous ones, the output signal is an electric wave train with a triangular envelop and a duration $2L/V_R$. It is the autocorrelation of the impulse response of either transducer Thus, the frequency response of a transducer is a $(\sin x)/x$ curve, that of the two transducers is a $[(\sin x)/x]^2$ curve. These results are also found when considering every pair of electrodes as equivalent to a set of alternate sign delta sources (or receivers).

In reality, this intuitive presentation leaves aside several problems such as the following ones :

- The input transducer transmits waves along two opposite directions
- Only a part of these waves is Rayleigh waves. The other part is bulk waves radiated toward the core of the solid.
- The Rayleigh wave velocity is slightly different according as the surface is or is not metallized.
- The fingers of either transducer cause reflexions and give rise to a partial conversion into bulk waves.
- The output transducer not only reflects but also transmits waves. A consequence of these phenomena is that the useful signal is followed by others, for example, the triple transit echo.

Theoretical analyses cannot take all these effects into account insomuch as the piezoelectric solid is anisotropic. The remedies studied for reducing these parasitic effects are, above all, technological. For example, the following modifications are adopted when it is necessary.

- In order to make it unidirectional, the input transducer is divided into several elements, three for instance, and these elements are fed through phasers in such a way that the waves they transmit interfere constructively only along the right direction.
- The fingers are split into two parts so that the inevitable reflexions take place at a frequency outside the useful band.
- The nature and cut of the materials are chosen whenever possible (there are other contraints imposed by specifications relating to temperature variation effects) for reducing the bulk wave transmission.

There are other specific solutions. In the next section one will be explained. Of course, the structure in Figure 2 constitutes a delay line. Let us summarize the main differences between the SAW delay lines and the bulk wave ones.

- The SAW frequency domain is about 50-1000 MHz. The finger width and interval are $\lambda/4$. Since the Rayleigh wave velocity is less than the bulk wave velocity the highest frequency is usually less than 1000 MHz : for ST cut quartz (X propagation), $V_R = 3200$ m/s $\rightarrow \lambda/4 = 0{,}8$ µm.

- Several output SAW transducers can be used. Thus multiple output delay lines can be built.
- Many SAW delay lines can be photoetched simultaneously on the same substrate.

FILTERING.

Signal filtering is based upon two principles : resonance (stationary waves), interference (progressive waves). Only the first is exploited with bulk waves, both are exploited with SAW.

Stationary wave filters.

We will not insist much on bulk wave resonators. They are known : they consist of piezoelectric plates (or bars) appropriately cut. Nevertheless, they are still the subject of researches. Let us indicate two recent improvements.

The first,which increases the time stability of the frequency,relates to the suspension and excitation of the resonator. This is made of a central active part connected by bridges with a ring (Figure 3). Only the central part, placed between two electrodes which do not touch it, can vibrate. The absence of contact between electrodes and resonating part prevents any diffusion of metal into the quartz. Thus the resonator does not suffer from the aging provoked by the diffusion which takes place in the long run when the crystal faces are metallized [3]

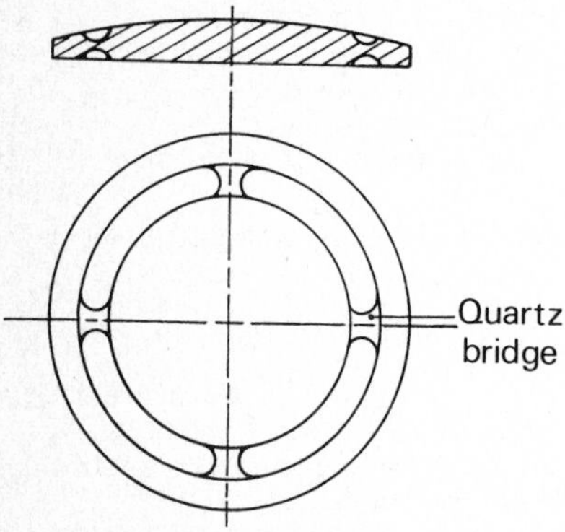

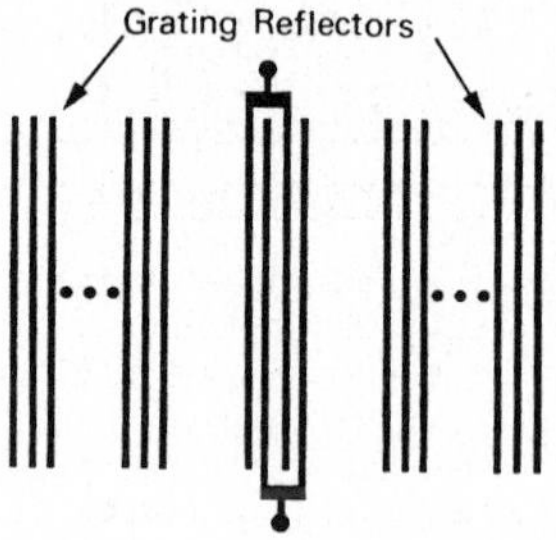

Figure 4 - SAW resonator.

Figure 3 - Resonator without metallic contact. The active part is the central disk. After [3].

The second improvement allows operation at higher frequencies. It consists in diminishing locally the thickness of the plates by ion etching. Resonators operating at a fondamental frequency of 500 MHz have been made. The thickness of a shear mode quartz resonator (AT cut) is 6.1 μm at 270 MHz [4].

Another technique for implementing high frequency resonators is under study. The matter is to deposit a very thin ZnO piezoelectric layer on a few μm thick silicon membrane.

Obviously a broad bandpass filter is made of several resonators appropriately coupled. The coupling can be external with discrete circuit element or internal by evanescent waves. In the latter case, the resonators have to be on the same substrate (monolithic filters).

The bulk wave resonator is more than sixty years old. The SAW resonator is only hardly ten years old.It comprises essentially two groove or metallic arrays which reflect the Rayleigh waves launched by a transducer located between the two arrays (Figure 4). A stationary regime builds in by itself at a frequency depending on the period and the spacing of the array-mirrors. Though the Q factors of these cavities (at the utmost 30 000) are less than those of bulk wave resonators, they present the advantages to operate easily at frequencies of several hundreds of MHz, to be not much affected by the mechanical mounting of the crystal, to authorize various associations, cascading, finger length or groove depth weighting. SAW resonators are now inserted into measurement instruments (for example f = 280 MHz → [5]).

Progressive wave filters.

All the wave interference filters use Rayleigh waves. Their operation lies in the fact that the impulse response of a SAW transducer is determined by the structure of the electrodes (figure 2 is only a particular case of figure 5) : each pair of fingers can be considered as equivalent to a source, the amplitude of which is defined by the finger overlapping and the instantaneous frequency by the interdigital interval.

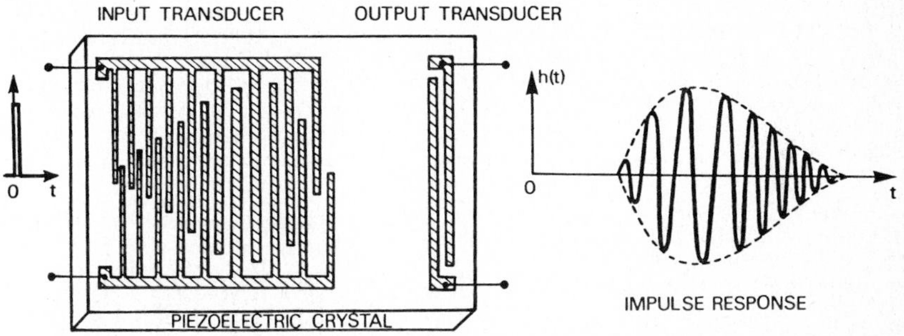

Figure 5 - The shape of the impulse response is defined by electrode finger overlapping and spacing.

For instance, a (sin x)/x shaped impulse response i.e., a rectangular frequency response (bandpass filter with central frequency f_0), is obtained with a constant interdigital interval transducer such as $d = V_R/2f_0$, the finger overlapping of which is proportional to $|(\sin x)/x|$. The receiver transducer of the filter, necessary to find again an electric voltage at the output, consists, in the simplest case, of a few fingers. The receiver finger length have to be at least equal to the largest transmitter source overlapping.

Numerous filters have been implemented according to this principle with improvements (compensation by inactive fingers of the phase shift caused by the uneven metallization) and various variations (filter with the transmitter transducer placed in between two receivers, weighting by transmitter and receiver). The most spectacular application in the consumer domain is the intermediary frequency TV filter.

A very ingenious device : the *multistrip coupler* [6] is often used in filter implementations and it is useful to describe its main properties.

A multistrip coupler is a set of metallic strips which couple two tracks A and B on a piezoelectric material (Figure 6.a). A Rayleigh wave launched in the track A induces an electric voltage between the strips. This voltage appearing also in track B generates a Rayleigh wave. The transfer of a track to the other is complete for a coupler length L_t : it is approximately determined by decomposing the signal which penetrates in A into a symmetric mode and an antisymetric mode propagating with different velocities. This transfer property is exploited on the one hand, to eliminate the bulk waves produced by a transducer placed in track A, the receiver being located in track B (the bulk waves go straight and cannot reach the receiver in B) and, on the other hand, to get in track B a uniform beam, whatever the dimensions of the sources in track A ; thus the diffraction effects of the small dimension sources are eliminated and the receiver can also assure a weighting (by variation of finger overlapping) since it is uniformly "illuminated" by the transmitter.

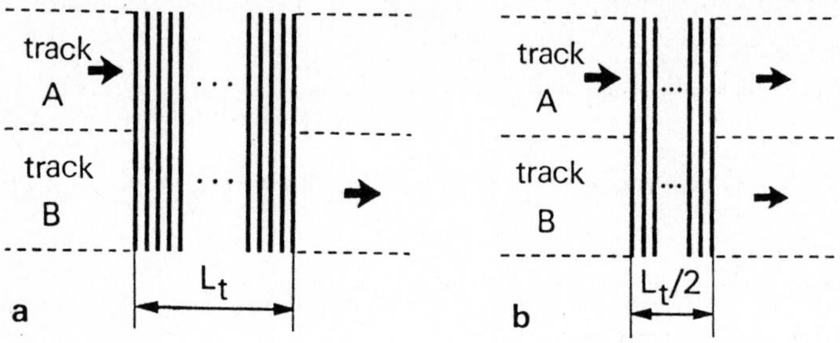

Figure 6 - Multistrip coupler (MSC) *a* - Complete transfert of a SAW beam from track A to track B. *b* - Splitting of a SAW beam into two parts.

The transfer is only partial if the coupler length is $L_t/2$: the beam penetrating in A comes out half by the track A, half by the track B. The B output is $\pi/2$ phase shifted with respect to the A output (Figure 6.b). This property is useful to get rid of echoes produced by reflection on the output transducer [7]. It suffices to add on the other track a second output transducer similar to the first one. The signals reflected by those two transducers come back towards the coupler with equal amplitudes and still $\pi/2$ phase shifted. The coupler recombines them into a unique signal coming out in B_1 where it is absorbed.

The partial transfer multistrip coupler is suitable to make high rejection filters. The configuration is presented in Figure 7. The geometry of the input A_1 and output B_2 transducers is chosen in order to get the desired passband shape. The coupler plays the three roles mentioned.

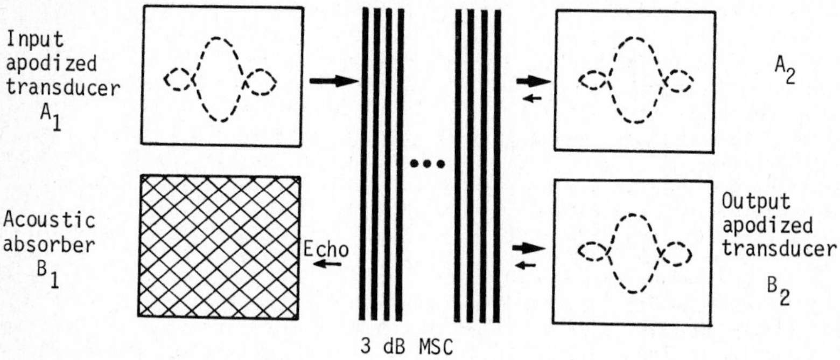

Figure 7 - MSC echo-trap. After [7].

Figure 8 is a remarkable example [8] of a frequency response obtained with this structure (except for a modification that we cannot give here).

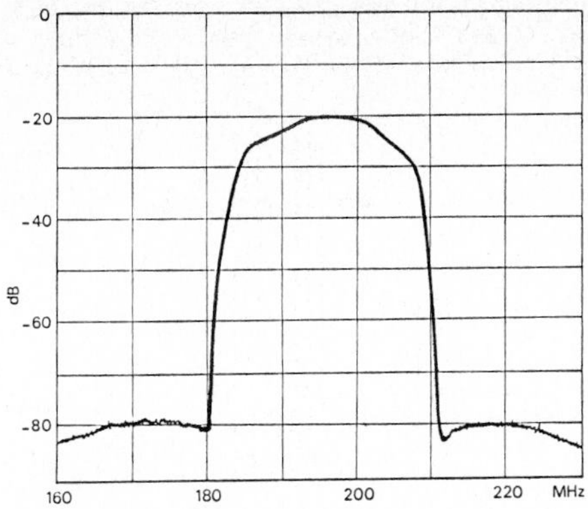

Figure 8 - Frequency response of a SAW-MSC bandpass filter. After [8].

CODING AND DECODING.

The electronics specialists and especially the radar experts often meet the difficulty of extracting a signal from noise. The difficulty is less when the signal, before being deteriorated by the noise, has been coded ; it can be more easily recognized if it traversed a decoding filter which is matched to it. This filter which performs the autocorrelation of the signal, provides a peak which raises above the noise level. It happens that the Rayleigh waves are suitable for generating coded signals and detecting them [9].

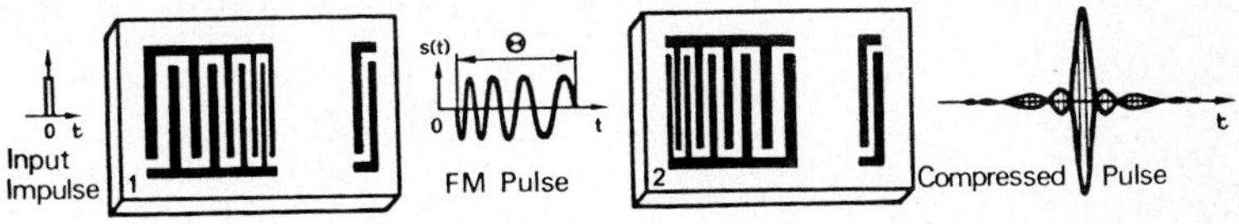

Figure 9 - Analog coding and decoding. Dispersive delay line 1 delivers a frequency modulated pulse which is time-compressed by matched filter 2.

Figure 9 refers to a signal often used by the radar experts : it is a pulse with a carrier, the frequency of which varies linearly. It is generated by an impulse applied to the dispersive delay line 1. Its instantaneous frequency is imposed by the interdigital spacing . Normally this pulse is sent to the atmosphere and then reflected by a target. Thus, it comes back distorted.However, we did not represent the noise which distorts it since the noise is not affected by the operation of correlation. Then, on its return, the pulse passes through filter 2 which is matched to it. It comes out time-compressed. The time-compression mechanism is understandable : the high frequency components of the signal s(t) which arrived first excite the narrow interdigital interval zones of the transducer ; the low frequency which come later excite the large interval zones at the instant the high frequency Rayleigh waves arrive. There is a concentration of energy at the output of the transmitter. The features of this kind of matched filters are of the following order :

Central frequency	:	$10 < f_0 < 500$ MHz
Relative frequency sweep	:	$B/f_0 < 50$ %
Signal duration	:	$1 < \theta < 20$ μs
Compression ratio $R=B\theta$	:	$10 < R < 1000$

The dispersive effect necessary to compress the pulse, produced in the foregoing principle by the displacement with the frequency of the transmitting zone, can be obtained in another way. Figure 10 illustrates the operation. The Rayleigh waves launched by a simple transducer are reflected at 90 degrees, by a first skew array of grooves in a zone where the wavelength matches the array step. As the period of the array increases from the input, the path of high frequency waves is shorter than that of low frequency waves. The second groove array plays a role similar to the first one and directs the waves toward the output transducer. This technique is well adapted to designing lines with compression ratio larger than 1000. [10].

Other recent methods resorting to simpler technology : dot arrays [11], 3 dB multistrip coupler [12] seem promising.

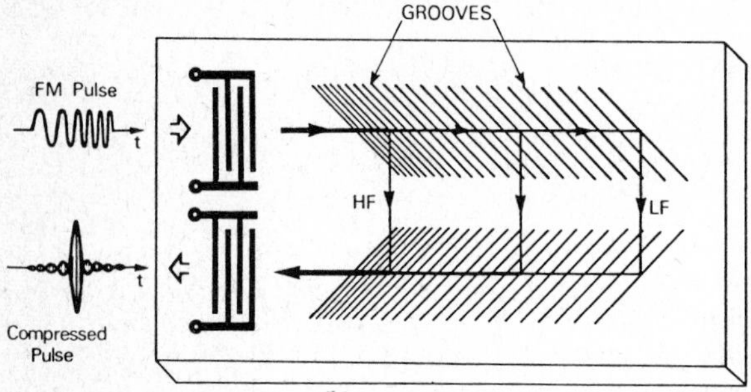

Figure 10 - Reflective array dispersive delay line.

FOURIER TRANSFORM.SPECTRUM ANALYSIS.

Two principles of spectrum analysis using acoustic waves can be mentioned. The first derives from optics and requires a time compression of the signal. The second is based on the spatial sorting of the components by acousto-optic interaction.

Time compression

The dispersive delay lines just described behave like time lenses. Thus, it is not surprising that optical experiments can be transposed in the acoustic domain. Actually, the Fourier transform S(f) of a signal s(t) can be written with the variable change $f = \mu t$

$$S(\mu t) = \exp\,(-\,j\pi\mu t^2)\int_{-\infty}^{+\infty}[s(\tau)\exp(-\,j\pi\mu t^2)]\exp[j\pi\mu(t-\tau)^2]d\tau$$

or symbolically, noting $D\pm = \exp\,(\pm\, j\pi\mu t^2)$

$$S = D_-\cdot[(s\cdot D_-) * D_+]$$

As shown in Figure 11 this operation can be performed with dispersive delay lines [13], the impulse responses of which are $D_+(t)$ or $D_-(t)$.

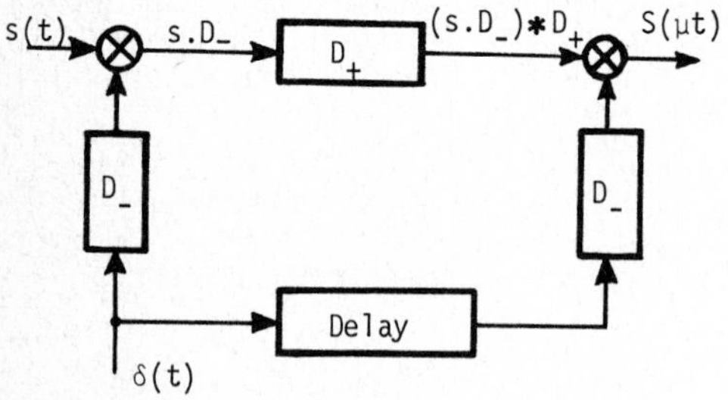

Figure 11 - Fourier transform processor using three SAW dispersive delay lines.

The finite duration of the impulse response of the first line imposes the analysis by time samples, for instance, of duration 20 μs. The resolving power is proportional to this duration. Taking into account the center frequency (a few hundred of MHz) and the length of the dispersive lines (> 10 cm), the spectrum domain extends over several tens of MHz, the resolving power being a few tens of kHz.

Of course, this device can perform other functions. Since the frequency components appear successively in the time domain, it is possible to modify them, to select them with a gate.

SPATIAL SORTING OF COMPONENTS.

This method is based on the interaction between a light beam and an acoustic beam. In short, here are the results of this interaction in the simplest case of longitudinal acoustic waves propagating in an isotropic bar and of a monochromatic laser beam. These waves, series of expansions and compressions, cause a periodic variation of the material index. With regard to the light waves, they behave like a phase array : any light beam (the width of which is large with respect to the acoustic wavelength) arriving at normal incidence is split into several diffracted beams. However, a light beam, correctly inclined at Bragg angle, generates only a single diffracted beam, the others being destroyed by interference (fig.12)

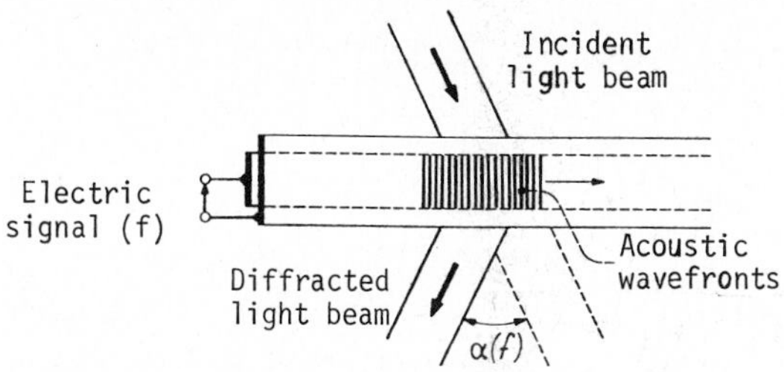

Figure 12 - Bragg angle acoustooptic interaction. The angle α of the diffracted beam depends on acoustic wave frequency f.

The properties of this single output beam are the following :
- The angle by which it is deflected is proportional to the acoustic wave frequency.
- Its intensity depends on the acoustic beam intensity.
- Its frequency contains the acoustic wave frequency since the index array, moving, involves a Doppler effect.

The third property has been used for delaying continuously a signal. The second is currently exploited for modulating a light beam (image reconstruction, laser printing system). The third we are here interested in for spectrum analysis is the object of great efforts, especially in order to sort signal radars by frequency in real time. Most of the specifications : bandwith at least 500 MHz, frequency resolution 1 MHz,dynamic range more than 40 dB are nearly satisfied. Difficulties about the implementation of high speed post detection signal processing hardware are still to be overcome [14].

Of course this acousto-optic interaction can also take place with SAW and a guided optical beam (integrated optics, semi-conductor laser source). Work is in progress in that field [15].

CONVOLUTION.

When a signal is distorted by noise it has to be received through a matched filter. We recalled this point previously and showed that Rayleigh waves are suitable to this purpose. The technique that we have illustrated by the example of the frequency modulated carrier pulse provides excellent results. However, it suffers from the drawback that a line is only matched to the signal for which it has been designed. Then it is necessary to have as many dispersive delay lines as signals to be processed. Of course, it would be more convenient to have at one's disposal a device able to match, if possible at will, several signals. It is with this aim in mind that elastic wave convolvers are studied. So far, it has implicitly been assumed that the strain amplitude remains below the threshold ($\sim 10^{-4}$) beyond which Hooke's law which relates proportionally strain, stress and electric field is no longer valid. This threshold has to be crossed in a convolver since the operation to perform, $s(t) = \int_{-\infty}^{+\infty} e_1(\tau)e_2(t - \tau)d\tau$, requires a non linear effect in order to generate the product of two signals. The first experiments were carried out with bulk waves [16] but Rayleigh waves are more suitable owing to the energy concentration near the surface.

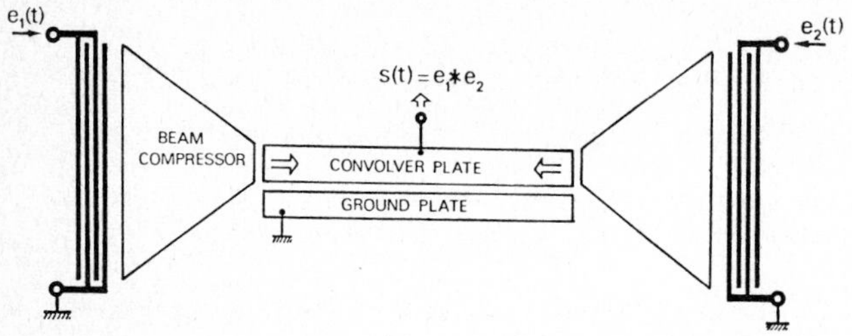

Figure 13.

Schematic of a SAW convolver

Figure 13 represents the principle of a Rayleigh wave convolver. The two signals e_1 and e_2 are carried by two wave trains of frequency f launched one toward the other by two transducers placed at the ends of a piezoelectric crystal (generally lithium niobate). In order to increase the power density, the width of each wave beam is compressed either by an asymmetrical multistrip coupler or by a horn guide. The non linear interaction between the two opposite wave vector beams gives rise to a stationary wave of frequency 2f , the amplitude of which is proportional to the product of the amplitudes of the two signals. This stationary wave is detected by the voltage that it induces between two electrodes. If the propagation time of the two wave trains below the central electrode is much larger than the duration of signals e_1 and e_2, the interaction duration can be considered as infinite and the induced voltage as the integration of the product of the two signals. In fact, these propagate along two opposite directions with the same velocity V, their relative velocity is 2V . It results that the output signal is time-compressed by a factor 2 as compared to s(t).

Since the induced electric power P_3 between the electrodes is proportional to powers P_1 and P_2 of e_1 and e_2, the convolver is characterized by the bilinearity factor $B = 10 \log (P_3/P_1P_2)$ The principle exposed is exploited with different variations with the purpose of increasing the efficiency (use of unidirectional transducers) and reducing the parasitic signals, particularly the signals reflected by transducers which can cause non desired convolution (use of two π phase shifted tracks to eliminate their effects by difference). The parameters of this kind of convolvers now leaving the laboratory are approximately the following [17, 18] :

bilinearity factor	: - 70 to - 60 dB
central frequency	: 150 to 300 MHz
integration time	: 10 to 20 μs
dynamic	: 70 dB
signal/noise ratio	: ∿ 50 dB

CONCLUSION.

The excellent features of acoustoelectric devices : compactness, reliability and their technology, partly similar to that of microelectronics, explain their regular development, especially in telecommunications and instrumentation. In the coming years, new functions will add to those described in this paper since much work is directed at designing other components such as sensors (position, acceleration, electric voltage, gas sensors). In parallel with those efforts for definite applications, researches are in progress in order to increase the number of usable materials (single crystals, ceramics and layers), to exploit other propagation modes (skimming bulk wave, edge guided mode, plate modes) and to generate elastic waves by new methods as, for example, with an intensity modulated light beam. This technique is interesting insomuch as optics, in certain applications, takes over electronics.

REFERENCES.

[1] SHIOSAKI, T., High-speed fabrication of high-quality sputtered ZnO thin-films for bulk and surface wave applications, IEEE Ultrason. Symp . Proc.,(1978) 100-110.

[2] LEMONS, R.A., and QUATE, C.F., Acoustic Microscopy, in Physical Acoustics, Vol. XIV (Ed. W.P. Mason and R.N. Thurston), Academic Press., (1979) 1-92.

[3] BESSON, R.J., Quartz crystal and superconductive resonators and oscillators, 10th Annual PTTI Meeting Proceedings (1978).

[4] BERTE, M., Acoustic bulk-wave resonators and filters operating in the fundamental mode at frequencies greater than 100 MHz, Electron. Lett., (1977) 248-250.

[5] ELLIOT, S., MIERZWINSKI, M. and PLANTING, P., The production of surface acoustic wave resonators, IEEE Ultrason. Symp. Proc. (1981) 89-93.

[6] MARSHALL, F.G. and PAIGE, E.G.S., Novel acoustic-surface-wave directional coupler with diverse applications, Electron. Lett. (1971) 460-462.

[7] MARSHALL F.G., New technique for the suppression of triple-transit signal in surface-acoustic-wave delay lines. Electron. Lett. (1972) 311-312.

[8] PORTER, W.A. and SMILOWITZ, B., Frequency sidelobes generated by apodized transducer pairs, IEEE Ultrason. Symp. Proc. (1982) 35-39

[9] DIEULESAINT, E., and ROYER, D., Elastic waves in solids. Applications to signal processing, J. Wiley (1980), chap. 9.

[10] WILLIAMSON, R.C., Properties and Applications of Reflective - Array devices, Proc. IEEE (1976) 702-710.

[11] SOLIE, L.P., The development of high performance RDA devices, IEEE Ultrason. Symp. Proc. (1979) 682-686.

[12] WOODS, R.C., Dispersive delay lines using 180° reflecting metal dot arrays, IEEE Ultrason. Symp. Proc., (1982) 88-91.

[13] JACK, M.A., GRANT P.M. and COLLINS, J.H., Theory, design and applications of surface acoustic wave Fourier transform processors, Proc. IEEE (1980) 450-468.

[14] HAMILTON, M.C., Acousto-optic spectrum analysis for electronic warfare applications, IEEE Ultrason. Symp. Proc., (1981) 714-720.

[15] JOSEPH, T.R., RAGANATH, T.R., LEE, J.Y. and PEDINOFF, M., Performance of the integrated optic spectrum analyser, IEEE Ultrason. Symp. Proc., (1981) 721-726.

[16] QUATE, C.F. and THOMSON, R.B., Convolution and correlation in real time with non linear acoustics, Appl. Phys. Lett., 16 (1970) 494.

[17] GAUTHIER, H. and MAERFELD, C., Wideband elastic convolvers, IEEE Ultrason. Symp. Proc. (1980) 30-36.

[18] HODGE, A.M. and LEWIS, M.F., Detailed investigations into SAW convolvers, IEEE Ultrason. Symp. Proc. (1982) 113-118.

The Mechanical Behavior of Electromagnetic Solid Continua
G.A. Maugin (editor)
Elsevier Science Publishers B.V. (North-Holland)

ELECTROMECHANICS OF ELECTRET TRANSDUCERS

Adalbert Prechtl

Institute for Electrical Machines
Technical University of Vienna
Vienna, Austria

Certain types of electret transducers are analyzed by means of a simple state function. The electrical response of an electret due to its elastic properties is briefly discussed.

Basically, an electret is a piece of solid dielectric material exhibiting (quasi) permanent electrification due to trapped surface charges, volume charges or remanent polarization or combinations of these. Recent advances in polymer science and technology have stimulated a new period of research in the field of electromechanical transducers based on electrostatic principles. Improved charge-storage capabilities and electric stability led to a wide variety of actual and potential new applications [1].

In the following contribution, the emphasis is placed on the basic analysis of simple yet practical arrangements. One-dimensional models are employed to display the more important aspects and to establish a basis for quantitative estimates.

BASIC MODEL

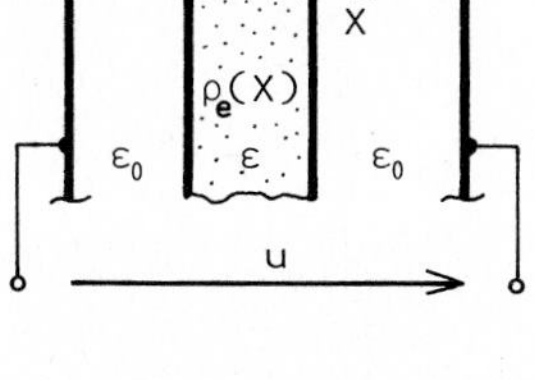

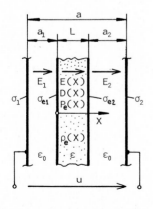

Figure 1

First, let us consider the basic model illustrated by Figure 1. A rigid layer of dielectric material carrying quasi-permanent surface charges of densities σ_{e1} and σ_{e2}, volume charges of density $\rho_e(X)$ and frozen-in polarization $P_e(X)$ is separated from planar electrodes by air gaps a_1 and a_2. Within the material we assume a constitutive equation of the form

$$D = \varepsilon E + P_e \, . \tag{1}$$

Denoting by

$$\sigma_e = \sigma_{e1} + \sigma_{e2} + \int_0^L \rho_e(X)\, dX \tag{2}$$

the total charge per unit area of the layer, the solution of $D'(X) = \rho_e(X)$ may be written as

$$D(X) = G(X) + \frac{\varepsilon_0}{2}(E_1 + E_2) \tag{3}$$

where

$$G(X) = -\frac{1}{2}\sigma_e + \sigma_{e1} + \int_0^X \rho_e(\bar{X})\, d\bar{X} \tag{4}$$

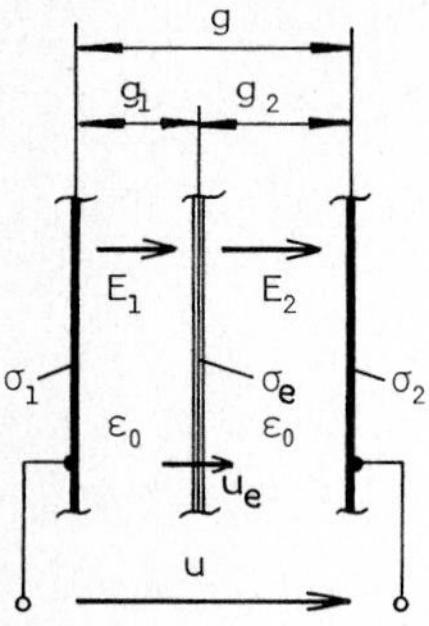

Figure 2

is regarded as prescribed, and

$$\varepsilon_0(-E_1+E_2) = \sigma_e . \quad (5)$$

To obtain the voltage drop accross the terminals, we have to integrate the electric field strength using (1) and (3),

$$u = g_1 E_1 + g_2 E_2 + u_e . \quad (6)$$

Here,

$$u_e = \frac{1}{\varepsilon} \int_0^L (G(X) - P_e(X))\, dX \quad (7)$$

is independent of E_1 and E_2,

$$g_1 = a_1 + \frac{1}{2} L\, \varepsilon_0/\varepsilon \;, \quad g_2 = a_2 + \frac{1}{2} L\, \varepsilon_0/\varepsilon \;, \quad (8)$$

and $g = g_1 + g_2 = a - L(\varepsilon-\varepsilon_0)/\varepsilon$. Expressions (5) and (6) yield the values

$$E_1 = (u - u_e - g_2\sigma_e/\varepsilon_0)/g \;,$$
$$E_2 = (u - u_e + g_1\sigma_e/\varepsilon_0)/g \;. \quad (9)$$

Note that, with respect to terminal quantities, the arrangement may be replaced by a sheet of vanishing thickness positioned at distances g_1 and g_2 between the electrodes (reduced total air gap g) (Figure 2). The sheet is characterized by the planar charge density σ_e and the constant voltage drop u_e of a double-layer.

STATE FUNCTIONS

Neglecting dissipative processes and imposing a global neutrality condition $\sigma_e + \sigma_1 + \sigma_2 = 0$ we assume the existence of a state function with values $w = \varepsilon_0(g_1 E_1^2 + g_2 E_2^2)/2 + \varepsilon_0(E_1 + E_2)u_e/2$ of the stored quasi-electrostatic energy per unit area (s.Figure 2). Together with $E_1 = \sigma_1/\varepsilon_0$ and $E_2 = (\sigma_1+\sigma_e)/\varepsilon_0$ we have

$$w(\sigma_1, g_1, g_2) = \frac{1}{2}(g_1\sigma_1^2 + g_2(\sigma_1+\sigma_e)^2)/\varepsilon_0 + (\sigma_1+\sigma_e/2)u_e \;, \quad (10)$$

where $u = \partial w/\partial\sigma_1 = (g_1\sigma_1 + g_2(\sigma_1+\sigma_e))/\varepsilon_0 + u_e$. In terms of variables u, g_1 and g_2 the state function with values $\tilde{w} = \sigma_1 u - w$,

$$\tilde{w}(u, g_1, g_2) = \frac{1}{2}\varepsilon_0(u - u_e - \sigma_e g_2/\varepsilon_0)^2/(g_1+g_2) - \frac{1}{2}(u_e + g_2\sigma_e/\varepsilon_0)\sigma_e \;, \quad (11)$$

is appropriate. It serves to calculate the surface charges $\sigma_1 = -\sigma_2 - \sigma_e = \partial\tilde{w}/\partial u$ as well as the forces in X-direction per unit area exerted on the left electrode, $f_1 = -\partial\tilde{w}/\partial g_1$, the right electrode, $f_2 = \partial\tilde{w}/\partial g_2$, and the electret slab, $f_e = -(f_1 + f_2)$, which have to be balanced by external or contact forces.

EXAMPLES

As a simple example we consider a transducer the principle of which is used in electret microphones (Figure 3: metallized electret foil, separated from backplate by variable air gap a_1). It differs from the classical electrostatic transducer in the sense that it does not require a dc bias. Even a net charge σ_e of the foil is not necessary. In this case and under (nearly) short-circuit conditions we obtain from (11) with $\sigma_e = 0$, (8), and $a_2 = 0$,

$$\sigma_1 = -\sigma_2 = \partial\tilde{w}/\partial u|_{u=0} = -\varepsilon_0 u_e/g\ , \quad g = a_1 + L\varepsilon_0/\varepsilon\ ,$$
$$f_{2e} = f_2 + f_e = -f_1 = \partial\tilde{w}/\partial g_1|_{u=0} = -\frac{1}{2}\varepsilon_0 (u_e/g)^2 \qquad (12)$$

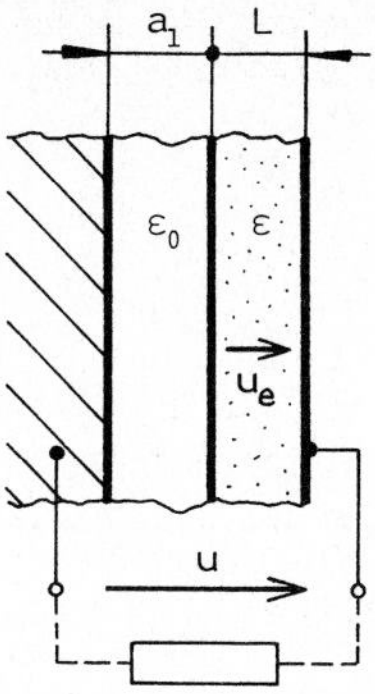

Figure 3

for the surface charges and the force per unit area of the metallized foil, respectively. Of course, the same result may be found more directly.

Another type of transducer is shown schematically in Figure 4. It consists of an electret located between two pairs of electrodes, which is movable in y-direction and carries no net charge ($\sigma_e = 0$). Neglecting edge effects we obtain from (11)

$$\tilde{W}(u_I, u_{II}, y) = \frac{1}{2}\varepsilon_0 d\left((b-c+y)u_I^2 + (b-c-y)u_{II}^2\right)/a + \\ + \frac{1}{2}\varepsilon_0 d\left((c-y)(u_I-u_e)^2 + (c+y)(u_{II}-u_e)^2\right)/g\ , \qquad (13)$$

where $g = a - L(\varepsilon-\varepsilon_0)/\varepsilon$, $|y| < b - c$, and d denotes the width(into page). Hence, the force in y-direction exerted on the electret is given by

$$F_e = \frac{\partial\tilde{W}}{\partial y} = -\frac{\varepsilon_0 L}{2\,a}\frac{\varepsilon-\varepsilon_0}{\varepsilon}\frac{d}{g}\left(u_I^2-u_{II}^2\right) + \varepsilon_0\frac{d}{g}u_e\left(u_I-u_{II}\right)\ . \qquad (14)$$

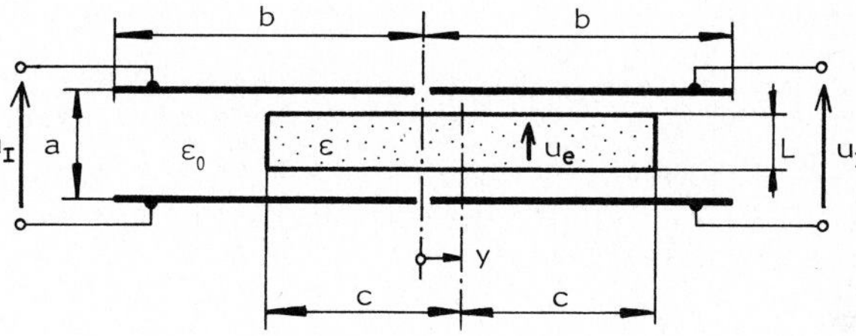

Figure 4

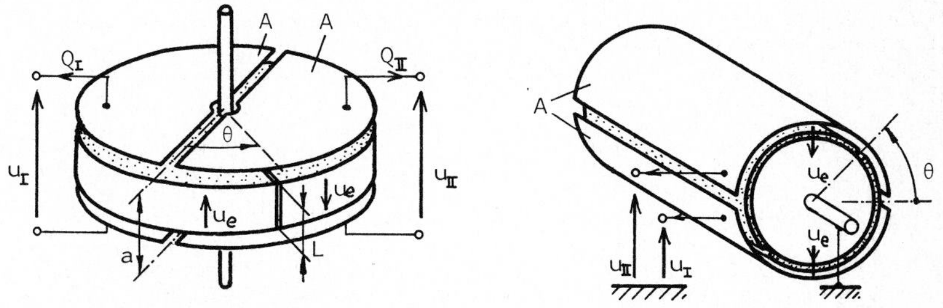

Figure 5

A similar transducer principle is applied in electret motors and generators which appear to have some applications in the fractional watt range, and for use in tachometers (Figure 5). Here, the state function $\tilde{W}$ assumes the form ($\sigma_e = 0$)

$$\tilde{W}(u_I, u_{II}, \theta) = \frac{1}{2}\frac{\varepsilon_0 A}{g\pi}\Big(\big(u_{II}-u_e\big)^2\theta + \big(u_{II}+u_e\big)^2(\pi-\theta) + \big(u_I+u_e\big)^2\theta + \big(u_I-u_e\big)^2(\pi-\theta)\Big) , \tag{15}$$

where $g = a - L(\varepsilon-\varepsilon_0)/\varepsilon$, $0 < \theta < \pi$, and A is half of the total capacitor area. The torque generated is given by

$$M_e = \frac{\partial\tilde{W}}{\partial\theta} = 2\frac{\varepsilon_0 A}{g\pi}u_e(u_I-u_{II}) , \tag{16}$$

while the expressions for the charges are obtained from $Q_I = \partial\tilde{W}/\partial u_I$, $Q_{II} = \partial\tilde{W}/\partial u_{II}$.

DEFORMABLE ELECTRET

To estimate the electrical response under applied forces of an electret due to its elastic properties we consider again a one-dimensional model (Figure 6). A slab of dielectric material is metallized on both sides and carries permanent Polarization $P_e(X)$, surface charges σ_{e1}, σ_{e2}, and volume charges of density $\rho_e(X)$ with respect to the undeformed state. The one-dimensional deformation $x = x(X)$ is characterized by the stretching $\lambda(X) = dx(X)/dX$. Imposing the neutrality condition $\sigma_1 + \sigma_2 + \sigma_e = 0$ with σ_e defined in (2) we have

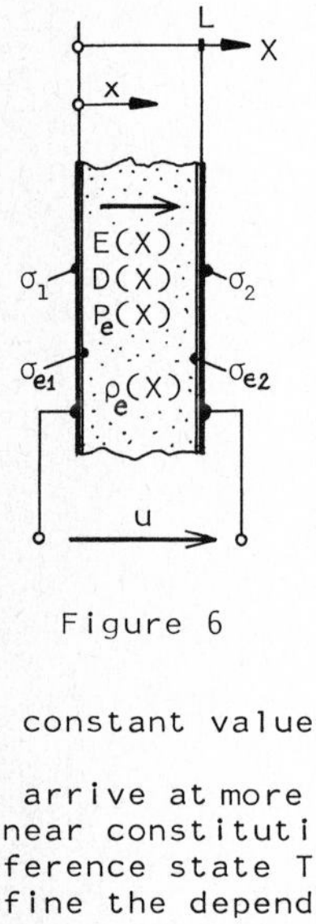

Figure 6

$$D(X) = G(X) + \sigma_e/2 + \sigma_1 , \tag{17}$$

where G(X) is given by (4). Let T denote the externally applied surface force (positive in tension) which retains its constant value within the slab (total stress). Then, in case of given constitutive equations $E = \mathcal{E}(D,T;X)$ and $\lambda = \Lambda(D,T;X)$, the voltage accross the terminals is related to T, σ_1 via (17) by

$$u = \int_0^L \lambda(x)E(X)\,dX = \hat{u}(T,\sigma_1) = \int_0^L \Lambda(D(X),T;X)\,\mathcal{E}(D(X),T;X)\,dX . \tag{18}$$

At constant values of u, $d\sigma_1/dT = -\big(\partial\hat{u}/\partial T\big)/\big(\partial\hat{u}/\partial\sigma_1\big)$.

To arrive at more explicit expressions we assume the electrically linear constitutive equation $\mathcal{E}(D,T;X) = \big(D - P_e(T;X)\big)/\varepsilon(T;X)$, choose a reference state $T = T_0$, $D(X) = P_e^0(X) = P_e(T_0;X)$, $\Lambda(P_e^0(X),T_0;X) = 1$, and define the dependences

$$\alpha_\lambda(X) = \frac{\partial\Lambda}{\partial T}(P_e^0(X),T_0;X) , \quad \beta_\lambda(X) = \frac{\partial\Lambda}{\partial D}(P_e^0(X),T_0;X) ,$$
$$\varepsilon^0(X) = \varepsilon(T_0;X) , \quad \alpha_\varepsilon(X) = \frac{\partial\varepsilon}{\partial T}(T_0;X)/\varepsilon^0(X) , \quad \alpha_p(X) = \frac{\partial P_e}{\partial T}(T_0;X) . \tag{19}$$

If, in addition, a weighted mean value is introduced by

$$\langle\cdot\rangle = \int_0^L (\cdot)\frac{dX}{\varepsilon^0}\Big/\int_0^L\frac{dX}{\varepsilon^0} , \tag{20}$$

a calculation based on (17) and (18) yields

$$\left.\frac{d\sigma_1}{dT}\right|_{T_0} = u\,\langle \alpha_\varepsilon - \alpha_\lambda + \alpha_p\beta_\lambda \rangle \Big/ \int_0^L dX/\varepsilon^0 + \\ + \left\langle \left(\left(\alpha_\varepsilon - \langle\alpha_\varepsilon\rangle\right) - \left(\alpha_\lambda - \langle\alpha_\lambda\rangle\right) + \left(\alpha_p\beta_\lambda - \langle\alpha_p\beta_\lambda\rangle\right)\right)\left(G - P_e^0\right) + \alpha_p \right\rangle \;, \tag{21}$$

which expresses the change in charge per unit area with applied stress at constant terminal voltage, and generalizes a formula given by Wada and Hayakawa [1, p.289]. In case of $u = 0$ (short circuit) and $P_e = 0$ (no extra polarization) we have, using (4),

$$\left.\frac{d\sigma_1}{dT}\right|_{T_0} = \left\langle \left(\left(\alpha_\varepsilon - \langle\alpha_\varepsilon\rangle\right) - \left(\alpha_\lambda - \langle\alpha_\lambda\rangle\right)\right) \int_0^X \rho_e(\bar{X})\, d\bar{X} \right\rangle \;. \tag{22}$$

This shows that permanent surface charges σ_{e1} and σ_{e2} play no role, while volume charges and inhomogeneities are crucial in obtaining electrical response.

REFERENCES

Extensive lists of references may be found in

[1] Sessler, G.M. (ed.), Electrets (Springer, Berlin Heidelberg New York, 1980)

The Mechanical Behavior of Electromagnetic Solid Continua
G.A. Maugin (editor)
Elsevier Science Publishers B.V. (North-Holland)

VELOCITY EQUATION FOR RAYLEIGH WAVES PROPAGATING IN CRYSTALS HAVING TWO PERPENDICULAR BINARY AXES OF SYMMETRY

D. Royer, J.M. Bonnet and E. Dieulesaint
Laboratoire d'Acoustoélectricité
Université Pierre et Marie Curie
10, rue Vauquelin, 75231 PARIS CEDEX 05
FRANCE

The Rayleigh wave velocity equation is established for configurations (direction of propagation and free surface) for which the equation of propagation and the boundary conditions simplify. Sixteen configurations in crystals belonging to the orthorhombic, tetragonal, cubic and hexagonal symmetry systems have been found. They include the three particular cases solved by Stoneley.

INTRODUCTION.

The operation at high frequency (f > 50 MHz) of surface acoustic wave devices especially those used for processing electronic signals requires low attenuation materials i.e crystals. Consequently, it is important to know the conditions of Rayleigh wave propagation in these anisotropic solids. They can be obtained by numerical techniques. However, it is generally fruitful to develop, when it is possible, analytical methods,above all if they are based on considerations of symmetry.

Since the discovery of surface waves by Rayleigh in 1885 [1] the conditions of their propagation in anisotropic media have been established - by Stoneley-only for three particular cases [2, 3].

We propose a more general method consisting of three steps : i) defining conditions on material stiffness constants for which the Christoffel's tensor and the boundary conditions on the free surface simplify, ii) stating the equation providing, with these conditions, the phase velocity V, iii) exploring systematically, in crystals having at least two perpendicular direct or inverse binary axes of symmetry, configurations for which this procedure is valid .

RAYLEIGH WAVE VELOCITY EQUATION.

In a non piezoelectric elastic solid the stress tensor components T_{ij} are related to the particle displacement components u_i by the stiffness tensor c_{ijkl}

$$T_{ij} = c_{ijkl} \frac{\partial u_l}{\partial x_k} \qquad (i,j,k,l = 1,2,3) \tag{1}$$

An elastic surface wave propagating in a semi-infinite medium, unbounded in the directions x_1 and x_3 (figure 1), must satisfy the equation of motion

$$\rho \frac{\partial^2 u_i}{\partial t^2} = \frac{\partial T_{ij}}{\partial x_j} = c_{ijkl} \frac{\partial^2 u_l}{\partial x_j \partial x_k} \qquad \text{for } x_2 > 0 \tag{2}$$

and the mechanical boundary conditions on the free surface

$$T_{i2} = c_{i2kl} \frac{\partial u_l}{\partial x_k} = 0 \qquad \text{for } x_2 = 0 \tag{3}$$

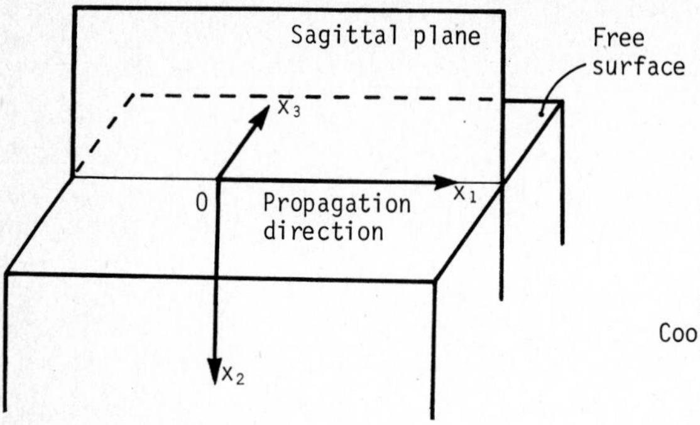

Figure 1.
Coordinate system.

A particle displacement of the form

$$u_i = {}^\circ u_i e^{-iqkx_2} \exp i(\omega t - kx_1) \qquad \text{with} \quad \mathrm{Im}[q] < 0 \tag{4}$$

describes a wave which propagates with a phase velocity $V = \omega/k$ along the direction x_1 and whose amplitude decreases with depth below the surface. Substituting this solution into the wave equation 2 leads to Christoffel's equations

$$\begin{pmatrix} \Gamma_{11}-\zeta & \Gamma_{12} & \Gamma_{13} \\ \Gamma_{12} & \Gamma_{22}-\zeta & \Gamma_{23} \\ \Gamma_{13} & \Gamma_{23} & \Gamma_{33}-\zeta \end{pmatrix} \begin{pmatrix} {}^\circ u_1 \\ {}^\circ u_2 \\ {}^\circ u_3 \end{pmatrix} = 0 \tag{5}$$

with

$$\zeta = \rho V^2 \tag{6}$$

and

$$\Gamma_{il} = c_{i1l1} + (c_{i12l} + c_{i21l})\, q + c_{i22l}\, q^2 \tag{7}$$

This system splits into two parts if Γ_{13} and Γ_{23} vanish i.e if the six elastic constants with a single index 3 are zero :

$$\begin{array}{lll} c_{1113} \equiv c_{15} = 0 & c_{1123} \equiv c_{14} = 0 & c_{2223} \equiv c_{24} = 0 \\ c_{1213} \equiv c_{65} = 0 & c_{2213} \equiv c_{25} = 0 & c_{1223} \equiv c_{64} = 0 \end{array} \tag{8}$$

In particular, it is the case when x_3 is parallel to a direct or an inverse diad axis of symmetry [4]. Then, the Rayleigh wave is polarized in the sagittal plane (${}^\circ u_3 = 0$) and the boundary conditions 3

$$c_{i211}\frac{\partial u_1}{\partial x_1} + c_{i212}\left(\frac{\partial u_1}{\partial x_2} + \frac{\partial u_2}{\partial x_1}\right) + c_{i222}\frac{\partial u_2}{\partial x_2} = 0 \qquad \text{for } x_2 = 0 \tag{9}$$

is always fulfilled when $i = 3$. The two other equations ($i = 1,2$) simplify if

$$c_{1211} \equiv c_{61} = 0 \qquad \text{and} \qquad c_{1222} \equiv c_{62} = 0 \tag{10}$$

In particular, it is the case when one of the axes x_1 or x_2 is along a direct or an inverse diad axis of symmetry.

In the following, conditions 8 and 10 are supposed to be satisfied, then the secular equation of system 5 reduces to

$$c_{22}c_{66}q^4 + [c_{22}(c_{11}-\zeta) + c_{66}(c_{66}-\zeta) - (c_{12}+c_{66})^2]q^2 + (c_{11}-\zeta)(c_{66}-\zeta) = 0 \tag{11}$$

Denoting q_1 and q_2 the two roots such that Im $[q_r] < 0$, the components of the corresponding eigenvectors $°u_1^{(r)}$ are given by Eq. 5 : $°u_1^{(r)} = 1$, $°u_3^{(r)} = 0$ and

$$°u_2^{(r)} = p_r = -\frac{c_{11}-\zeta + c_{66}q_r^2}{(c_{12}+c_{66})q_r} \qquad r = 1,2 \tag{12}$$

The general solution is a linear combination of these two partial waves propagating at the same velocity V :

$$\begin{cases} u_1 = (A_1\, e^{-iq_1kx_2} + A_2\, e^{-iq_2kx_2}) \exp i(\omega t - kx_1) \\ u_2 = (A_1p_1\, e^{-iq_1kx_2} + A_2p_2\, e^{-iq_2kx_2}) \exp i(\omega t - kx_1) \end{cases} \tag{13}$$

The weighting factors A_1, A_2 and the velocity V are determined by the boundary conditions 9 that expressed as

$$(q_1 + p_1)\, A_1 + (q_2 + p_2)\, A_2 = 0 \tag{14}$$

$$(c_{21} + c_{22}q_1p_1)\, A_1 + (c_{21} + c_{22}q_2p_2)\, A_2 = 0 \tag{15}$$

These two equations are compatible providing the determinant is zero

$$(p_1-p_2)(c_{12}-c_{22}q_1q_2) + (q_1-q_2)(c_{12}-c_{22}p_1p_2) = 0$$

Substituting the values of p_1-p_2 deduced from (12) yields

$$(c_{11}-\zeta-c_{66}q_1q_2)(c_{12}-c_{22}q_1q_2) + (c_{12}+c_{66})(c_{12}q_1q_2-c_{22}p_1p_2q_1q_2) = 0 \tag{16}$$

It appears the factor

$$c_{22}p_1p_2q_1q_2 = \frac{c_{22}(c_{11}-\zeta)^2 + (c_{11}-\zeta)c_{22}c_{66}S + c_{22}c_{66}^2P}{(c_{12}+ c_{66})^2}$$

that expresses as a function of the sum $S = q_1^2 + q_2^2$ and the product $P = q_1^2q_2^2$ of the roots of Eq. 11 :

$$c_{22}c_{66}S = (c_{12}+c_{66})^2 - c_{22}(c_{11}-\zeta) - c_{66}(c_{66}-\zeta) \tag{17}$$

$$c_{22}c_{66}P = (c_{11}-\zeta)(c_{66}-\zeta) \tag{18}$$

Then

$$c_{22}p_1p_2q_1q_2 = c_{11}-\zeta \tag{19}$$

and Eq. 16 reduces to

$$q_1q_2 = -\zeta\,(c_{11}-\zeta)/[c_{22}(c_{11}-\zeta) - c_{12}^2] \tag{20}$$

Squaring and equating to the value of P (Eq. 18) leads to the equation

$$c_{22}c_{66}\zeta^2(c_{11}-\zeta) = (c_{66}-\zeta)[c_{22}(c_{11}-\zeta) - c_{12}^2]^2 \tag{21}$$

which gives the Rayleigh wave velocity $V = (\zeta / \rho)^{1/2}$.

This equation is similar to that established by Stoneley [2] for Rayleigh wave propagation in the basal plane of hexagonal crystals. For cubic crystals ($c_{22}=c_{11}$) it includes the two cases (propagation in [001] plane along direction [100] and [110]) also studied by Stoneley [3]. However, as we have shown, this equation is more general and can be applied to other configurations. For example we have checked it for an orthorhombic crystal : $Ba_2\,Na\,Nb_5O_{15}$, class 2mm . With the same elastic constants [5], the velocities given by Eq. 21 are identical to those computed by Slobodnik [6] :

Y cut, X propagation	V = 3460,4 m/s
X cut, Y propagation	V = 3464,9 m/s

In the following paragraph, the different possible configurations are systematically sought for.

THE CONFIGURATIONS SATISFYING THE REQUIRED CONDITIONS.

The configurations i.e the dispositions of the axes x_1, direction of propagation and x_3, normal to the sagittal plane, with respect to the crystallographic axes X, Y, Z, for which the foregoing analysis is valid have now to be determined. The conditions 8 and 10 on the elastic constants must be satisfied by these configurations. Since they are verified when x_1 and x_3 are both parallel to a (direct or inverse) diad axis let us consider the crystals belonging to the orthorhombic, tetragonal, cubic and hexagonal symmetry systems.

1 - Orthorhombic system.

Each of the three crystallographic axes is a direct or an inverse diad axis. Six equivalent combinations are possible. The elastic constants of Eq. 21 are directly linked to the constants c^R_{ijkl} referred to the crystallographic axes :

$$c_{11} = c^R_{iiii} \qquad c_{22} = c^R_{jjjj} \qquad c_{12} = c^R_{iijj} \qquad c_{66} = c^R_{ijij}$$

i is the index (1 for X, 2 for Y, 3 for Z) of the crystallographic axis parallel to the direction of propagation and j the index of the axis perpendicular to the free surface.

2 - Quadratic, Cubic and Hexagonal Systems.

The table of the stiffness constants given in the crystallographic system is the following.

$$c^R_{\alpha\beta} = \begin{vmatrix} c^R_{11} & c^R_{12} & c^R_{13} & 0 & 0 & c^R_{16} \\ c^R_{12} & c^R_{22} & c^R_{13} & 0 & 0 & -c^R_{16} \\ c^R_{13} & c^R_{13} & c^R_{33} & 0 & 0 & 0 \\ 0 & 0 & 0 & c^R_{44} & 0 & 0 \\ 0 & 0 & 0 & 0 & c^R_{55} & 0 \\ c^R_{16} & -c^R_{16} & 0 & 0 & 0 & c^R_{66} \end{vmatrix}$$

The axis Z contains always a diad axis. Three methods of investigation using rotations around this axis are examined (Figure 2).

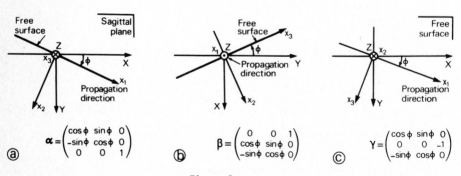

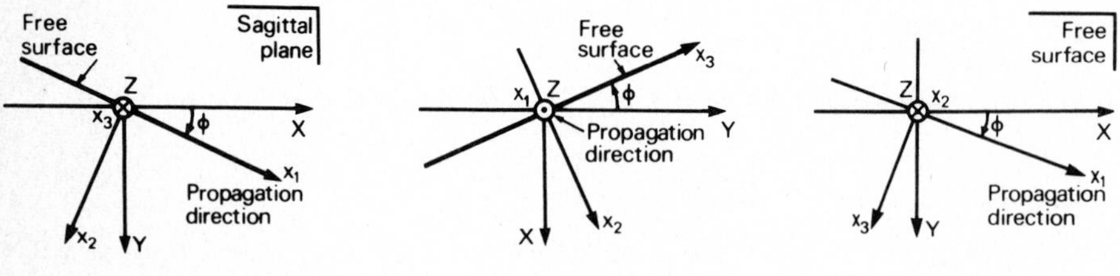

ⓐ $$\alpha = \begin{pmatrix} \cos\phi & \sin\phi & 0 \\ -\sin\phi & \cos\phi & 0 \\ 0 & 0 & 1 \end{pmatrix}$$

ⓑ $$\beta = \begin{pmatrix} 0 & 0 & 1 \\ \cos\phi & \sin\phi & 0 \\ -\sin\phi & \cos\phi & 0 \end{pmatrix}$$

ⓒ $$\gamma = \begin{pmatrix} \cos\phi & \sin\phi & 0 \\ 0 & 0 & -1 \\ -\sin\phi & \cos\phi & 0 \end{pmatrix}$$

Figure 2.

Rotation around *a)* the normal to the sagittal plane, *b)* the direction of propagation, *c)* the normal to the free surface.

a - The sagittal plane x_1x_2 is perpendicular to Z (fig. 2*a*).

Z being a diad axis, the condition 8 is verified : the constants with a single index 3 are zero, whatever the direction of propagation x_1 in the plane X Y.. Moreover, since $c_{26} = - c_{16}$ the two conditions 10 reduce to one. Denoting α_i^j the elements of the matrix of rotation of angle ϕ around Z,it expresses as :

$$c_{16} \equiv c_{1112} = \alpha_1^p \alpha_1^q \alpha_1^r \alpha_2^s c^R_{pqrs} = 0 \tag{22}$$

b - The direction of propagation x_1 is parallel to Z (fig. 2*b*).

The constants with a single index 1 or with three indices 1 are zero. Eq. 10 and Eq.8 regarding c_{15}, c_{25} and c_{64} are satisfied whatever the angle ϕ defining the crystal cut. Since in the matrix of rotation β_i^j : $\beta_1^i = \delta_{i3}$ and β_2^3 and β_3^3 are zero, $c_{14} \equiv c_{1123}$ and $c_{65} \equiv c_{1213}$ are also zero whatever ϕ :

$$c_{1123} = \beta_2^r \beta_3^s c^R_{33rs} = 0 \qquad \text{since } c^R_{36} = 0 \text{ and } c^R_{32} = c^R_{31} \tag{23}$$

$$c_{1213} = \beta_2^q \beta_3^s c^R_{3q3s} = 0 \qquad \text{since } c^R_{54} = 0 \text{ and } c^R_{55} = c^R_{44} \tag{24}$$

The unique condition on ϕ is given by :

$$c_{24} \equiv c_{2223} = \beta_2^p \beta_2^q \beta_2^r \beta_3^s c^R_{pqrs} = 0 \tag{25}$$

c - The normal x_2 to the free surface is parallel to Z (fig. 2*c*).

The constants with a single index 2 (c_{14}, c_{65}, c_{61}) and those with three indices 2(c_{24}, c_{62}) are zero whatever ϕ defining the direction of propagation x_1. The permutation of indices 1 and 2 in Eq.23 and 24 yields :

$$c_{2213} \equiv c_{25} = 0 \qquad \text{and} \qquad c_{2123} \equiv c_{64} = 0 \qquad \forall\ \phi$$

The only condition on the angle ϕ is

$$c_{15} \equiv c_{1113} = \gamma_1^p \gamma_1^q \gamma_1^r \gamma_3^s c^R_{pqrs} = 0 \tag{26}$$

In fact $\alpha_1^p = \beta_2^p = \gamma_1^p$ and $\alpha_2^p = \beta_3^p = \gamma_3^p$ whatever p : Eq.22 , 25 and 26 are identical. Since the coefficients α_1^3 and α_2^3 are zero, only the constants c^R_{pqrs} without any index 3, which express as functions of $c^R_{11} = c^R_{22}$, c^R_{12}, c^R_{66} and $c^R_{16} = - c^R_{26}$, appear in their development :

$$(-c^R_{11}+c^R_{12}+2c^R_{66})\sin\phi \cos\phi \ (\cos^2\phi-\sin^2\phi)$$
$$+c^R_{16}[\cos^2\phi(\cos2\phi-2\sin^2\phi)-\sin^2\phi(\cos2\phi+2\cos^2\phi)] = 0 \tag{27}$$

or

$$\frac{\gamma}{2} \sin4\phi + c^R_{16} \cos4\phi = 0 \tag{28}$$

where

$$\gamma = c^R_{66} - \frac{c^R_{11} - c^R_{12}}{2} \tag{29}$$

- For the crystals of classes 4, $\bar{4}$, 4/m of the tetragonal system, solutions of Eq. 28 are

$$\phi_0 = - \frac{1}{4} \tan^{-1}\left(\frac{2c^R_{16}}{\gamma}\right) \qquad \text{and} \qquad \phi_1 = \phi_0 + \frac{\pi}{4} \tag{30}$$

- For the crystals of classes 422, 4mm, $\bar{4}$2m, 4/mmm and of all cubic classes : $c^R_{16} = 0$, the solutions are

$$\phi_0 = 0 \qquad \text{and} \qquad \phi_1 = \frac{\pi}{4} \tag{31}$$

- For crystals of the hexagonal system (and isotropic solids) c_{16} and γ are zero. Eq. 28 is true for any ϕ.

The table 1 sums up the results. It comprises 16 configurations. The cases solved by Stoneley are indicated.

Crystal symmetry	ORTHORHOMBIC	TETRAGONAL (4, $\bar{4}$, 4/m)					
Propagation direction	A_2	$\perp Z$		//Z		$X+\phi_0$	$X+\phi_0+\frac{\pi}{4}$
Normal to the free surface	A_2'	$X+\phi_0$	$X+\phi_0+\frac{\pi}{4}$	$X+\phi_0$	$X+\phi_0+\frac{\pi}{4}$	//Z	

Crystal symmetry	HEXAGONAL			TETRAGONAL (422, 4mm, $\bar{4}$2m, 4/mmm) or CUBIC					
Propagation direction	$\perp Z$	//Z	$\perp Z$	$\perp Z$		//Z		X	$X+\frac{\pi}{4}$
Normal to the free surface	$\perp Z$	$\perp Z$	//Z	X	$X+\frac{\pi}{4}$	X	$X+\frac{\pi}{4}$	//Z	
Cases solved by Stoneley			[2]					[3]	[3]

Table 1 - The sixteen configurations for which the velocity equation is valid.

The constants c_{11}, c_{22}, c_{12} and c_{66}, necessary to calculate from Eq. 21 the velocity in one of those configurations, are to be determined by applying the corresponding rotation around Z of matrix α or β or γ to the tensor c^{R}_{pqrs}.

CONCLUSION.

After noticing that the equation of propagation and the boundary conditions simplified under the conditions that some elastic constants vanished we have established the equation giving the velocity of the Rayleigh waves. We sought for the configurations for which these conditions are fulfilled and numbered sixteen cases of application in crystals belonging to the orthorhombic, tetragonal, hexagonal and cubic systems. They include the three configurations treated independently by Stoneley.

REFERENCES.

[1] Strutt, J., (Lord Rayleigh), On waves propagated along the plane surface of an elastic solid, Proc. London Math. Soc., 17 (1885) 4 - 11.

[2] Stoneley, R., The seismological implications of aelotropy in continental structures, Mon. Not. Roy. Astr. Soc. Geophys. Suppl., 5 (1949) 343 - 353.

[3] Stoneley, R., The propagation of surface elastic waves in cubic crystal, Proc., Roy. Society, 232 A(1955) 447 - 458.

[4] Dieulesaint, E. and Royer, D., Elastic waves in solids (J. Wiley, London,1980) 143.

[5] Warner, A.W., Coquin, G.A. and Fink, J.L., Elastic and piezoelectric constants of $Ba_2NaNb_5O_{15}$, J. Appl. Phys. 40 (1969) 4353 - 4356.

[6] Slobodnik, A.J., Conway, E.D. and Delmonico, R.T., Microwave acoustics handbook, Vol 1A, Surface wave velocities, AFCRL Report n° 73-0597. Air Force Cambridge Research Labs., Bedford, Mass., U.S.A. (1973) 162 - 170.

The Mechanical Behavior of Electromagnetic Solid Continua
G.A. Maugin (editor)
Elsevier Science Publishers B.V. (North-Holland)

TRANSDUCTEURS INTERDIGITES DANS UNE STRUCTURE COUCHE MINCE PIEZOELECTRIQUE/SUBSTRAT AMORPHE

Jean DUCLOS et Michel LEDUC

U.E.R. Sciences et Techniques
B.P. 4006   76077 Le HAVRE cedex

Nous abordons le problème de l'excitation d'ondes de surface de type Rayleigh dans des structures stratifiées (couche mince piezoélectrique sur substrat amorphe) par des transducteurs interdigités métalliques. La notion de constante diélectrique effective facilite l'écriture du couplage existant entre le potentiel et les charges électriques. Une méthode numérique est proposée, conduisant aux caractéristiques de l'onde et à l'admittance électrique du transducteur.

INTRODUCTION

La propagation des ondes de surface sur des solides est bien connue depuis Lord RAYLEIGH. Les substrats piezoélectriques sont très utilisés maintenant pour propager des ondes de surface à des fréquences élevées (jusqu'à 1 GHz). La création d'ondes sur des matéri aux amorphes tels que la silice, a pu être réalisée en les recouvrant d'une couche mince piezoélectrique (sulfure de cadmium, oxyde de zinc). Les transducteurs interdigités, formés de languettes métalliques soumises à un potentiel H.F., constituent le meilleur dispositif d'émission-réception de ces ondes de surface [1].

ONDES DE SURFACE

Nous proposons une breve description des ondes susceptibles de se propager dans le milieu stratifié décrit par la figure 1.

Les ondes recherchées sont planes sinusoïdales et se propagent avec une vitesse de phase v selon Ox_1.

Elles combinent une vibration mécanique (déplacement $\vec{u}$) et un potentiel φ électrostatique qui doivent tendre vers 0 lorsqu'on s'éloigne des interfaces [2] et [3].

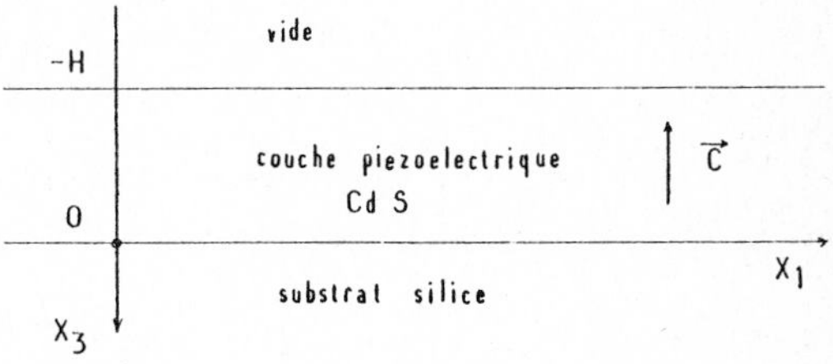

Ces grandeurs sont reliées dans la couche par les lois de la piezoélectricité, vérifient dans chaque solide le principe de Newton et dans chaque milieu le théorème de Gauss. Enfin les interfaces air/couche (x_3=-H) et couche/substrat (x_3=0) imposent des conditions aux limites:

- continuité des déplacements u_1 et u_3 en x_3=0
- continuité des contraintes T_{31} et T_{33} en x_3=0 et x_3=-H
- continuité des composantes E_1 et D_3 en x_3=0 et x_3=-H ou nullité de φ

On aboutit à un système de 8 équations linéaires homogènes dont les 8 inconnues sont les coefficients de pondération de solutions élémentaires.

Pour une valeur donnée du paramètre $kH=2\pi H/\Lambda$ (Λ désignant la longueur d'onde), il existe une ou plusieurs valeurs de la vitesse v annulant le déterminant de ce système. Le résultat apparaît sur la figure 2 qui met en évidence le caractère dispersif de la propagation et la présence de modes. Dans la pratique, seuls les modes I et II sont utilisés, car ils correspondent aux valeurs de kH les plus accessibles et sont plus efficaces que les autres.

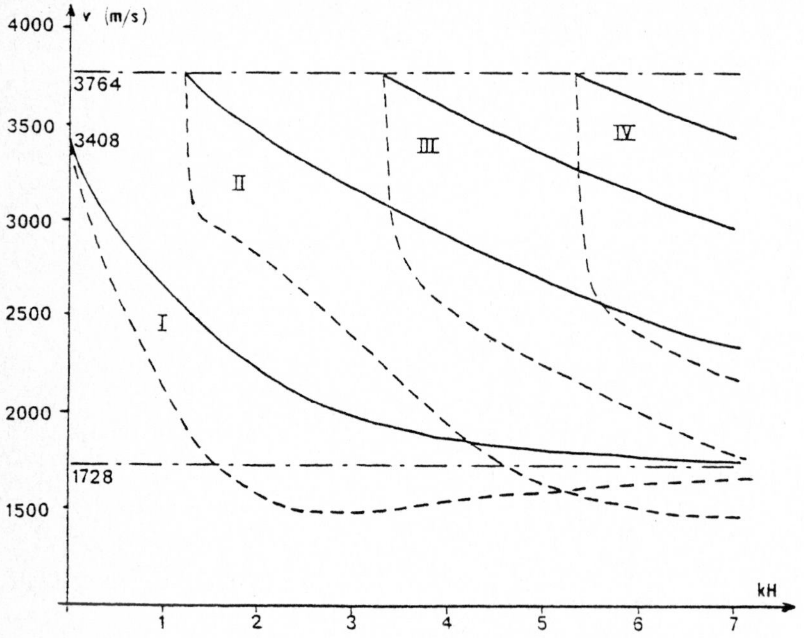

Figure 2

Les grandeurs physiques associées à l'onde étant déterminées à un coefficient multiplicatif près, on considère les déplacement et champ électrique normalisés:

$$u_{ci}\sqrt{\frac{k}{P}} \quad \text{et} \quad E_i\sqrt{\frac{1}{kP}}$$

Dans ces expressions, seules les valeurs 1 et 3 de l'indice i sont considérées et P désigne le flux de puissance mécanique et électrique traversant une bande plane unitaire perpendiculaire à Ox_1.

Sur la figure 3 sont tracées, en fonction de kH, les variations de $|E_1|$ à l'interface où se trouve le transducteur pour réaliser la conversion électrique-mécanique. Les lettres A, B, C et D affectées aux modes I et II précisent la configuration utilisée [4]

cas	interface $x_3=0$	interface $x_3=-H$
A	transducteur	rien
B	transducteur	métallisation
C	rien	transducteur
D	métallisation	transducteur

La valeur prise par E_1 permet de prévoir quelle sera l'efficacité d'un transducteur interdigité disposé à l'interface considérée. A ce propos, soulignons quelques résultats significatifs:
- dans le cas du 1° mode, E_1 dépend beaucoup de kH ainsi que de la configuration; un maximum important est atteint pour kH de l'ordre de 2,5 (difficile à réaliser). Aux valeurs voisines de 0,5 la métallisation de l'interface sans transducteur est très efficace.
- pour le 2° mode, IIB et IID présentent des maxima importants correspondant à des configurations réalisables (kH de l'ordre de 1,3)
- les valeurs obtenues avec un substrat piezoélectrique de même nature que la couche, sont 2 à 3 fois moindres que les maxima obtenus ici.

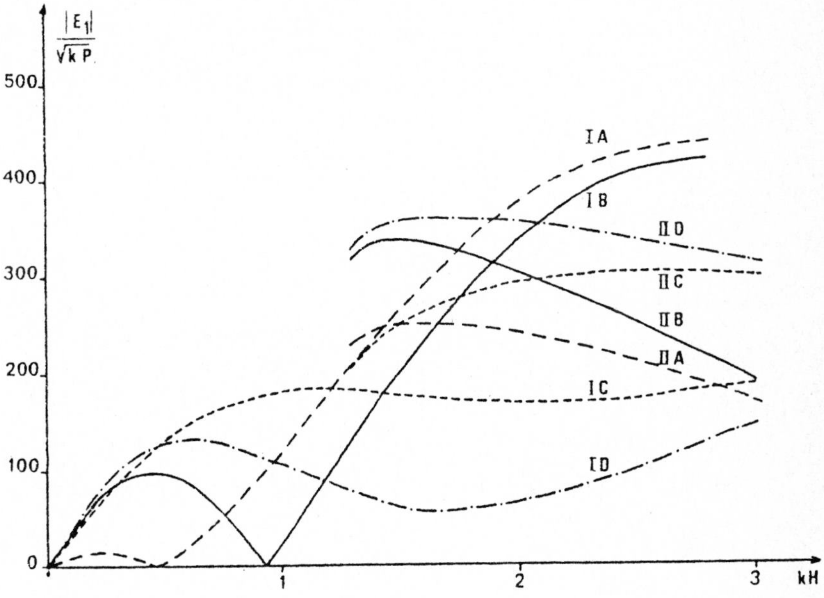

Figure 3

TRANSDUCTEURS

L'émission et la réception des ondes de surface sont réalisées par des électrodes métalliques alternées, appelées peignes interdigités, d'extension infinie selon x_2; nous n'envisageons ici que l'émission par un transducteur situé à l'interface $x_3=0$. L'onde créée n'est plus une onde plane; chaque fonction s'exprime par une intégrale de Fourier. En particulier, le potentiel électrique en un point de l'interface contenant les électrodes, s'écrit:

$$\varphi(x_1,x_3=0,t) = \int_{-\infty}^{+\infty} \tilde{\varphi}(k,x_3=0)\, \exp(j(kx_1-\omega t))\, dk$$

où $\tilde{\varphi}(k,x=0)$ désigne la transformée de Fourier de $\varphi(x_1,x_3=0)$.

On peut ainsi exprimer que les valeurs du potentiel sur les électrodes sont fixées (+ ou - V suivant la polarité).
La mise en équation du problème est semblable à celle exposée dans la première partie, mais diffère sur un point essentiel: la discontinuité de D_3 à l'interface couche/substrat, qui s'exprime en introduisant la densité de charge $\sigma(x_1)$ sur les électrodes ou sa transformée de Fourier $\tilde{\sigma}(k)$.

Le système linéaire possède alors un second membre proportionnel à: $\frac{\tilde{\sigma}(k)}{|k|}$
et sa résolution conduit à la relation: $\tilde{\varphi}(k) = \frac{\tilde{\sigma}(k)}{|k|\,\varepsilon(k)}$
où la fonction $\varepsilon(k)$ a la dimension d'une constante diélectrique.
C'est une fonction paire, qui résume l'ensemble des propriétés des 3 milieux au niveau $x_3=0$ pour la fréquence choisie.

La figure 4 représente ses variations dans le cas d'une fréquence de 100 MHz, d'une couche de CdS d'épaisseur 5 μm, l'interface air/couche n'étant pas métallisée.

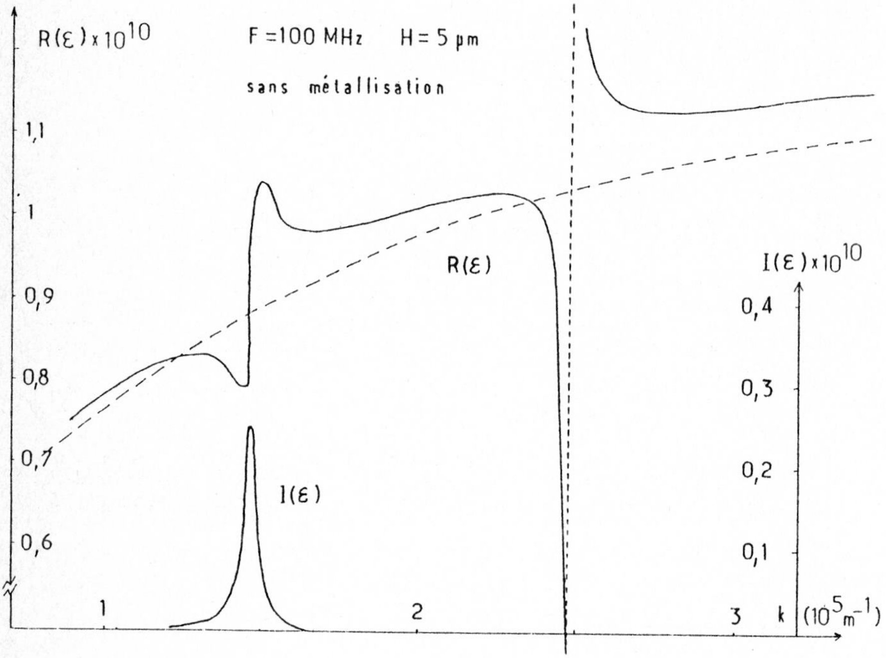

Figure 4

On distingue plusieurs domaines:

si $k < \frac{2\pi F}{V_T}$ (où V_T=3764 m/s est la vitesse de l'onde transversale dans la silice)
$\varepsilon(k)$ est complexe (partie imaginaire relativement faible) et peut présenter de brusques variations correspondant aux ondes de volume possibles dans le substrat.

si $\frac{2\pi F}{V_T} < k < \frac{2\pi F}{V_R}$ (où V_R=1728 m/s est la vitesse des ondes de Rayleigh sur un monocristal de CdS)
$\varepsilon(k)$ est réelle; ses zéros correspondent aux vitesses des ondes de surface, lorsque l'interface $x_3=0$ n'est pas métallisée ($\tilde{\sigma}$ nulle); ses pôles, d'allure hyperbolique, correspondent aux vitesses de ces ondes lorsque cette interface est métallisée ($\tilde{\varphi}$ nulle).

si $k > \frac{2\pi F}{V_R}$ il n'y a plus d'onde de surface et la fonction $\varepsilon(k)$ tend vers une limite ε_∞ différente de celle obtenue lorsque la piezoélectricité n'est pas prise en compte.

Le calcul de la capacité d'un ensemble périodique d'électrodes interdigitées [5] montre que la fonction $\sigma(x_1)$ tend vers l'infini aux bords de chaque électrode; on préfère donc utiliser sa primitive $q(x_1)$ s'annulant hors du transducteur qui est une fonction continue. L'expression de $\tilde{\varphi}(k)$ devient alors:

$$\tilde{\varphi}(k) = j \frac{k\,\tilde{q}(k)}{|k|\,\mathcal{E}(k)}$$

La détermination de $\varphi(x_1,t)$ et $q(x_1,t)$ est alors primordiale puisque la première fonction est une mesure directe de l'efficacité du dispositif émetteur, alors que la seconde permet la détermination de l'admittance du transducteur.
Des solutions à ce problème ont déjà été apportées dans le cas des substrats piezo-électriques (milieux non stratifiés) [6] et [7]; nous proposons la méthode suivante.
On subdivise les électrodes en segments de largeurs Δx_J inégales, tenant compte de l'allure de $q(x_1)$; en considérant les accroissements Δq_J de q sur ces segments, on obtient une expression approchée simple de $\tilde{q}(k)$ et, par transformation de Fourier, il vient:

$$\varphi(x) = \frac{4}{\pi} \sum_J \frac{\Delta q_J}{\Delta x_J} \int_0^{\infty} \frac{\sin(kx_J)\,\sin(k\Delta x_J/2)\,\sin(kx)}{k^2\,\mathcal{E}(k)}\,dk$$

On aboutit ainsi à un système linéaire dont les premières équations expriment la nullité de $\tilde{q}(k)$ pour chaque zéro de la fonction $\mathcal{E}(k)$; c'est une condition nécessaire à la convergence de l'intégrale donnant $\varphi(x)$. Les équations suivantes traduisent la constance du potentiel sur chaque électrode (valeurs prises en des points régulièrement espacés); leurs coefficients sont des intégrales dont le calcul exploite la nullité de $\tilde{q}(k)$ aux zéros de $\mathcal{E}(k)$.
Les accroissements Δq_J obtenus comme solutions de ce système, permettent de reconstituer la fonction $q(x)$ et, éventuellement, d'évaluer le potentiel en tout point de l'interface $x_3=0$.
Leur connaissance fournit aussi, immédiatement, l'admittance du transducteur.
En effet, la charge Q des électrodes vaut: $\frac{jI}{\omega} = (\frac{jG}{\omega} + C)\ V$

d'après le schéma équivalent, faisant apparaître la partie réelle G et la partie imaginaire $-j\omega C$ de l'admittance; or, Q étant la somme des accroissements de la fonction $q(x_1)$ sur les électrodes, on obtient une relation simple entre C , G et les parties réelle et imaginaire de $q(x_1)$.

CONCLUSION

Notre méthode présente l'avantage de ne pas s'écarter de la représentation physique du problème.
L'utilisation de la fonction $q(x)$ qui s'était révélée fructueuse dans le calcul des capacités, s'avère encore judicieuse puisqu'elle permet d'alléger les calculs et conduit directement au schéma équivalent du transducteur.
Le calcul des intégrales finales reste toutefois délicat, étant donné l'allure de la fonction $\mathcal{E}(k)$; il nécessite le recours à un ordinateur puissant.
La notion de constante diélectrique effective à l'interface, peut s'étendre:
- aux cas où le nombre d'interfaces est supérieur à 2;
- à d'autres fonctions, afin de conduire directement aux déplacements.

REFERENCES

[1] Dieulesaint, E. et Royer, D.: Ondes élastiques dans les solides (Masson, Paris, 1974)

[2] Slobodnick, A.J., Conway, E.D. et Delmonico, R.T.: Microwave acoustic handbook, vol. 1A, Surface wave velocities, scientific report N.T.I.S. (1973)

[3] Duclos, J., Leduc, M. et Delestre, P.: Propagation d'ondes de surface sur une structure couche mince piezoélectrique/substrat amorphe; revue du CETHEDEC 71 (1982) 189-207

[4] Hickernell, F.S.: Piezoelectric film surface wave transducers; in: Acoustic surface wave and acoustooptic devices (Optosonic Press, New-York, 1971)

[5] Duclos, J. et Leduc, M.: Capacité statique des transducteurs interdigités à couches minces, répartition des charges et des potentiels; Revue Phys. Appl. 17 (1982) 75-82

[6] Milsom, R.F., Reilly, N.H.C. et Redwood, M.: Analysis of generation and detection of surface and bulk acoustic waves by interdigital transducers; I.E.E.E. Trans. Sonics Ultrasonics, SU 24, 3 (1977) 147-166

[7] Hussein, A.M. et Ristic, V.M.: The evaluation of the input admittance of SAW interdigital transducers; J. Appl. Phys. 50-7 (1979) 4794-4801.

The Mechanical Behavior of Electromagnetic Solid Continua
G.A. Maugin (editor)
Elsevier Science Publishers B.V. (North-Holland)

SYMMETRY BREAKING AND DYNAMICAL ELECTROMAGNETIC-ELASTIC COUPLINGS

G.A.Maugin

Université Pierre-et-Marie-Curie
Laboratoire de Mécanique Théorique Associé au C.N.R.S.,
Tour 66, 4 Place Jussieu, 75230 Paris Cedex 05,
France

The aim of this contribution is to exhibit the important role played by bias and spontaneous fields in the dynamical properties of magnetized and/or electrically polarized deformable media. The general effect of initial (or permanent) electric or magnetic dipoles, insofar as dynamical properties are concerned, is ,through a so-called symmetry breaking, to allow for couplings which would not necessarily exist in a natural field-free configuration, providing thus much richer possibilities for electromechanical and magnetomechanical interactions. The review is not analytical. It rather aims at introducing some basic notions and general properties, sophisticated illustrations being essentially provided by examples from ferroelectric and ferromagnetic or antiferromagnetic media previously, or currently, worked out in detail by the author and co-workers.

INTRODUCTION:THE EXAMPLE OF ELECTROMAGNETIC OPTICS

Because of the small tensorial order of the involved effects, the simplest example of symmetry breaking is provided by the phenomenon of <u>anisotropy inducement</u> in crystalline optics:certain optically isotropic or cubic crystals become birefringent on being subjected to an electric field; some transparent plastics which were optically isotropic in their mechanically free state, become anisotropic in their optical properties when subjected to a system of mechanical loads and this results in the celebrated Maxwell-Neumann-Brewster photo-mechanical effect. Clearly, these two examples show that properties and phenomena which were not allowed in a certain configuration of the material, become feasible after an external stimulus has caused a change in the symmetry in the physical property of interest. The electric Kerr effect, the Faraday effect and the Voigt-Cotton-Mouton effect [1] belong in the class of phenomena related to optical anisotropy inducement. An important point here is to note that the existence of such effects in material media is direct evidence that the equations for the electromagnetic fields in a general material continuum, unlike the equations for them in vacuum, are <u>nonlinear</u> since the sum of two solutions generally fails to be a solution. An analysis of electro-magneto-optical effects, though based on nonlinear equations for which all but the simplest exact solutions are difficult to exhibit, may be simplified and made tractable by assuming that the dynamical part of the solution - the light wave - has such a weak intensity that the vectors which describe it may be treated as infinitesimal. Then,in a quite general approach to this class of phenomena, it may generally be assumed that the displacement and magnetic intensity fields are <u>functions</u> (in the usual sense) of the bias electric and magnetic induction fields and <u>linear functional</u> of the small electric and magnetic induction fields. This duality in functional behavior can be originally produced but by a <u>fully nonlinear behavior</u>. Only a deductive approach, from the fully nonlinear version to the linearized one, can be envisaged from the mathematical point of view.+ We therefore witness the close relationship

+ Approaches to nonlinear descriptions which simply consist in adding ad hoc terms to the linear description are doomed to yield false and insufficient generalizations.

between anisotropy inducement and the original nonlinear behavior. Indeed, for the sake of illustration consider the following general nonlinear relationship in a dielectric: $\underline{D} = \hat{D}(\underline{E})$. Call $\underline{E}_o$ a bias electric field of strong intensity and $(\underline{d},\underline{e})$ the small-amplitude fields corresponding to a light wave. Initially, we have $\underline{D}_o = \hat{D}(\underline{E}_o)$ while after linearization we have $\underline{d} = L\hat{D}(\underline{E}_o).\underline{e}$ where $L\hat{D}$ is a linear operator of $\mathcal{V}$ onto $\mathcal{V}$ (the space of real polar vectors in $\mathbb{R}^3$) that we write $\underline{\varepsilon}(\underline{E}_o)$, the dielectric tensor. If the medium was initially isotropic, then we can check that

$$\underline{\varepsilon}(\underline{0}) = \varepsilon_o \underline{I} \quad , \quad \underline{\varepsilon}(\underline{E}_o \neq \underline{0}) = \bar{\varepsilon}_o \underline{I} + \varepsilon_E (\underline{E}_o \otimes \underline{E}_o / |\underline{E}_o|^2) \qquad .(1)$$

These two formulas encapsulate the phenomenon of anisotropy inducement in a typical manner . The symmetry group for the dynamical process in the absence of $\underline{E}_o$ is full isotropy, O_3 ,while with $\underline{E}_o \neq \underline{0}$ the symmetry group has become the one of so-called transverse isotropy or cylindrical symmetry with respect to $\underline{E}_o$ (the effects of $\underline{E}_o$ and $-\underline{E}_o$ cannot be distinguished). In addition, the isotropic value ε_o is certainly modified ($\bar{\varepsilon}_o$) , but to a small extent, by the bias field. Another more general example involving <u>photoelasticity</u> can be given thus. Instead of the previous constitutive equation consider

$$\underline{D} = \hat{D}(\underline{E},\underline{C}) \qquad ,$$

where $\underline{C}$ is a strain tensor. In general we shall have

$$\underline{d} = \underline{\varepsilon}_1(\underline{E}_o,\underline{C}_o).\underline{e} + \underline{\varepsilon}_2(\underline{E}_o,\underline{C}_o).\underline{c} \qquad ,(2)$$

where $\underline{c}$ is a dynamical perturbation in strain. The usual photoelastic-effect hypothesis consists in taking $\underline{E}_o = \underline{0}$ and $\underline{c} = \underline{0}$, hence considering that mechanical processes contribute only statically. Then it remains $\underline{d} = \underline{\varepsilon}_1(\underline{0},\underline{C}_o).\underline{e}$ with $\underline{\varepsilon}_1(\underline{0},\underline{0}) = \varepsilon_o \underline{I}$, and $\underline{\varepsilon}_1(\underline{0},\underline{C}_o)$ will take the form

$$\underline{\varepsilon}_1(\underline{0},\underline{C}_o) = \bar{\varepsilon}_o \underline{I} + K \underline{C}_o \qquad ,(3)$$

so that $\underline{\varepsilon}_1$ will have the same principal directions as the strain tensor $\underline{C}_o$ and the originally isotropic material will behave like a uniaxial or biaxial crystal depending on the state of strain. Streaming birefringence due to a shear motion in a transparent fluid can be approached in the like manner.

SYMMETRY BREAKING AND PHASE TRANSITION

The above-given example, especially eqns.(1),exhibits a striking analogy with the problem of broken symmetry at a phase transition in solid state physics [2]-[3]. There, the field (analogous to $\underline{E}_o$) which will initiate the breaking of symmetry in certain conditions of temperature is the so-called <u>transition</u> or <u>order parameter</u>.We are thus led to enunciating the celebrated <u>Curie principle</u> of superposition of symmetries: The group of symmetries of two or more objects, regarded as a whole, is the highest common subgroup of the symmetry groups of these objects, determined taking into account the relative arrangement of their elements of symmetry [4]. In other words [5], this can also be expressed as: In the case of a combination of two or more <u>unequal</u> symmetrical figures into a single composite whole, the latter will comprise only those elements of symmetry which are common to all components of the figure, taking into account the real distribution of the figures in space. Then, extending this statement from geometrical figures to physical phenomena, it follows that

> "if we denote by Γ the symmetry group of the transition parameter (choosing a definite orientation for it in representation space), the symmetry G of the low-temperature phase would be only the maximal subgroup of G_o (the symmetry of the high-temperature phase) formed by the intersection of G_o

and Γ :

$$G = G_o \cap \Gamma \qquad ,(4)$$

and this can also be restated as

"the symmetry group G of the lower phase is the maximal subgroup of the symmetry group G_o of the upper phase which leaves the transition parameter invariant".

Equation (1) is easily illustrated in Figure 1 by considering the ferroelectric phase transition occuring in $BaTiO_3$ at 403 K . In this case G_o = m3m (or O_h). The

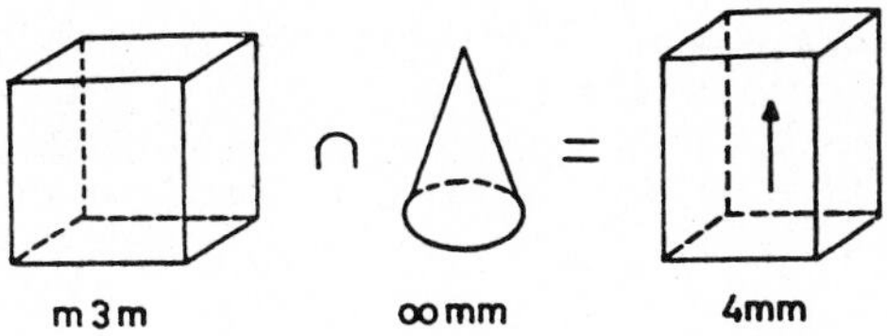

Figure 1.- A Curie-principle interpretation of the cubic-to-tetragonal phase transition in $BaTiO_3$ (after [6]).

electric polarization has the same symmetry transformation properties as a cone, so that $\Gamma = \infty mm$(or $C_{\infty v}$). If the polarization vector is directed along any of the edges of the cubic unit cell, we have $G = m3m \cap \infty mm = 4mm$ (or C_{4v}), which is indeed the observed symmetry for $BaTiO_3$ (cf.Figure 2a) with the spontaneous polarization vector pointing along the tetragonal axis. In the case of $BaTiO_3$ the symmetry group G allows for the local existence of electromechanical couplings in the form of piezoelectricity (linear coupling between electric and mechanical perturbations). However, while the local (individual-domain ;see below) symmetry has actually been reduced from G_o to G,macroscopically, the crystal specimen as a whole should have an average symmetry G_o even below the transition temperature because, macroscopically, it is a scalar, the temperature, which governs the transition phenomenon. Therefore,the statement made above and the last one can be reconciled only if the crystal splits into domains, assuming of course that all the possible domains are able to develop uniformly and to the same extent. Then one understands why the application of an intense electric bias field, favoring the growth of domains aligned with its direction to the expense of others, is necessary to re-obtain a one-domain sample exhibiting globally the G symmetry and the accompanying piezoelectric coupling, as experimentally discovered by B.Vul in the forties (cf.[7]). In the process both the viewpoints of phase transition and anisotropy inducement join together to provide a globally ordred phase. We shall develop this point below when one accounts for strains. Notice that ferroelectrics of the $BaTiO_3$ type are of the displacive type (spontaneous polarization related to the displacement of the central atome;cf. Figure 2a). Another prototype of ferroelectric crystals is provided by $NaNO_2$ (Figure 2b) where the local spontaneous polarization is connected with a central molecular group (here NO_2); its crystalline structure is centered orthorhombic of symmetry group 1m2m (or C_{2v}^{20}) in the ferroelectric phase and there exists a so-called incommensurate phase between the high-temperature disordered phase and the low-temperature ferroelectric one.

LINEAR ELECTRO- AND MAGNETO-ACOUSTICS

The procedure involved in the examples of eqns.(1) and (2) is extremely simple and clearly amounts to a usual first-order variation in which no geometrical background plays a role. The situation is quite different when stress-strain processes are in-

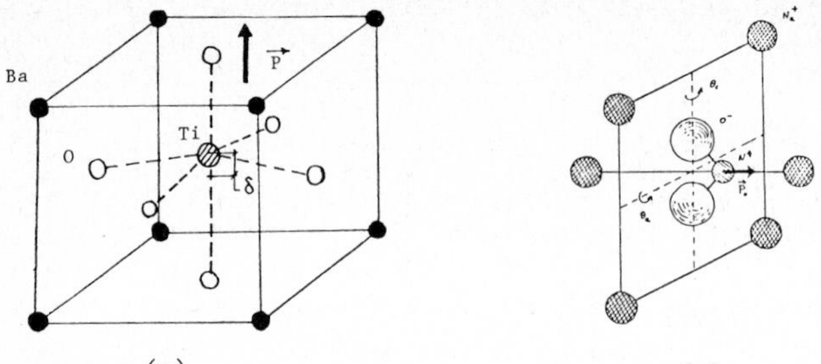

(a) (b)

Figure 2.- Crystalline structure of $BaTiO_3$(a) and $NaNO_2$(b) in their ferroelectric phase.

volved because the consideration of a nonlinear theory then implies the recognition of various configurations of the material with mappings between these configurations. A variational operation aiming at deducing equations governing small fields superimposed on bias fields (of any tensorial nature, but of the scalar type) requires cautions which were altogether ignored until now. This is already quite familiar in pure nonlinear elasticity [8],[9], but a small group of authors starting with Toupin [10] and Tiersten [11] also recognized this fact in the framework of the electrodynamics of deformable continua. Indeed, let us call K_t the current configuration at time t, K_o the initial configuration corresponding to the application of bias fields and K_R some ideal field-free configuration used as a reference at time t=0. For the sake of example we consider two nonlinear processes, one mechanical, the other electromagnetic (electric or magnetic) such that we have the nonlinear , supposedly "exact", constitutive equations

$$\underline{T}_\alpha = \hat{F}_\alpha(\underline{C}_\beta) \ , \quad \alpha,\beta = 1,2 \qquad .(5)$$

Because of the notion of configuration, the $\underline{T}_\alpha$'s and $\underline{C}_\beta$'s are geometrical objects. Assume that there exist initial (bias) fields $\underline{C}_{o\beta}$ such that, initially, eqns.(5) read $\underline{T}_{o\alpha} = \hat{F}(\underline{C}_{o\beta})$. The variational operation which yielded (1) now is complicated by the fact that only geometrical objects in the same configuration can be compared. Therefore, for example, the variation of the $\underline{T}_\alpha$'s between K_o and K_t is formally necessarily of the form

$$\underline{t}_\alpha = \delta \underline{T}_\alpha = \overset{K_R \to K_t}{\mathcal{C}} \left[\overset{K_R \leftarrow K_t}{\mathcal{C}} (\underline{T}_\alpha) - \overset{K_R \leftarrow K_o}{\mathcal{C}} (\underline{T}_{o\alpha}) \right] \qquad ,(6)$$

where it is understood that K_t remains in a neighborhood of K_o , $K_t \subset N(K_o)$ in some precise topological sense and the $\mathcal{C}$'s indicate so-called convections or transports from a configuration to another one, by which it must be understood not passive transformations of co-ordinates , but tensorial transformations based on a deformation mapping itself (motion). In the limit eqn.(6) provides the definition of a so-called convected derivative of which the Lie derivative of differential geometry is an example. That is, the operator δ is a derivative in the sense that it is linear and it obeys Leibnitz product rule. Unfortunately, the right-hand side of eqns.(5), in order to satisfy the basic invariance requirement of continuum mechanics (so-called "objectivity"), has a general form such that it contains operations which do not commute with δ . As a consequence the result of the δ-operation will be rather complicated and this, in turns, greatly complicates the phenomenon of symmetry breaking. To fix the ideas let us call $\mathcal{V}$ the set of vectors of E^3 , $\mathcal{C}$ the set of general second-order tensors and $\mathcal{C}_S$ the set of symmetric second-order tensors. $\mathcal{L}(A,B)$ denotes the set of linear applications of A in B and $\mathcal{L}_S(A,B)$ its symmetric part. Lower case letters will denote the corresponding small-amplitude dynamical fields. We call $L_{\alpha\beta} = L_\alpha \hat{F}_\beta$ the linear operators generalizing the operator $\hat{LD}$ of the introduction. In performing the perturbation of (5) about K_o with the help of (6), in full generality the following symbolic equations obtain

$$t_\alpha = \sum_\beta \underline{L}_{\alpha\beta}(\underline{C}_{o\gamma},\underline{T}_{o\gamma}(\underline{C}_{o\eta})).\underline{\tilde{c}}_\beta \qquad ,(7)$$

where $\underline{\tilde{c}}_\beta$ has not necessarily the same exact tensorial character as $\underline{C}_\beta$. To be more explicit, consider the electro-elastic case where $\underline{T}_1$ is the symmetry stress tensor, $\underline{C}_1$ is a symmetric (finite)strain tensor, $\underline{T}_2$ is a vector field such as an electric field and $\underline{C}_2$ is a vector field such as an electric polarization, and assume that $\underline{C}_{o1} = \underline{0}$ (K_o is unstrained). Then, in general, $\underline{T}_{o\alpha} = \hat{\underline{F}}_\alpha(\underline{0},\underline{C}_{o2}) \neq \underline{0}$, $\alpha = 1,2$. Let $\underline{T}_1 = \tilde{F}_1(\underline{C}_1,\underline{0})$ be the purely nonlinear elastic law associated with eqns.(5). Then $\underline{T}_{o1} = \hat{F}_1(\underline{0},\underline{C}_{o2}\neq\underline{0})$ are internal stresses created by the bias field $\underline{C}_{o2}$ (here an electric polarization) and, assuming that $\tilde{F}_1$ admits a unique inverse, there are associated with $\underline{T}_{o1}$ <u>internal</u> strains $\underline{C}_{o1}^{int}$ such that

$$\underline{C}_{o1}^{int} = \tilde{F}_1^{-1} \circ \underline{T}_{o1}(\underline{0},\underline{C}_{o2}) \qquad .(8)$$

Since we are in a fully nonlinear theory for the K_o solution, these strains are compatible if and only if the associated Riemmann-Christoffell curvature tensor, $\mathbb{R}(\underline{C}_{o1}^{int})$, vanishes. This will be identically the case if $\underline{C}_{o2}$ is spatially uniform. As to $\underline{t}_1$ and $\underline{t}_2$, they are obtained in the form

$$\begin{aligned} \underline{t}_1 &= \underline{L}_{11}(\underline{0},\underline{C}_{o2},\underline{T}_{o1}(\underline{0},\underline{C}_{o2}),\underline{T}_{o2}(\underline{0},\underline{C}_{o2}).\underline{\tilde{c}}_1 + \underline{L}_{12}(.,.,.,.,.,.).\underline{c}_2 \\ \underline{t}_2 &= \underline{L}_{21}(.,\ .\ ,\quad .\quad ,\quad .\quad).\underline{\tilde{c}}_1 + \underline{L}_{22}(.,.,.,.,.,.).\underline{c}_2 \end{aligned} \qquad ,(9)$$

where (this is shown in each particular case of eqns.(5))

$$\begin{array}{c|c} \underline{L}_{11} \in \mathcal{L}(\mathscr{C}, \mathscr{C}_S) & \underline{L}_{12} \in \mathcal{L}(\mathscr{V}, \mathscr{C}_S) \\ \hline \underline{L}_{21} \in \mathcal{L}(\mathscr{C}, \mathscr{V}) & \underline{L}_{22} \in \mathcal{L}(\mathscr{V}, \mathscr{V}) \end{array} \qquad (10)$$

and $\underline{\tilde{c}}_1 = \nabla\underline{u}$ is the full displacement gradient and <u>not</u> only its symmetric part $\underline{c}_1$. Equations (10) provide the analog of eqn.$(1)_2$ for electromechanical phenomena. Now the analog of eqn.$(1)_1$ is obtained by making $\underline{C}_{o2} = \underline{0}$ so that, assuming no hysteresis effects, hence $\hat{\underline{F}}_\alpha(\underline{0},\underline{0}) = \underline{0}$,α=1,2 ,it can be shown that the $\underline{L}_{\alpha\beta}(\underline{0},\underline{0},\underline{0},\underline{0}) = \underline{L}^o_{\alpha\beta}$ become such that eqn.(10) is replaced by (T = transpose)

$$\begin{array}{c|c} \underline{L}^o_{11} \in \mathcal{L}_S(\mathscr{C}_S, \mathscr{C}_S) & \underline{L}^o_{12} \in \mathcal{L}(\mathscr{V}, \mathscr{C}_S) \\ \hline \underline{L}^o_{21} = \underline{L}^{oT}_{12} \in \mathcal{L}(\mathscr{C}_S, \mathscr{V}) & \underline{L}^o_{22} \in \mathcal{L}_S(\mathscr{V}, \mathscr{V}) \end{array} \qquad . (11)$$

Consequently , $\underline{\tilde{c}}_1$ is replaced by $\underline{c}_1$ (linearized symmetric strain tensor), $\underline{L}^o_{11}$ admits at most 21 independent components while $\underline{L}^o_{12}$ and $\underline{L}^o_{22}$ will have at most 18 and 6 independent components, respectively. The Voigt notation of linear piezoelectricity underlies the scheme (11) for $\mathscr{C}_S \to \mathbb{R}^6$ and $\mathscr{V} \to \mathbb{R}^3$. The properties of symmetry and transposition of the operators in (11) follow from the fact that $F_1 \sim \partial\Sigma / \partial\underline{C}_1$ and $F_2 \sim \partial\Sigma / \partial\underline{C}_2$ if Σ is the energy density and

$$\underline{L}^o_{11} = \left.\frac{\partial^2\Sigma}{\partial\underline{C}_1 \otimes \partial\underline{C}_1}\right|_0 \quad , \quad \underline{L}^o_{12} = \left.\frac{\partial^2\Sigma}{\partial\underline{C}_1 \otimes \partial\underline{C}_2}\right|_0 \quad , \quad \underline{L}^o_{22} = \left.\frac{\partial^2\Sigma}{\partial\underline{C}_2 \otimes \partial\underline{C}_2}\right|_0 \qquad .(12)$$

The two results (10) and (11) exhibit all the differences between a linearization about a natural free state ($K_o = K_R$) —eqns.(11) — and the linearization about a bias state — eqns.(10). Indeed, once a material symmetry has been chosen with respect to K_R, from the theory of invariants of a scalar-valued function one knows on which set of invariants formed from $\underline{C}_1$ and $\underline{C}_2$ does the energy depend. The expressions of $\underline{L}^o_{\alpha\beta}$ and $\underline{L}_{\alpha\beta}(\underline{C}_{o2} \neq \underline{0})$ are easily computed but, whereas $\underline{L}^o$ contains only second-order derivatives of $\dot{}$ with respect to the $\underline{C}_\beta$'s, $\underline{L}_{\alpha\beta}(\underline{C}_{o2} \neq \underline{0})$ will contain first-order derivatives as well, thus exhibiting the influence of initial fields. In addition, eqns.(10) show that the rotation will appear along with the strain in the linearized constitutive equations while the initial field will modify the value of the elasticity coefficients. The important point here, obviously, is the existence or nonexistence of the coupling terms $\underline{L}_{12}$ and $\underline{L}_{21}$. In the electroelastic case, if K_R presents no centre of symmetry, we may have $\underline{L}^o_{12}$ nonzero for some of its components and, consequently, piezoelectricity. The important parameter then is the electromechanical coupling factor usually defined by [12]

$$K^2 = (L^o_{12})^2 / \rho_o L^o_{11} L^o_{22} \quad (\rho_o\text{:matter density}), \quad (13)$$

where we have considered typical components of the $\underline{L}_{\alpha\beta}$'s. Consequences of a nonzero K^2 are the stiffening of elasticity coefficients and the alteration in wave speeds (direct electroacoustic effect) and, for crystals whose axis of order six is set orthogonal to the sagittal plane, the possibility to propagate SH surface waves at the free or shorted surface of a piezoelectric substrate in the form of Bleustein-Gulyaev waves [12]-[13]. If the initial symmetry group is full isotropy or centered cubic, then obviously $\underline{L}^o_{12} = \underline{L}^o_{21} = \underline{0}$ and one must eventually go one step further to obtain electroelastic couplings of the quadratic type (electrostriction). The situation is quite different in the presence of an intense bias field where (10) applies since we may have both $\underline{L}_{12}$ and $\underline{L}_{21}$ nonzero in spite of the high degree of symmetry of the material in its free state. We may have biased piezoelectricity. If $\underline{E}_o$ is the bias field, we shall have, using a definition of the same type as (13), $K^2 \propto E_o^2$ so that, in theory at least, the alterations brought in the dynamical behavior can be controlled by the intensity of the bias field, provided the latter is sufficiently intense to have effects of significant magnitude. Furthermore, while in the case $\underline{E}_o = \underline{0}$ the propagation of Bleustein-Gulyaev waves may be forbidden, it may be allowed for a field $\underline{E}_o \neq \underline{0}$ set orthogonal to the sagittal plane and it is checked that the speed and penetration depth of this surface mode are controlled by $|\underline{E}_o|$ and that the wave degenerates into a face-shear mode for the vanishing of the bias field. This class of phenomena is illustrated for piezoelectric semiconductors in [14] and for paramagnets exhibiting induced linear magnetoelastic couplings (biased piezomagnetism) when subjected to an intense magnetic field in [15]. Insofar as the acoustic waves are concerned, the symmetry breaking may have for effect to split the double root for transverse elastic waves in two distinct roots, thus causing a linear birefringence which is the acoustic analog of the Voigt-Cotton Mouton effect mentioned in the Introduction (see [15] for an example). In the presence of dissipative effects we may have the acoustic analog of the optical Faraday effect (rotation of the plane of polarization). Obviously, the same formalism as the one in eqns.(10) will have to be used when studying the electro- or magneto-elastic stability of structures in bias electromagnetic fields. An example of this is given by the case of soft-ferromagnetic elastic plates in a transverse bias magnetic field where the anisotropy induced by the bias field causes additional couplings between the essentially two-dimensional problem for the flexion of the plate and the three-dimensional magnetostatic problem. The same conclusions regarding the symmetry breaking are reached whenever one studies the parametric excitation of the said plate by a spatially uniform, but time-varying, bias magnetic field (see [16]-[17]).

Ponderomotive force and couple.- Until now we have ignored these quantities. These are a priori quadratic in the electromagnetic entities and, as such, never appear in "apriori" linear theories, i.e., theories such as (11) and Voigt's theory of piezoelectricity where the linearization is perfomed about a free state. In the case of the scheme (10), it can be shown that these two concepts can be introduced in $\underline{t}_1$

and they cause additional alterations in $\underline{L}_{11}$ and $\underline{L}_{12}$.
In conclusion of this point , the present treatment proves the necessity and usefulness of "exact" nonlinear , rotationally invariant, theories of deformable electromagnetic continua (e.g.,in [18]-[20]) eventhough one may be finally interested only in linear electro- and/or magneto-acoustics since all subtleties contained in eqn. (10) would certainly escape any simpler approach.

ELECTRICALLY AND MAGNETICALLY ORDERED MATERIALS

As already noticed in the second section above, in the case of electromagnetically ordered crystals in their low-temperature phase a structure in domains is most common and one has to apply a strong bias field to recover a spatially homogeneous situation such as a one-domain structure. The effect of both ordering and bias field necessitates the primary construction of nonlinear theories of deformable ferroelectrics and ferromagnets and antiferromagnets. Such nonlinear theories using the notions of configurations and such ingredients as (6) have been constructed for deformable ferromagnets [21],[22], deformable antiferromagnets [23],[24] and deformable ferroelectrics [25],[26] in a continuum framework. The corresponding linear magneto- or electro-acoustic theories, using the scheme (6) — cf.Figure 3 — and producing

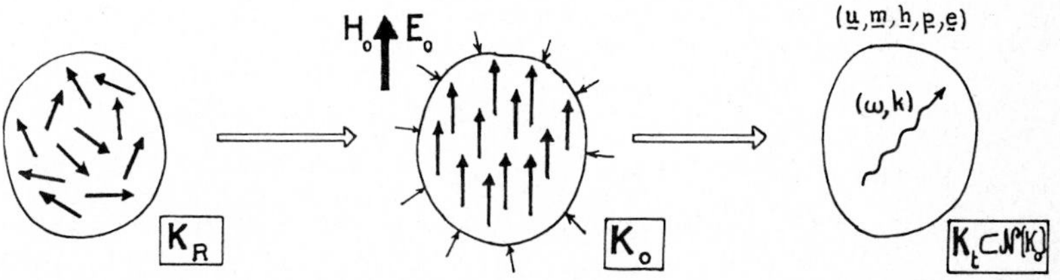

Figure 3.- Superimposition of small dynamical fields on a configuration rendered of the one-domain type by application of an intense bias field.

equations of the form of eqns.(10) have been deduced in various papers([27],[24], [26]). But there are essential differences with the case described by eqns.(10). First, additional independent variables are involved (magnetization and polarization gradients allowing for the phenomenological representation of ordering effects) and, more important, while $\underline{t}_2$ in eqns.(9) is a field which can eventually be eliminated after use of Maxwell's equations to yield a pure mechanical problem (hence only mechanical vibration modes are involved), in the case of electromagnetically ordered media the field $\underline{t}_2$ is not a Maxwellian field but a material field (magnetic anisotropy field in one case, local electric field in another) which contributes directly in a"non-mechanical" field equation (the spin-precession equation,or equations, in the case of ferromagnets and antiferromagnets, the equation governing the electric polarization in the ferroelectric case) which gives rise to new modes of propagation , the spin-wave modes or magnons in magnetically ordered crystals [28] and the ferroelectric soft modes in ferroelectrics [6]. In a linear analysis these new elementary oscillations couple with acoustic phenomena, for wavelengths which are still in the range of phenomenological physics,at resonance points [29],[30] on the real dispersion diagram with a typical repulsion of coupled branches at those points. The same holds true even if dissipative effects are considered in ferroelectrics [31] or in ferromagnets [32],[33].If a rather high symmetry, such as isotropy, is considered in the field-free state K_R, then the linearization of nonlinear equations allows one to show that the couplings result from the bias field, and the observed repulsion for bulk coupled modes (Figure 4a) is directly related to the magnitude of the bias field. Illustration of these for ferroelectrics and ferromagnets are given in Figure 4b and 4c, respectively. Direct consequences of the symmetry breaking by the bias field then are a birefringence effect for polaritons and an acoustical activity in ferroelectrics such as $BaTiO_3$ [31] , while a magnetoacoustic Faraday effect, an exchange of relaxation at the resonance point and the possibility of a magnetoacoustic dichroïsm are exhibited in the case of ferromagnets[27].

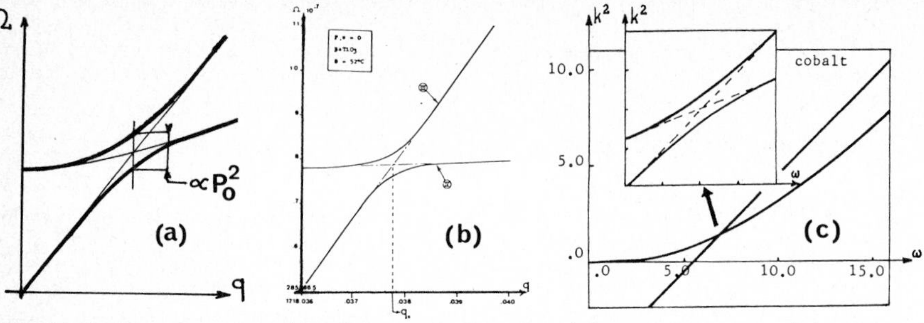

Figure 4.-Resonance coupling between bulk modes:(a)general picture,(b) in a ferroelectric, (c) in a ferromagnet (after [31] and [32]).

In the case of antiferromagnets [34]-[36], for certain settings of the bias magnetic field $\underline{H}_o$ one obtains a splitting of the spin wave branch in high and low magnon branches which is directly related to $|\underline{H}_o|$ (Figure 5a), with accompanying resonances and magnetoacoustic Faraday effects [36], while in other settings the breaking is complete in that it produces two distinct transverse acoustic modes which both couple with two magnon branches (Figure 5b). For the sake of illustration an experimentally measured dispersion relation is also given in Figure 5c.

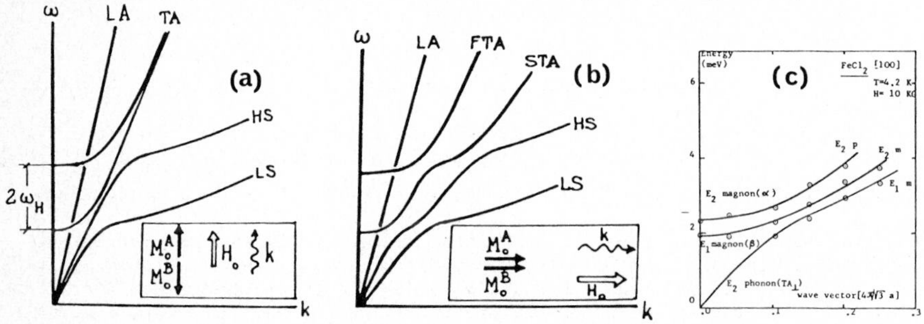

Figure 5.-Magnetoacoustic resonance in antiferromagnets:(a)energy gap between magnons proportional to H_o,(b)complete breaking,(c)$FeCl_2$(after [37]).

Insofar as coupled surface waves are concerned, the application of an intense bias electric field orthogonally to the sagittal plane has allowed us [38] to prove the existence of a strongly dispersive (in reason of its coupling with the ferroelectric soft mode) SH surface mode of the Bleustein-Gulyaev type in ferroelectrics such as $BaTiO_3$ even when the field-free configuration has been assumed of a rather high symmetry. In addition, the penetration depth of the mode is controlled by the intensity of the bias electric field and boundary-layer effects complicate the picture as a result of the ordering inherent in ferroelectrics (cf.Figures 6). Correspondingly, Parekh [39] was the first to recognize that whenever the bias magnetic field is set orthogonal to the sagittal plane, an essentially SH surface wave, akin to the Bleustein-Gulyaev mode of piezoelectricity, could be maintained at the free surface of a ferromagnetic substrate. Due to the axial nature of magnetic fields, one direction of propagation and its opposite are not equivalent for this process, that is, SH magnetoacoustic waves propagating to the right and the left of the bias magnetic field have differing dispersion relations [40] — Figures 7. It was indeed proven (see Section VII in Ref.[13]) that the generalized Rayleigh problem for magnetoelastic waves at the free surface of a ferromagnet splits exactly into a classical Rayleigh problem (governing the elastic component parallel to the sagittal plane) and the above SH surface mode whenever isotropy is considered in the field-free state in the nonlinear theory and the symmetry is bkoken by an intense bias magnetic field set orthogonally to the sagittal plane.

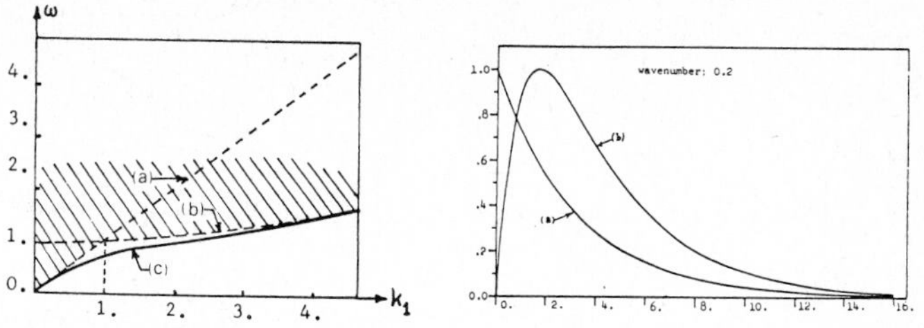

(a) (b)

Figure 6.- (a)Dispersion relation for Bleustein-Gulyaev waves in ferroelectrics (grounded boundary[38]);(b)Behavior of amplitude with depth (after [38]).

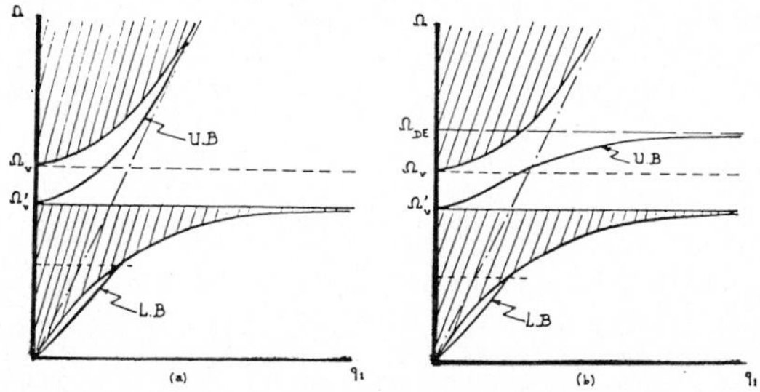

Figure 7.- Nonreciprocity of SH surface magnetoelastic waves in elastic ferromagnets(a)left to right,(b)right to left (after [40] and [13]).

NONLINEAR PROBLEMS

The equations of the nonlinear theory of ferroelectrics linearized about a bias electric field can be fully recovered by taking the long wavelength limit of a one-dimensional lattice model [41] based on the proptotype picture provided by the crystalline structure of $NaNO_2$(cf.Figure 2a).The model reproduces well the stiffening of elasticity coefficients by the permanent electric dipoles and the resonance coupling between the ferroelectric soft mode and a transverse acoustic mode, the coupling parameter being, in effect, directly related to the square of the spontaneous polarization. Whenever one extends this model so as to account for nonlinearities in the ferroelectric order parameter (the electric polarization, a vector, rigidly attached to the central molecular group in the crystalline structure) then, without introducing elastic nonlinearities, single- and multiple-soliton solutions can be exhibited which couple to the stress field with an effective electromechanical coupling factor. The system obtained can be shown to be governed by a double sine-Gordon equation (for the orientation of dynamical electric dipoles) in the form [42]

$$\frac{\partial^2 \phi}{\partial \tau^2} - \frac{\partial^2 \phi}{\partial X^2} - \sin\phi + \frac{\eta^2}{2}\left[\frac{Q^2}{(\Omega^2 - \Omega_T^2)}\right]\sin(2\phi) = 0, \quad (14)$$

with a transverse elastic displacement v given by

$$\frac{dv}{d(QX - \Omega\tau)} = -\frac{\eta Q \sin\phi}{(\Omega^2 - \Omega_T^2)}, \quad \Omega_T^2 = \hat{V}_T^2 Q^2 \quad ,(15)$$

where Q and Ω are a pseudo wavenumber and a pseudo frequency, $\hat{V}_T$ is the transverse elastic disturbance speed altered by permanent dipoles and $\eta \propto P_0^2$. A difficult mathematical and numerical study [42], which is here irrelevant, allows one to exhibit in a picturesque manner the incidence of the η-parameter on electro-elastic solutions which physically represent the motion of a wall in a ferroelectric (Figure 8) with the accompanying radiation of elastic solitons through η, as well as the soliton-antisoliton collision accompanied by the radiation of elastic solitons and the radiation of harmonic electric waves (Figure 9).The latter case represents two walls moving toward one another and then separating and leaving polarization states which have been interchanged. This rather sophisticated example where the initial configuration K_o is not spatially homogeneous , concludes the present review. In this case,

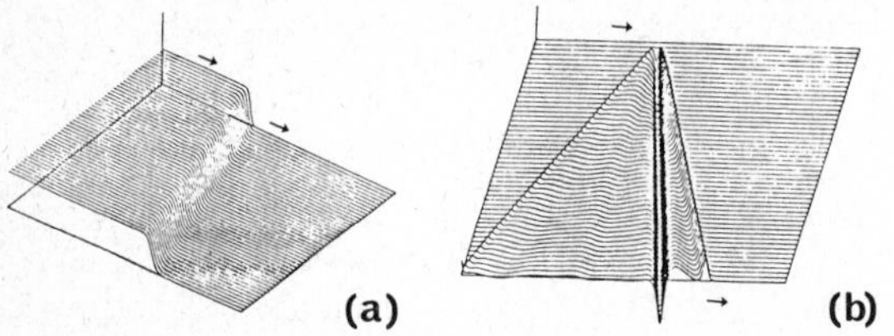

Figure 8.-Single soliton in $NaNO_2$:(a) rotation of the polarization for a 180°wall, (b)accompanying radiation of elastic soliton (after [42]).

because of the inhomogeneity of the initial configuration, we may say that the symmetry breaking is only local . Other sophisticated examples of such a local symmetry breaking superimposed on inhomogeneous initial configurations are provided by the study of Bloch and Néel wall vibrations in elastic ferromagnets where magnetostriction causes a local anisotropy inducement within the wall [43], as also by the study of electroacoustic echoes in a resonant piezoelectric powder of which each grain is equipped with its own piezoactive axis [44]. Further problems along the same line of thought could consider materials such as certain compounds which are simultaneously ferroelectric and ferromagnetic (cf. the linearized rigid case in [45]).

Acknowledgment : The author expresses his debt to his co-workers and friends A.Askar, B.Collet, N.Daher, A.Fomèthe, C.Goudjo, A.Hakmi, S.Motogi, J.Pouget and J.Sioké-Rainaldy.

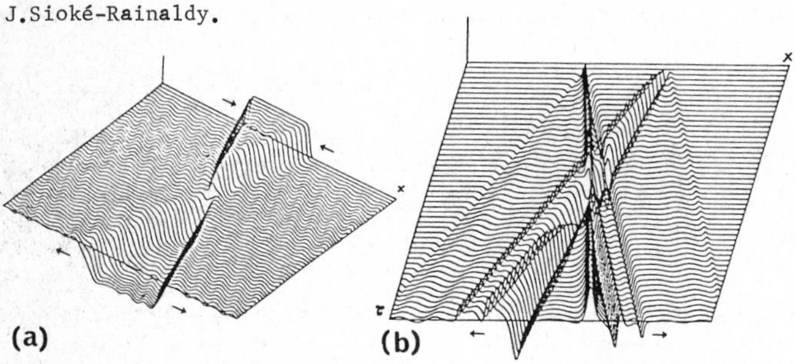

Figure 9.-Soliton-antisoliton collision in $NaNO_2$(a);(b)accompanying radiation of elastic solitons (after [42]).

REFERENCES

[1] Landau L.D. and Lifshitz E.M.,Electrodynamics of Continua(Pergamon,Oxford,1960)
[2] Boccara N., Symétrie brisée (Hermann, Paris, 1972).
[3] Boccara N.(Editor),Symmetries and Broken Symmetries in Condensed Matter Physics, (Proc.Pierre Curie Colloquium,Paris,1980;I.D.S.E.T,Paris,1981).
[4] Wadhwan V.K., Ferroelasticity and related properties of crystals, Phase Transitions,3(1982) 3-103.
[5] Zheludev I.S., Ferroelectricity and symmetry, in Solid State Physics,Vol.26, pp.429-464,eds.H.F.Ehrenreich, F.Seiz and D.Turnbull (Academic Press, New New , 1971).
[6] Blinc R. and Zeks B., Soft modes in ferroelectrics and antiferroelectrics, (North Holland Publ.Co.,Amsterdam, 1974).
[7] Tareev B., Physics of Dielectric crystals (MIR Publishers,Moscow, 1975).
[8] Zorski H.,On the equations describing small deformations superposed on finite deformation, in Proc.Intern.Symposium on Second-order effects, ed.H.Reissner, pp.109-128 (Pergamon, Oxford, 1962).

[9] Suhubi E.S.,Thermoelastic Solids, in Continuum Physics, Vol.II,pp.173-265, ed.A.C.Eringen (Academic Press, New York, 1975).
[10] Toupin R.A., A Dynamical Theory of dielectrics, Int.J.Engng.Sci.,1(1963)101-106.
[11] Baumhauer J.C. and Tiersten H.F.,Nonlinear electroacoustic equations for small fields superimposed on a bias, J.Acoust.Soc;Amer.,54(1973)1017-1034.
[12] Dieulesaint E. and Royer D.,Ondes élastiques dans les solides:Application au traitement du signal (Masson, Paris, 1974).
[13] Maugin G.A.,Elastic surface waves with transverse horizontal polarization, in Advances in Applied Mechanics, Vol.23,pp.373-434,ed.J.W.Hutchinson (Academic Press, New York, 1983).
[14] Daher N.,Waves in elastic semiconductors in a bias field, in these proceedings.
[15] Hakmi A. and Maugin G.A.,Wave propagation in paramagnetic bodies exhibiting induced linear magnetoelastic couplings, in these proceedings.
[16] Maugin G.A., and Goudjo C., The equations of soft-ferromagnetic plates, Int.J.Solids Structures, 18(1982) 889-912.
[17] Goudjo C. and Maugin G.A., On the static and dynamic stability of soft-ferromagnetic plates, J.Méca.Théor.Appl., 2(1983,in print).
[18] Nelson D.F., Electric, optic and acoustic interactions in dielectrics, (J.Wiley-Interscience, New York,1979).
[19] Eringen A.C., Mechanics of Continua, 2nd Edition(Chapter 10 added)(Krieger, New York,1980).
[20] Maugin G.A., The principle of virtual power in continuum mechanics:Application to coupled fields, Acta Mechanica, 35(1980)-1-70 [partial print of "The Foundations of the Electrodynamics of Nonlinear Continua",Prace I.P.P.T-P.A.N., Warsaw,1976].
[21] Tiersten H.F.,Coupled magnetomechanical equations for magnetically-saturated insulators, J.Math.Phys., 5(1964) 1298-1318.
[22] Maugin G.A., Micromagnetism, in Continuum Physics, Vol.III,pp.213-312, ed. A.C.Eringen,(Academic Press, New York, 1976).
[23] Maugin G.A., A continuum theory of deformable ferrimagnetic bodies-I,II, J.Math.Phys., 17(1976)1727-1751.
[24] Sioké-Rainaldy J. and Maugin G.A.,Magnetoelastic equations for antiferromagnetic insulators of the easy-axis type, J.Appl.Phys.,54(1983)1490-1506.
[25] Maugin G.A.,Deformable dielectrics-I, Arch.Mech.(Poland),28(1976)679-692; Parts II and III, ibid,29(1977)143-159, 251-258.
[26] Maugin G.A. and Pouget J.,Electroacoustic equations in one-domain ferroelectric bodies, J.Acoust.Soc.Amer., 67(1980)575-587.
[27] Maugin G.A., A continuum Approach to magnon-phonon couplings-I,II, Int;J;Engng.Sci., 17,(1979)1073-1091, 1093-1108.
[28] Turov A.E., Physical properties of magnetically ordered crystals (Academic Press, New York, 1965).
[29] Maugin G.A., Elastic-electromagnetic resonance couplings in electromagnetically ordered media, in Theoretical and applied mechanics, pp.345-355, eds.F.P.J. Rimrott and B.Tabarrok (North Holland Publ.Co.,Amsterdam, 1980).
[30] Maugin G.A., Wave motion in magnetizable deformable solids:Recent advances, Int.J.Engng.Sci., 19(1981)321-388.
[31] Pouget J. and Maugin G.A., Coupled acoustic-optic modes in deformable ferroelectrics, J.Acoust.Soc.Amer.,68(1980) 588-601.
[32] Maugin G.A. and Pouget J., A continuum approach to magnon-phonon couplings-III, Int.J.Engng.Sci., 19(1981) 479-492.
[33] Fomèthe A. and Maugin G.A., Influence of dislocations on magnon-phonon couplings,Phenomenological approach, Int.J.Engng.Sci.,20(1982) 1125-1144.
[34] Morgenthaler F.R., Dynamic magnetoelastic coupling in ferromagnets and antiferromagnets, I.E.E.E.Trans.Mag.,8(1972)130-151.
[35] Maugin G.A. and Sioké-Rainaldy J., Caractérisation de milieux magnétiques par l'étude de la résonance magnétoacoustique, Revue du CETHEDEC,No.71(1982)209-218.

[36] Maugin G.A. and Sioké-Rainaldy J., Magnetoacoustic resonance in antiferromagnetic insulators in weak magnetic fields, J.Appl.Phys.,54(1983)1507-1518.
[37] Lovesey S.W.,Magnon-phonon hybridization in magnetic salts with unquenched orbital angular momentum, Commun.Solid State Phys.,7(1974)117-124.
[38] Pouget J. and Maugin G.A., Bleustein-Gulyaev Surface modes in elastic ferroelectrics, J.Acoust.Soc;Amer.,69(1981)1304-1318.
[39] Parekh J., in Proceedings of the I.E.E.E. Ultrasonics Symposium(Boston,1972); p.333 (I.E.E.E.Publ.,New York,1972)
[40] Scott R.Q. and Mills D.L., Propagation of surface magnetoelastic waves on ferromagnetic crystal substrate, Phys.Rev.,B15 (1977) 3545-3557.
[41] Maugin G.A. and Pouget J., Nonlinear elastic waves generated by electric solitons in ferroelectric crystals, in Nonlinear Deformation Waves, pp.410-417, eds.U.Nigul and J.Engelbrecht (Springer-Verlag,Berlin,Heidelberg,1983).
[42] Pouget J. and Maugin G.A.,Solitons and electroacoustic interactions -I- Single soliton and domain walls,II-Interactions of solitons and radiations (Preprints, 1983).
[43] Motogi S. and Maugin G.A.,Effects of magnetostriction on vibrations of Bloch and Néel walls (Preprint,1983);see also in these proceedings.
[44] Pouget J. and Maugin G.A.,A continuum approach to electroacoustic echoes in piezoelectric powders, J.Acoust.Soc.Amer.,70(1983,in print);see also Pouget J. in these proceedings.
[45] Maugin G.A., Dynamic magnetoelectric couplings in ferroelectric ferromagnets, Phys.Rev.,B23(1981)4608-4614.

The Mechanical Behavior of Electromagnetic Solid Continua
G.A. Maugin (editor)
Elsevier Science Publishers B.V. (North-Holland)

MODELES ET SOLUTIONS EFFECTIVES EN ELECTROMAGNETO-ELASTICITE

Parton V.Z.

Institut des constructions mécaniques
pour l'industrie chimique
Moscou, URSS

La principale propriété des piézo-électriques, celle de transformer l'énergie mécanique en énergie électrique et inversément trouve aujourd'hui des applications de plus en plus larges dans divers domaines de la technique. Ce progrès est, avant tout, dû à la position correcte et à l'élaboration des méthodes de solution des problèmes d'électro-élasticité qui se basent sur la description phénoménologique de l'effet piézo-électrique [1-4].

1. Equations d'état de l'électro-élasticité linéaire. La forme la plus usitée des équations d'état d'un milieu piézo-électrique se déduit de l'expression pour l'enthalpie électrique H_2 qui est une fonction des déformations ε_{ij} et du champ électrique E_i .
En utilisant la représentation de la fonction H_2 sous la forme

$$H_2 = \frac{1}{2} C_{ijk\ell} \varepsilon_{ij} \varepsilon_{k\ell} - \ell_{ijk} E_i \varepsilon_{jk} - \frac{1}{2} \varepsilon^S_{ij} E_i E_j \tag{1.1}$$

et les expressions des contraintes σ_{ij} et de l'induction électrique D_i au moyen de H_2

$$\sigma_{ij} = \frac{\partial H_2}{\partial \varepsilon_{ij}} , \quad D_i = -\frac{\partial H_2}{\partial E_i} , \tag{1.2}$$

on obtient des relations linéaires

$$\sigma_{ij} = C_{ijk\ell} \varepsilon_{k\ell} - \ell_{mij} E_m , \quad D_i = \ell_{ik\ell} \varepsilon_{k\ell} + \varepsilon^S_{ij} E_j , \tag{1.3}$$

où $C_{ijk\ell}$ sont des modules d'élasticité (tenseur de rang 4), ℓ_{mij} , des constantes piézo-électriques (tenseur de rang 3), ε^S_{ij} , des constantes diélectriques (tenseur de rang 2) mesurés à déformations constantes.

2. Equations de mouvement. Il est d'usage, en Electro-élasticité de négliger les forces massiques qui peuvent apparaître lors d'interaction des courants et polarisations induits dans le matériau avec les champs électriques et magnétiques. Ainsi donc, la solution des problèmes d'électro-élasticité fait appel aux équations de mouvement connues sous la forme

$$\sigma_{ij,j} = \rho \frac{\partial^2 u_i}{\partial t^2} \tag{2.1}$$

où u_i est le vecteur déplacement, ρ , la densité du matériau.

De plus, en position linéaire on peut négliger les contraintes électromagnétiques surfaciques. Par conséquent, les conditions aux limites pour les composantes mécaniques du champ électro-élastique dans les piézo-électriques ont la même forme que dans la théorie d'élasticité classique.

3. Champ électromagnétique. En l'absence du courant de conduction et des charges libres le champ électromagnétique dans un milieu piézo-électrique doit satisfaire aux équations de Maxwell

$$\varepsilon_{ijk} E_{j,k} + \frac{\partial B_i}{\partial t} = 0, \qquad \varepsilon_{ijk} H_{j,k} - \frac{\partial D_i}{\partial t} = 0, \tag{3.1}$$
$$D_{i,i} = 0, \quad B_{i,i} = 0,$$

où ε_{ijk} est le tenseur unité antisymétrique, $B_i = \mu_0 H_i$, le vecteur induction magnétique.

En éliminant B_i des équations de Maxwell il vient

$$E_{k,ki} - E_{i,kk} = -\mu_0 \frac{\partial^2 D_i}{\partial t^2}. \tag{3.2}$$

Ce système d'équations avec les équations d'état sont utilisées pour résoudre les problèmes portant sur le rayonnement des ondes électromagnétiques par un milieu piézo-actif.

Dans la plupart de problèmes sur le fonctionnement des transducteurs piézo-électriques et sur la propagation des ondes électroacoustiques il est possible de négliger les effets magnétiques et d'utiliser l'approximation quasistatique pour le champ électrique, i.e. les équations d'électrostatique

$$\varepsilon_{ijk} E_{j,k} = 0, \quad D_{i,i} = 0. \tag{3.3}$$

Si l'on introduit le potentiel électrique $E_i = -\varphi_{,i}$ le système d'équations complet pour un milieu piézo-électrique linéaire s'obtient par la substitution des équations d'état dans les équations d'électrostatique et de mouvement

$$c_{ijkl} \cdot u_{k,lj} + e_{mij}\, \varphi_{,mj} = \rho \frac{\partial^2 u_i}{\partial t^2},$$
$$e_{ikl}\, u_{k,li} - \varepsilon^S_{ij}\, \varphi_{,ji} = 0. \tag{3.4}$$

Notons que dans un cas général où la surface d'un corps électrique est limitrophe d'un certain milieu extérieur il y a lieu d'ajouter aux équations (3.4) les équations de Maxwell pour le milieu donné et de tenir compte des conditions aux limites pour les composantes du champ électromagnétique. Si le milieu extérieur est le vide ou l'air pour lequel la permittivité diélectrique est sensiblement inférieure aux constantes ε^S_{ij} de la plupart des matériaux piézo-actifs la condition de la continuité de la composante normale du vecteur induction électrique peut être remplacée par une égalité approchée

$$n_i D_i = 0, \tag{3.5}$$

où n_i est le vecteur unité de la normale à la surface du corps.

Certains problèmes d'importance pratique de vibration des corps piézo-électriques se laissent résoudre en partant des principes variationnels d'électroélasticité. L'équation variationnelle fondamentale pour le milieu piézo-électrique, une généralisation du principe de Hamilton peut être écrite sous la forme

$$\delta\int_{t_0}^{t_1} dt\int_V (K-H_2)dV+\int_{t_0}^{t_1} dt\int_S (t_i\delta u_i+\sigma_0\,\delta\varphi)dS=0. \tag{3.6}$$

Ici $K=\frac{1}{2}\rho\dot{u}_i\dot{u}_i$ est l'énergie cinétique, H_2, la densité d'enthalpie électrique, $t_i=\sigma_{ij}n_j$, les forces surfaciques, $\sigma_0=n_iD_i$, les charges électriques surfaciques, t_0, t_1 les dates fixes.

Une telle position permet de poser et de résoudre des problèmes de concentration des contraintes, de résistance, de fiabilité et de rupture, i.e. de développer la théorie d'élasticité, la théorie des plaques et des enveloppes pour de tels matériaux que sont les piézo-électriques. Considérons quelques exemples caractéristiques.

4. <u>Rupture fragile des matériaux piézo-électriques</u>. La présence, dans les matériaux piézo-électriques réels, des défauts du types fissure, entaille, cavités étroites, etc., peut causer parfois leur rupture. Dans ce cas aussi l'approche energétique de Griffits s'est trouvée fort fructueuse pour l'étude théorique des divers aspects du processus de rupture et lors d'élaboration des méthodes pratiques de calcul de résistance des éléments de construction en matériaux piézo-électriques.

Considérons un piézo-électrique occupant un volume V et caractérisé par une fissure de forme quelconque. Soit $\sigma_{ij}^{(0)}$, $\varepsilon_{ij}^{(0)}$, $u_i^{(0)}$, $\varphi^{(0)}$ les composantes des tenseurs des contraintes, des déformations, du vecteur déplacement et le potentiel électrique dans un certain état initial "0" du corps et $\sigma_{ij}^{(1)}$, $\varepsilon_{ij}^{(1)}$, $u_i^{(1)}$, $\varphi^{(1)}$ les quantités correspondantes dans l'état "1" tel qu'une partie de la surface bilatérale des fissures reçoit un accroissement $\Delta\Sigma$. On écrit les conditions aux limites sur la surface $\Sigma+\Delta\Sigma$ conformément à la position linéarisée du problème. En écrivant l'expression de l'énergie interne du corps en états "0" et "1"

$$U^{(\kappa)}=\frac{1}{2}\int_V\left[\sigma_{ij}^{(\kappa)}\varepsilon_{ij}^{(\kappa)}+E_j^{(\kappa)}D_j^{(\kappa)}\right]dV,\quad (\kappa=0,1) \tag{4.1}$$

et en calculant l'accroissement de l'énergie interne ΔU au passage de l'état "0" à l'état "1" nous obtenons pour le flux d'énergie lors de formation de la discontinuité, l'expression suivante

$$\Delta A_{\Delta\Sigma}=\frac{1}{2}\int_{\Delta\Sigma}\left[\sigma_{ij}^{(0)}n_ju_i^{(1)}+D_j^{(0)}n_j\varphi^{(1)}\right]dS. \tag{4.2}$$

Conférons à la condition de cheminement de la fissure la forme suivante:

$$\gamma\left[\Delta\Sigma_1+\Delta\Sigma_2\right]=-\,\Delta A_{\Delta\Sigma}, \tag{4.3}$$

où γ est la densité de l'énergie surfacique; $\Delta\Sigma_1,\Delta\Sigma_2$ les surfaces de la discontinuité supplémentaire.

Rapportons le milieu piézo-électrique à un repère trirectangle d'orientation arbitraire $x_k(k=1,2,3)$ et supposons que la fissure rectiligne de largeur 2ℓ ($|x_1|<\ell$) se trouve dans le plan $x_2=0$. En supposant l'état électro-élastique indépendant de la coordonnée x_3 et les bords de la fissure exempts de charges libres et de charge mécanique présentons la condition de cheminement de la fissure sous la forme

$$2\gamma=\frac{1}{2}\lim_{\Delta\ell\to 0}\frac{1}{\Delta\ell}\int_{\ell}^{\ell+\Delta\ell}\Big\{\sigma_{i2}^{+}(x_1,0,\ell)\big[u_i(x_1,0,\ell+\Delta\ell)\big]+ \quad (4.4)$$
$$+D_2^{+}(x_1,0,\ell)\big[\varphi(x_1,0,\ell+\Delta\ell)\big]\Big\}\,dx_1 .$$

Ici $\sigma_{i2}^{+}(x_1,0,\ell), D_2^{+}(x_1,0,\ell)$ sont des composantes des contraintes et la composante normale du vecteur induction électrique sur le prolongement de la fissure $[u_i]$, $[\varphi]$ le saut du vecteur déplacements et du potentiel électrique sur la fissure de largeur $2(\ell+\Delta\ell)$.

Il découle des raisonnements physiques que le bord de la fissure est une sorte de déversoir d'énergie libérée reparti le long du contour de la fissure. En introduisant le vecteur densité d'écoulement d'énergie $\vec{\Gamma}$ (Γ_1, Γ_2, Γ_3) pour un milieu piézo-électrique dans les conditions de statique il est possible de démontrer à l'aide de relations

$$\Gamma_k=\oint_C\big[Hn_i-\sigma_{ij}n_iu_{j,k}+D_in_iE_k\big]dS,\quad (k=1,2,3) \qquad (4.5)$$

que les Γ_k ne dépendent pas du choix du contour fermé C entourant l'extrémité de la fissure au cas où sur la surface de la fissure sont vérifiées les conditions

$$D_in_i=0\quad \sigma_{ij}n_i=0. \qquad (4.6)$$

Dans (4.5) $H = U - E_iD_i$ est l'enthalpie électrique, h_k, les composantes du vecteur unité de la normale à C.

Si le développement de la fissure dans le milieu piézo-électrique se produit dans son plan primitif, $x_2 = 0$ la condition de rupture revêt la forme

$$\vec{t}\cdot\vec{\Gamma}=2\gamma. \qquad (4.7)$$

La condition (4.7) est valable pour les fissures contenues dans le plan isotrope du matériau; dans le cas général d'anisotropie du matériau on a

$$\Gamma(\theta)=\Gamma_1\cos\theta+\Gamma_2\sin\theta=2\gamma(\theta) \qquad (4.8)$$

où $\gamma(\theta)$ est une fonction de θ obtenue expérimentalement.

5. Enveloppes piézo-électriques. Une large application d'éléments piézo-électriques à parois minces comme transducteurs d'énergie basse fréquence a nécessité la création des théories appliquées.

Soit une enveloppe en céramique piézo-électrique d'épaisseur h polarisée suivant la normale à la surface médiane et rapportée au repère trirectangle α_1, α_2, z . Supposons que les courbes coordonnées α_1, α_2 coïncident avec les lignes de courbures principales de la surface médiane et l'axe des z avec la direction de polarisation préalable. Sur les surfaces $z = \pm h/2$ de l'enveloppe il y a de fines électrodes indéfinies aux quelles est appliqué un potentiel

$\varphi(\pm h/2) = \pm V_0$. L'adoption des hypothèses de Kirchhoff-Love conduit, on le sait, aux équations ordinaires d'équilibre d'un élément d'enveloppe et à la répartition linéaire des déplacements tangentiels et des déformations suivant l'épaisseur de l'enveloppe. En tenant compte des hypothèses admises et de la forme des tenseurs C^E_{ijkl}, ℓ_{mij}, ε^S_{ij} pour les classes de symétrie ∞m et 6 mm conférons aux équations d'état (1.3) la forme suivante

$$\sigma_{11} = C^*_{11}\varepsilon_{11} + C^*_{12}\varepsilon_{22} - \ell^*_{31}E_3,\quad \sigma_{22} = C^*_{12}\varepsilon_{11} + C^*_{11}\varepsilon_{22} - \ell^*_{31}E_3, \tag{5.1}$$
$$\sigma_{12} = \frac{1}{2}\left(C^E_{11} - C^E_{12}\right)\varepsilon_{12},\quad D_j = \varepsilon^S_{11}\cdot E_j,\ D_3 = \ell^*_{31}(\varepsilon_{11}+\varepsilon_{22}) + \varepsilon^*_{33}E_3, (j=1,2).$$

La seconde équation (3.4), équation d'électrostatique, compte tenu de (5.1), pour les classes de symétrie indiquées, s'écrit sous forme symbolique [5]

$$\frac{d^2\varphi}{dz^2} + p^2\varphi = \frac{\ell^*_{31}}{\varepsilon^*_{33}}\left(\varkappa_1 + \varkappa_2\right), \tag{5.2}$$
$$p^2 = \frac{1}{A_1A_2}\frac{\varepsilon^*_{11}}{\varepsilon^*_{33}}\left[\frac{\partial}{\partial\alpha_1}\frac{A_2}{A_1}\frac{\partial(\ldots)}{\partial\alpha_1} + \frac{\partial}{\partial\alpha_2}\frac{A_1}{A_2}\frac{\partial(\ldots)}{\partial\alpha_2}\right],$$

où $\varkappa_1, \varkappa_2$ sont les variations de courbures principales, A_j, les paramètres de Lamé de la surface médiane de l'enveloppe. En introduisant les caractéristiques intégrales de champ électrique à l'aide de relations

$$\Phi_j = \int_{-h/2}^{h/2} z^{j-1}\varphi(\alpha_1,\alpha_2,z)\,dz, \tag{5.3}$$

représentons la solution générale de (5.2) sous la forme

$$\varphi = \frac{p\,h/2\cos pz}{h\sin p\,h/2}\Phi_1 + \frac{2(p\,h/2)^2\cdot\sin pz}{h^2\sin p\,h/2\,(1 - p\,h/2\,\mathrm{ctg}\,p\,h/2)}\Phi_2 - \tag{5.4}$$
$$-\frac{\ell^*_{31}h^2\cos pz}{4\varepsilon^*_{33}\,p\,h/2\sin p\,h/2}\left(\varkappa_1+\varkappa_2\right) + \frac{\ell^*_{31}}{\varepsilon^*_{33}p^2}\left(\varkappa_1+\varkappa_2\right).$$

En satisfaisant la condition pour φ nous obtenons les équations d'ordre indini. En n'y retenant que les termes à $(p\,h/2)^2$ près on obtient les équations approchées par rapport à Φ_1, Φ_2 :

$$\left[1 - \frac{(ph^2)}{12}\right]\Phi_1 = -\frac{\ell^*_{31}h^3}{12\varepsilon^*_{33}}\left(\varkappa_1+\varkappa_2\right), \tag{5.5.}$$
$$\left[1 - \frac{(ph^2)}{60}\right]\Phi_2 = \frac{V_0h^2}{6}.$$

En introduisant les caractéristiques intégrales (efforts et moments) et en attirant les relations connues [6] il vient

$$T_1 = C^*_{11}h(\varepsilon_1 + \nu_*\varepsilon_2) + 2\ell^*_{31}V_0,\quad T_2 = C^*_{11}h(\nu_*\varepsilon_1 + \varepsilon_2) + 2\ell^*_{31}V_0,$$
$$S = \frac{1}{2}\left(C^E_{11} - C^E_{12}\right)h\omega,\quad H = \frac{1}{24}\left(C^E_{11} - C^E_{12}\right)h^3\tau, \tag{5.6}$$
$$M_1 = \frac{C^*_{11}h^3}{12}\left(\varkappa_1 + \nu_*\varkappa_2\right) - \ell^*_{31}\Phi_1,\quad M_2 = \frac{C^*_{11}h^3}{12}\left(\nu_*\varkappa_1 + \varkappa_2\right) - \ell^*_{31}\Phi_1,$$

où $\varepsilon_j, \varkappa_j, \omega, \tau$ sont des composantes des déformations de la surface médiane de l'enveloppe.

Les conditions aux limites posées au contour de l'enveloppe comportent les conditions ordinaires mécaniques et la condition pour le potentiel

$$\left(\frac{\cos\gamma}{A_1}\frac{\partial}{\partial\alpha_1}+\frac{\sin\gamma}{A_2}\frac{\partial}{\partial\alpha_2}\right)\varphi=-\frac{D}{\varepsilon_{11}^*} \qquad (5.7)$$

dont on tire, compte tenu de (5.3), pour Φ_j

$$\left(\frac{\cos\gamma}{A_1}\frac{\partial}{\partial\alpha_1}+\frac{\sin\gamma}{A_2}\frac{\partial}{\partial\alpha_2}\right)\Phi_j=-\Phi_j^*, \qquad (5.8)$$

où $\Phi_j^*=\frac{1}{\varepsilon_{11}^*}\int_{-h/2}^{h/2}z^{j-1}D\,dz$, γ est l'angle entre la normale au contour de la surface médiane et la courbe coordonnée α_1.

Notons que la déduction des équations approchées (5.5) n'utilise point la condition d'épaisseur constante de l'enveloppe. En admettant, dans (5.5) et (5.6) que $h=h(\alpha_1,\alpha_2)$ nous obtenons les équations pour la détermination des caractéristiques intégrales du potentiel du champ électrique et les équations pour les déplacements de la surface médiane des enveloppes d'épaisseur variable.

6. Enveloppes multicouches en céramique piézoélectrique à polarisation en épaisseur des couches séparées l'une de l'autre par des électrodes. Considérons une enveloppe composée, de 2m+1 couches piézoélectriques séparées l'une de l'autre par des électrodes infiniment minces portées à des potentiels donnés. L'on suppose que les couches symétriques par rapport à la surface médiane de la couche médiane ont des épaisseurs égales et des propriétés physicomécaniques identiques. On applique à la surface métallisée de chaque couche une différence de potentiel de $2V_0$.

Hypothèses principales utilisées pour la mise en oeuvre d'une théorie des enveloppes multicouches: 1) hypothèse des normales non déformables admise pour tout l'ensemble du paquet de l'enveloppe; 2) en tout point de toute couche les contraintes σ_{i3} peuvent être admises négligeables par rapport aux σ_{11}, σ_{22}, σ_{12}.

On sait que la première hypothèse conduit aux relations cinématiques coïncidant avec la théorie de l'enveloppe à une couche alors que les équations d'équilibre en efforts et moments ont la forme ordinaire.

En supposant que les couches de l'enveloppe travaillent sans glissement on peut montrer que les conditions mécaniques de contact sont remplies en vertu de la première hypothèse et de l'équation d'équilibre. Comme les surfaces de face de chaque couche sont métallisées alors, dans les limites de la k-ème couche sont valables les équations d'Electrostatique (3.3). Introduisons, dans chaque couche, un repère local α_1, α_2, z lié à la surface médiane de la k-ème couche (δ_k étant l'épaisseur de cette couche). Le potentiel électrique aux électrodes étant symétrique gauche on obtient 2(m+1) équations de forme (5.5) pour la détermination des caractéristiques intégrales. Les conditions aux limites mécaniques ont la forme ordinaire celles électriques dans les limites de la k-ème couche sont données sous la forme (5.8).

En introduisant de manière ordinaire les efforts et moments [6] on obtient

$$T_1 = B_{11}\varepsilon_1 + B_{12}\varepsilon_2 + 2E_{31}V_0\ , \quad S = B\omega\ ,$$
$$T_2 = B_{12}\varepsilon_1 + B_{11}\varepsilon_2 + 2E_{31}V_0\ , \quad H = D\tau\ , \tag{6.1}$$
$$M_1 = D_{11}\varkappa_1 + D_{12}\varkappa_2 + M_Э\ , \quad M_2 = D_{12}\varkappa_1 + D_{11}\varkappa_2 + M_Э\ ,$$

où

$$B_{1j} = 2\sum_{S=1}^{m} C_{1j}^{*(s)}\cdot\delta_S + C_{1j}^{*(m+1)}\cdot\delta_{m+1}\ ,$$

$$D_{1j} = \frac{2}{3}\sum_{S=1}^{m} C_{1j}^{*(s)}\cdot\left(h_S^3 - h_{S+1}^3\right) + \frac{2}{3}C_{1j}^{*(m+1)}h_{m+1}^3\ , \tag{6.2}$$

$$B = \sum_{S=1}^{m}\left(C_{11}^{E(s)} - C_{12}^{E(s)}\right)\delta_S + \left(C_{11}^{E(m+1)} - C_{12}^{E(m+1)}\right)\delta_{m+1}\ ,$$

$$D = \frac{1}{3}\sum_{S=1}^{m}\left(C_{11}^{E(s)} - C_{12}^{E(s)}\right)\left(h_S^3 - h_{S+1}^3\right) + \frac{1}{3}\left(C_{11}^{E(m+1)} - C_{12}^{E(m+1)}\right)h_{m+1}^3$$

$$E_{31} = 2\sum_{S=1}^{m}(-1)^S\, \ell_{31}^{*(s)} + \ell_{31}^{*(m+1)}\ , \quad M_Э = -2\sum_{S=1}^{m}\ell_{31}^{*(s)}\Phi_1^{(s)} - \ell_{31}^{*(m+1)}\cdot\Phi_1^{(m+1)}\ .$$

En vertu des relations (6.1) le système d'équations de solution pour l'enveloppe composée de 2m+1 couches sera lié aux équations correspondantes pour la détermination des caractéristiques intégrales du potentiel du champ électrique.

On peut traiter de la façon analogue l'enveloppe composée de nombre paire, 2m, de couches. Cependant, comme V_0 =const et les épaisseurs des couches le sont aussi le système d'équations de solution pour une enveloppe de 2m couches sera non lié ce qui facilite notablement sa solution.

En conclusion notons que l'approche exposée de calcul des enveloppes piézoélectriques multicouches permet de prendre en compte la rigidité des électrodes entre couches ce qui s'avère fort utile dans nombre de cas. En l'occurence, seules les relations du type (6.2) changeront.

7. Exemples. En utilisant les équations obtenues considérons le problème d'oscillations longitudinales stationnaires d'une console à épaisseur linéairement variable en supposant que les oscillations soient excitées par une différence de potentiel $\pm V_0 e^{i\Omega t}$. La loi de variation de l'épaisseur est donnée par $h(x) = a + x\,\mathrm{tg}\,\alpha$ où $\mathrm{tg}\,\alpha = (a - b)/L$ (L est la longueur de la console, a, b les hauteurs de l'extrémité encastrée et de l'extrémité libre). On suppose que l'épaisseur de la console dans la direction de l'axe des y est inférieure à l'épaisseur minimale de sorte que l'on peut poser $T_2 = 0$. On suppose également que l'extrémité x=0 de la console est encastrée, celle, x=L, libre. Les conditions aux limites électriques correspondent au cas où la console est en contact avec

le vide à x=0. En utilisant les équations (5.5) et en admettant que le potentiel du champ varie le long de l'épaisseur de l'enveloppe suivant une loi quadratique on obtient de (5.4) l'expression suivante pour le potentiel

$$\varphi = 2V_0 z / h(x) \tag{7.1}$$

La solution générale de l'équation de mouvement

$$\frac{1}{h(x)}\frac{d}{dx}\left[h(x)\frac{dU}{dx}\right] + \varkappa^2 U = 0, \quad \left(\varkappa^2 = \rho\Omega / C_{11}^*(1-\nu_*^2)\right), \tag{7.2}$$

écrite en composante d'amplitude du déplacement et satisfaisant aux conditions aux limites a la forme

$$U(x) = \frac{1}{\Delta}\left\{\Delta_1 J_0(\varkappa h(x)/tg\alpha) + \Delta_2 Y_0(\varkappa h(x)/tg\alpha)\right\} \tag{7.3}$$

où

$$\Delta = J_0(\varkappa a/tg\alpha)Y_1(\varkappa b/tg\alpha) - J_1(\varkappa b/tg\alpha)Y_0(\varkappa a/tg\alpha),$$

$$\Delta_1 = -\frac{2\ell_{31}^* V_0 J_1(\varkappa b/tg\alpha)Y_0(\varkappa a/tg\alpha)}{C_{11}^*(1+\nu_*)\varkappa b},$$

$$\Delta_2 = \frac{2\ell_{31}^* V_0 J_0(\varkappa a/tg\alpha)Y_1(\varkappa b/tg\alpha)}{C_{11}^*(1+\nu_*)\varkappa b}.$$

La composante normale du vecteur induction électrique

$$D_n = D_3\cos\alpha + D_1\sin\alpha = \left[\ell_{31}^*(1-\nu_*)\frac{dU}{dx} - \right.$$
$$\left. - \frac{2V_0(1+\kappa_*^2)\varepsilon_{33}^*}{h(x)}\left(1 - \frac{\varepsilon_{11}^* tg^2\alpha}{\varepsilon_{33}^*(1+\kappa_*^2)}\right)\right]\cos\alpha. \tag{7.4}$$

En calculant la charge électrique totale aux armatures $D_\Sigma = \int_S D_n dS$. trouvons le courant traversant le circuit comme vitesse de variation de la charge totale dans le temps:

$$I = \frac{2V_0 i\Omega e^{i\Omega t}(\ell_{31}^*)^2(1-\nu_*)}{C_{11}^*(1+\nu_*)\varkappa b\Delta}\left\{J_1(\varkappa b/tg\alpha)Y_0(\varkappa a/tg\alpha)\cdot\right. \tag{7.5}$$

$$\left.[J_0(\varkappa b/tg\alpha) + \mu(\kappa,\delta)(\varkappa L)] - J_0(\varkappa a/tg\alpha)Y_1(\varkappa b/tg\alpha)[Y_0(\varkappa b/tg\alpha) + \mu(\kappa,\delta)(\varkappa L)]\right\}$$

où

$$\mu(\kappa,\delta) = \frac{\kappa \ln\kappa(1+\kappa_*^2)(1+\nu_*)}{(1-\kappa)(1-\nu_*)\kappa_*^2}\left[1 - \frac{\varepsilon_{11}^* a^2(1-\kappa)}{\varepsilon_{33}^* L^2(1+\kappa_*^2)}\right], \quad \kappa_*^2 = \frac{(\ell_{31}^*)^2}{\varepsilon_{33}^* C_{11}^*}, \kappa = \frac{b}{a}, \delta = \frac{a}{L}.$$

Il découle de (7.5) qu'aux fréquences satisfaisant à l'équation

$$J_0(\varkappa a/tg\alpha)Y_1(\varkappa b/tg\alpha) - J_1(\varkappa b/tg\alpha)Y_0(\varkappa a/tg\alpha) = 0, \tag{7.6}$$

le courant traversant le circuit devient infiniment grand, i.e. il y a résonance. Si la fréquence est la racine de l'équation

$$J_1(xb/tg\alpha)Y_0(xa/tg\alpha)\left[J_0(xb/tg\alpha)+\mu(\kappa,\delta)(xL)\right]- \\ -J_0(xa/tg\alpha)Y_1(xb/tg\alpha)\left[Y_0(xb/tg\alpha)+\mu(\kappa,\delta)(xL)\right]=0, \qquad (7.7)$$

le courant est nul dans le circuit, i.e. on assiste à l'anti-résonance.

Dans la table 1 ci-après sont données les valeurs des trois premières fréquences de résonance $(xL)_n\pi^{-1}$ en fonction de $k=b/a$ (la dernière colonne de la table 1 correspond à la console d'épaisseur constante).

Table 1

n \ k	0,1	0,2	0,3	0,4	0,5	0,6	0,7	0,8	0,9	1
1	0,7	0,663	0,62	0,593	0,573	0,558	0,538	0,525	0,512	1/2
2	1,636	1,587	1,540	1,537	1,528	-	-	1,508	1,504	3/2
3	2,606	2,56	2,54	2,526	2,518	2,512	2,508	2,505	2,5025	5/2

Pour la console fabriquée en céramique PZT-4 les premières recines de l'équation (7.7) sont données dans la table 2 ($\delta=0,01$).

Table 2

$(xL)_a$ \ k	0,1	0,8	1
1	2,21	1,67	1,7

L'analyse du calcul numérique effectué montre qu'avec la diminution du paramètre k les racines de résonance et d'antirésonance tendent à s'approcher l'une de l'autre. Si l'on applique, pour rendre le fonctionnement du transducteur plus efficace, la formule de Mason [7] $K_3^2=(f_a^2-f_p^2)/f_a^2$ on arrive à la conclusion qu'un transducteur à épaisseur linéairement variable est moins efficace qu'un transducteur à épaisseur constante.

Les raisonnements analogues permettent d'étudier l'influence de la rigidité des électrodes sur les fréquences de résonance et d'antirésonance (par exemple, dans le cas d'une plaque ronde en céramique piézoélectrique couverte d'électrodes d'épaisseurs égales).

8. En concluant il convient de noter que jusqu'à présent on n'a pas encore commencé des études expérimentales de large envergure sur la mécanique des matériaux piézoélectriques. Le problème de toute première importance pour les expérimentateurs est celui de détermination pour les vastes classes de piézoélectriques (céramiques piézoélectriques, entre autres): des modèles élastiques, des constantes piézoélectriques et diélectriques mesurées à déformations constantes. Il est nécessaire de vérifier expérimentalement les équations d'état et des résultats des diverses approches théoriques de descroption du comportement de ces matériaux sous charge; de déterminer les caractéristiques de résistance à la fissuration, la densité d'énergie surfacique et les longueurs des fissures; de déterminer les fréquences propres, de résonance et d'antirésonance etc. De telles recherches permettront de jeter des bases théoriques solide de la mécanique des matériaux piézoélectriques.

REFERENCES

1. Parton V.Z., Fracture mecanics for piezoelectric materials, Acta Astronautica, Pergamon Press, 1976, vol. 3, pp.671-683.

2. Партон В.З., Кудрявцев Б.А., Механика разрушения при наличии электрических полей. Физ.-хим. материалов, 1982, №5,т.18,с.3-15.

3. Партон В.З., Сеник Н.А. О применении метода символического интеррования в теории пьезокерамических оболочек. ПММ, 1983, т. 46, вып. 2, с. 257-262.

4. Партон В.З., Сеник Н.А. Соотношения электроупругости для многослойных пьезооболочек. Изв. АН СССР, МТТ, 1983, №3.

5. Лурье А.И. Теория упругости. М.: Наука, 1970. 940 с.

6. Амбарцумян С.А. Общая теория анизотропных оболочек. М.: Наука, 1974. 448 с.

7. Berlincourt D.A., Curran D.R., and Jaffe H., Piezoelectric and piezo magnetic materials and their function in transducers. In: Physical Acoustics (W.P. Mason and R.N. Thurston, eds) vol. 1, A, chap. 3. New York: Academic Press (1964).

The Mechanical Behavior of Electromagnetic Solid Continua
G.A. Maugin (editor)
Elsevier Science Publishers B.V. (North-Holland)

SURFACE WAVE PROPAGATION WITH FINITE AMPLITUDE : RESONATORS AND CONVOLVERS

Michel PLANAT

Laboratoire de Physique et Métrologie des Oscillateurs du CNRS
associé à l'Université de Franche-Comté-Besançon
32, avenue de l'Observatoire - 25000 Besançon - France

Nonlinear elastic wave propagations are mainly characterized by the distorsion of the shape of the wave front and by interaction phenomena. For a resonator, the first phenomenon leads to the amplitude-frequency effect which is to be lowered in frequency control application. On the contrary, nonlinear interaction can be used for signal processing in devices like convolvers or correlators.

INTRODUCTION

Depuis la mise au point du transducteur métallique à électrodes interdigitées par White en 1965, les ondes de surface ont connu un développement important. Dans la gamme de fréquences 10 MHz - 2 GHz, leurs performances les ont rendues attrayantes aussi bien dans le domaine de la métrologie, par la réalisation d'oscillateurs ou de capteurs de haute résolution, que dans le domaine du traitement du signal.

Dans ces deux domaines, l'étude des effets de propagation non linéaire est prépondérante : d'une part parce qu'ils introduisent un couplage entre l'onde et le milieu extérieur et sont donc responsables de la sensibilité des dispositifs à la température, la pression, aux forces, aux accélérations [1], d'autre part parce qu'ils sont la cause des phénomènes d'interaction onde-onde à l'origine de la génération d'harmoniques, de l'anisochronisme des résonateurs, de l'intermodulation dans les filtres et des produits de convolution.

Nous nous intéresserons dans cet article à ce deuxième type de problème. Les termes non linéaires de l'équation d'onde sont dûs à la déformation finie du cristal et à ses constantes fondamentales d'ordre supérieur. La résolution de cette équation sera effectuée en considérant ces termes non linéaires comme des excitations dont il s'agira d'étudier la réponse.

La propagation des harmoniques d'une onde de Rayleigh a fait l'objet de nombreuses études expérimentales [2][3][4], principalement par l'utilisation d'une sonde optique. La propagation non linéaire a également pour effet de distordre le profil de l'onde et s'accompagne de la modification de la vitesse de l'onde, ou de la fréquence du résonateur : c'est le phénomène d'anisochronisme [5].

Dans les milieux anisotropes, ces phénomènes dépendent de la configuration cristalline particulière et du cristal choisis, ainsi que de la géométrie des transducteurs.

La formation de produits de convolution, lorsque deux ondes de même fréquence et de sens opposés interagissent, a été démontrée par Chaban [6]. Une gamme de composants pour le traitement du signal, fonctionnant sur ce principe, a été introduite. Leur intérêt est de travailler en temps réel, avec un volume réduit et une large bande passante. Un des convoluteurs les plus performants utilise un milieu de propagation de niobate de lithium où l'onde est guidée par un film mince d'aluminium [7]. De très hautes densités d'énergie et une grande dynamique sont obtenues.

Dans cet article, nous effectuons le calcul des effets non linéaires pour ces deux types d'application, en utilisant la méthode des perturbations.

A/ ANALYSE NON LINEAIRE DE LA LIGNE A RETARD A ONDES DE RAYLEIGH

On considère un milieu de propagation semi-infini et faiblement piézoélectrique (quartz) limité par une surface libre $a_2 = 0$. Sur cette surface se trouvent deux transducteurs métalliques interdigités utilisés pour l'excitation et la détection d'ondes de Rayleigh. Le coefficient de couplage électromécanique étant faible, il est possible de séparer l'analyse en deux parties [8] : une partie où le milieu est considéré comme purement élastique et une partie résiduelle électrostatique.

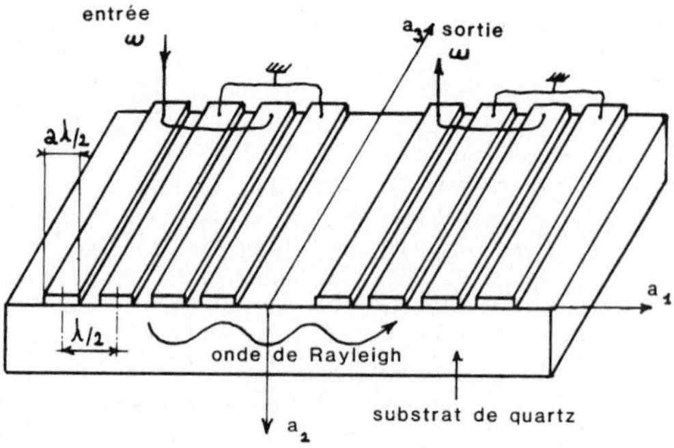

Fig. 1 : Ligne à retard à ondes de Rayleigh

Pour le milieu purement mécanique, les équations de propagation et la condition aux limites associée qui exprime l'annulation de la contrainte sur la surface $a_2=0$ sont :

$$\rho_o \ddot{u}_j = P_{ij,i} \qquad \text{et} \qquad P_{2j} = 0 \text{ pour } a_2 = 0 \tag{1}$$

P_{ij} est le tenseur de Piola-Kirchoff donné par l'expression [5] :

$$P_{ij} = C_{ijk\ell}\, u_{k,\ell} + (\tfrac{1}{2} C_{ijn\ell}\, \delta_{km} + C_{ink\ell}\, \delta_{jm} + \tfrac{1}{2} C_{ijk\ell mn})\, u_{k,\ell}\, u_{m,n} \tag{2}$$

où u_j désigne le déplacement mécanique, ρ_o est la masse de l'unité de volume et les "$C_{ijk\ell}$" et $C_{ijk\ell mn}$" sont les coefficients élastiques du 2e et du 3e ordre du cristal étudié. δ est le symbole de Kronecker.

a) Etude linéaire

L'onde de Rayleigh de fréquence ω, solution du système (1)-(2), est la superposition de vibrations se propageant à la même vitesse V_o suivant la direction Oa_1 et évanescentes suivant Oa_2 :

$$\overset{o}{u}_j = U_1 \sum_{r=1}^{3} C_j^{(r)}\, e^{i(\omega t - k_o a_1 - k_o n_2^{(r)} a_2)} \tag{3}$$

$k_o = \omega/V_o$ est le nombre d'onde et la partie imaginaire des coefficients de pénétration $n_2^{(r)}$ est négative.

Pour les dispositifs électroacoustiques, la grandeur intéressante n'est pas le déplacement mécanique mais la tension électrique détectée sur le transducteur de sortie. Sa détermination nécessite la prise en compte du couplage piézoélectrique et des conditions aux limites électriques. La tension électrique recueillie peut alors être reliée [8] aux caractéristiques géométriques de ce transducteur, aux propriétés piézoélectriques du cristal et à la terminaison électrique utilisée.

b) **Etude non linéaire**

Le système non linéaire (1)-(2) est résolu en utilisant la méthode des perturbations. La solution s'écrit sous la forme :

$$u_j = \overset{0}{u}_j + \overset{1}{u}_j + \overset{2}{u}_j \qquad (4)$$

où $\overset{1}{u}_j$ et $\overset{2}{u}_j$ sont des termes correctifs d'ordre un et deux de la solution linéaire

En reportant cette solution dans l'équation (1), on en déduit aux différents ordres d'approximation :

1) l'équation de propagation et la condition aux limites linéaires (1). Sa solution est $\overset{0}{u}_j$ donnée par la relation (3).

2) un système d'équation inhomogène dont le premier membre linéaire est en $\overset{1}{u}_j$ et dont le second membre non linéaire est fonction de $\overset{0}{u}_j$. En reportant $\overset{0}{u}_j$ donnée par (3) dans ce second membre quadratique, on voit que celui-ci contient une vibration à l'harmonique deux et une contribution statique. La solution de ce système est [5] :

$$\overset{1}{u}_j = k_o U_1^2 \{ \sum_{r,s} \alpha_j^{(r,s)} e^{2i\phi^{(r,s)}} + \sum_{r\neq s} \beta_j^{(r,s)} e^{2i\Psi^{(r,s)}} + \lambda \sum_r C_j^{(r)} [1 + \frac{\Delta C_j^{(r)}}{C_j^{(r)}} - 2i\,\Delta k_o\, a_1 - 2i\,\Delta(k_o n_2^{(r)}) a_2]\, e^{2i\Psi^{(r,r)}} \qquad (5)$$

$$\text{où} \quad \phi^{(r,s)} = k_o[n_2^{(r)*} - n_2^{(s)}]a_2 \;;\quad \Psi^{(r,s)} = \omega t - k_o a_1 - 1/2\, k_o(n_2^{(r)} + n_2^{(s)})a_2 \qquad (6)$$

et les "α", "β", "$\Delta C/C$", Δk_o et $\Delta(k_o n_2^{(r)})$ sont des constantes.

Le premier terme est la solution particulière de l'équation de propagation correspondant au terme statique ; le second terme est la solution à la fréquence 2ω qui ne vibre pas en phase avec la solution du système homogène. Ces deux solutions ont donc été choisies sous une forme analogue aux termes excitateurs. Lorsque l'excitation vibre en phase avec la solution du système homogène, la solution (3e terme de la relation (5)) est choisie en perturbant le nombre d'onde, les coefficients de pénétration et l'amplitude de la solution homogène et en la développant au premier ordre d'approximation. Cette dernière solution possède en outre les degrés de liberté nécessaires pour satisfaire la condition aux limites.

La solution générale comprend des termes proportionnels à la distance de propagation a_1, à la profondeur de pénétration a_2 ou des termes constants. Comme la distance de propagation de l'onde de Rayleigh dans la ligne à retard est généralement supérieure à 100 longueurs d'onde, la profondeur de pénétration n'excédant pas 3 longueurs d'onde, le terme proportionnel à a_1 est prépondérant.

Le taux d'harmonique deux défini comme le rapport entre les amplitudes U_2 et U_1 des solutions aux fréquences 2ω et ω est donc :

$$U_2/U_1 = \lambda' \, k_o^2 \, U_1^2 \, a_1 \qquad (7)$$

où $\lambda' = 2i\lambda\, \Delta k_o/k_o$ est une constante sans dimension dépendant des caractéristiques fondamentales du cristal, particulièrement de ses constantes élastiques du 3e ordre.

A l'ordre suivant d'approximation est obtenu un système homogène dont le premier membre est formé des parties linéaires des relations (1) mais d'inconnue u_j^2. Son second membre non linéaire est calculé en fonction de u_j^0 et u_j^1. Il contient une vibration à la fréquence de l'harmonique trois et une vibration à la fréquence fondamentale. Les solutions sont obtenues par la même méthode que précédemment mais contiennent des termes dont l'amplitude est proportionnelle à a_1^2, a_2^2, $a_1 a_2$, a_1, a_2 ou constants, le terme en a_1^2 étant prépondérant à une distance suffisante de la source.

La vibration à la fréquence fondamentale joue un rôle extrêmement important dans les dispositifs métrologiques. En effet, en s'ajoutant à la solution linéaire (3), elle a pour effet de distordre le profil de l'onde au cours de la propagation et d'introduire une modification de son amplitude et de sa vitesse.

Pour la ligne à retard, l'anisochronisme δ peut être défini par la variation relative de la vitesse

$$\delta = \Delta V/V = \pi\, f_1\, n\, k_o^2\, U_1^2 \qquad (8)$$

où n est le nombre de longueurs d'onde de propagation et f_1 une constante réelle sans dimension calculée en fonction des constantes du cristal. Cet anisochronisme peut être exprimé en fonction du courant électrique I détecté sur le transducteur de sortie. On définit ainsi la constante d'anisochronisme k par la relation :

$$\delta = kI^2 \quad \text{où} \quad k = \frac{f_1\, n\, K'^2(s)}{V_o^2\, W^2\, N^2\, \pi\, |\Phi_s|^2 (\varepsilon_{22}\varepsilon)^2} \qquad (9)$$

où a est la largeur relative de recouvrement des électrodes (fig.1), $s = \sin \frac{\pi}{2} a$, K'(s) est l'intégrale elliptique de 1ère espèce du module complémentaire $s' = \sqrt{1-s^2}$, W est l'ouverture du transducteur, N le nombre de paires de doigts, ε_{22} la constante diélectrique du cristal, ε une constante diélectrique relative caractérisant l'anisotropie et $|\Phi_s|$ le potentiel électrostatique introduit à la référence (8).

Comparaison avec l'expérience

Afin de vérifier la validité de nos résultats théoriques, nous avons d'abord étudié des lignes à retard à quartz fonctionnant à la fréquence de 100 MHz (n = 192 ; N = 150 ; W = 6.6 mm ; a = 0.5). L'anisochronisme calculé pour différentes orientations cristallines est toujours inférieur à $k = 10^{-8}/A^2$. Les expériences d'anisochronisme mises en oeuvre sur ces lignes ont montré que cette constante restait inférieure à $k = 10^{-5}/A^2$. Une plus grande précision de mesure est difficile à obtenir en raison du faible coefficient de qualité de ces lignes (de l'ordre de 500). Il était donc nécessaire d'utiliser de très fortes puissances d'excitation (supérieures à 1 W) qui ont pour effet d'échauffer considérablement le cristal et d'entraîner une dérive thermique.

Ces deux inconvénients ont été éliminés en mesurant l'anisochronisme de résonateurs à quartz à ondes de surface dont le coefficient de qualité est supérieur à 10 000. La dérive thermique a été éliminée en effectuant une mesure dynamique pour séparer les effets de la température et les effets non linéaires élastiques de constantes de temps très différentes [5].

La constante d'anisochronisme k_R du résonateur peut être déduite de celle de la ligne à retard en utilisant leurs schémas équivalents électriques [10].

Nous avons obtenu $k_R = k\,(R_0/R_1)^2$ avec $R_0/R_1 = 4Q/n\pi$ où R_0 est la résistance de rayonnement des transducteurs, R_1 l'impédance du matériau à la résonance, Q le coefficient de qualité du résonateur et n le nombre de longueurs d'onde de la cavité résonnante. Pour des résonateurs de coupe ST fonctionnant à 100 MHz (n = 190 ; N = 25 ; W = 2.5 mm ; a = 0.4 ; Q = 40 000) nous avons calculé $k_R = 1.6.10^{-3}/A^2$ (valeur mesurée : $k_R = 1.8.10^{-3}/A^2$). La constante obtenue pour des résonateurs fonctionnant à fréquence plus élevée (380 MHz) est $20.10^{-3}/A^2$ et s'explique par la diminution de l'ouverture W des transducteurs (W = 0.9 mm).

La comparaison avec des résonateurs à quartz à ondes de volume de coupe AT fonctionnant sur le partiel 5 aux fréquences de 5 à 100 MHz peut être effectuée : on trouve un anisochronisme de l'ordre de $2.10^{-1}/A^2$ [14]. Les ondes de Rayleigh semblent donc extrêmement intéressantes à ce point de vue.

B/ ANALYSE D'UN CONVOLUTEUR A GUIDE D'ONDE

Le convoluteur étudié (fig. 2) se compose de deux transducteurs d'entrée à électrodes métalliques interdigitées et d'éléments de compression capables de transmettre de très hautes densités d'énergie sur une largeur de quelques longueurs d'onde. Les ondes sont guidées par une électrode centrale et détectées sur des électrodes de masse situées de part et d'autre.

La structure effectivement analysée est représentée en coupe sur la figure 3.

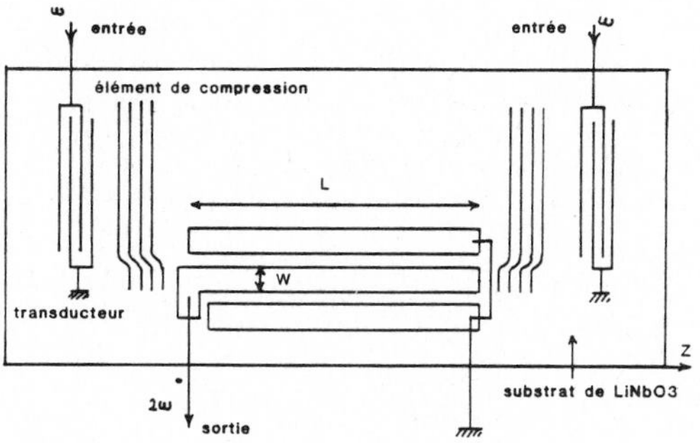

Fig. 2
Convoluteur à ondes guidées

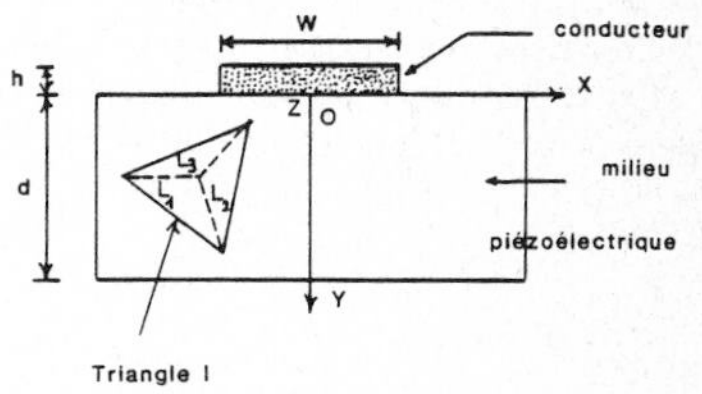

Fig. 3
Section transverse du guide d'ondes

Elle se compose d'un substrat piézoélectrique et d'un élément métallique conducteur, de largeur W et de hauteur h, déposé sur le milieu de propagation. La vibration se propage perpendiculairement au plan de la figure avec le nombre d'onde k_3. Elle est multimode et dispersive.

Pour résoudre ce problème, nous utilisons une méthode variationnelle aux éléments finis.

a) **Etude linéaire**

Nous introduisons un lagrangien dont la variation au premier ordre est nulle lorsque l'ensemble des équations de propagation, des conditions aux limites et à l'interface entre les deux solides sont satisfaites [11] :

$$\mathcal{L} = \frac{1}{2} \{\int_V \frac{1}{2} (-\rho_0 \omega^2 u_j u_j^* + C_{ijk\ell} u_{k,\ell} u_{j,i}^* + 2e_{ik\ell} u_{k,\ell} \phi_{,i}^* - \varepsilon_{i\ell} \phi_{,\ell} \phi_{,i}^*) dV$$
$$- \int_V F_j u_j^* dV - \int_A Q_j u_j^* dA + \int_V G\phi^* dV - \int_A \sigma \phi^* dA \qquad (10)$$
$$- \int_{A_1} (\phi - \phi_0) [n_i (e_{ik\ell} u_{k,\ell}^* - \varepsilon_{i\ell} \phi_{,\ell}^*) - \sigma] dA + c.c.\}$$

où u_j et ϕ sont les déplacements mécaniques et potentiel électrique d'essai à déterminer.

Ce lagrangien correspond à un milieu de propagation piézoélectrique de masse volumique ρ_0, de constantes élastiques "c", piézoélectriques "e" et diélectriques "ε". Il est le siège de forces mécaniques de volume F_j, de surface Q_j ainsi que de charges électriques de volume G et de surface σ. Ce milieu peut être également en contact avec une électrode conductrice au potentiel ϕ_0, le long d'une surface A_1 de normale n_i. Ce lagrangien s'interprète comme la différence entre l'énergie cinétique de la vibration et l'enthalpie, quantité à laquelle on ajoute les travaux des forces extérieures et de contact. Le dernier terme de cette expression est ajouté afin de pouvoir satisfaire la condition de potentiel constant ϕ_0 sur l'électrode indépendamment du choix des fonctions d'essai.

Dans le cas de la propagation linéaire de l'onde guidée, il n'y a pas de forces de volume F_j ni de charges libres G. D'autre part, le champ électrique qui accompagne la propagation de l'onde sous l'électrode est court-circuitée par cette électrode et donc les charges électriques de surface σ et le potentiel électrique sont également nuls.

Le lagrangien pour la structure complète de la fig. 3 se compose du lagrangien correspondant au milieu piézoélectrique et de celui correspondant au conducteur qui est simplement élastique. Les forces de contact Q_j entre ces deux milieux sont des vecteurs opposés et par conséquent disparaissent de l'expression de $\mathcal{L}$.

Le déplacement mécanique u_j et le potentiel ϕ de l'onde linéaire guidée sont choisis sous la forme d'une vibration de fréquence ω se propageant suivant la direction Oz avec le nombre d'onde k_3. A l'aide de cette hypothèse, on est ramené à un lagrangien à deux dimensions x et y.

La méthode des éléments finis est alors appliquée en effectuant un maillage dans la tranche du guide (fig. 3). La section du guide est décomposée en triangles élémentaires I et chaque triangle comprend 6 noeuds situés à ses sommets et aux milieux de ses côtés.

Le lagrangien obtenu en effectuant cette décomposition est ensuite intégré sur la surface de chaque triangle en utilisant les nouvelles coordonnées de surface L_1, L_2 et L_3, puis minimisé par rapport aux seuls degrés de liberté indépendants que sont les déplacements mécaniques $U^{(a)}$ et potentiel électrique $\phi^{(b)}$ aux noeuds indépendants de la structure. On obtient :

$$\begin{bmatrix} K(k_3) - \omega^2 M & L(k_3) \\ L^t(k_3) & J(k_3) \end{bmatrix} \begin{bmatrix} U^{(a)} \\ \phi^{(b)} \end{bmatrix} = \begin{bmatrix} 0 \end{bmatrix} \qquad (11)$$

où K, M, L et J sont respectivement les matrices élastiques, de masse, piézoélectriques et diélectriques caractérisant le matériau et le maillage choisi. L^t désigne la transposée de L. L'annulation du déterminant de ce système conduit à la relation de dispersion $k_3 = f(\omega)$. Sa résolution permet d'obtenir les champs linéaires qui vont maintenant être utilisés dans le convoluteur.

b) **Etude non linéaire**

Pour l'étude non linéaire, on peut se ramener au maillage précédent en introduisant en chaque noeud de la structure des sources d'excitation provenant des effets non linéaires.

Les forces mécaniques F_j, Q_j et les charges électriques d'excitation G et σ peuvent être calculées en fonction des constantes fondamentales du cristal étudié [12]. Le milieu de propagation (niobate de lithium, PZT) ayant un coefficient de couplage électromécanique élevé, ces excitations font intervenir non seulement les constantes élastiques du 3e ordre, mais aussi les constantes électroélastiques, d'électrostriction et les constantes diélectriques du 3e ordre.

L'approche variationnelle du convoluteur est obtenue de la manière suivante : les vibrations linéaires dans chaque triangle sont la superposition de deux ondes de même fréquence ω et de nombres d'ondes opposés k_3 et $-k_3$. Les solutions reportées dans l'expression des termes non linéaires F_j, Q_j, G et σ engendrent en particulier une composante d'excitation à la fréquence 2ω et constante suivant Oz. Cette composante d'excitation est située dans la tranche du guide ; sa réponse est recherchée à la même fréquence. En utilisant la même méthode aux éléments finis que dans l'étude linéaire, mais cette fois en présence des sources $F^{(a)}$ et $G^{(b)}$ dues aux effets non linéaires, on en déduit que les déplacements mécaniques $\overset{1}{U}{}^{(a)}$ et les potentiels $\overset{1}{\phi}{}^{(b)}$ aux noeuds indépendants sont solution du système :

$$\begin{bmatrix} K(k_3=0) - 4\omega^2 M & L(k_3=0) \\ L^t(k_3=0) & J(k_3=0) \end{bmatrix} \begin{bmatrix} \overset{1}{U}{}^{(a)} \\ \overset{1}{\phi}{}^{(b)} \end{bmatrix} = \begin{bmatrix} F^{(a)} \\ G^{(b)} \end{bmatrix} \qquad (12)$$

On définit un facteur de mérite du convoluteur par le rapport entre la valeur efficace du potentiel sur l'électrode de guidage relativement au flux par unité de largeur des ondes incidentes :

$$M = \phi_0^{rms} / (\text{flux}/W) \qquad (13)$$

c) **Résultats**

Le calcul a d'abord été testé dans le cas d'une onde de Rayleigh afin de comparer les résultats de la méthode d'éléments finis avec ceux donnés par un calcul analytique. Pour cela, des conditions de symétrie ont été introduites afin de représenter un guide de largeur infinie.

A l'aide d'un maillage non uniforme à 55 noeuds, une précision meilleure que 1‰ a été obtenu sur la vitesse de phase. Le facteur de mérite du convoluteur dégénéré que nous avons obtenu est de $1.3\ 10^{-4}$ Vm/W, qui peut être comparé au facteur de mérite mesuré : 0.9 à $1.2\ 10^{-4}$ Vm/W ou au facteur de mérite calculé par Ganguly [13] : $2.6\ 10^{-4}$ Vm/W par la méthode analytique. Ganguly attribue la différence entre ce qu'il a calculé et les résultats expérimentaux à l'imprécision des valeurs des constantes fondamentales d'ordre supérieur.

Dans un deuxième temps, des maillages ont été développés pour l'étude d'une onde se propageant dans le niobate de lithium sous une électrode de guidage d'aluminium. Les courbes de dispersion obtenues sont en excellent accord avec l'expérience lorsque le maillage choisi couvre l'ensemble des régions de l'espace où les champs ne sont pas nuls [12] : il doit s'étendre sur une profondeur d'environ 3 longueurs d'onde de profondeur et sur une largeur de 2 longueurs d'onde de part et d'autre de l'électrode conductrice.

Pour l'étude non linéaire, le maillage utilisé est représenté sur la figure 4. Un demi guide d'ondes est maillé à l'aide de 46 noeuds en introduisant une condition de symétrie sur l'axe du guide. Avec une précision de l'ordre de 5 ‰ sur la vitesse de phase, un facteur de mérite de $0.47\ 10^{-4}$ Vm/W a été obtenu, en bon accord avec les résultats expérimentaux [7] (0.4 à $0.5\ 10^{-4}$ Vm/W).

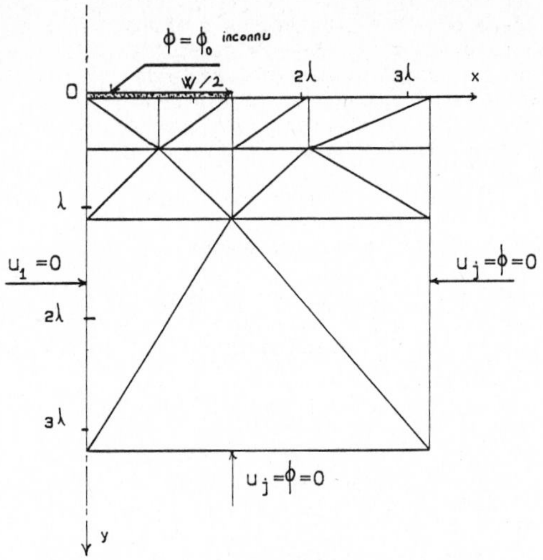

Fig. 4
Domaine pour l'étude aux éléments finis du guide d'ondes

Sur les figures 5 et 6 sont représentées respectivement les distributions linéaires et non linéaires des champs mécaniques (courbes en pointillés) et du potentiel (courbes en traits pleins) en fonction de la profondeur y au sein du substrat et de la distance x à l'axe du guide pour un guide sans épaisseur de 2.75 longueurs d'onde.

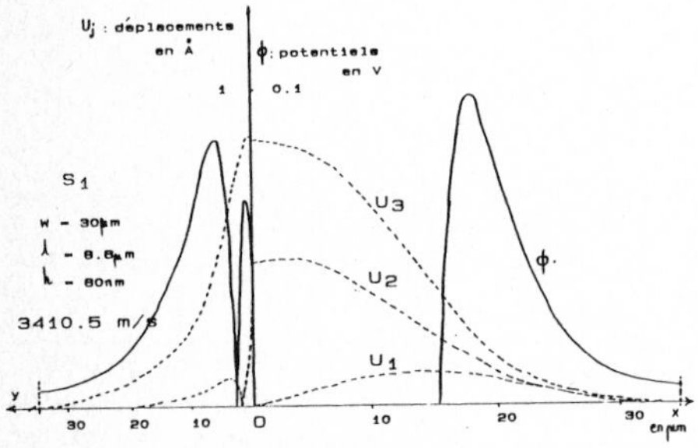

Fig. 5
Champs linéaires

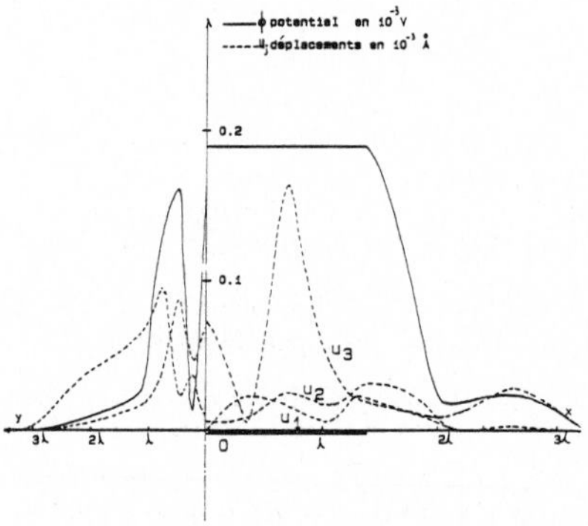

Fig. 6
Champs non linéaires

CONCLUSION

La méthode des perturbations a été développée pour étudier les effets non linéaires de propagation d'ondes de surface dans un milieu piézoélectrique. Dans l'optique de la métrologie des fréquences et pour un milieu faiblement piézoélectrique comme le quartz, l'anisochronisme d'une onde de Rayleigh a été obtenu en fonction des constantes élastiques du 3e ordre et de la géométrie des transducteurs utilisés.

Dans l'optique du traitement du signal, a été calculé le facteur de mérite d'un convoluteur à guides d'onde aluminium sur niobate de lithium, en utilisant la méthode variationnelle aux éléments finis.

Ces études, valables pour un matériau ou une géométrie, des transducteurs ou du guide d'onde arbitraires peuvent être utilisées soit pour diminuer les effets non linéaires dans le cas des oscillateurs, ou pour les optimaliser dans le cas du convoluteur.

REFERENCES

[1] Hauden D., Rousseau S., Gagnepain J.J., 34th Annual Symposium on Frequency Control, Philadelphie, 1980.

[2] Lopen P.O., J. Appl. Phys., 39, 5400 (1968).

[3] Lean E.G. and Tseng C.C., J. Appl. Phys., 41 (10), 3912 (1970).

[4] Alippi A., Plama A., Palmieri L. and Socino G., J. Appl. Phys., 48, 2812 (1977).

[5] Planat M., Theobald G., Gagnepain J.J., L'Onde Electrique, 60, n° 11 (1980).

[6] Chaban A.A., Sov. Phys. Sld. St., 9, 2622 (1968).

[7] Defranould P., Maerfeld C., Proc. IEEE, 64, 748 (1976).

[8] Coquin G.A. and Tiersten H.F., J. Acoust. Soc. Am., 41 (4), 924 (1967).

[9] Gagnepain J.J. et Besson R., Physical Acoustics, vol. XI, Academic Press, New-York (1975).

[10] Schmidt R.V. and Coldren L.A., IEEE Trans. Sonics and Ultrasonics, SU-22, 115-122 (1975).

[11] Holland R. and Eernisse E.P., Research Monograph, n° 56, MIT Press, Cambridge and London (1969).

[12] Planat M., Vanderbock G., Gautier M. and Maerfeld C., IEEE Ultrasonics Symposium San Diego (1982).

[13] Ganguly A.K. and Davis K.L., J. Appl. Phys., 51, 920-926 (1980).

[14] Schoenwald J.S., Schreve W.R. and Rosenfeld R.C., 29th Annual Symposium on Frequency Control, 150 (1975).

The Mechanical Behavior of Electromagnetic Solid Continua
G.A. Maugin (editor)
Elsevier Science Publishers B.V. (North-Holland)

NONLINEAR BEHAVIOR OF QUARTZ AND $LiNbO_3$

M. A. Breazeale and Paul Jerry Latimer

Department of Physics
The University of Tennessee
Knoxville, Tennessee
U.S.A.

The ultrasonic harmonic generation technique has been extended to the measurement of trigonal crystals. An initially sinusoidal longitudinal ultrasonic wave distorts as it propagates along pure mode directions. Measurement of the amplitudes of the fundamental and the second harmonic allows one to calculate third-order elastic constants. Measures values of C_{111} and C_{333} for quartz agree with published values. The values for $LiNbO_3$ disagree with published values. New data with better samples are reported.

INTRODUCTION

We have used nonlinear distortion of finite amplitude ultrasonic waves to measure the third-order elastic constants of quartz and $LiNbO_3$. One reason we chose to make the measurements was that the measurements of $LiNbO_3$ which had been made using stress dependence of ultrasonic wave velocity (1) led to errors in C_{333} for $LiNbO_3$ which appear to be anomalously large (±200%). Our technique gives C_{111} and C_{333} directly, and hence our values are considerably more accurate than this. A second reason is that recent developments in growth techniques have improved $LiNbO_3$ sample quality, and one really needs the values corresponding to the most perfect samples. Further, one would like to evaluate the effect of piezoelectric stiffening on the third-order elastic constants. In fact, it turns out that we are not able to do this, but we have some preliminary indications that such an evaluation is possible.

We measured values of C_{111} and C_{333} for quartz and compared with other measurements. Knowing that our quartz values compared favorably with those of Thurston, McSkimin, and Andreatch (2), we were able to be much more confident of our $LiNbO_3$ values which differ considerably from those of Nakagawa. Such a procedure was necessary since we report the first set of measurements of trigonal symmetry crystals taken by the harmonic generation technique.

THEORY

Nonpiezoelectric Directions. The theory (3) previously used for description of the nonlinear behavior of cubic crystals can be extended directly to trigonal symmetry; however, proper attention must be paid to the effect of piezoelectricity in the [100] direction in quartz and the [001] direction in $LiNbO_3$. For nonpiezoelectric directions one can use the strain energy

$$\phi = \frac{1}{2} C_{ijk\ell}\eta_{jk}\eta_{k\ell} + \frac{1}{3!} C_{ijk\ell mn}\eta_{ij}\eta_{k\ell}\eta_{mn} + \cdots \tag{1}$$

in the equation of motion

$$\rho_0\ddot{u}_i = \frac{\partial}{\partial a_k}\left[\frac{\partial x_i}{\partial a_m}\frac{\partial \phi}{\partial \eta_{km}}\right] \tag{2}$$

To derive the nonlinear equation for longitudinal wave pure mode propagation directions in crystals of trigonal symmetry with the result that the wave equation for the [100] direction is

$$\rho_0 \ddot{u} = C_{11} \frac{\partial^2 u}{\partial a_1^2} + (3C_{11} + C_{111}) \frac{\partial u}{\partial a_1} \frac{\partial^2 u}{\partial a_1^2} , \tag{3}$$

which is identical in form to that for a cubic crystal. For the [001] direction the longitudinal wave equation is similarly straightforward:

$$\rho_0 \ddot{w} = C_{33} \frac{\partial^2 w}{\partial a_3^2} + (3C_{33} + C_{333}) \frac{\partial w}{\partial a_3} \frac{\partial^2 w}{\partial a_3^2} . \tag{4}$$

(The [010] direction is not a pure mode direction for a longitudinal wave.) Since these two equations have the same form, the solutions have the same form. For example, for the [100] direction, an initially sinusoidal ultrasonic wave at a = 0 is described by

$$u(a,t) = A_1 \sin(ka-\omega t) - \frac{k^2A^2a}{8} \beta_{[100]} \cos 2(ka-\omega t) + \ldots \tag{5}$$

where a is a measure of the distance in the [100] direction and

$$\beta_{[100]} = - \frac{3C_{11} + C_{111}}{C_{11}} \tag{6}$$

is the nonlinearity parameter which contains the third-order elastic constant C_{111} to be measured. For the [001] direction the nonlinearity parameter is

$$\beta_{[001]} = - \frac{3C_{33} + C_{333}}{C_{33}} . \tag{7}$$

<u>Piezoelectric Directions</u>. The general equation of motion, Eq. 2, can be used to describe the propagation of finite amplitude longitudinal waves in piezoelectric directions if one reinterprets the thermodynamic tension

$$\frac{\partial \phi}{\partial \eta_{km}} = t_{km} . \tag{8}$$

For piezoelectric directions the strain energy ϕ no longer accounts for all results of a deformation. Rather, the strain energy ϕ is replaced by the thermodynamic potential H_2, the electric enthalpy, defined by

$$\begin{aligned} H_2 &= \phi - E_i D_i \\ &= \frac{1}{2} C^{(E)}_{ijk\ell} \eta_{ij} \eta_{k\ell} + \frac{1}{3!} C^{(E)}_{ijk\ell mn} \eta_{ij} \eta_{k\ell}\, \eta_{mn} - \frac{1}{2} \varepsilon^{(S)}_{ij} E_i E_j \\ &\quad - \frac{1}{6} \varepsilon^{(S)}_{ijk} E_i E_j E_k - e_{ijk} E_i \eta_{jk} - \frac{1}{2} d_{ijk\ell} E_i E_j \eta_{k\ell} - \frac{1}{2} f_{ijk\ell m} E_i \eta_{jk} \eta_{\ell m} + \ldots \end{aligned} \tag{9}$$

Using the thermodynamic tension

$$t_{km} = \frac{\partial H_2}{\partial \eta_{km}} \tag{10}$$

and Maxwell's equations in the equation of motion, Eq. 2, one finds that the propagation of a finite amplitude wave in a piezoelectric direction is described by nonlinear equations of the form of Eq. 3 or Eq. 4 if the elastic constants are interpreted as the effective, or electrically "stiffened" elastic constants. For the [001] direction, the direction of interest for $LiNbO_3$, the "stiffened" constant is:

$$C_{333} = C_{333}^{(E)} + \varepsilon_{333}^{(S)} (\rho_{33}/\varepsilon_{22}^{(S)})^3 + \frac{3f_{333}e_{33}}{\varepsilon_{33}^{(S)}} - e_{333}^2 d_{333} \left(\frac{2}{\varepsilon_{33}^{(S)}} + \frac{1}{\varepsilon_{33}^{(S)2}}\right) . \quad (11)$$

This expression differs from that of McMahon (4) in the final term. McMahon made an algebraic error which will need to be corrected at the time one is able to measure all of the coefficients in this equation; at the moment one cannot isolate $C_{333}^{(E)}$ because of lack of data on f_{333}.

EXPERIMENTAL TECHNIQUE

The technique of measuring the third-order elastic constants of trigonal crystals is the same as has been used previously to measure cubic crystals. Optically flat and parallel samples are mounted as shown in Fig. 1. A 30 MHz quartz crystal at the top of the sample generates a sinusoidal wave of finite amplitude which produces its own harmonics as it propagates downward and impinges on the bottom surface which has been coated with a conductor. The receiving electrode and the end of the sample form a parallel plate capacitor with a spacing of approximately 5 microns which is biased by approximately 150 volts. The electrical signal generated by the ultrasonic pulse as it impinges on the end of the sample is proportional to the amplitude of the ultrasonic wave. Appropriate filtering allows one to measure the amplitude of the fundamental and of the second harmonic of the ultrasonic wave. This information, along with Eq. 5, allows one to determine the nonlinearity parameter β. An additional measurement of the velocity of ultrasonic waves allows one to calculate the second-order elastic constant and hence to determine the third-order elastic constant.

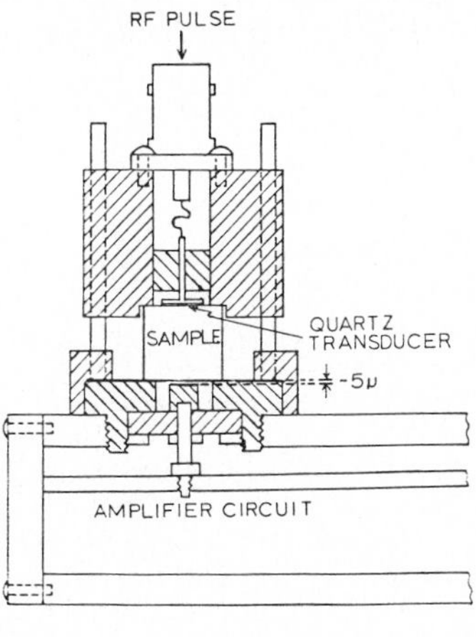

Fig. 1. Capacitive Receiver for Measuring Harmonic Generation.

RESULTS

Results of the measurements can be presented in two forms. For example, a plot of A_2 vs. A_1^2 for Z-cut quartz as shown in Fig. 2 illustrates the point that the measurements were made in the range of validity of the perturbation solution, Eq. 5, which predicts that the amplitude of the second harmonic is proportional to the square of the fundamental amplitude. The slope of this curve is a measure of β_Z for quartz. A much more sensitive plot which allows one to extrapolate to zero amplitude is given in Fig. 3. The extrapolation should be made in such a manner that the curve approaches the ordinate with zero slope (5). In addition, correction for diffraction can be made (6) to give the values of TOE constants in Table I. The agreement among the three sets of measurements reassured us about the use of the harmonic generation technique with trigonal crystals. As explained by Thurston et al. (2), the difference between the "piezoelectrically stiffened" and the "unstiffened" constants is smaller than the experimental uncertainty. Thus, we have not made that distinction in Table I. Measured values of the nonlinearity parameters of $LiNbO_3$ are presented in Table II. By measuring the velocity of ultrasonic waves in the samples we obtained the values of C_{11} listed in Table III. Using these data and relationships of the form

$$\beta_X = - \frac{3C_{11} + C_{111}}{C_{11}} \quad (12)$$

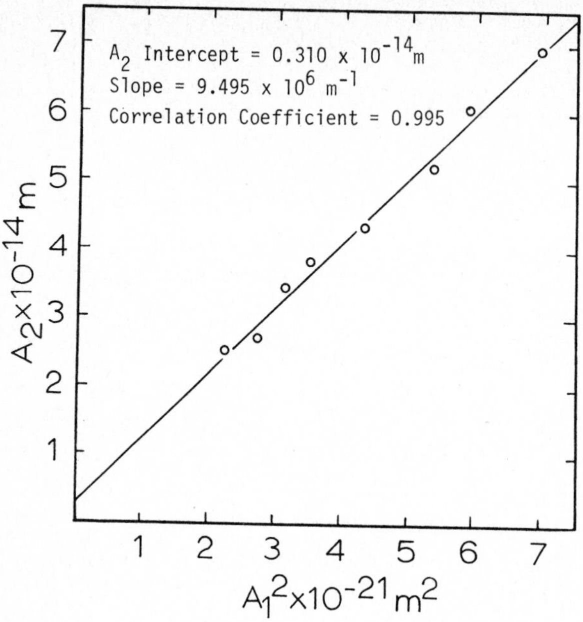

Fig. 2. Graph of A_2 vs. A_1^2 for Z-Cut Quartz.

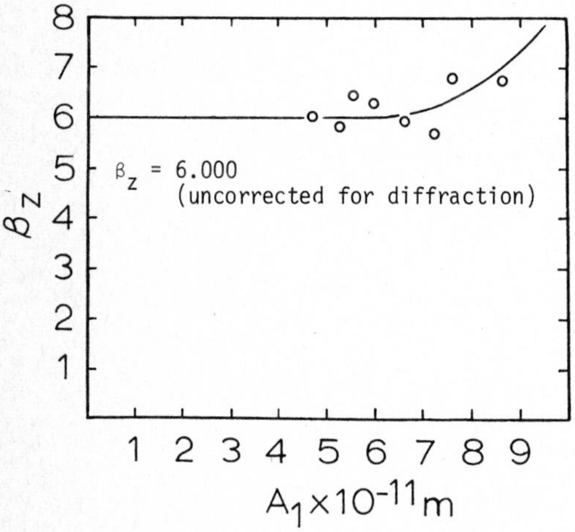

Fig. 3. Graph of β_z vs. A_1 for Z-Cut Quartz.

we calculated the TOE constants given in Table IV. For comparison we present earlier data from our laboratory as well as the data of Nakagawa (1). The Z-direction (corresponding to measured values of C_{333}) is the piezoelectric direction. Even though we list Nakagawa's value for C^E_{333}, his data still have an undetermined coefficient f_{333} buried in it, so it is not, strictly speaking, C^E_{333}, but is a quantity somewhere between C^E_{333} and C^D_{333}. At the moment we do not have values of f_{333}, so we can only list our value C^D_{333}, which is barely within Nakagawa's uncertainty limits of ±200%. Our value of C^E_{111} is outside the limits of Nakagawa's stated uncertainty limits. In part, this difference may be the result of refinement of techniques for growing samples, so that in fact our values should differ from those of Nakagawa because our samples are different. We assume that the more recently grown samples are purest and most defect-free, however.

AN EXPERIMENTAL POSSIBILITY

During the course of our measurements we discovered that the capacitive receiver produces a measurable signal in a piezoelectric direction even without a bias voltage. An example of the fundamental signal with and without bias voltage is shown in Fig. 4. Obviously this signal is produced by piezoelectric coupling, since it is not observed in nonpiezoelectric directions. However, the relationship between the magnitude of this signal and the corresponding piezoelectric coupling coefficients is yet to be evaluated. We hope that future analysis will allow us to evaluate such quantities as f_{333} for $LiNbO_3$.

TABLE I

SECOND-ORDER ELASTIC CONSTANTS AND THIRD-ORDER ELASTIC CONSTANTS FOR QUARTZ

Sample Orientation	SOE Constants x 10^{11} dynes/cm^2 McSkimin et al. (Ref. 7)	TOE Constants x 10^{12} dynes/cm^2 Present Experiment	Thurston et al. (Ref. 2)	Stern and Smith (Ref. 8)
X	C_{11} = 8.680	C_{111} = -2.15 ± .06	C_{111} = -2.10 ± .07	C_{111} = -2.18
Z	C_{33} = 10.575	C_{333} = -8.34 ± .55	C_{333} = -8.15 ± .18	C_{333} = -8.37

TABLE II

VALUES OF THE NONLINEARITY PARAMETER β FOR LITHIUM NIOBATE (CORRECTED FOR DIFFRACTION)

Sample Orientation	β
X	6.01
Z	0.359

TABLE III

SOE CONSTANTS OF $LiNbO_3$ USING DENSITY ρ = 4.644 gm/cm^3

	C^E_{11} x 10^{12} dynes/cm^2	C^D_{33} x 10^{12} dynes/cm^2
Present experiment	1.98	2.48
Nakagawa et al. (1973)	1.98	2.48
Philip and Breazeale (1982)	1.964	

TABLE IV

THIRD ORDER ELASTIC CONSTANTS OF $LiNbO_3$

	C^E_{111} x 10^{12} dynes/cm^2	C^D_{333} x 10^{12} dynes/cm^2	C^E_{333} x 10^{12} dynes/cm^2
Present experiment	-17.83 ± .35	-8.33 ± .15	
Nakagawa et al. (1973)	- 5.12 ± 1.94		-3.63 ± 6.90*
Philip and Breazeale (1982)	-16.1		

*This value includes one term in addition to the constant field coefficient.

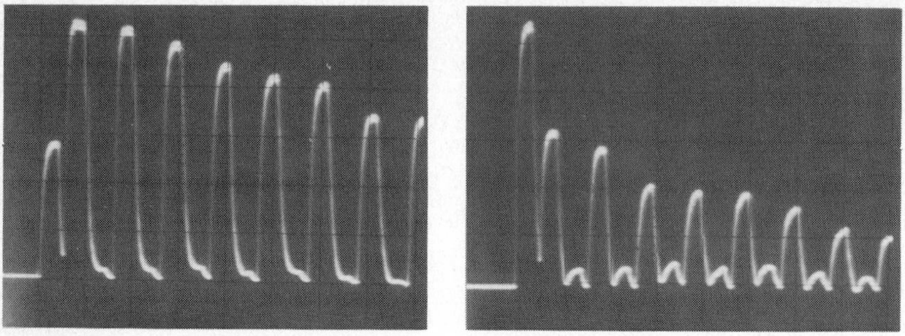

(a) Fundamental with bias voltage 1 V/cm

(b) Fundamental without bias voltage .5 V/cm

Fig. 4. A Comparison of the Fundamental Ultrasonic Signal from $LiNbO_3$ with Bias Voltage Applied and without Bias Voltage Applied.

ACKNOWLEDGMENT

This research was sponsored by the Office of Naval Research

REFERENCES

[1] Nakagawa, Y., K. Yamanouchi, and K. Shibayama, "Third-Order Elastic Constants of Lithium Niobate," J. Appl. Phys. 44, 3969 (1973).

[2] Thurston, R. N., H. J. McSkimin, and P. Andreatch, Jr., "Third-Order Elastic Coefficients of Quartz," J. Appl. Phys. 37, 267 (1966).

[3] Holt, A. C., and J. Ford, "Theory of Ultrasonic Pulse Measurements of Third-Order Elastic Constants for Cubic Crystals," J. Appl. Phys. 38, 42 (1967); Thurston, R. N., "Wave Propagation in Fluids and Normal Solids," in Physical Acoustics, edited by Warren P. Mason (Academic Press, New York, 1964), Vol. IA, pp. 2-110.

[4] McMahon, D. H., "Acoustic Second-Harmonic Generation in Piezoelectric Crystals," J. Acoust. Soc. Am. 44, 1007 (1968).

[5] Yost, W. T., and M. A. Breazeale, "Ultrasonic Nonlinearity Parameters and Third-Order Elastic Constants of Germanium between 300 and 77°K," Phys. Rev. B 9, 510 (1974).

[6] Breazeale, M. A., and Bruce Blackburn, "Measurement of Nonlinearity Parameters in Small Solid Samples by the Harmonic Generation Technique," Ultrasonics International 1979 Conference Proceedings, Graz, Austria, p. 500.

[7] McSkimin, H. J., P. Andreatch, Jr., and R. N. Thurston, "Elastic Moduli of Quartz versus Hydrostatic Pressure at 25° and -195.8°C," J. Appl. Phys. 36, 1624 (1965).

[8] Stern, R., and R. T. Smith, "On the Third-Order Elastic Moduli of Quartz," J. Acoust. Soc. Am. 44, 640 (1968).

The Mechanical Behavior of Electromagnetic Solid Continua
G.A. Maugin (editor)
Elsevier Science Publishers B.V. (North-Holland)

REFLECTION AND TRANSMISSION OF WEAK DISCONTINUITY WAVES IN ELECTRO-MAGNETOELASTIC MATERIALS

Eiji Matsumoto

Department of Aeronautical Engineering
Kyoto University
Kyoto 606
Japan

A weak discontinuity wave is reflected and transmitted at the boundary between two distinct electro-magnetoelastic materials. When different kinds of reflected and transmitted waves from the incident wave are generated, an interaction occurs between the jumps of the mechanical and electromagnetic fields. It is shown that the total of the jumps of a field is conserved in the collision of the incident wave. A relation connecting the speeds and the jumps of the incident, reflected and transmitted waves is obtained. The result is applied to the boundary between two piezoelectric materials.

1. INTRODUCTION

A weak discontinuity wave is defined as a propagating surface across which physical quantities suffer discontinuities weaker than those across shock waves. The wave may be reflected from and transmitted (refracted) through the boundary between two dissimilar materials. For elastic materials the subject has been studied by Chen and Gurtin (1973), Wesołowski (1975), Suhubi and Jeffrey (1976), and Borejko (1979), and for elastic-plastic materials by Jasman (1974). Even in a single material such phenomena occur at a surface where mechanical properties change discontinuously. For example, Cizek and Ting (1978) considered reflected waves from an elastic-plastic boundary, and Strumia (1979) and Boillat and Ruggeri (1979) investigated reflection and transmission of the waves at shock surfaces. This paper starts out from a set of generalized balance laws which includes basic equations for electro-magnetoelastic materials as a special case. A mathematical foundation is given to the reflection and transmission of weak discontinuity waves in the case of one-dimensional motion. As an application of the result, we consider the boundary between two piezoelectric materials. When a transverse wave collides with the boundary, electromagnetic waves are generated in addition to reflected and transmitted transverse waves.

2. GENERALIZED BALANCE LAWS AND WEAK DISCONTINUITY WAVES

In many cases, the basic equations for homogeneous continuous media can be expressed in the integral form in the reference configuration:

$$\frac{d}{dt}\int_V Y^\alpha \, dV + \oint_S \boldsymbol{\Phi}^\alpha \cdot d\boldsymbol{S} + \int_V F^\alpha \, dV = 0 \, , \tag{1}$$

$$\frac{d}{dt}\int_S \boldsymbol{Z}^\Gamma \cdot d\boldsymbol{S} + \oint_C \boldsymbol{\Psi}^\Gamma \cdot d\boldsymbol{L} + \int_S \boldsymbol{G}^\Gamma \cdot d\boldsymbol{S} = 0 \, , \tag{2}$$

$$\oint_S \boldsymbol{Z}^\Gamma \cdot d\boldsymbol{S} + \int_V H^\Gamma \, dV = 0 \, , \tag{3}$$

where Y^α (α=1,..,M) are scalars, and $\underset{\sim}{Z}^\Gamma$ (Γ=1,..,N) vectors. F^α and H^Γ are scalar-valued functions of Y^α and $\underset{\sim}{Z}^\Gamma$, and $\underset{\sim}{\Phi}^\alpha$, $\underset{\sim}{\Psi}^\Gamma$ and $\underset{\sim}{G}^\Gamma$ are vector-valued functions of them. The local form of (1)-(3) becomes

$$\dot{Y}^\alpha + \mathrm{DIV}\, \underset{\sim}{\Phi}^\alpha + F^\alpha = 0 \,, \tag{4}$$

$$\dot{\underset{\sim}{Z}}^\Gamma + \mathrm{ROT}\, \underset{\sim}{\Psi}^\Gamma + \underset{\sim}{G}^\Gamma = \underset{\sim}{0} \,, \tag{5}$$

$$\mathrm{DIV}\, \underset{\sim}{Z}^\Gamma + H^\Gamma = 0 \,. \tag{6}$$

We are concerned with a plane weak discontinuity wave Σ which has the properties:

(1) $\underset{\sim}{a} \equiv (Y^\alpha, \underset{\sim}{Z}^\Gamma)$ is continuous everywhere, and its first derivatives are continuous except on Σ, where they are discontinuous.

(2) $\underset{\sim}{a}$ is uniformly constant over Σ at each instant.

The compatibility conditions for a continuous function $\zeta(\underset{\sim}{X}, t)$ across Σ are

$$[\dot{\zeta}] = - U \bar{\zeta} \,, \quad [\partial_i \zeta] = \bar{\zeta} N_i \,, \quad \bar{\zeta} \equiv [\partial_i \zeta] N^i \,, \tag{7}$$

where ∂_i denotes partial derivative with respect to X^i, $\underset{\sim}{N}$ the unit normal vector to Σ, U the normal velocity, and $[\cdot]$ the jump of a quantity within it. Applying the compatibility conditions (7) to (4)-(6), we get

$$- U \bar{Y}^\alpha + A^{\alpha i}{}_\beta N_i \bar{Y}^\beta + B^{\alpha i}{}_{\Delta k} N_i \bar{Z}^{\Delta k} = 0 \,, \tag{8}$$

$$- U \bar{Z}^{\Gamma i} + \varepsilon^{ijk} C^\Gamma{}_{k\beta} N_j \bar{Y}^\beta + \varepsilon^{ijk} D^\Gamma{}_{k\Delta l} N_j \bar{Z}^{\Delta l} = 0 \,, \tag{9}$$

$$\underset{\sim}{N} \cdot \bar{\underset{\sim}{Z}}^\Gamma = 0 \,, \tag{10}$$

where

$$A^{\alpha i}{}_\beta \equiv \frac{\partial \Phi^{\alpha i}}{\partial Y^\beta} \,, \quad B^{\alpha i}{}_{\Delta k} \equiv \frac{\partial \Phi^{\alpha i}}{\partial Z^{\Delta k}} \,, \quad C^{\Gamma i}{}_\beta \equiv \frac{\partial \Psi^{\Gamma i}}{\partial Y^\beta} \,, \quad D^{\Gamma i}{}_{\Delta k} \equiv \frac{\partial \Psi^{\Gamma i}}{\partial Z^{\Delta k}} \,. \tag{11}$$

Equations (8) and (9) can be written as

$$(- U \delta^{\Gamma'}{}_{\Delta'} + Q^{\Gamma'}{}_{\Delta'}) \bar{a}^{\Delta'} \qquad (\Gamma', \Delta' = 1, \ldots, M+3N) \,, \tag{12}$$

where

$$\underset{\sim}{Q} \equiv \begin{pmatrix} A^{\alpha i}{}_\beta N_i & B^{\alpha i}{}_{\Delta l} N_i \\ \varepsilon^{ijk} C^\Gamma{}_{k\beta} N_j & \varepsilon^{ijk} D^\Gamma{}_{k\Delta l} N_j \end{pmatrix} \,. \tag{13}$$

Thus the normal velocity U is a proper vector of $\underset{\sim}{Q}$, and the jump $\bar{\underset{\sim}{a}}$ is a proper vector belonging to U. We call $\bar{\underset{\sim}{a}}$ the amplitude (vector) of the wave.

3. REFLECTION AND TRANSMISSION OF WEAK DISCONTINUITY WAVES

Let B be the plane boundary between two dissimilar homogeneous media, and $\underset{\sim}{N}$ its unit normal vector. Then the functional forms of $\underset{\sim}{\Phi}^\alpha$, $\underset{\sim}{\Psi}^\Gamma$, F^α, $\underset{\sim}{G}^\Gamma$, and H^Γ on one side of B are, respectively, different from those on the other side. The boundary conditions are derived from (1)-(3):

$$\underset{\sim}{N} \cdot [\underset{\sim}{\Phi}^\alpha]_B + F^\alpha{}_S = 0 \,, \tag{14}$$

$$\underset{\sim}{N} \times [\underset{\sim}{\Psi}^\Gamma]_B + \underset{\sim}{G}^\Gamma{}_S = \underset{\sim}{0} \,, \tag{15}$$

$$\underset{\sim}{N} \cdot [\underset{\sim}{Z}^\Gamma]_B + H^\Gamma{}_S = 0 \,, \tag{16}$$

where $F^\alpha{}_S$, $\underset{\sim}{G}^\Gamma{}_S$, and $H^\Gamma{}_S$ are surface quantities of F^α, $\underset{\sim}{G}^\Gamma$, and H^Γ such as the surface charge and the surface current. Let us consider the one-dimensional wave motion. That is, suppose that a plane weak discontinuity wave normally incident to B generates plane weak discontinuity waves and that all the quantities are transversally uniform.

Henceforth the subscripts + and - denote, respectively, the limits of a quantity on the $\underset{\sim}{N}$ and $-\underset{\sim}{N}$ sides of B. Let U_I and $\bar{\underset{\sim}{a}}_I$ be the normal velocity and the amplitude of the incident wave. Let p be the number of all negative proper values of $\underset{\sim}{Q}^-$, and q be that of all positive proper values of $\underset{\sim}{Q}^+$. The velocities of the reflected waves are written as $U_R^{(1)} <\ldots< U_R^{(p)}$, and those of the transmitted waves are written as $U_T^{(1)} <\ldots< U_T^{(q)}$. The corresponding amplitudes are denoted by similar super- and subscripts. Figure 1 shows the plane of the N-direction and the time, where $N = 0$ means the boundary and $t = 0$ the time of the collision of the incident wave. The vicinity of the origin can be divided into $p + q + 3$ regions by the loci of the waves. Suppose that in each region $\underset{\sim}{a}$ is differentiable and the derivatives have unique limits at the origin. Then it follows from an identity that

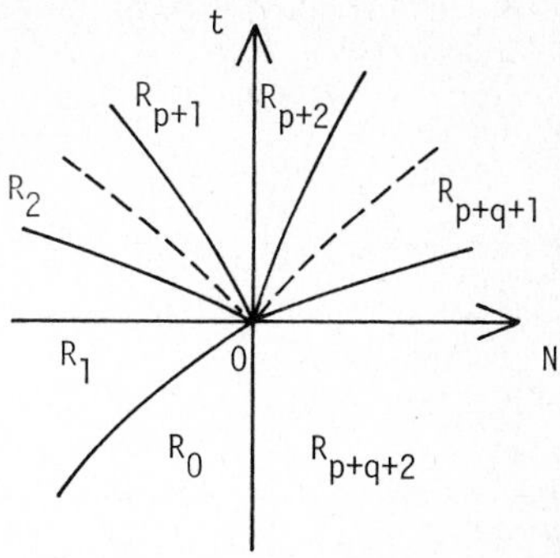

Figure 1 Loci of Waves

$$\bar{\underset{\sim}{a}}_I + \bar{\underset{\sim}{a}}_B = \sum_{i=1}^{p} \bar{\underset{\sim}{a}}_R^{(i)} + \sum_{j=1}^{q} \bar{\underset{\sim}{a}}_T^{(j)} + \bar{\underset{\sim}{a}}_{B'} \ , \tag{17}$$

where the subscripts B or B' denotes a quantity at the boundary for $t = -0$ or $t = +0$, respectively. Equation (17) implies that the total of the amplitudes of all the waves including the jump across the boundary is conserved just before and just after the collision of the incident wave.

To determine the amplitudes of the reflected and transmitted waves, we apply the compatibility conditions (7) to each wave and combine the results with the time derivatives of the boundary conditions (14)-(16). The obtained equations are

$$U_I^2 \, \bar{\underset{\sim}{a}}_I + \dot{\underset{\sim}{b}}_S = \sum_{i=1}^{p} \{U_R^{(i)}\}^2 \, \bar{\underset{\sim}{a}}_R^{(i)} + \sum_{j=1}^{q} \{U_T^{(j)}\}^2 \, \bar{\underset{\sim}{a}}_T^{(j)} + \dot{\underset{\sim}{b}}_S{}' \ , \tag{18}$$

$$\dot{H}^{\Gamma}{}_S = \dot{H}^{\Gamma}{}_S{}' \ , \tag{19}$$

where

$$\underset{\sim}{b}_S \equiv (F^{\alpha}{}_S \, , \, \underset{\sim}{G}^{\Gamma}{}_S) \ , \tag{20}$$

and a surface quantity without or with prime denotes its value for $t = -0$ or for $t = +0$, respectively. In many theories of electro-magnetoelastic materials, the surface quantity $\underset{\sim}{b}_S$ linearly depends on $H^{\Gamma}{}_S$. Then (18) reduces to

$$U_I^2 \, \bar{\underset{\sim}{a}}_I = \sum_{i=1}^{p} \{U_R^{(i)}\}^2 \, \bar{\underset{\sim}{a}}_R^{(i)} + \sum_{j=1}^{q} \{U_T^{(j)}\}^2 \, \bar{\underset{\sim}{a}}_T^{(j)} \ . \tag{21}$$

Let S_I be the subspace of $M + 3N$-dimensional space spanned by $\bar{\underset{\sim}{a}}_I$, and S_R be the subspace spanned by all the proper vectors belonging to negative proper values of $\underset{\sim}{Q}^-$. Similarly, S_T denotes the subspace spanned by all the proper vectors belonging to positive proper values of $\underset{\sim}{Q}^+$. Then in view of (21) we see that

$$S_I \subset S_R \oplus S_T \ . \tag{22}$$

If all the amplitudes of the transmitted waves vanish, (22) implies $S_I \in S_R$, which contradicts the fact that proper vectors belonging to different proper values are linearly independent. Thus we conclude that at least one transmitted wave exists. When

$$S_R \cap S_T = \{\underset{\sim}{0}\} \ , \tag{23}$$

the amplitudes of the reflected and transmitted waves are uniquely determined by (21). When

$$S_I \subset S_T \ , \tag{24}$$

no reflected wave is excited.

Across the boundary there may exsist a strong discontinuity in $\underset{\sim}{a}$. The kinematical compatibility condition for $\underset{\sim}{a}$ across B is given by

$$\frac{d}{dt}[\underset{\sim}{a}]_B = [\dot{\underset{\sim}{a}}]_B \quad . \tag{25}$$

By a similar process to derivation of (18), we obtain

$$[\dot{\underset{\sim}{a}}]_{B'} - [\dot{\underset{\sim}{a}}]_B = U_I \bar{\underset{\sim}{a}}_I - \sum_{i=1}^{p} U_R^{(i)} \bar{\underset{\sim}{a}}_R^{(i)} - \sum_{j=1}^{q} U_T^{(j)} \bar{\underset{\sim}{a}}_T^{(j)} \ , \tag{26}$$

which means that $[\dot{\underset{\sim}{a}}]$ changes discontinuously at time $t = 0$. From this fact and (25) we see that the strong discontinuity in $\underset{\sim}{a}$ across the boundary changes roughly when the incident wave collides with the boundary.

4. EXAMPLE: PIEZOELECTRIC MATERIAL WITH CUBIC SYMMETRY

According to Dunkin and Eringen (1963) and Pao (1978), the basic equations for deformable media in slow motion are given by

$$\frac{d}{dt}\int_V \rho \, dv = 0 \ , \tag{27}$$

$$\frac{d}{dt}\int_V \rho \underset{\sim}{v} \, dv - \oint_S \underset{\sim}{t}\cdot d\underset{\sim}{s} - \int_V \underset{\sim}{f} \, dv = \underset{\sim}{0} \ , \tag{28}$$

$$\frac{d}{dt}\int_V \rho_e \, dv + \oint_S \underset{\sim}{J}_e \cdot d\underset{\sim}{s} = 0 \ , \tag{29}$$

$$\frac{d}{dt}\int_S \underset{\sim}{B}\cdot d\underset{\sim}{s} + \oint_C (\underset{\sim}{E} + \underset{\sim}{v}\times\underset{\sim}{B})\cdot d\underset{\sim}{l} = 0 \tag{30}$$

$$\frac{d}{dt}\int_S \underset{\sim}{D}\cdot d\underset{\sim}{s} - \oint_C (\underset{\sim}{H} - \underset{\sim}{v}\times\underset{\sim}{D})\cdot d\underset{\sim}{l} + \int_S \underset{\sim}{J}_e\cdot d\underset{\sim}{s} = 0 \ , \tag{31}$$

$$\oint_S \underset{\sim}{B}\cdot d\underset{\sim}{s} = 0 \ , \qquad \oint_S \underset{\sim}{D}\cdot d\underset{\sim}{s} - \int_V \rho_e \, dv = 0 \ , \tag{32}$$

where ρ is the material density, $\underset{\sim}{v}$ the velocity, $\underset{\sim}{t}$ the Cauchy stress, $\underset{\sim}{f}$ the external force including electromagnetic one, ρ_e the electric charge, $\underset{\sim}{J}_e$ the effective electric current, $\underset{\sim}{B}$ the magnetic induction, $\underset{\sim}{H}$ the magnetic field intensity, $\underset{\sim}{D}$ the electric displacement, $\underset{\sim}{E}$ the electric field intensity. We consider two linear piezoelectric materials with cubic symmetry, and suppose that one of the lattice points coincides with the origin of the co-ordinate and the edges with the axes. Then the constitutive equations for each material are given by

$$T_{\kappa}^{\ ij} = \lambda e_{kk} \delta^{ij} + 2\mu e^{ij} + \beta \sum_{m=1}^{3} \delta^{im} \delta^{jm} e_{mm} + \theta^{ij}{}_k (E^k + \varepsilon^{kpq} v_p B_q) \ , \tag{33}$$

$$D^i = \varepsilon E^i - \theta^{ijk} e_{jk} + \alpha\varepsilon^{ijk} v_j H_k \ , \tag{34}$$

$$B^i = \kappa H^i - \alpha\varepsilon^{ijk} v_j E_k \ , \tag{35}$$

$$J_e^{\ i} \equiv J^i - \rho_e v^i = \sigma(E^i + \varepsilon^{ijk} v_j B_k) + \omega^{ijk} e_{jk} \ , \tag{36}$$

$$f^i = \rho_e E^i + \varepsilon^{ijk} J_j B_k \ , \tag{37}$$

where cf. Thomas (1966) and

$$e_{ij} = \frac{1}{2}(\partial_j u_i + \partial_i u_j) \ , \tag{38}$$

$$\alpha \equiv \kappa\varepsilon - \kappa_0 \varepsilon_0 \ , \qquad \theta^{ijk} \equiv \theta|\varepsilon^{ijk}| \quad . \tag{39}$$

The basic equations (27)-(32) can be transformed into the integral form in the reference configuration. To complete the basic equations in the same form as (1)-(3), the compatibility condition for $\underset{\sim}{v}$ and $\underset{\sim}{F}$ is necessary:

$$\frac{d}{dt}\int_V e\, dV - \oint_S \underset{\sim}{R}\cdot d\underset{\sim}{S} = \underset{\sim}{0} \quad , \qquad R^{ijk} \equiv \frac{1}{2}(v^i\, \delta^{jk} + v^j\, \delta^{ki}) \quad . \tag{40}$$

For the obtained set of basic equations, the vector $\underset{\sim}{a}$ becomes $(\rho_\kappa \underset{\sim}{v}, \underset{\sim}{e}, J\rho_e, J\underset{\sim}{B}\,\underset{\sim}{F}^{-T}, J\underset{\sim}{D}\,\underset{\sim}{F}^{-T})$.

In order to emphasize the interaction in the reflection and transmission of weak discontinuity waves, we consider a simple case where the propagation direction equals (1, 1, 1) and where $\underset{\sim}{v} = \underset{\sim}{B} = \underset{\sim}{D} = \underset{\sim}{0}$ and $\underset{\sim}{F} = \underset{\sim}{I}$ ahead of the wave. Taking the jump of the local form of (27)-(32) and (40) across a wave and using (33)-(40), we get

$$(\rho_\kappa U^2 - \mu - \frac{\beta}{3})\bar{\underset{\sim}{v}} - (\lambda + \mu + \frac{4\theta^2}{3\varepsilon})(\underset{\sim}{N}\cdot\bar{\underset{\sim}{v}})\underset{\sim}{N} = \underset{\sim}{0} \quad , \tag{41}$$

$$-U\,\bar{\rho}_e + \underset{\sim}{N}\cdot\bar{\underset{\sim}{J}}_e = 0 \quad , \tag{42}$$

$$-\bar{\underset{\sim}{B}} + \underset{\sim}{N}\times\bar{\underset{\sim}{E}} = \underset{\sim}{0} \quad , \qquad U\bar{\underset{\sim}{D}} + \underset{\sim}{N}\times\bar{\underset{\sim}{H}} = \underset{\sim}{0} \quad , \tag{43}$$

$$\underset{\sim}{N}\cdot\bar{\underset{\sim}{B}} = 0 \quad , \qquad \underset{\sim}{N}\cdot\bar{\underset{\sim}{D}} = 0 \quad , \tag{44}$$

$$\bar{\underset{\sim}{D}} = \varepsilon\bar{\underset{\sim}{E}} + \frac{1}{U}\underset{\sim}{\theta}\,(\bar{\underset{\sim}{v}}\otimes\underset{\sim}{N}) \quad , \qquad \bar{\underset{\sim}{B}} = \kappa\bar{\underset{\sim}{H}} \quad . \tag{45}$$

Then the following three kinds of weak discontinuity waves can exist.

A. TRANSVERSE WAVE. When $\underset{\sim}{N}\times\underset{\sim}{v} \neq \underset{\sim}{0}$, (41) imposes

$$U_T^{\,2} = (\mu + \frac{\beta}{3})/\rho_\kappa \quad . \tag{46}$$

In this case it follows that

$$\bar{\underset{\sim}{B}} = -\frac{\kappa\theta}{\sqrt{3}}\underset{\sim}{N}\times\bar{\underset{\sim}{v}} \quad , \qquad \bar{\underset{\sim}{D}} = -\frac{\theta}{\sqrt{3}U_T}\bar{\underset{\sim}{v}} \quad . \tag{47}$$

B. LONGITUDINAL WAVE. When $\underset{\sim}{N}\cdot\bar{\underset{\sim}{v}} \neq 0$, (41) imposes

$$U_L^{\,2} = (\lambda + 2\mu + \frac{\beta}{3} + \frac{4\theta^2}{3\varepsilon})/\rho_\kappa \quad . \tag{48}$$

Neglecting smaller terms than $U_L{}^2/c^2$, we have $\bar{\underset{\sim}{B}} = \bar{\underset{\sim}{D}} = \underset{\sim}{0}$ for the wave,

C. ELECTROMAGNETIC WAVE. For a wave different from the above ones, we have $\bar{\underset{\sim}{v}} = \underset{\sim}{0}$, and (43)-(45) imply

$$U_{EM}^{\;2} = 1/(\varepsilon\kappa) \quad . \tag{49}$$

Thus for the longitudinal wave the electromagnetic components of the amplitude vanish, and for the electromagnetic wave the mechanical components vanish. On the other hand, the transverse wave has non-vanishing components of the both fields. Note that in general for the longitudinal wave $\bar{\underset{\sim}{E}}$ does not vanish, cf. $(45)_1$.

We next consider the case where the incident wave is a transverse wave. By applying the result in the previous section, we can uniquely determine all the amplitudes of the excited waves. In a result, the reflected and transmitted waves consist of transverse waves and electromagnetic waves. Let $\bar{\underset{\sim}{v}}_I$ be the jump for the incident wave, then (21) implies

$$\bar{\underset{\sim}{v}}_R = \frac{1 - \xi\eta}{1 + \xi\eta}\bar{\underset{\sim}{v}}_I \quad , \qquad \bar{\underset{\sim}{v}}_T = \frac{2}{\eta(1 + \xi\eta)}\bar{\underset{\sim}{v}}_I \quad , \tag{50}$$

where $\bar{\underset{\sim}{v}}_R$ and $\bar{\underset{\sim}{v}}_T$ are, respectively, the jumps for the reflected and transmitted transverse waves and

$$\xi \equiv \rho_\kappa^+/\rho_\kappa^- \quad , \qquad \eta \equiv U_T^+/U_T^- \quad . \tag{51}$$

Similarly, from (21) and (43)-(45) we can obtain the jumps $\bar{\underset{\sim}{B}}_R$, $\bar{\underset{\sim}{D}}_R$ and $\bar{\underset{\sim}{B}}_T$, $\bar{\underset{\sim}{D}}_T$ for the backward and forward electromagnetic waves. Furthermore, by use of the expressions we can calculate the following jumps which are more convenient for measurement:

$$[\dot{\underset{\sim}{H}}]_R = \frac{2U_T^-\,[\theta]_B}{\sqrt{3}(1 + \xi\eta)\{1 + (\varepsilon^+/\varepsilon^-)(U_{EM}^+/U_{EM}^-)\}}\;\underset{\sim}{N}\times\bar{\underset{\sim}{v}}_I \quad , \tag{52}$$

$$[\dot{\underset{\sim}{E}}]_R = \frac{-2U_T^- \kappa^+ U_{EM}^+ [\theta]_B}{\sqrt{3}(1+\xi\eta)\{1+(\kappa^+/\kappa^-)(U_{EM}^+/U_{EM}^-)\}} \bar{\underset{\sim}{v}}_I \quad , \tag{53}$$

$$[\dot{\underset{\sim}{H}}]_T = \frac{2U_T^- [\theta]_B}{\sqrt{3}(1+\xi\eta)\{1+(\kappa^+/\kappa^-)(U_{EM}^+/U_{EM}^-)\}} \underset{\sim}{N}\times\bar{\underset{\sim}{v}}_I \quad , \tag{54}$$

$$[\dot{\underset{\sim}{E}}]_T = \frac{2U_T^- \kappa^- U_{EM}^- [\theta]_B}{\sqrt{3}(1+\xi\eta)\{1+(\varepsilon^+/\varepsilon^-)(U_{EM}^+/U_{EM}^-)\}} \bar{\underset{\sim}{v}}_I \quad , \tag{55}$$

where we have omitted smaller terms. Thus we find that the intensities of the excited electromagnetic waves linearly depend on $[\theta]_B$, the difference of the electromechanical coupling constants between the two materials.

In the other case where the incident wave is a longitudinal wave or an electromagnetic wave, we can also obtain the amplitudes of the reflected and transmitted waves. In each case the incident wave does not excite different kinds of waves.

REFERENCES

[1] Chen, P.J. and Gurtin, M.E., On the propagation of one-dimensional acceleration waves in laminated composites, J. Appl. Mech. 40 (1973) 1055-1060.
[2] Wesołowski, Z., Linear independence of amplitudes of reflected acceleration waves, Bull. Acad. Polon. Sci., Serie Sci. Tech. 23 (1975) 547-552.
[3] Suhubi, E.S. and Jeffrey, A., Propagation of weak discontinuities in a layered hyperelastic half-space, Proc. R. Soc. Edinburgh Sect. A 75 (1976) 209-221.
[4] Borejko, P., Reflection and refraction of an acceleration wave at boundary between two nonlinear elastic materials, Arch. Mech. 31 (1979) 373-384.
[5] Jahsman, W.E., Reflection and refraction of weak elastic-plastic waves, J. Appl. Mech. 41 (1974) 117-123.
[6] Cizek, J.C. and Ting, T.C.T., Reflection of acceleration waves in an elastic-plastic medium, J. Appl. Mech. 45 (1978) 51-59.
[7] Strumia, A., Evolution law of a weak discontinuity crossing a non characteristic shock in a nonlinear dielectric medium, Meccanica 14 (1979) 67-71.
[8] Boillat, G. and Ruggeri, T., Reflection and transmission of discontinuity waves through a shock wave, Proc. R. Soc. Edinburgh Sect. A 83 (1979) 17-24.
[9] Dunkin, J.W. and Eringen, A.C., On the propagation of waves in an electromagnetic elastic solids, Int. J. Engng Sci. 1 (1963) 461-495.
[10] Pao, Y.-H., Electromagnetic forces in deformable continua, in: Nemat-Nasser, S. (ed.), Mechanics Today IV (Pergamon Press, New York, 1978).
[11] Thomas, T.Y., On the stress-strain relations for cubic crystals, Proc. Natl. Acad. Sci. 55 (1966) 235-239.

The Mechanical Behavior of Electromagnetic Solid Continua
G.A. Maugin (editor)
Elsevier Science Publishers B.V. (North-Holland)

A ONE-DIMENSIONAL PROBLEM IN THE GENERALIZED THEORY OF THERMOPIEZOELASTICITY

E.Bassiouny[+] and A.F.Ghaleb[++]

[+]Department of Mathematics,Faculty of Education,Cairo University, Fayum, Egypt.

[++]Department of Mathematics, Faculty of Science, Cairo University, Cairo,Egypt.

A simple model of generalized thermopiezoelasticity is used to investigate the motion of a semi-infinite piezoelectric rod under a thermal shock at its end. Fourier's law for heat conduction assumes the form $q + t_1\dot{q} = -k\,\partial T/\partial x$. For the one-dimensional case under consideration, the electric displacement vanishes, thus allowing for a solution of the thermoelastic problem independently of the electric field. The latter may then be obtained from the equation of state. Due to the presence of the thermal relaxation time t_1, the temperature, stress and electric field distributions in the rod suffer two discontinuities which propagate along the rod with different velocities. This provides a simple electrical means for the determination of the relaxation time experimentally.

INTRODUCTION

Due to its numerous applications, the study of the mechanical behaviour of piezoelectric media has received a growing interest in the past few years. In a different scope, a new trend in research over dynamic problems in thermoelasticity has been observed since Lord and Schulman [1] first introduced their theory of generalized thermoelasticity. This was further developed by Sherief and Dhaliwal [2]. The principle feature of this theory resides in a modification of Fourier's law for heat conduction in order to account for the thermal relaxation phenomenon.

To the authors' knowledge, there have been very few attempts to apply the generalized theory of thermoelasticity to the study of polarizable and magnetized media in general , and piezoelectrics more specially. We note here the work of Smirnov [3] devoted to the generalized thermoelasticity of Cosserat media. The aim of the present work is to investigate the effect of the thermal relaxation time on the behaviour of a piezoelectric medium subjected to thermal disturbances. The one-dimensional problem, for its simplicity, is most suitable for our purpose.

THREE-DIMENSIONAL EQUATIONS

The three-dimensional equations of generalized thermopiezoelasticity, from which the one-dimensional equations are to be deduced, are the same as for the usual theory, with the exception of Fourier's law for heat conduction. In a system of orthogonal Cartesian coordinates, these equations are

Equations of motion :

$$\rho \ddot{u}_i = \sigma_{ij,j} \quad ,$$

Equations of electrostatics:

$$D_{i,i} = 0 \quad , \quad E_i = -V_{,i} \quad ,$$

Equation for entropy production:

$$T\dot{s} = -q_{i,i}$$

Constitutive equations:

$$\sigma_{ij} = C_{ijk\ell}\,\varepsilon_{k\ell} - e_{kij}\,D_k - a_{ij}T$$

$$E_i = -\,e_{ijk}\,\varepsilon_{jk} + b_{ij}\,D_j - c_iT$$

$$s = a_{ij}\,\varepsilon_{ij} + c_i\,D_i + a\,T$$

Fourier's law for heat conduction:

$$q_i + A_{ij}\,\dot{q}_j = -k_{ij}\,T_{,j}$$

Here, (A_{ij}) denotes the thermal relaxation tensor. Other symbols have their usual meaning.

ONE-DIMENSIONAL EQUATIONS:

Consider a semi-infinite piezoelectric rod occupying the region $0 \leq x < \infty$ and initially at rest at temperature T_o. At the near end of the rod a thermal shock is given, which raises the temperature of this end to a prescribed value which is then maintained fixed all the time. The one-dimensional equations can be put in a more convenient form by using the following set of dimensionless parameters:

$$\xi = (C_{1111}/\rho_o)^{\frac{1}{2}}(aT_o/k_{11})\,x \quad , \quad \tau = (C_{1111}/\rho_o)\,(aT_o/k_{11})\,t$$

$$\tau_1 = (C_{1111}/\rho_o)\,(aT_o/k_{11})\,A_{11} \quad ,$$

$$\theta = (T-T_o)/T_o \quad , \quad u = (C_{1111}/\rho_o)^{\frac{1}{2}}(aT_o/k_{11})u_1 \quad ,$$

$$\sigma = \sigma_{11}/C_{1111} \quad , \quad E = E_1/c_1T_o ,$$

where ρ_o denotes the initial density of the rod. After some manipulations, the linearized one-dimensional equations reduce to the following:

$$\partial_\xi^2\theta = (\partial_\tau\theta + \tau_1\,\partial_\tau^2\theta) + g\,(\partial_\xi\partial_\tau u + \tau_1\,\partial_\tau^2\partial_\xi u) \quad , \tag{1}$$

$$\partial_\tau^2 u = \partial_\xi^2 u - \varepsilon\,\partial_\xi\theta \quad , \tag{2}$$

$$\sigma = \partial_\xi u - \varepsilon\,\theta \quad , \tag{3}$$

$$E = -e\partial_\xi u - \theta \quad , \tag{4}$$

where

$$\varepsilon = a_{11}T_0/C_{1111} \quad , \quad g = a_{11}/aT_0 \quad , \quad e = e_{111}/c_1T_0 .$$

Equations (1)-(4) will be solved under the following set of limiting conditions:

$$\theta(\xi,\tau)|_{\xi=0} = \theta_0 \quad , \quad \theta(\xi,\tau)|_{\xi=\infty} = 0 \qquad \text{for} \quad \tau \geqslant 0$$

$$\sigma(\xi,\tau)|_{\xi=o} = 0 \quad , \quad \sigma(\xi,\tau)|_{\tau=\infty} = 0 \quad \text{for} \quad \tau \geqslant 0 ,$$

$$\theta(\xi,\tau)|_{\tau=o} = 0 \quad \text{for} \quad \xi > 0 ,$$

$$\sigma(\xi,\tau)|_{\tau=o} = 0 \quad , \quad \partial_\xi \sigma(\xi,\tau)|_{\tau=o} = 0 \quad \text{for} \quad \xi > 0 ,$$

$$u(\xi,\tau)|_{\tau=o} = 0 \quad , \quad \partial_\tau u(\xi,\tau)|_{\tau=o} = 0 \quad \text{for} \quad \xi > 0 .$$

Following Sherief and Dhaliwal [4], we introduce the thermoelastic potential function ϕ by the relation

$$u = \partial_\xi \phi \quad .$$

The other unknowns of the problem are expressed through ϕ as follows:

$$\theta = (1/\varepsilon)(\partial^2_\xi\phi - \partial^2_\tau\phi) \quad , \quad \sigma = \partial^2_\tau\phi \quad , \quad E = -e\,\partial^2_\xi\phi - \theta \quad .$$

It may be easily verified that function ϕ satisfies the fourth order partial differential equation

$$[\partial^4_\xi - (1+\tau_1+\varepsilon_1\tau_1)\partial^2_\xi\partial^2_\tau - (1+\varepsilon_1)\partial^2_\xi\partial_\tau + \partial^3_\tau + \tau_1\,\partial^4_\tau]\phi = 0 \quad , \tag{5}$$

where $\varepsilon_1 = g\,\varepsilon$, and the boundary conditions

$$\partial^2_\xi\phi(\xi,\tau)|_{\xi=o} = \varepsilon\,\theta_o \quad , \quad \partial^2_\xi\phi(\xi,\tau)|_{\xi=\infty} = 0 \quad \text{for} \quad \tau > 0 ,$$

$$\partial^2_\tau\phi(\xi,\tau)|_{\xi=o} = 0 \quad , \quad \partial^2_\tau\phi(\xi,\tau)|_{\xi=\infty} = 0 \quad \text{for} \quad \tau > 0 . \tag{6}$$

The problem thus reduces to that of solving equation (5), together with conditions (6). The Laplace transform technique is used for that purpose. The details of the procedure will not be given, since this can be found somewhere else [4]. We only note that the inverse transformations are carried out by analysing the behaviour of the transforms in the complex plane, near the point at infinity, and then by making use of a well-known theorem on the expansion of the Laplace transform at the point at infinity. Only the first few terms of such expansions are retained, implying that the resulting formulae for the originals are valid for small time values. Herebelow we give the final expressions for temperature, stress and electric field distributions in the rod:

$$= \theta_o\Big[H(x_1)\,e^{-\beta_{11}\xi}\sum_{\ell=o}^{3} m_{1\ell}x_1^{\ell/2}J_\ell(z_1) - H(x_2)\,e^{-\beta_{21}\xi}\sum_{\ell=o}^{3} m_{2\ell}\,x_2^{\ell/2}\,I_\ell(z_2)\Big]$$

$$\sigma = \varepsilon\theta_o\Big[H(x_1)\,e^{-\beta_{11}\xi}\sum_{\ell=o}^{3}\beta_\ell\,x_1^{\ell/2}\,J_\ell(z_1) - H(x_2)\,e^{-\beta_{21}\xi}\sum_{\ell=o}^{3}\beta_\ell\,x_2^{\ell/2}\,I_\ell(z_2)\Big]$$

$$E = \theta_o\Big[H(x_1)\,e^{-\beta_{11}\xi}\sum_{\ell=o}^{3} n_{1\ell}x_1^{\ell/2}\,J_\ell(z_1) - H(x_2)\,e^{-\beta_{21}\xi}\sum_{\ell=o}^{3} n_{2\ell}\,x_2^{\ell/2}\,I_\ell(z_2)\Big], \tag{7}$$

where

$$x_i = \nu(\tau - \beta_{io}\xi)/\beta_{i2}\xi \quad , \quad z_i = 2\nu\beta_{i2}\xi x_i^{\frac{1}{2}} \quad , \quad \nu = (-1)^{i+1} \quad (i=1,2)$$

and $\beta_{12} > 0, \beta_{22} < 0$. Coefficients appearing in formulae (7) are listed in the appendix. The range of validity of these formulae may be extended to larger time values by including more terms in the series.

NUMERICAL RESULTS AND DISCUSSION

The obtained formulae show that each of the unknown functions of the problem suffer two discontinuities that start propagating from the end of the rod at $\tau = 0$, with velocities $v_1 = 1/\beta_{10}$ and $v_2 = 1/\beta_{20}$, respectively. It may be verified that $v_2 > v_1$. For small values of the relaxation time, $\tau_1 \leq 1$, the following approximate expressions hold:

$$v_1 \simeq 1 - \varepsilon_1 \tau_1/2 \quad \text{and} \quad v_2 \simeq \tau_1^{-\frac{1}{2}} (1 + \varepsilon_1 \tau_1/2) \quad .$$

It is thus clear that the thermal relaxation mainly affects the greatest of the two velocities of propagation of discontinuities. Ahead of the fast shock wave, the rod is still in its initial state and has not yet felt the thermal disturbance. This is in contrast to the coupled theory of thermoelasticity, where disturbances propagate instantaneously.

At time moment τ, the jumps occur at points of the rod with abcissae $\xi_1 = \tau/\beta_{10}$ and $\xi_2 = \tau/\beta_{20}$. For the electric field, the magnitudes of these jumps are

$$[E]_1 = -n_{10}\, \theta_0\, e^{-\beta_{11}\xi} \quad \text{and} \quad [E]_2 = n_{20}\, \theta_0\, e^{-\beta_{21}\xi} \quad ,$$

respectively. A measurement of either of these may be used to evaluate the constant e, which is directly related to the piezoelectric constant. Again, a measurement of the electric field at a fixed point of the rod, ξ_0 say, makes it possible to determine the time interval $\Delta\tau$ that elapses between the passage of the two shock waves through that point:

$$\Delta\tau/\xi_0 = (\beta_{10} - \beta_{20}) = \{\tfrac{1}{2}\, 1 + (\varepsilon_1 + 1)\tau_1 + A\}^{\frac{1}{2}} - \{\tfrac{1}{2}\, 1 + (\varepsilon_1 + 1)\tau_1 - A\}^{\frac{1}{2}}$$

This relation may be used to evaluate τ_1. For $\tau_1 \leq 1$, it reduces to

$$\varepsilon_1 \tau_1 - 2\tau_1^{\frac{1}{2}} - 2(\Delta\tau/\xi_0 - 1) = 0 \quad ,$$

whence

$$\tau_1^{\frac{1}{2}} = (1/\varepsilon_1)\{1 - [1 + 2\varepsilon_1(\Delta\tau/\xi_0 - 1)]^{\frac{1}{2}}\} \quad .$$

We have obtained some numerical results for the following values of the constants:

$$\varepsilon = \varepsilon_1 = 0.003 \quad \text{and} \quad e = 100/3.$$

Figures (1)-(3) show the temperature, stress and electric field distributions in the rod for $\tau = 0.25$ and for two values of the relaxation time: $\tau_1 = 0.05$ and $\tau_1 = 0.1$.

REFERENCES

[1] Lord H.W. and SHULMAN Y.,J.Mech.Phys.Solids,15(1967),5.
[2] Sherief H.H. and Dhaliwal R.S., Quart.Appl;Math.,38(1980),1.
[3] Smirnov V.N.,J.Engrg.Phys., 39(1980),4.
[4] Sherief H.H. and Dhaliwal R.S.,"A generalized one-dimensional thermal shock problem for small times"(to appear).

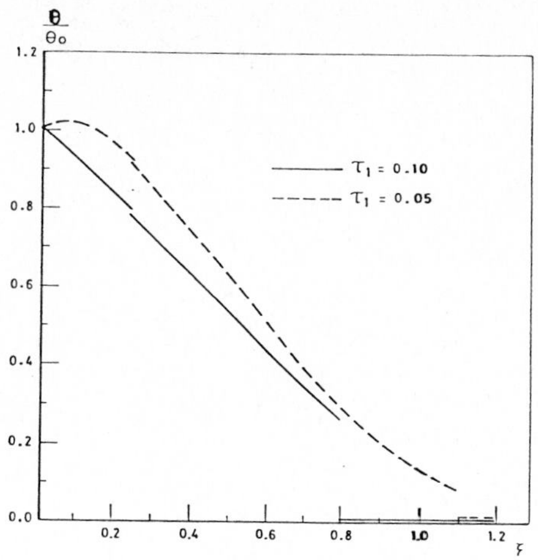

Fig. 1. Temperature at t=0.25.

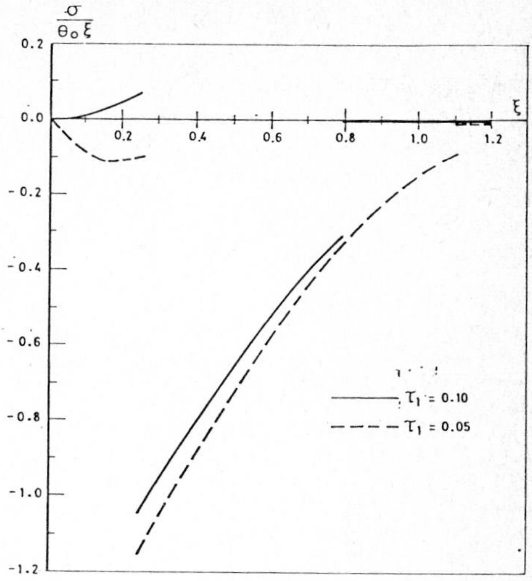

Fig. 2. Stress at t=0.25

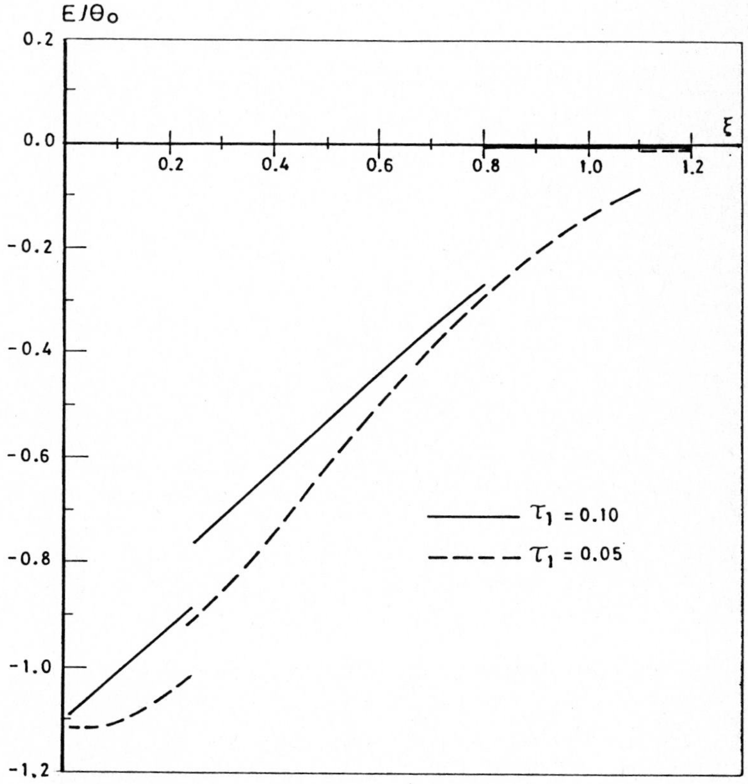

Fig. 3. Electric field at t=0.25.

APPENDIX

$\beta_{io}=\sqrt{\alpha_{io}}$, $\beta_{i1}=\alpha_{i1}/(2\beta_{io})$, $\beta_{i2}=(4\alpha_{i2}\,\alpha_{io}-\alpha_{i1}^{2})/(8\,\beta_{io}^{3})$, i=1,2,

$\alpha_{1o}=\frac{1}{2}\{1+(\varepsilon_1+1)\tau_1+A\}$, $\alpha_{11}=\frac{1}{2}\{\varepsilon_1+1+[\varepsilon_1-1+(\varepsilon_1+1)^2\tau_1]/A\}$,

$\alpha_{12}=-\alpha_{22}=\varepsilon_1/A^3$, $\alpha_{13}=-\alpha_{23}=-\varepsilon_1\{\varepsilon_1-1+(1+\varepsilon_1)^2\tau_1\}/A^5$,

$\alpha_{2o}=\frac{1}{2}\{1+(\varepsilon_1+1)\tau_1-A\}$, $\alpha_{21}=\frac{1}{2}\{\varepsilon_1+1-[\varepsilon_1-1+(\varepsilon_1+1)^2\tau_1]/A\}$,

$A=\{1+2(\varepsilon_1-1)\tau_1+(\varepsilon_1+1)^2\tau_1^2\}^{\frac{1}{2}}$,

$m_{1o}=\beta_o(\alpha_{1o}-1)$, $m_{11}=\beta_o\alpha_{11}+\beta_1(\alpha_{1o}-1)$,

$m_{12}=\beta_o(\beta_{11}^2+2\beta_{1o}\beta_{12})+\beta_1\alpha_{11}+\beta_2(\alpha_{1o}-1)$,

$m_{13}=2\beta_o\beta_{11}\beta_{12}+\beta_1(\beta_{11}^2+2\beta_{1o}\beta_{12})+\beta_2\alpha_{11}+\beta_3(\alpha_{1o}-1)$,

$m_{2o}=-\beta_o(\alpha_{2o}-1)$, $m_{21}=-\beta_o\alpha_{21}-\beta_1(\alpha_{2o}-1)$,

$m_{22}=-\beta_o(\beta_{21}^2+2\beta_{2o}\beta_{22})-\beta_1\alpha_{21}-\beta_2(\alpha_{2o}-1)$,

$m_{23}=-2\beta_o\beta_{21}\beta_{22}-\beta_1(\beta_{21}^2+2\beta_{2o}\beta_{22})-\beta_2\alpha_{21}-\beta_3(\alpha_{2o}-1)$,

$n_{1o}=-e\,\varepsilon_1\beta_o\alpha_{1o}-m_{1o}$, $n_{11}=-e\varepsilon_1(\beta_o\alpha_{11}+\beta_1\alpha_{1o})-m_{11}$,

$n_{12}=-e\,\varepsilon_1\{\beta_o(\beta_{11}^2+2\beta_{1o}\beta_{12})+\beta_1\alpha_{11}+\beta_2\alpha_{1o}\}-m_{12}$,

$n_{13}=-e\,\varepsilon_1\{2\beta_o\beta_{11}\beta_{12}+(\beta_1+\beta_3)(\beta_{11}^2+2\beta_{1o}\beta_{12})+\beta_2\alpha_{11}\}-m_{13}$,

$n_{2o}=-e\,\varepsilon_1\beta_o\alpha_{2o}-m_{2o}$, $n_{21}=-e\varepsilon_1(\beta_o\alpha_{21}+\beta_1\alpha_{2o})-m_{21}$,

$n_{22}=-e\,\varepsilon_1\{\beta_o(\beta_{21}^2+2\beta_{2o}\beta_{22})+\beta_1\alpha_{21}+\beta_2\alpha_{2o}\}-m_{22}$,

$n_{23}=-e\,\varepsilon_1\{2\beta_o\beta_{21}\beta_{22}+(\beta_1+\beta_3)(\beta_{21}^2+2\beta_{2o}\beta_{22})+\beta_2\alpha_{21}\}-m_{23}$.

$\beta_o=1/A$, $\beta_1=-\{\varepsilon_1-1+(\varepsilon_1+1)^2\,\tau_1\}/A^3$,

$\beta_2=\{\varepsilon_1^2-4\varepsilon_1+1+2(\varepsilon_1-1)(\varepsilon_1+1)^2\,\tau_1+(\varepsilon_1+1)^4\,\tau_1^2\}/A^5$

$$\beta_3=-\{(\varepsilon_1-1)(\varepsilon_1^2-8\varepsilon_1+1)+3(\varepsilon_1+1)^2(\varepsilon_1^2-4\varepsilon_1+1)\,\tau_1+3(\varepsilon_1-1)(\varepsilon_1+1)^4\tau_1^2+(\varepsilon_1+1)^2\tau_1^3\}/A^7$$

Part 2

ACOUSTOOPTICS AND PHOTOELASTICITY

The Mechanical Behavior of Electromagnetic Solid Continua
G.A. Maugin (editor)
Elsevier Science Publishers B.V. (North-Holland)

ACOUSTIC WAVE GENERATION BY THE CONVERSE ELASTOOPTIC EFFECT

D. F. Nelson

Bell Laboratories
Murray Hill, New Jersey 07974

A general Lagrangian theory of acoustic wave generation by the converse elastooptic effect is presented. In the process of solution the wave equations for the two optical waves and the one acoustic wave are converted to coupled mode equations. Taking the two optical waves as inputs, we obtain a general expression for the intensity of the output acoustic wave. In contrast to previous derivations, this one applies explicitly to all orientations of the three waves (provided they are near or at phase matching) and to a crystal of any symmetry, contains all needed (and new) anisotropy factors, and derives the material interaction coefficient, its symmetry, and dispersion. The material interaction coefficient includes the normal elastooptic effect, the rotooptic effect, and the indirect elastooptic effect consisting of the converse electrooptic and piezoelectric effects in succession.

INTRODUCTION. The elastooptic effect, also called the photoelastic or piezooptic effect, has traditionally been studied by measuring the effect of a mechanical deformation on an input light wave. In early studies the mechanical deformation was a static, homogeneous deformation.[1] Later the mechanical deformation used was a thermally excited acoustic wave in Brillouin scattering[2,3] or a coherently generated acoustic wave in acoustooptic diffraction.[4,5] In all three of these variations both the mechanical deformation and the light wave are inputs to the interaction and an altered light wave is the output.

A new mode of study of the elastooptic effect was made possible by the use of an intense, coherent light wave from a laser as the single input wave. In this manner Chiao, Townes, and Stoicheff[6] in 1964 inverted the interaction and produced both an output hypersonic wave and an output light wave in a process they called stimulated Brillouin scattering.

The last of the three possible modes of study of the elastooptic effect was first realized by Korpel, Adler, and Alpiner[7] also in 1964 when they produced an ultrasonic wave in water at the difference frequency of the two input light waves. Later Caddes, Quate, and Wilkinson[8] produced hypersonic waves in crystals by this process.

Optical difference frequency generation of acoustic waves was not only the last permutation of input and output waves of the elastooptic effect to be studied, it has also been the least studied of the three. The only other experimental studies apparently were those carried out by Cachier[9] and recently by Nelson and coworkers[10,11]. Theoretical studies were published by Kastler[12] and Kroll[13] for bulk generation, by Yang, Richards, and Shen[14] for superlattice generation, and by Chang[15] for surface generation. Others[16–18] have considered coupled mode equations with elastooptic coupling without specifically considering acoustic wave generation from optical difference frequency mixing.

The theoretical treatments have, in general, been phenomenological and applicable only to simple media. None, for example, derives the elastooptic coupling tensor with its symmetry and dispersion. None includes the rotooptic effect. None obtains the appropriate anisotropy factors which apply to waves travelling in arbitrary directions in low symmetry crystals. Only one[17] considers, and then in only a special case, the contribution of the indirect elastooptic effect to the generation, that is, the succession of optical difference frequency generation of a propagating, longitudinal, forced electric wave and the piezoelectric generation of an acoustic wave by that electric wave. These deficiencies of previous treatments are all remedied in the derivation presented here.

The present treatment is based on a Lagrangian theory[19,20] of a dielectric crystal of arbitrary symmetry, anisotropy, structural complexity, and nonlinearity in interaction with the electromagnetic field. All long-wavelength modes of mechanical motion of the crystal are accounted for including both acoustic and optic modes of vibration. If the primitive unit cell of the crystal contains N particles, then there are 3N-3 optic modes and 3 acoustic modes. If all N particles are taken to be atoms, then the optic modes are all ionic modes with resonant frequencies typically in the infrared. If some of the N particles are taken to be valence electrons (one or two are generally enough[21]), then the corresponding optic modes are electronic modes with resonant frequencies typically in the ultraviolet. Inclusion of the optic modes allows this theory to give a correct description of interaction phenomena in crystals for electromagnetic frequencies up to the ultraviolet region. In the present calculation it allows the derivation of the frequency dispersion of the elastooptic tensor.

A few remarks on nomenclature seem worthwhile at this point. Many of the above cited papers refer to acoustic wave generation from optical difference frequency mixing as "electrostrictive mixing" or "generation by electrostriction". Such usage is quite natural since electrostriction refers to the deformation of a body which is proportional to the square of an applied electric field. Furthermore, it was believed for decades that the electrostriction tensor was equal to the low frequency limit of the elastooptic tensor. However, recently the traditionally measured electrostriction tensor was shown[22,23] to differ from the low frequency limit of the elastooptic tensor by the Maxwell stress tensor because the two tensors are referred to different frames, material and spatial, respectively, as a result of their methods of measurement. For this reason it seems wise to avoid the term "electrostriction" in the present context and use a term unencumbered with historical connotations. Thus we propose to call the interaction studied here the "*converse elastooptic effect*".

LAGRANGIAN. The total spatial frame Lagrangian density of a lossless dielectric crystal of arbitrary symmetry, anisotropy, structural complexity, and nonlinearity in interaction with the electromagnetic field is given by[24]

$$\mathscr{L} = \frac{\rho}{2} C_{KL} \frac{\partial X_K}{\partial t} \frac{\partial X_L}{\partial t} + \sum_{\nu} \frac{\rho^\nu}{2} \left(\frac{\partial y_i^{T\nu}}{\partial t} - x_{j,K} \frac{\partial X_K}{\partial t} y_{i,j}^{T\nu} \right)^2 \tag{1}$$

$$- \rho\Sigma\,(E_{AB},\Pi_C^\nu) + \frac{1}{J} \sum_{\nu} q^\nu y_i^{T\nu} (\vec{E} + \dot{\vec{x}} \times \vec{B})_i + \frac{\epsilon_o}{2} \vec{E}\cdot\vec{E} - \frac{1}{2\mu_o} \vec{B}\cdot\vec{B}$$

with the stored energy $\rho\Sigma$ given by[25]

$$\rho\Sigma(E_{AB},\Pi_c^\nu) = \frac{1}{J} \Big({}^{02}M_{ABCD} E_{AB} E_{CD} + \sum_{\nu\mu} {}^{20}M_{CD}^{\nu\mu} \Pi_C^\nu \Pi_D^\mu \tag{2}$$

$$+ \sum_{\nu\mu\lambda} {}^{30}M_{CDE}^{\nu\mu\lambda} \Pi_C^\nu \Pi_D^\mu \Pi_E^\lambda + \sum_{\nu} {}^{11}M_{CAB}^{\nu} \Pi_C^\nu E_{AB} + \sum_{\nu\mu} {}^{21}M_{CDAB}^{\nu\mu} \Pi_C^\nu \Pi_D^\mu E_{AB} \Big).$$

Here X_K is the material coordinate, x_i the spatial coordinate, $x_{j,K} \equiv \partial x_j / \partial X_K$ the deformation gradient, u_i the displacement, such that $x_i = X_J \delta_{Ji} + u_i$, $y_i^{T\nu}$ the ν–th total internal coordinate, Y_J^ν its natural state value and y_i^ν its variable part, such that $y_i^{T\nu} = Y_J^\nu \delta_{Ji} + y_i^\nu$, $C_{KL} \equiv x_{i,K} x_{i,L}$, $E_{KL} \equiv (C_{KL} - \delta_{KL})/2$, $\Pi_J^\nu \equiv X_{J,i} y_i^{T\nu} - Y_J^\nu$, ρ and ρ^0 the deformed and undeformed mass densities, $\rho = \rho^0/J$, $J(x/X)$ the Jacobian, ρ^ν and m^ν the deformed and undeformed mass densities of the ν– sublattice, $\rho^\nu \equiv m^\nu/J$,. q^ν the charge density associated with the ν - sublattice, $\vec{E}$ the electric field given in terms of the vector and scalar potentials $\vec{A},\Phi$ by $\vec{E} = -\vec{\nabla}\Phi - \partial\vec{A}/\partial t$, and $\vec{B} = \vec{\nabla} \times \vec{A}$ the magnetic induction.

FOURIER EXPANSION. Since the elastooptic effect involves two optical fields and one acoustic field, two and three field terms which contain at most one displacement field should be kept in the Lagrangian. We perform a Fourier expansion within the Lagrangian in order to select out the desired field components. Consider the frequency relation, $\omega_B = \omega_O + \omega_A$, where ω is an angular frequency and B,O,A refer to the Brillouin field, the optical field, and the acoustic field. For this purpose it is sufficient to express y_i^ν, $u_{i,j}$, E_i, and B_i in a form like

$$y_i^\nu = \frac{1}{2} (y_i^{\nu B} e^{i\omega_B t} + y_i^{\nu O} e^{i\omega_O t} + y_i^{\nu A} e^{i\omega_A t} + \text{c.c.}) \tag{3}$$

and retain only the time independent terms in the Lagrangian.

The internal coordinates can now be eliminated from the Lagrangian with the use of the Lagrange equation for the internal coordinates. That equation for $y_i^{\nu A*}$ yields

$$0 = \omega_A^2 m^\nu (y_i^{\nu A} - u_{i,J}^A Y_J^\nu) + q^\nu E_i^A - {}^{11}M_{iab}^\nu u_{a,b}^A \tag{4}$$

$$- 2\sum_{\mu} {}^{20}M_{id}^{\nu\mu} (y_d^{\mu A} - u_{d,k}^A Y_k^\mu) - 3\sum_{\mu\lambda} {}^{30}M_{ide}^{\nu\mu\lambda} y_d^{\mu O*} y_e^{\lambda B}.$$

We regard the linear terms here as driven by the bilinear terms.

NORMAL COORDINATES. Though internal coordinates, being vectors, are useful for the stored energy expansion, use of normal coordinates is preferred for interpreting the derived interaction tensors. Normal coordinates are scalar coordinates that are the linear combinations of internal coordinates that uncouple the internal motions in the absence of perturbations into the optic modes of the crystal. They are defined by

$$\eta^k \equiv \sum_\mu (m^\mu)^{1/2} \vec{i}^{k\mu} \cdot \vec{y}^\mu \tag{5}$$

(with a similar definition of H^k in tems of $\vec{Y}^\mu$), where the normal mode eigenvectors satisfy

$$2\sum_\mu \frac{i_a^{k\mu}\ {}^{20}M_{ab}^{\mu\nu}}{(m^\mu m^\nu)^{1/2}} = \Omega_k^2 i_b^{k\nu}, \tag{6}$$

and Ω_k^2 are the eigenvalues found from this equation and are squares of the transverse optic mode frequencies. They satisfy

$$\sum_\nu \vec{i}^{k\nu} \cdot \vec{i}^{l\nu} = \delta^{kl}, \quad \sum_k i_i^{k\mu} i_j^{k\nu} = \delta_{ij}\delta^{\mu\nu}. \tag{7}$$

It is also helpful to introduce

$$\vec{c}^k \equiv \sum_\mu \frac{q^\mu \vec{i}^{k\mu}}{(m^\mu)^{1/2}}, \quad {}^{11}N^k{}_{ab} \equiv \sum_\nu \frac{i_c^{k\nu}\ {}^{11}M_{cab}^\nu}{(m^\nu)^{1/2}} \tag{8}$$

and similarly for ${}^{20}M_{ab}^{\nu\mu}$, ${}^{30}M_{abc}^{\nu\mu\lambda}$, etc. With these definitions (4) can be solved for

$$\eta^{kA} = \frac{c_i^k E_i^A}{\Omega_k^2 - \omega_A^2} + \sum_{\nu l} i_i^{k\nu} u_{i,j}^A i_j^{l\nu} H^l - {}^{11}N^k{}_{ab} \frac{u_{a,b}^A}{\Omega_k^2 - u_A^2} - 3\sum_{lm} \frac{{}^{30}M^{klm}\eta^{lB}\eta^{mO*}}{\Omega_k^2 - \omega_A^2}. \tag{9}$$

Similarly at the optical frequency ω_O we get

$$\eta^{kO} = \frac{c_i^k E_l^O}{\Omega_k^2 - \omega_O^2} - \frac{c_i^k E_i^B u_{j,j}^{A*}}{2(\Omega_k^2 - \omega_O^2)} + \frac{(\Omega_k^2 - \omega_B^2)}{2(\Omega_k^2 - \omega_O^2)} \eta^{kB} u_{j,j}^{A*} \tag{10}$$

$$+ \frac{1}{2}\sum_{l\nu} \frac{(\Omega_l^2 i_d^{l\nu} u_{d,e}^{A*} i_e^{k\nu} + \Omega_k^2 i_d^{k\nu} u_{d,e}^{A*} i_e^{l\nu})\eta^{lB}}{\Omega_k^2 - \omega_O^2}$$

$$- 3\sum_{lm} {}^{30}N^{klm}\eta^{lB}(\eta^{mA*} - \sum_{n\xi} i_f^{m\xi} u_{f,g}^{A*} i_g^{n\xi} H^n) - \sum_l \frac{{}^{21}N^{kl}{}_{ab}\ \eta^{lB}\ u_{a,b}^{A*}}{\Omega_k^2 - \omega_O^2}.$$

A similar expression for η^{kB} can also be obtained.

MODIFIED LAGRANGIAN. Elimination of η^{kA}, η^{kB}, η^{kO} from the Lagrangian now yields

$$L = \frac{\rho^0 \omega_A^2}{8} u_e^A u_e^{A*} - \frac{1}{4} c_{abcd} u_{a,b}^A u_{c,d}^{A*} + \frac{1}{4} e_{jlm} E_j^A u_{l,m}^{A*} \tag{11}$$

$$+ \frac{\epsilon_o}{8} E_k^A \kappa_{kl}(\omega_A) E_l^{A*} - \frac{1}{8\mu_o} B_k^A B_k^{A*} + \frac{\epsilon_o}{8} E_k^O (\kappa_{kl}(\omega_O) + \frac{\omega_O}{2} \frac{\partial \kappa_{kl}}{\partial \omega_O}) E_l^{O*}$$

$$- \frac{1}{8\mu_o} B_e^O B_e^{O*} + \frac{\epsilon_o}{8} E_k^B (\kappa_{kl}(\omega_B) + \frac{\omega_B}{2} \frac{\partial \kappa_{kl}}{\partial \omega_B}) E_l^{B*} - \frac{1}{8\mu_o} B_l^B B_l^{B*}$$

$$+ \frac{\epsilon_o}{4} b_{k\,j\,l}^{\omega_B \omega_O \omega_A} E_k^B E_j^{O*} E_l^{A*} + \frac{\epsilon_o}{4} \chi_{k\,j\,lm}^{\omega_B \omega_O \omega_A} E_k^B E_j^{O*} u_{l,m}^{A*},$$

where the elastic stiffness tensor, the piezoelectric stress tensor, the dielectric tensor, the optical mixing tensor, and the elastooptic tensor are given respectively by

$$c_{Abcd} \equiv 2^{02}M_{abcd} - \sum_k \frac{{}^{11}N^k{}_{ab}\ {}^{11}N^k{}_{cd}}{\Omega_k^2 - \omega_A^2}, \tag{12}$$

$$e_{abc} \equiv -\sum_k \frac{c_a^k\ {}^{11}N^k{}_{bc}}{\Omega_k^2 - \omega_A^2}, \quad \epsilon_o \kappa_{ij}(\omega) \equiv \epsilon_o + \sum_k \frac{c_i^k c_j^k}{\Omega_k^2 - \omega^2}, \tag{13}$$

$$\epsilon_o b_{i\,j\,k}^{\omega_B \omega_O \omega_A} \equiv -\frac{3}{2}\sum_{lmn} \frac{c_i^l c_j^m c_k^n\ {}^{30}N^{lmn}}{(\Omega_l^2 - \omega_B^2)(\Omega_m^2 - \omega_O^2)(\Omega_n^2 - \omega_A^2)} \tag{14}$$

$$\epsilon_o \chi^{\omega_B \omega_O \omega_A}_{i\ j\ kl} \equiv -\sum_{mn} \frac{c_i^m c_j^{n21} N^{mn}{}_{kl}}{(\Omega_m^2-\omega_B^2)(\Omega_n^2-\omega_O^2)} \tag{15}$$

$$+3\sum_{mnp} \frac{c_i^m c_j^{n30} N^{mnp11} N^p{}_{kl}}{(\Omega_m^2-\omega_B^2)(\Omega_n^2-\omega_O^2)(\Omega_p^2-\omega_A^2)} + \frac{1}{2}\sum_{mn\nu} \frac{c_i^m c_j^n (i_k^{m\nu} i_l^{n\nu}\omega_B^2 + i_k^{n\nu} i_l^{m\nu}\omega_O^2)}{(\Omega_m^2-\omega_B^2)(\Omega_n^2-\omega_O^2)}$$

$$+\frac{\epsilon_o}{2}\chi_{il}(\omega_B)\,\delta_{jk} + \frac{\epsilon_o}{2}\chi_{jl}(\omega_O)\delta_{ik} - \frac{\epsilon_o}{4}\chi_{ij}(\omega_B)\delta_{kl} - \frac{\epsilon_o}{4}\chi_{ij}(\omega_O)\delta_{kl}.$$

The antisymmetric part of the elastooptic tensor,

$$\chi^{\omega_B\omega_O\omega_A}_{i\ j\ [kl]} = -\frac{1}{2}\chi_{i[k}(\omega_B)\delta_{l]j} - \frac{1}{2}\chi_{j[k}(\omega_O)\delta_{l]i}, \tag{16}$$

represents the rotooptic effect.[26] Note that the frequency dependence of the measurable interaction tensors (12) - (16) is predicted; this results from inclusion of the internal coordinates in the description.

ACOUSTIC WAVE EQUATION. The Lagrange equation for $\vec{u}^{A*}$ yields

$$-\rho^0\omega_A^2 u_l^A = c_{lmab} u^A_{a,bm} - e_{jlm} E^A_{j,m} - \epsilon_o \chi^{\omega_B\omega_O\omega_A}_{i\ j\ lm} (E_i^B E_j^{O*})_{,m} \tag{17}$$

while the Lagrange equation for Φ^{A*}, when applied to a plane wave, yields

$$\Phi^A = -\frac{e_{efg} u^A_{f,ge} + 2\epsilon_o b^{\omega_B\omega_O\omega_A}_{k\ l\ m} (E_k^B E_l^{O*})_{,j}}{\epsilon_o k_A^2 \vec{s}^A \cdot \overleftrightarrow{\kappa}(\omega_A) \cdot \vec{s}^A} \tag{18}$$

where $\vec{s}^A = \vec{k}_A/k_A$ is a unit propagation vector. Combining these two equations leads to

$$-\rho^0\omega_A^2 u_l^A - \bar{c}_{lmab} u^A_{a,bm} = \frac{2e_{jlm} b^{\omega_B\omega_O\omega_A}_{k\ i\ n} (E_k^B E_i^{O*})_{,mjn}}{k_A^2 \vec{s}^A \cdot \overleftrightarrow{\kappa}(\omega_A) \cdot \vec{s}^A} - \epsilon_o \chi^{\omega_B\omega_O\omega_A}_{k\ j\ lm} (E_k^B E_j^{O*})_{,m} \tag{19}$$

where

$$\bar{c}_{lmab} = c_{lmab} + \frac{s_k^A e_{klm} s_c^A e_{cab}}{\epsilon_o \vec{s}^A \cdot \overleftrightarrow{\kappa}(\omega_A) \cdot \vec{s}^A} \tag{20}$$

is the piezoelectrically stiffened stiffness tensor.

OPTICAL WAVE EQUATION. The Lagrange equation for the vector potential $\vec{A}^{O*}$ yields

$$(\vec{\nabla}\times\vec{B}^O)_i + \frac{i\omega_O}{c^2}\left[\kappa_{il}(\omega_O) + \frac{\omega_O}{2}\frac{\partial\kappa_{il}}{\partial\omega_O}\right]E_l^O + 2d^{\omega_B\omega_O\omega_A}_{k\ i\ m} E_k^B E_m^{A*} + \chi^{\omega_B\omega_O\omega_A}_{k\ i\ lm} E_k^B u^{A*}_{l,m} = 0. \tag{21}$$

From the definition of the electric field in terms of the potentials we also have $(\vec{\nabla}\times\vec{E}^O)_j - i\omega_O B_j^O = 0$. Combining these yields the driven optical wave equation,

$$\left\{\frac{c}{\omega_O{}^2}\vec{\nabla}\times(\vec{\nabla}\times\vec{E}^O) - \left[\overleftrightarrow{\kappa}(\omega_O) + \frac{\omega_O}{2}\frac{\partial\overleftrightarrow{\kappa}}{\partial\omega_O}\right]\cdot\vec{E}^O\right\}_j = 2b^{\omega_B\omega_O\omega_A}_{i\ j\ k} E_i^B E_k^{A*} + \chi^{\omega_B\omega_O\omega_A}_{i\ j\ kl} E_i^B u^{A*}_{k,l}. \tag{22}$$

A similar equation at frequency ω_B also results.

COUPLED MODE EQUATIONS. When waves interact by a weak nonlinearity, it is convenient to characterize the interaction by approximate first order differential equations governing the scalar amplitudes of the waves. The interaction will be significant only if phase matching, $\vec{k}_B = \vec{k}_O + \vec{k}_A$, is at least approximately met among the wave vectors.

The waves can be represented as $\vec{u}^A = \vec{b}^A U(z)\exp(i\vec{k}^A\cdot\vec{z})$, $\vec{E}^B = \vec{e}^B E_B(z)\exp(i\vec{k}^B\cdot\vec{z})$, and $\vec{E}^O = \vec{e}^O E_O(z)\exp(i\vec{k}^O\cdot\vec{z})$ where $\vec{b}^A$, $\vec{e}^B$, $\vec{e}^O$ are unit eigenvectors of the uncoupled waves, and $z = \vec{n}\cdot\vec{z}$, $\vec{n}$ being the unit normal to the planar surface at which the nonlinear interaction begins. The amplitudes are taken as slowly varying, that is, $\partial U/\partial z << k_A U$, $\partial E_B/\partial z << k_B E_B$, and $\partial E_O/\partial z << k_O E_O$.

The coupled mode equation for U is found by forming the scalar product of (19) with $\vec{b}^A$ dropping $\partial^2 U/\partial z^2$, subtracting off the undriven acoustic wave equation, and introducing the group velocity

$$(v_g^A)_i = \frac{1}{\rho^0 v_A}\bar{c}_{ijkl} b_j^A b_k^A s_l^A, \tag{23}$$

where v_A is the acoustic phase velocity. The result is

$$\frac{\partial U}{\partial z}=\frac{\epsilon_o k_A \chi E_B E_O^* e^{i\Delta k_n z}}{2\omega_A \rho^0 \vec{n}\cdot\vec{v}_g^A} \tag{24}$$

where

$$\chi\equiv\chi_{i\ j\ kl}^{\omega_B\omega_O\omega_A}\, e_i^B e_j^O b_k^A s_l^A-\frac{2b_{i\ j\ k}^{\omega_B\omega_O\omega_A}\, e_i^B e_j^O s_k^A s_e^A e_{efg} b_f^A s_g^A}{\epsilon_o \vec{s}^A\cdot\overleftrightarrow{\kappa}(\omega_A)\cdot\vec{s}^A}, \tag{25}$$

and $\Delta k_n=(\vec{k}_B-\vec{k}_O-\vec{k}_A)\cdot\vec{n}$ is the phase mismatch.

The coupled mode equation for E_O is found by an analogous procedure to be

$$\frac{\partial E_O}{\partial z}=\frac{k_A \chi E_B U^* e^{i\Delta k_n z}}{2k_O \cos\delta_O \vec{n}\cdot\vec{t}_O} \tag{26}$$

where $\vec{t}_O=[\vec{s}^O-(\vec{e}^O\cdot\vec{s}^O)\vec{e}^O]\sec\delta_O$ is the unit vector in the direction of the group velocity and δ_O is the angle between it and the unit wave vector $\vec{s}^O$. An analogous equation for E_B,

$$\frac{\partial E_B}{\partial z}=-\frac{k_A \chi E_O U\ e^{-i\Delta k_n z}}{2k_B \cos\delta_B \vec{n}\cdot\vec{t}_B}, \tag{27}$$

can be obtained similarly.

MANLEY-ROWE RELATIONS. With the energy flow vectors of the three waves denoted by $\vec{S}_A$, $\vec{S}_O$, and $\vec{S}_B$ the Manley-Rowe relations,

$$\frac{\partial}{\partial z}\left(\frac{<\vec{S}_A\cdot\vec{n}>}{\omega_A}\right)=\frac{\partial}{\partial z}\left(\frac{<\vec{S}_O\cdot\vec{n}>}{\omega_O}\right)=-\frac{\partial}{\partial z}\left(\frac{<\vec{S}_B\cdot\vec{n}>}{\omega_B}\right), \tag{28}$$

can be derived from (24), (26), (27). Here $<>$ refers to a time average and each of the terms of (28) is equal to $\epsilon_o k_A \chi \mathcal{R}\{ E_B E_O^* U^* \exp i\Delta k_n z \} /2$.

CONVERSE ELASTOOPTIC EFFECT. The growth of an acoustic wave produced by optical difference frequency generation can be found by integration of (24) in the regime of negligible pump depletion (E_B,E_O constant) with the condition $U=0$ at $z=0$ to be

$$U=\frac{\epsilon_o z k_A \chi E_B E_O^* \Phi^{1/2}(\Delta k_n z/2)}{2\omega_A \rho^0 \vec{n}\cdot\vec{v}_g^A}, \tag{29}$$

where $\Phi(\sigma)=\sin^2\sigma/\sigma^2$ is the phase matching function. This produces an acoustic energy flow vector of

$$<\vec{S}_A>=\frac{(\omega_A z\chi)^2<S_B><S_O>\Phi(\Delta k_n z/2)\vec{t}_A}{2\rho^0 v c^2 n_B n_O \cos\delta_B \cos\delta_O \cos\delta_A (\vec{n}\cdot\vec{v}_g^A)^2} \tag{30}$$

where n_B and n_O are refractive indices and δ_A is the angle between the acoustic phase and group velocities. A discussion of the new aspects of this expression appears in the INTRODUCTION.

I wish to thank M. Lax and H. F. Tiersten for helpful conversations.

[1] F. Pockels, *Lehrbuch der Kristalloptik* (Leipzig, Teubner, 1906).

[2] L. Brillouin, Ann. Phys. (Paris) *17*, 88 (1922).

[3] E. Gross, Nature *126*, 201, 400, 603 (1930).

[4] P. Debye and F. W. Sears, Proc. Natl. Acad. Sci. U.S. *18*, 409 (1932).

[5] R. Lucas and P. Biquard, J. Phys. Radium *3*, 464 (1932).

[6] R. Y. Chiao, C. H. Townes, and B. P. Stoicheff, Phys. Rev. Lett. *12*, 592 (1964).

[7] A. Korpel, R. Adler, and B. Alpiner, Appl. Phys. Lett. *5*, 86 (1964).

[8] D. E. Caddes, C. F. Quate, and C. D. W. Wilkinson, Appl. Phys. Lett. *28*, 309 (1966); in *Modern Optics*, J. Fox, Ed. (Polytechnic Press, Brooklyn, NY, 1967), p. 219.

[9] C. Cachier, J. Acoust. Soc. Am. *49*, 974 (1971).

[10] K. A. Nelson, D. R. Lutz, and M. D. Fayer, Phys. Rev. B *24*, 3261 (1981).

[11] K. A. Nelson, R. J. D. Miller, D. R. Lutz, and M. D. Fayer, J. Appl. Phys. *53*, 1144 (1982).

[12] A. Kastler, C. R. Acad. Sci. (Paris) *259*, 4233, 4535 (1964); *260*, 77 (1965).

[13] N. M. Kroll, J. Appl. Phys. *36*, 34 (1965).

[14] K. H. Yang, P. L. Richards, and Y. R. Shen, J. Appl. Phys. *44*, 1417 (1973).

[15] M. S. Chang, Appl. Opt. *16*, 1960 (1977).

[16] A. Yariv, IEEE J. Quant. Electron. *QE-1*, 28 (1965).

[17] D. A. Sealer and H. Hsu, IEEE J. Quant. Electron. *QE-1*, 116 (1965).

[18] G. N. Burlak and N. Ya. Kotsarenko, Fiz. Tverd. Tela *19*, 2648 (1977); English transl.: Soviet Phys. - Solid State *19*, 1551 (1977).

[19] M. Lax and D. F. Nelson, Phys. Rev. B*4*, 3694 (1971); Phys. Rev. B *13*, 1759 (1976).

[20] D. F. Nelson, *Electric, Optic, and Acoustic Interactions in Dielectrics* (Wiley, New York, 1979).

[21] S. H. Wemple and M. DiDomenico, Jr., Phys. Rev. Lett. *23*, 1156 (1969); Phys. Rev. B *1*, 193 (1970).

[22] Ref. 20, p. 428.

[23] D. F. Nelson, in *Basic Optical Properties of Materials* , A. Feldman, Ed. (U.S., Dept. of Commerce, Washington, D.C., 1980), p. 209.

[24] Ref. 20, p. 80.

[25] Ref. 20, p. 112.

[26] Ref. 20, p. 289.

The Mechanical Behavior of Electromagnetic Solid Continua
G.A. Maugin (editor)
Elsevier Science Publishers B.V. (North-Holland)

OPTICAL ACTIVITY IN A LINEAR NON-DISSIPATIVE ANISOTROPIC DIELECTRIC OF THE RATE TYPE

Ph.BOULANGER and G.MAYNE

Université Libre de Bruxelles - Département de Mathématique
Campus Plaine, C.P.218 - Boulevard du Triomphe
1050 Bruxelles, Belgium

The most general form of the constitutive equation for a linear anisotropic non-dissipative dielectric of the rate type is derived in the case of two time-derivatives of the electric field and four of the polarization vector. For such dielectrics with uniaxial symmetry the propagation of plane harmonic waves in the direction of the axis is studied.

INTRODUCTION

A usual way of describing the electromagnetic behaviour of dielectrics with several kinds of charges is to introduce partial polarization vectors. In the case of two partial polarization vectors $P_{(1)}$, $P_{(2)}$, for a linear theory including dispersion effects, it is natural to formulate the constitutive equations in the form

$$\begin{cases} E = a_o^{(1)} P_{(1)} + a_1^{(1)} \dot{P}_{(1)} + a_2^{(1)} \ddot{P}_{(1)} \\ E = a_o^{(2)} P_{(2)} + a_1^{(2)} \dot{P}_{(2)} + a_2^{(2)} \ddot{P}_{(2)} \end{cases} , \qquad (1)$$

where E denotes the electric field, and the superimposed dots denote time derivatives. Equations (1) can be regarded as equations of motion for the two kinds of charges of the dielectric. However, if one wishes to formulate a single constitutive equation in terms of the total macroscopic polarization vector

$$P = P_{(1)} + P_{(2)} , \qquad (2)$$

one may eliminate $P_{(1)}$, $P_{(2)}$ from equations (1) (2). This leads to a constitutive equation in the form

$$b_o E + b_1 \dot{E} + b_2 \ddot{E} = a_o P + a_1 \dot{P} + a_2 \ddot{P} + a_3 \dddot{P} + a_4 \ddddot{P} . \qquad (3)$$

Using the terminology of [1], we call the equation (3) constitutive equation of the rate type. In a former paper [2], we studied such constitutive equations for linear isotropic dielectrics : requirements of dissipation, passivity, causality and the propagation of plane harmonic waves were investigated. In this paper, we present some results concerning (3) in the case when the dielectric is anisotropic. In what follows, the coefficients $b_o, \dots, a_o, \dots, a_4$ have thus to be regarded as second order tensors.

NON-DISSIPATION - INTERNAL ENERGY

We here derive the most general form that equation (3) may have in order to describe a non-dissipative dielectric. The absence of dissipation is expressed by requiring the existence of an internal energy function (here a quadratic form)

$$\begin{aligned}\sigma = {} & \frac{1}{2}\,E\mathcal{A}E + \frac{1}{2}\,P\mathcal{B}P + \frac{1}{2}\,\dot{E}\mathcal{C}\dot{E} + \frac{1}{2}\,\dot{P}\mathcal{D}\dot{P} + \frac{1}{2}\,\ddot{P}\mathcal{E}\ddot{P} + \frac{1}{2}\,\dddot{P}\mathcal{F}\dddot{P} \\ & + E\mathcal{G}P + E\mathcal{J}\ddot{P} + P\mathcal{N}\ddot{P} + \dot{E}\mathcal{P}\dot{P} + \dot{E}\mathcal{R}\dddot{P} + \dot{P}\mathcal{T}\dddot{P} \\ & + (E\mathcal{H} + P\mathcal{L})\dot{E} + (E\mathcal{K} + P\mathcal{O})\dddot{P} \\ & + (E\mathcal{I} + P\mathcal{M})\dot{P} + (\dot{E}\mathcal{Q} + \dot{P}\mathcal{S})\ddot{P} + \ddot{P}\mathcal{U}\dddot{P} \quad ,\end{aligned} \tag{4}$$

such that

$$\dot{\sigma} - E.\dot{P} = 0 \tag{5}$$

for every $E(t)$, $P(t)$ compatible with (3).

The two last lines of (4) consist of terms that change sign when the odd derivatives of E and P change sign ("not invariant under time-reversal"). The others terms of (4) are invariant when the odd derivatives of E and P change sign.

At an arbitrary given time, the left-hand side of (5) is a quadratic form in E, P and their derivatives up to order 2 and 4. These vectors cannot take arbitrary values since they have to satisfy the constitutive equation (3). However, introducing a vector valued multiplier

$$\lambda = \lambda_o E + \lambda_1 \dot{E} + \lambda_2 \ddot{E} + \mu_o P + \mu_1 \dot{P} + \mu_2 \ddot{P} + \mu_3 \dddot{P} + \mu_4 \ddddot{P} \quad , \tag{6}$$

we transform our requirement into the equivalent requirement that

$$\dot{\sigma} - E.\dot{P} - \lambda.(b_o E + b_1 \dot{E} + b_2 \ddot{E} - a_o P - a_1 \dot{P} - a_2 \ddot{P} - a_3 \dddot{P} - a_4 \ddddot{P}) = 0 \quad , \tag{7}$$

for every vector values of E, $\dot{E}$, $\ddot{E}$, P, $\dot{P}$, $\ddot{P}$, $\dddot{P}$, $\ddddot{P}$.

This yields an algebraic system of matrix equations relating the coefficients of (3), (4), (6). After several algebraic manipulations we could show from this system that the coefficients of the constitutive equations (3) have to be given as functions of four symmetric tensors A_o, A_2, S, T and three skew-symmetric tensors W, W', Ω, by

$$\begin{aligned} & b_o = 1 + A_o(S - \Omega A_2 \Omega) \quad , \quad b_1 = (W + A_o \Omega A_2)S - (1 + W\Omega)A_2\Omega \quad , \\ & \qquad b_2 = (1 + W\Omega)A_2 S \quad , \\ & a_o = A_o \quad , \quad a_1 = W + A_o \Omega A_2 - b_o W' \quad , \quad a_2 = (1 + W\Omega)A_2 - b_1 W' + b_o T \\ & \qquad a_3 = b_1 T - b_2 W' \quad , \quad a_4 = b_2 T \quad . \end{aligned} \tag{8}$$

In deriving this result use has been made of the following remark : equation (3) is unchanged when all the tensor coefficients are multiplied to the left by the same invertible tensor. Note that when $T = W' = S = \Omega = 0$, the constitutive equation (3) reduces to an usual one [3], where W is regarded as a gyration coefficient.

The expressions of the coefficients of σ are then given by

$$\begin{aligned} & \mathcal{A} = (S - \Omega A_2 \Omega) b_o \ , \ \mathcal{B} = A_o \ , \ \mathcal{C} = S\tilde{A}_2 S \ , \ \mathcal{D} = \tilde{A}_2 + T - W'\mathcal{A}W' + A_2\Omega b_o W' + W' b_o^T \Omega A_2 , \\ & \mathcal{E} = T\mathcal{A}T - W'S\tilde{A}_2 SW' - W'SA_2 b_o T - Tb_o^T \Omega A_2 SW' \ , \ \mathcal{F} = TS\tilde{A}_2 ST \ , \ \mathcal{G} = -\,(S - \Omega A_2 \Omega)A_o \ , \\ & \mathcal{J} = -\,\mathcal{A}T + b_o^T \Omega A_2 SW' \ , \ \mathcal{N} = A_o ST - A_o \Omega A_2 (SW' + \Omega T) \ , \ \mathcal{P} = -\,S(\tilde{A}_2 + A_2 \Omega b_o W') \ , \end{aligned}$$

$$R = -\tilde{A}_2 ST\ ,\ T = (\tilde{A}_2 + A_2\Omega b_o W')^T T\ ,\ H = b_o^T \Omega A_2 S\ ,\ L = -A_o \Omega A_2 S\ ,\ K = -HT\ , \quad (9)$$

$$O = -LT\ ,\ I = b_o^T [SW' - \Omega A_2(1 + \Omega W')]\ ,\ M = -A_o [SW' - \Omega A_2(1 + \Omega W')]\ ,$$

$$S = P^T W - [SW' - \Omega A_2(1 + \Omega W')]^T T\ ,\ Q = S\tilde{A}_2 SW' + SA_2\Omega b_o T\ ,\ U = Q^T T\ ,$$

where the notation $\tilde{A}_2 = A_2 - A_2\Omega A_o \Omega A_2$ has been used.

This result shows that the terms of σ that are not invariant under time-reversal vanish in the case when $W' = \Omega = 0$. The coefficients H, L, K, O vanish with Ω, while the coefficients I, M, S, Q, U vanish when $W' = \Omega = 0$. Note that the skew-symmetric tensor W does not enter the expression of the internal energy σ.

WAVE PROPAGATION

It seems interesting to analyze the influence of the skew-symmetric tensors W, W', Ω of the non-dissipative anisotropic model (3), (8), on wave propagation in this model. For plane harmonic waves in an infinite medium $E(x,t)$, $P(x,t)$ are given by

$$E = \mathrm{Re}\{e \exp i\omega(t - \frac{n}{c} s.x)\}\ , \quad P = \mathrm{Re}\{p \exp i\omega(t - \frac{n}{c} s.x)\}\ . \quad (10)$$

Substituting (10) into (3), and introducing the amplitude $d = e + p$ of the electric displacement vector, one obtains the constitutive equation for the amplitudes in the form

$$d = \kappa(\omega) e\ , \quad (11)$$

where $\kappa(\omega)$ called complex permittivity tensor is expressed in terms of the coefficients of the constitutive equation (3). Using the expressions (8) of these coefficients, one can show that this tensor is hermitian for every frequency, which means that there is no absorption of the waves.

For the sake of simplicity, we now assume that the anisotropy group of the model is such that the symmetric and skew-symmetric tensors are in the form

$$\begin{pmatrix} A_o \\ A_2 \\ S \\ T \end{pmatrix} = \begin{pmatrix} a'_o \\ a'_2 \\ s' \\ t' \end{pmatrix} 1 + \begin{pmatrix} a''_o \\ a''_2 \\ s'' \\ t'' \end{pmatrix} 1_z \otimes 1_z\ , \quad \begin{pmatrix} W \\ W' \\ \Omega \end{pmatrix} = \begin{pmatrix} \alpha \\ \beta \\ \gamma \end{pmatrix} (1_x \otimes 1_y - 1_z \otimes 1_x), \quad (12)$$

where 1_x, 1_y, 1_z are orthonormal basis vectors (uniaxial symmetry with axis 1_z). The complex permittivity tensor $\kappa(\omega)$ then reads

$$\kappa(\omega) = \kappa_{11} 1 + (\kappa_{33} - \kappa_{11})\, 1_z \otimes 1_z + i\kappa_{12}(1_x \otimes 1_y - 1_y \otimes 1_x)\ , \quad (13)$$

and, for propagation of waves in the direction $s = 1_z$, the refractive index and amplitude vector e are given by

$$n^2 = \kappa_{11} \pm i\kappa_{12}\ , \qquad e_x = \pm i e_y\ . \quad (14)$$

There is thus in general circular birefringence (optical activity) for waves propagating in the $s = 1_z$ direction. The expressions of κ_{11}, κ_{12} in terms of the constitutive coefficients a'_o, a''_o, ..., β, γ can be obtained from (8) and (12). Four special cases are studied.

o) W = W' = Ω = 0 (no gyration - σ invariant under time-reversal).

In this case,

$$\kappa_{12} = 0 \quad , \tag{15}$$

$$\kappa_{11} - 1 = \frac{1 + a_o's' - \omega^2 a_2's'}{a_o' - \omega^2(a_2' + t' + a_o's't') + \omega^4 a_2's't'} \equiv \frac{\omega_*^2 - \omega^2}{t'(\omega^2 - \omega_1^2)(\omega^2 - \omega_2^2)} \tag{16}$$

There is no optical activity, and the model exhibits two optical resonance frequencies ω_1, ω_2.

1) W' = Ω = 0 (W-gyration - σ invariant under time-reversal).

In this case

$$\kappa_{12} = \frac{-\omega\alpha}{(a_2's't')^2(\omega^2 - \omega_1^2)^2(\omega^2 - \omega_2^2)^2 - \omega^2\alpha^2(1 - \omega^2 s't')^2} \tag{17}$$

There is optical activity, non vanishing, except for $\omega = 0$.

2) W = Ω = 0 (W'-gyration).

In this case,

$$\kappa_{12} = \frac{\omega(\omega_*^2 - \omega^2)^2\beta}{t'^2(\omega^2 - \omega_1^2)^2(\omega^2 - \omega_2^2)^2 - \omega^2\beta^2(\omega_*^2 - \omega^2)}$$

There is optical activity vanishing for $\omega = \omega_*$. Since κ_{12} does not change sign for varying ω, the optical activity does not change orientation.

3) W = W' = 0 (Ω-gyration - no vanishing term in σ)

In this case

$$\kappa_{12} = \frac{-\omega(2a_o/a_2 - a_o^2\gamma^2 - \omega^2)\gamma}{s'^2t'^2(\omega^2 - \omega_1^2)^2(\omega^2 - \omega_2^2)^2 - \omega^2\gamma^2[a_o' - \omega^2(a_o's' - 1)t']^2} \tag{18}$$

There is optical activity vanishing for $\omega^2 = 2a_o/a_2 - a_o^2\gamma^2$. Since κ_{12} change sign at this value of ω, the optical activity changes orientation.

Note that for cases (1) (2) (3) the values of κ_{11} are the same as in case (o) up to second order terms in α, β, γ. In each case, the effect of these terms is a splitting of each resonance frequency (ω_1, ω_2) into two resonance frequencies $(\omega_1^{\pm}, \omega_2^{\pm})$.

Further investigation of the signification of the skew-symmetric tensors W, W', Ω and their connections with the internal energy and the partial polarization models is still needed.

REFERENCES

[1] Truesdell, C. and Noll, W., The non linear Field Theories of Mechanics, Handbuch der Physik III/3 (Springer, Berlin, Heidelberg, New York, 1965).

[2] Boulanger, Ph. et Mayné, G., Modèles intégro-différentiels en électrodynamique linéaire des continus, Bull.Acad.Roy.Belg.Cl.Sci. 64 (1978) 153-168.

[3] Toupin, R.A., A dynamical Theory of Elastic Dielectrics, Int.J.Eng.Sci. 1 (1963), 101-126.

Part 3

ACOUSTOELECTRICITY OF DEFORMABLE SEMICONDUCTORS

The Mechanical Behavior of Electromagnetic Solid Continua
G.A. Maugin (editor)
Elsevier Science Publishers B.V. (North-Holland)

ELECTRIC FIELDS, DEFORMABLE SEMICONDUCTORS AND PIEZOELECTRIC DEVICES

H.F. Tiersten
Department of Mechanical Engineering,
Aeronautical Engineering & Mechanics
Rensselaer Polytechnic Institute
Troy, New York 12181
U.S.A.

In the nonlinear interaction of the quasi-static electric field with deformable semiconductors the condition of rotational invariance has an important influence on the description, just as it has in the case of deformable insulators. The description is obtained by means of a systematic application of the laws of continuum physics to a well-defined macroscopic model consisting of interpenetrating solid continua, which interact by means of defined local material fields causing forces and couples to be exerted between the continua. Electric polarization results from the relative displacement of the bound electronic charge continuum with respect to the lattice continuum and electrical conduction arises from the motion of the charged conduction-electronic and hole fluids. Although fundamentally each conducting fluid interacts with itself by means of a pressure, it is convenient to transform to the chemical potentials and obtain the resulting force equations per unit charge. The integral forms associated with each of these latter equations results in boundary conditions at an abrupt macroscopic surface of a deformable semiconductor, the need for which has not been appreciated in other existing work. The new (or rather missing) boundary conditions relate the jump discontinuities in the chemical potentials across the interface to the forces exerted by the lattice continuum on the charged fluids, which prevent the charge from leaving the semiconductor. When the deformation is omitted the resulting boundary conditions have important implications in the description of rigid semiconductor interfaces. The linear equations for small fields superposed on a bias are presented in the absence of conduction. The equation for the perturbation of the eigenfrequency of the piezoelectric solution due to a bias, which is obtained from the linear equations, is presented. The change in frequency resulting from any bias may readily be calculated from the perturbation equation when the linear piezoelectric solution and biasing state are known. The importance of the use of reference coordinates, which can be used in the presence of a bias only with the nonlinear rotationally invariant description, in the simplest case of essentially stress-free thermal deformation, which is just about always present in technological devices, is shown.

1. INTRODUCTION

Although the importance of rotational invariance on the description of the interaction of the electric field with deformable solids originally shown by Toupin[1] is by now quite widely appreciated[2-6], awareness of it is far from universal. In the usual description[7-11] of deformable semiconductors the equations of linear piezoelectricity are coupled to the nonlinear diffusion-drift current equations from semiconductor physics[12]. Although this theory has been somewhat useful in the description of the behavior of piezoelectric semiconductors for small fields under a variety of circumstances, it is not rotationally invariant, lacks consistency in certain other respects also and, most importantly, does not provide proper semiconduction boundary conditions, which are required when surfaces are present, as, e.g., in the case of surface wave propagation. Moreover, this latter criticism can be leveled at the standard description of semiconductors that is employed[12-14] when the electron number density varies appreciably[15], as it often does near boundary surfaces, or in cases of inhomogeneous doping. This standard macroscopic description, consisting of diffusion-drift current equations, the charge balance equations, and the electrostatic constitutive equations, is a continuum field theory, i.e., a system of partial differential equations containing dependent macroscopic field variables as a function of space and time. However, a complete field theory must also have a set of consistent boundary conditions. These have been obtained[16] from integral forms of the governing equations just as in electromagnetism, where electromagnetic boundary conditions are obtained from the integral forms of Maxwell's equations[17,18]. The integral forms have been shown[16] to yield the conventional diffusion-drift current differential equations plus the associated boundary conditions across the surface of the semiconductor, which are missing in the usual macroscopic semiconductor theory[12-14].

Here, following earlier work[19], the differential equations and boundary conditions describing the interaction of a finitely deformable, polarizable, heat conducting and electrically semiconducting continuum with the quasi-static electric field are obtained by means of a systematic application of the laws of continuum physics to a well-defined macroscopic model consisting of five suitably defined interpenetrating continua. As in Ref.19 the five continua consist of the positively charged lattice continuum coupled to four distinct charged continua. The four electrically charged continua are referred to as the bound electronic continuum, the conduction-electronic continuum, the hole continuum and the impurity continuum, respectively. The negatively charged bound electronic continuum can displace slightly with respect to the positively charged lattice continuum and, thereby, produce the electric polarization[20]. The impurity continuum, which can be positively or negatively charged and is required for the general balance of electric charge in the semiconductor, is rigidly attached to the lattice continuum. Both the conduction-electronic continuum and hole continuum are inertialess charged fluids that can move with respect to the lattice continuum while experiencing a force of resistance. In addition, each conducting fluid interacts with neighboring elements of the same fluid by means of defined fluid pressure forces. Naturally, the two conducting fluids are allowed to exchange charge with each other and with the impurity continuum in order to allow for recombination-generation phenomena.

As in all such descriptions, the application of the appropriate equations of balance of charge and momentum yields the equations of motion, which with the equations of electrostatics, constitutes an underdetermined system. The application of the equation of conservation of energy to the combined material continuum results in the first law of thermodynamics which, with the aid of the second law of thermodynamics[21-23] and the principle of material objectivity[24], enables the determination of the constitutive equations. These constitutive equations along with the aforementioned equations of motion and electrostatics and the thermodynamic dissipation equation result in a properly determined system. In order to complete the system of equations, jump (or boundary) conditions across moving surfaces of discontinuity are determined from the appropriate integral forms of the field equations, which are taken to be valid even when the differential forms from which they were obtained are not. The integral forms of the force equations per unit charge[25,16] for each of the conducting fluids results in semiconduction boundary conditions across the surface of the semiconductor, which have been missing from the usual descriptions. These boundary conditions relate the jump discontinuities in the electronic and hole chemical potentials across the surface to the forces per unit area per unit charge densities exerted by the lattice continuum on the respective fluids that keep the electrons from leaving the solid. Moreover, the usual work[9-11] in this area tacitly assumes that the electric surface charge density vanishes at an interface between a semiconductor and the surrounding space. Since the material is a semiconductor and not an insulator, this is a restrictive assumption. By virtue of the aforementioned semiconduction boundary conditions, this restrictive assumption on electric surface charge density does not exist in the description of the deformable semiconductor presented here and in Ref.19.

When the deformation is omitted the resulting description, including the new (or rather missing) semiconduction boundary conditions, is applicable to ordinary semiconductors. In particular, the differential forms of the equations of motion for the conduction electronic and hole fluids along with linear constitutive assumptions yield[16] precisely the usual diffusion-drift current equations containing the material coefficients of mobility and diffusivity. The new boundary conditions at the surface of the semiconductor, which are required for the solution of semiconduction problems involving bounded media, clearly have been obtained from integral forms of the semiconduction equations. As with all the other differential equations and their associated boundary conditions, this procedure ensures consistency between the semiconduction differential equations and the semiconduction boundary conditions. The expressions for the forces per unit area per unit charge densities, which are presented, contain material surface coefficients which are to be determined from measurements just as the mobility and diffusivity coefficients in the differential equations for the semiconductor are often measured. Presumably, the material surface coefficients could be calculated from a more fundamental quantum-mechanical model by means of electronic surface structure calculations[26,27] but the available results seem to indicate that for quantitative detail such calculations would be prohibitively complicated.

The missing boundary conditions have not deterred workers from treating problems concerning semiconductors with boundaries analytically[28]. To be sure, many authors have circumvented the difficulty by imposing various boundary conditions in order to provide the additional condition required for the solution of the boundary-value problem. For example, some have simply disregarded the existence of surface charge on the semiconductor surface[9]. Others have assumed a priori the value of the surface charge[29-31]. Although some of these procedures are based on reasonable assumptions over certain ranges, in a number of cases results have been obtained which are clearly at variance with experiment[32]. In addition, none of those approaches have obtained the boundary conditions from fundamental principles. In the procedure employed in Refs.16 and 33 and presented here, when the newly defined material surface coefficients have been found from measurements of a particular material surface, the new boundary conditions enable the solution of semiconduction boundary-value problems. Furthermore, it has recently been shown[34] that the electric field terms appearing in the expressions for the surface forces, which arise naturally in the description presented here and do not appear in conventional work[35] in this area, are essential for the accurate description of the MOS capacitor in weak inversion.

In the absence of conduction the nonlinear equations reduce to those for the electroelastic insulator. The linear equations for small dynamic fields superposed on a static bias, which are obtained from these latter equations, are presented. The equation for the perturbation of the eigenfrequency of the piezoelectric solution due to a bias, which is obtained from the linear equations, is presented. The change in frequency resulting from any bias may readily be calculated from the perturbation equation when the linear piezoelectric solution and biasing state are known. Piezoelectric devices are subject to many different types of biases and the results of some calculations are presented. However, the associated piezoelectric solution and biasing state are omitted in each instance for the sake of brevity. Almost all piezoelectric devices are subject to biasing states resulting from a change in temperature. When an anisotropic crystal is subject to a change in temperature all line elements not along principal axes shear or skew, i.e., rotate, in addition to extending or contracting. This shearing or skewing of the axes was omitted in all work[36-38] on piezoelectric devices that employed the conventional linear description, which must be referred to the variable temperature dependent intermediate position rather than the fixed reference position. Since the nonlinear rotationally invariant description employed here is referred to the fixed reference state at the reference temperature, the density and geometry do not change. Consequently, the aforementioned skewing is automatically accounted for as are all other effects.

2. DEFORMABLE SEMICONDUCTORS

The application of the procedures employed in Secs.2-5 of Ref.19 to the model introduced there and shown in Figs.1-6 of Ref.19 in the absence of all magnetic quantities results in the system of differential equations

$$\nabla\cdot\tau - \nabla p^e - \nabla p^h + P\cdot\nabla E + \mu E = \rho\, dv/dt\,, \tag{2.1}$$

$$-\nabla p^e + \mu^e(E + E^e) = 0\,,\quad -\nabla p^h + \mu^h(E + E^h) = 0\,, \tag{2.2}$$

$$\nabla\cdot(\mu^e v^e) + \partial\mu^e/\partial t = \gamma^e\,,\quad \nabla\cdot(\mu^h v^h) + \partial\mu^h/\partial t = \gamma^h\,, \tag{2.3}$$

$$\nabla\cdot(\mu^i v) + \partial\mu^i/\partial t = \gamma^i\,,\quad \nabla\cdot(\mu^r v) + \partial\mu^r/\partial t = 0\,, \tag{2.4}$$

$$\mu = \mu^r + \mu^i + \mu^e + \mu^h\,,\quad J = (\mu^r + \mu^i)v + \mu^e v^e + \mu^h v^h\,, \tag{2.5}$$

$$\gamma^e + \gamma^h + \gamma^i = 0\,,\quad \nabla\cdot J + \partial\mu/\partial t = 0\,,\quad \nabla\cdot D = \mu\,, \tag{2.6}$$

$$D = \epsilon_0 E + P\,,\quad E = -\nabla\varphi\,,\quad v = dy/dt\,,\quad P = \rho\pi\,, \tag{2.7}$$

$$-\mu^e E^e\cdot(v^e - v) - \mu^h E^h\cdot(v^h - v) - \left(\mathcal{E}^e + \frac{p^e}{\mu^e}\right)\gamma^e - \left(\mathcal{E}^h + \frac{p^h}{\mu^h}\right)\gamma^h - \nabla\cdot q = \rho\theta\,\frac{d\eta}{dt}\,, \tag{2.8}$$

where the variables are defined in Ref.19, but for clarity we repeat the definitions with minor modifications here. Accordingly, τ denotes the local mechanical stress tensor; v, v^e, v^h, E, E^e, E^h, D, P, J and q denote the velocity of the solid (lattice continuum), velocity of the conduction-electronic fluid, velocity of the hole fluid, Maxwell electric field, local material electric field exerted on the conduction-electronic fluid, local material electric field exerted on the hole fluid, electric displacement, polarization, total current and heat flux vector, respectively; ρ, p^e, p^h, μ^e, μ^h, μ^i, μ^r, μ, γ^e, γ^h, γ^i, $\mathcal{E}^e$, $\mathcal{E}^h$, θ and η denote the mass density, conduction electronic pressure, hole pressure, electronic charge density, hole charge density, residual lattice charge density, net charge density, charge source densities in the electronic, hole and impurity continua, which permit exchange of charge, electronic energy density per unit charge, hole energy density per unit charge, absolute temperature and entropy per unit mass, respectively. From Sec.5 of Ref.19 the associated recoverable constitutive equations take the form

$$\tau_{ij} = \rho y_{i,L} y_{j,M} \frac{\partial\chi}{\partial E_{LM}} + \rho y_{i,L} \frac{\partial\chi}{\partial W_L} E_j\,,\quad \pi_i = -y_{i,L}\,\partial\chi/\partial W_L\,, \tag{2.9}$$

$$\eta = -\partial\chi/\partial\theta\,,\quad p^e = (\mu^e)^2\partial\mathcal{E}^e/\partial\mu^e\,,\quad p^h = (\mu^h)^2\partial\mathcal{E}^h/\partial\mu^h\,, \tag{2.10}$$

where

$$\chi = \chi(y_{j,M}, E_i, \theta) = \chi(E_{LM}, W_L, \theta)\,, \tag{2.11}$$

and the dependence on the far right arises from the first dependence as a result of rotational invariance and

$$E_{LM} = \frac{1}{2}(y_{k,L}y_{k,M} - \delta_{LM}), \quad W_L = y_{k,L}E_k = -\varphi_{,L}, \tag{2.12}$$

in which $y_i = y_i(X_L, t)$ denotes the present coordinates of material points and X_L, the reference coordinates and t denotes the time. Clearly $v_i = dy_i/dt$ where d/dt is the time derivative with X_L fixed and we have introduced the usual conventions and Cartesian notation. Also from Sec.5 of Ref.19, but with one minor but important modification, the dissipative constitutive equations take the rotationally invariant forms

$$q_i = y_{i,K}L_K, \quad E_i^e = y_{i,K}\Omega_K^e, \quad E_i^h = y_{i,K}\Omega_K^h, \tag{2.13}$$

where

$$L_K = L_K(\theta_{,M}, \mu^e, \mu^h, w_L^e, w_L^h, E_{LM}, W_L, \theta),$$

$$\Omega_K^{e/h} = \Omega_K^{e/h}(\theta_{,M}, \mu^{e/h}, w_L^{e/h}, E_{LM}, W_L, \theta),$$

$$\gamma^{e/h} = (\theta_{,M}, \mu^{e/h}, w_L^{e/h}, E_{LM}, W_L, \theta), \tag{2.14}$$

in which

$$w_L^e = y_{i,L}(v_i^e - v_i), \quad w_L^h = y_{i,L}(v_i^h - v_i). \tag{2.15}$$

In addition to the foregoing we have the conservation of mass

$$\rho J = \rho_o, \quad J = \det y_{i,K}, \tag{2.16}$$

where ρ_o is the reference mass density. We now note that this is a deterministic system of equations that can readily be reduced to 13 equations in the 13 dependent variables y_i, φ, μ^e, μ^h, v_i^e, v_i^h and θ. The 13 equations consist of (2.1) - (2.3), $(2.6)_3$ and (2.8).

Before proceeding to the boundary conditions, we note that it is advantageous to transform Eqs.(2.2) for the semiconducting fluids from the forms containing the fluid pressures p^e and p^h to more convenient and, indeed, more conventional forms containing the chemical potentials $\varphi^e = \varphi^e(\mu^e, \theta)$ and $\varphi^h = \varphi^h(\mu^h, \theta)$. As usual, the chemical potentials are defined in terms of the energy densities $\mathcal{E}^e$ and $\mathcal{E}^h$ by[39]

$$\varphi^e = \frac{\partial}{\partial\mu^e}(\mu^e\mathcal{E}^e), \quad \varphi^h = \frac{\partial}{\partial\mu^h}(\mu^h\mathcal{E}^h). \tag{2.17}$$

Then, assuming no material inhomogeneity, from (2.17) with $(2.10)_{2-3}$ we obtain

$$\frac{1}{\mu^e}\nabla p^e = \nabla\varphi^e, \quad \frac{1}{\mu^h}\nabla p^h = \nabla\varphi^h, \tag{2.18}$$

the substitution of which in (2.2) yields

$$-\nabla(\varphi^e + \varphi) + E^e = 0, \quad -\nabla(\varphi^h + \varphi) + E^h = 0, \tag{2.19}$$

which are very convenient forms of the differential equations for the semiconducting fluids. In order to obtain the boundary conditions for our system of differential equations we note that the integral forms which yield (2.19) when φ, φ^e

and φ^h are differentiable may be written

$$\int_S \underset{\sim}{n}(\varphi^e + \varphi)\,dS + \int_V \underset{\sim}{E}^e\,dV = 0\,, \quad - \int_S \underset{\sim}{n}(\varphi^h + \varphi)\,dS + \int_V \underset{\sim}{E}^h\,dV = 0\,. \tag{2.20}$$

From either Sec.6 of Ref.19 or the integral forms that yield (2.1), (2.3), $(2.4)_1$, $(2.6)_3$ and the conservation of energy[40], we obtain

$$\underset{\sim}{n} \cdot [\underset{\sim\sim}{\tau} - (p^e + p^h)\underset{\sim}{I} + \underset{\sim}{T}^E] = 0\,, \tag{2.21}$$

$$\underset{\sim}{n} \cdot [\mu^e(\underset{\sim}{v}^e - \underset{\sim}{v})] + \frac{\partial \sigma^e}{\partial t} = \Gamma^e\,, \quad \underset{\sim}{n} \cdot [\mu^h(\underset{\sim}{v}^h - \underset{\sim}{v})] + \frac{\partial \sigma^h}{\partial t} = \Gamma^h\,, \tag{2.22}$$

$$\partial \sigma_i/\partial t = \Gamma_i\,, \quad \underset{\sim}{n} \cdot [\underset{\sim}{D}] = \sigma\,, \quad \underset{\sim}{n} \cdot [\underset{\sim}{q}] + \mathscr{A}_s = 0\,, \tag{2.23}$$

where

$$\underset{\sim}{T}^E = \underset{\sim\sim}{D\,E} - \frac{1}{2}\,\epsilon_o \underset{\sim}{E} \cdot \underset{\sim}{E}\,\underset{\sim}{I}\,, \quad \int_S \sigma\,dS = \lim_{V \to 0} \int_V \mu\,dV\,,$$

$$\sigma = \sigma^r + \sigma^i + \sigma^e + \sigma^h\,, \quad \Gamma^i + \Gamma^e + \Gamma^h = 0\,, \tag{2.24}$$

and relations similar to $(2.24)_2$ hold for σ^r, σ^i, σ^e and σ^h, Γ^i, Γ^e and Γ^h are the appropriate surface charge source densities and $\mathscr{A}_s$ represents a collection of terms resulting from the semiconductor surface[40], which we do not bother to write here. In addition to the boundary conditions in (2.21) - (2.23) we have the additional conditions

$$[\underset{\sim}{y}] = 0\,, \quad [\varphi] = 0\,, \quad [\theta] = 0\,, \tag{2.25}$$

where we have introduced the conventional notation [a] for $(a^+ - a^-)$ across a material surface of discontinuity in which $\underset{\sim}{n}$ is directed from the - to the + side of the surface. The new (or rather missing) semiconduction boundary conditions are obtained by applying (2.20) to an arbitrary limiting pill-box region exactly in the usual manner in which $(2.23)_2$ is obtained and taking the limits as $\underset{\sim}{E}^e$ and $\underset{\sim}{E}^h$ become unbounded, with the result[16]

$$[\varphi^e] = \underset{\sim}{n} \cdot \underset{\sim}{\mathfrak{F}}^e\,, \quad [\varphi^h] = \underset{\sim}{n} \cdot \underset{\sim}{\mathfrak{F}}^h\,, \tag{2.26}$$

where

$$\int_S \underset{\sim}{\mathfrak{F}}^e\,dS = \lim_{V \to 0} \int_V \underset{\sim}{E}^e\,dV\,, \quad \int_S \underset{\sim}{\mathfrak{F}}^h\,dS = \lim_{V \to 0} \int_V \underset{\sim}{E}^h\,dV\,, \tag{2.27}$$

and we note that $\underset{\sim}{n} \cdot \underset{\sim}{\mathfrak{F}}^e$ and $\underset{\sim}{n} \cdot \underset{\sim}{\mathfrak{F}}^h$ are the forces per unit charge densities per unit area exerted by the lattice continuum on the conduction-electronic and hole fluids, respectively, that prevent the electrons from leaving the semiconductor. Since the constitutive equations for $\underset{\sim}{\mathfrak{F}}^e$ and $\underset{\sim}{\mathfrak{F}}^h$ must be rotationally invariant[19], they may be written in the form

$$\mathfrak{F}^e_j = y_{j,K} G^e_K(\theta_{,M}, \mu^e, w^e_L, E_{LM}, W_L, \theta)\,,$$

$$\mathfrak{F}^h_j = y_{j,K} G^h_K(\theta_{,M}, \mu^h, w^h_L, E_{LM}, W_L, \theta)\,. \tag{2.28}$$

Since the system of equations is referred to the present coordinates, which are unknown, we now transform them in order to refer them to the known reference coordinates X_L. In doing this we restrict ourselves to intrinsic semiconductors for which $\mu^i = \gamma^i = 0$. The procedures for making these transformations are well known (see, e.g., Ref.25) and we simply present the resulting equations and boundary conditions that are needed in the solution of a problem. After the transformation the differential equations that replace (2.1), $(2.6)_3$, (2.3) and (2.2), respectively, take the forms

$$K_{Lj,L} - \mathcal{P}_{Lj,L} = \rho^o \frac{dv_j}{dt}, \quad \mathcal{D}_{L,L} = J\mu = \bar{\mu}, \tag{2.29}$$

$$\mathcal{J}^e_{L,L} + \frac{d}{dt}\bar{\mu}^e = \bar{\gamma}^e, \quad \mathcal{J}^h_{L,L} + \frac{d}{dt}\bar{\mu}^h = \bar{\gamma}^h, \tag{2.30}$$

$$C_{LK}\Omega^e_K = (\varphi + \varphi^e)_{,L}, \quad C_{LK}\Omega^h_K = (\varphi + \varphi^h)_{,L}, \tag{2.31}$$

where ρ^o is the reference mass density and we have the constitutive equations

$$K_{Lj} = \rho^o y_{j,K} \frac{\partial X}{\partial E_{LK}} + JX_{L,i}T^{EF}_{ij}, \quad \mathcal{D}_L = -\varepsilon_o JC^{-1}_{LM}\varphi_{,M} - \rho^o \frac{\partial X}{\partial W_L}, \tag{2.32}$$

$$\mathcal{J}^e_L = C^{-1}_{LK}\bar{\mu}^e w^e_K, \quad \mathcal{J}^h_L = C^{-1}_{LK}\bar{\mu}^h w^h_K, \tag{2.33}$$

and

$$\mathcal{P}_{Lj} = JX_{L,j}(p^e + p^h), \quad T^{EF}_{ij} = \varepsilon_o\left(E_iE_j - \frac{1}{2}E_kE_k\delta_{ij}\right), \tag{2.34}$$

$$C_{KL} = y_{i,K}y_{i,L}, \quad C^{-1}_{LM} = X_{L,i}X_{M,i}, \quad \bar{\mu}^{e/h} = J\mu^{e/h}, \quad \bar{\gamma}^{e/h} = J\gamma^{e/h}, \tag{2.35}$$

and we note that all other variables have been defined previously. We further note that

$$\mathcal{J}_L = \mathcal{J}^e_L + \mathcal{J}^h_L, \quad \mathcal{J}_{L,L} + d(\mathcal{D}_{L,L})/dt = 0\,. \tag{2.35}$$

The associated boundary conditions across material surfaces of discontinuity take the forms

$$\underset{\sim}{N_L[K_{Lj} - \mathcal{P}_{Lj}]} = 0, \quad \underset{\sim\sim\sim}{[y]} = 0, \tag{2.36}$$

$$\underset{\sim}{[\varphi^e]} = N_L C_{LK}G^e_K, \quad \underset{\sim}{[\varphi^h]} = N_L C_{LK}G^h_K, \tag{2.37}$$

$$N_L\underset{\sim}{\left[\mathcal{J}_L + \frac{d}{dt}\mathcal{D}_L\right]} = 0, \quad \underset{\sim\sim}{[\varphi]} = 0, \quad \underset{\sim}{N_L[\mathcal{D}_L]} = \bar{\sigma}, \tag{2.38}$$

in which N_L denotes the unit normal to the reference position of the material surface. If the boundary separates matter from free-space, the boundary conditions take the form

$$N_L(K^f_{Lj} - K_{Lj} + \mathcal{P}_{Lj}) = 0, \tag{2.39}$$

$$-\varphi^e = N_L C_{LK}G^e_K, \quad -\varphi^h = N_L C_{LK}G^h_K, \tag{2.40}$$

$$N_L\left[\frac{d}{dt}\mathcal{D}^f_L - \mathcal{J}_L - \frac{d}{dt}\mathcal{D}_L\right] = 0, \quad \varphi = \varphi^f, \tag{2.41}$$

where

$$K^f_{Lj} = JX_{L,i}\epsilon_o\left(E^f_i E^f_j - \frac{1}{2} E^f_k E^f_k \delta_{ij}\right), \quad \mathcal{D}^f_L = JX_{L,i}\epsilon_o E^f_i , \tag{2.42}$$

and

$$\varphi^f(P^B) = \varphi^f(R^B) + u_M \varphi^f_{,M}(R^B) + \frac{1}{2} u_M u_L \varphi^f_{,ML}(R^B) + \ldots ,$$

$$E^f_L(P^B) = E^f_L(R^B) + u_M E^f_{L,M}(R^B) + \frac{1}{2} u_M u_K E^f_{L,MK}(R^B) + \ldots , \tag{2.43}$$

in which P^B and R^B represent the unknown present and known reference position of a point of a material boundary, respectively, u_M is the unknown displacement of the point of the boundary $E^f_i = \delta_{Li} E^f_L(P^B)$ and $E^f_L(R^B) = -\varphi^f_{,L}(R^B)$. In addition, we have the differential equation on φ^f for points of free-space, which takes the form

$$\varphi^f_{,LL} = 0 , \tag{2.44}$$

and we note that for brevity we have ignored heat conduction in referring the equations to reference coordinates.

3. RIGID SEMICONDUCTORS

When the deformation is suppressed the resulting equations hold for rigid (or rather typical) semiconductors[33]. In this section we present the differential equations and boundary conditions for semiconductors themselves that arise from either the reduction of the deformable description or the model of semiconduction without deformation at the outset. Since when deformation is omitted the mass density ρ does not occur and it is conventional to use ρ for charge density in work on semiconductors, we replace all μ by ρ in this section. Then the resulting differential equations for the semiconductor may be written in the form

$$\nabla(\varphi + \varphi^e) = \mathbf{E}^e , \quad \nabla(\varphi + \varphi^h) = \mathbf{E}^h , \tag{3.1}$$

$$\nabla \cdot \mathbf{J}^e + \frac{\partial \rho^e}{\partial t} = \gamma^e , \quad \nabla \cdot \mathbf{J}^h + \frac{\partial \rho^h}{\partial t} = \gamma^h , \tag{3.2}$$

$$\mathbf{J}^e = \rho^e \mathbf{v}^e , \quad \mathbf{J}^h = \rho^h \mathbf{v}^h , \quad \nabla \cdot \mathbf{D} = \rho , \quad \mathbf{D} = \epsilon \mathbf{E} , \tag{3.3}$$

$$\mathbf{E} = -\nabla\varphi , \quad \rho = \rho^e + \rho^h + \rho^i , \quad \gamma^e + \gamma^h = 0 , \tag{3.4}$$

and we note that we have assumed the usual simple linear isotropic constitutive relation between $\mathbf{P}$ and $\mathbf{E}$. If we assume the constitutive equations for $\mathbf{E}^e$ and $\mathbf{E}^h$ to be of the important limiting linear form usually chosen, we have

$$\mathbf{E}^e = \mathbf{v}^e / \mu^e , \quad \mathbf{E}^h = -\mathbf{v}^h / \mu^h , \tag{3.5}$$

where μ^e and μ^h are the measured mobilities of the respective fluids, then the semiconduction equations (3.1) can be written in the form[16]

$$\mathbf{J}^e = -\mu^e \rho^e \mathbf{E} - D^e \nabla\rho^e , \quad \mathbf{J}^h = \mu^h \rho^h \mathbf{E} - D^h \nabla\rho^h , \tag{3.6}$$

where

$$D^e = -\mu^e \rho^e \partial\varphi^e / \partial\rho^e , \quad D^h = \mu^h \rho^h \partial\varphi^h / \partial\rho^h , \tag{3.7}$$

are the usual diffusion coefficients. Equations (3.6) are the conventional diffusion drift-current equations for the respective fluids. If we take as constitutive choices for $\mathcal{E}^e$ and $\mathcal{E}^h$ the forms for the Maxwell gas[16], then we obtain

$$\varphi^e = -\frac{kT}{q}\log\left(\frac{\rho^e}{-qN_c}\right), \quad \varphi^h = \frac{kT}{q}\log\left(\frac{\rho^h}{qN_v}\right), \tag{3.8}$$

where in this section we have used T for absolute temperature instead of θ, q is the charge of an electron and N_c and N_v are the quasi-microscopically defined "effective densities of states" in the conduction and valence bands, respectively. A common selection for γ^e in n-type material is

$$\gamma^e = (\rho_1^h/\tau_p) - G, \tag{3.9}$$

where ρ_1^h is the dynamic portion of ρ^h which is superposed on a static bias ρ_0^h, τ_p is the "minority-carrier lifetime" and G is the charge source density due to, say, illumination. We now note that although Eqs.(3.1) result in the well-known conventional description when the conventional constitutive assumptions are made, Eqs.(3.1) hold under all other circumstances when drift inertia is negligible, simply by making different appropriate constitutive assumptions, e.g., for the degenerate electron gas.

At all interfaces we have the usual electrical boundary conditions

$$[\varphi] = 0, \quad \mathbf{n}\cdot\left[\frac{\partial\mathbf{D}}{\partial t} + \mathbf{J}\right] = 0, \quad \mathbf{n}\cdot[\mathbf{D}] = \sigma, \tag{3.10}$$

the latter of which serves to define the surface charge σ and where

$$\mathbf{J} = \mathbf{J}^e + \mathbf{J}^h. \tag{3.11}$$

In addition we have the new (or rather missing) semiconduction boundary conditions

$$[\varphi^e] = f^e(\rho^e, \mathbf{v}^e, \mathbf{E}), \quad [\varphi^h] = f^h(\rho^h, \mathbf{v}^h, \mathbf{E}), \tag{3.12}$$

where for a semiconductor-insulator interface we have taken the constitutive equations for the surface forces in the form[16,33]

$$f^e = A_o^R + A_1^R\,\mathbf{n}\cdot\mathbf{E}_{ins} + A_2^R\,\mathbf{n}\cdot\mathbf{E}_{sem} + A_3^R\log(\rho^e/\rho_b^e)$$
$$+ A_4^R\log^2\left(\frac{\rho^e}{\rho_b^e}\right) + \ldots + \int_{-\infty}^{t} E(t-s)\,\frac{\partial\rho^e(s)}{\partial s}\,ds + \mathbf{A}^D\cdot\mathbf{v}^e + \ldots, \tag{3.13}$$

$$f^h = D_o^R + D_1^R\,\mathbf{n}\cdot\mathbf{E}_{ins} + D_2^R\,\mathbf{n}\cdot\mathbf{E}_{sem} + D_3^R\log(\rho^h/\rho_b^h)$$
$$+ D_4^R\log^2\left(\frac{\rho^h}{\rho_b^h}\right) + \ldots + \int_{-\infty}^{t} H(t-s)\,\frac{\partial\rho^h(s)}{\partial s}\,ds + \mathbf{D}^D\cdot\mathbf{v}^h + \ldots, \tag{3.14}$$

where, from thermodynamic arguments[41],

$$\mathbf{A}^D > 0, \quad \mathbf{D}^D < 0. \tag{3.15}$$

In Eqs.(3.13) and (3.14) the material surface coefficients with the superscript R are called recoverable coefficients because they do not produce entropy and influence static as well as dynamic cases. The material surface coefficients with the superscript D are called dissipative coefficients because they produce entropy and influence dynamic cases only. The memory functionals are required for the description of high frequency behavior.

The recoverable surface coefficients have been evaluated[16] for silicon from quasi-static capacitance-voltage measurements on MOS capacitors and the dissipative coefficients have been evaluated[33] from low-frequency small-signal (superposed on a quasi-static bias) admittance measurements on the same structure. In the case of photogenerated D.C. an asymptotic outer expansion of the small-signal (superposed on a bias) equations within the charge neutrality approximation results in a single condition to be applied at the edge of the charge neutral region of the same form as the Shockley recombination velocity condition with the recombination velocity (heretofore measured directly) expressed in terms of the now known material coefficients and static solution. The calculated[33] recombination velocity as a function of surface potential has the same shape as the measured curve. The small signal equations have been applied in the analytical description of the admittance of the MOS capacitor and the resulting linear equations are very similar to the conventional[35] equivalent circuit description. This is as expected since the small signal description obtained from the equations presented here constitutes the macroscopic linear field theory underlying existing lumped parameter representations. However, the terms that arise in the small signal description resulting from the electric field terms in (3.13) and (3.14), i.e., those containing A_1^R, A_2^R, D_1^R and D_2^R, which arise naturally in the description presented here, do not appear in any of the existing lumped parameter descriptions. The necessity for the existence of these electric field terms in the description of an MOS capacitor in weak inversion has recently been verified experimentally[34]. A curve taken from Ref.34 exhibiting this agreement is shown in Fig.1, which contains measurements of a type of semiconductor capacitance versus a type of semiconductor response frequency in inversion. The measurements were made at 1.7 Hz for a number of temperatures. It can be seen from the figure that all curves have the same nonzero intercept on the vertical axis. Since this nonzero intercept exists only[34] when the electric field terms appear in (3.13) and (3.14), the experimental results unequivocally demonstrate the existence of the electric field terms in (3.13) and (3.14).

4. NONLINEAR EFFECTS IN PIEZOELECTRIC DEVICES

In the absence of electrical conduction the equations presented in Sec.2 reduce to those for the deformable insulator[42]. When heat conduction is omitted the linear electroelastic equations for small fields superposed on a static bias have been obtained[43]. These equations may be written in the form

$$\tilde{K}_{L\gamma,L} = \overset{\circ}{\rho}\ddot{u}_\gamma, \quad \tilde{\mathcal{D}}_{L,L} = 0, \tag{4.1}$$

where

$$\tilde{K}_{L\gamma} = \underset{1}{G}_{L\gamma M\delta}u_{\delta,M} + \underset{2}{G}_{ML\gamma}\tilde{\hat{\varphi}}_{,M}, \quad \tilde{\mathcal{D}}_L = \underset{1}{R}_{LM\alpha}u_{\alpha,M} + \underset{2}{R}_{LM}\tilde{\hat{\varphi}}_{,M}, \tag{4.2}$$

and the effective coefficients are defined by

$$\underset{1}{G}_{L\gamma M\delta} = \underset{2}{c}_{L\gamma M\delta} + \hat{c}_{L\gamma M\delta}, \quad \underset{2}{R}_{LM} = -\epsilon_{LM} - \hat{\epsilon}_{LM}, \quad \underset{2}{G}_{ML\gamma} = \underset{1}{R}_{ML\gamma} = e_{ML\gamma} + \hat{e}_{ML\gamma},$$

$$\hat{c}_{L\gamma M\delta} = \overset{1}{T}_{LM}\delta_{\gamma\delta} + \underset{3}{c}_{L\gamma M\delta AB}\overset{1}{E}_{AB} + \underset{2}{c}_{L\gamma KM}w_{\delta,K} + \underset{2}{c}_{LKM\delta}w_{\gamma,K} - \underset{1}{k}_{AL\gamma M\delta}\overset{1}{\varphi}_{,A} + \overset{1}{g}_{L\gamma M\delta},$$

$$\hat{e}_{ML\gamma} = -\underset{1}{k}_{ML\gamma BC}\overset{1}{E}_{BC} + e_{MLK}w_{\gamma,K} - b_{AML\gamma}\overset{1}{\varphi}_{,A} + \overset{2}{g}_{ML\gamma},$$

$$\hat{\epsilon}_{LM} = b_{LMCD}\overset{1}{E}_{CD} - \underset{2}{\chi}_{LMC}\overset{1}{\varphi}_{,C} - 2\epsilon_o\overset{1}{J}\overset{1}{E}_{ML}, \tag{4.3}$$

and $\overset{1}{g}_{L\gamma M\delta}$ and $\overset{2}{g}_{ML\gamma}$, which for brevity we do not bother to write[44], depend on the biasing electric field E^1_ζ and static deformation. In obtaining these equations we have defined the intermediate coordinates $\xi_\alpha = \xi_\alpha(X_L)$ so that we have[43,44]

$$y_i = \bar{y}_i(\xi_\alpha, t) = y_i(X_L, t) , \tag{4.4}$$

and the static and dynamic small field displacement fields $w_K(X_L)$ and $u_\alpha(X_L, t)$ are defined by

$$\underset{\sim}{w} = \underset{\sim}{\xi} - \underset{\sim}{X} , \quad \underset{\sim}{u} = \underset{\sim}{y} - \underset{\sim}{\xi} . \tag{4.5}$$

Hence, we have

$$\underset{\sim}{y}(X_L, t) = \underset{\sim}{X} + \underset{\sim}{w}(X_L) + \underset{\sim}{u}(X_L, t) , \tag{4.6}$$

and

$$\hat{\varphi}(X_L, t) = \hat{\varphi}^1(X_L) + \tilde{\hat{\varphi}}(X_L, t) , \quad \varphi^f(X_L, t) = \varphi^{f1}(X_L) + \tilde{\varphi}^f(X_L, t) , \tag{4.7}$$

where $\hat{\varphi}^1$ and φ^{f1} denote the biasing electric potentials at the reference positions of material points and points of free-space, respectively. Equations (4.1) are the small-field stress equations of motion and charge equation of electrostatics referred to the reference coordinates X_L. Equations (4.2) are the linear electro-elastic constitutive equations. In (4.1) - (4.3) ρ_0 denotes the reference mass density, and $\underset{2}{c}_{L\gamma M\alpha}$, $e_{ML\gamma}$ and ϵ_{LM} denote the second-order elastic, piezoelectric, and dielectric constants, respectively. The symbols T^1_{LM}, E^1_{AB}, and E^1_j denote the components of the static biasing stress, strain, and electric field, respectively, and J^1 is the Jacobian of the static deformation. The biasing variables satisfy the appropriate static equations given in (66) - (72) of Ref.43, or the equivalent equations using reference coordinates as independent variables. In (4.3) ${}_3c_{L\gamma M\alpha AB}$, $b_{AML\gamma}$, ${}_2\chi_{LMC}$, and ${}_1k_{ML\gamma BC}$ denote the third-order elastic, electrostrictive, third-order electric permeability, and first-order electroelastic constants, respectively. To the foregoing equations we must adjoin the dynamic boundary conditions, which we do not present here for the sake of brevity[44].

From the foregoing equations the equation for the perturbation of the eigenfrequency of the piezoelectric solution due to a bias has been obtained[44] and may be written in the form

$$\Delta_\mu = H_\mu / 2\omega_\mu , \quad \omega = \omega_\mu - \Delta_\mu , \tag{4.8}$$

where ω_μ and ω are the unperturbed and perturbed eigenfrequencies, respectively, and for many boundary conditions

$$H_\mu = - \int_V [\tilde{K}^n_{L\gamma} g^\mu_{\gamma,L} + \tilde{\mathscr{D}}^n_L \hat{f}^\mu_{,L}] \, dV , \tag{4.9}$$

in which the normalized mechanical displacement field and electric potential have been defined by

$$g^\mu_\gamma = \frac{u^\mu_\gamma}{N_\mu} , \quad \hat{f}^\mu = \frac{\tilde{\varphi}^\mu}{N_\mu} , \quad N^2_\mu = \int_V \rho u^\mu_\gamma u^\mu_\gamma \, dV . \tag{4.10}$$

The fields in (4.9) and (4.10) are obtained from solutions of the linear piezoelectric equations[44]

$$\tilde{K}^\ell_{L\gamma,L} = \rho\ddot{u}_\gamma , \quad \tilde{\mathscr{D}}^\ell_{L,L} = 0 , \quad \tilde{K}^\ell_{L\gamma} = \underset{2}{c}_{L\gamma M\alpha} u_{\alpha,M} + e_{ML\gamma}\tilde{\varphi}_{,M} , \quad \tilde{\mathscr{D}}^\ell_L = e_{LM\gamma} u_{\gamma,M} - \epsilon_{LM}\tilde{\varphi}_{,M} . \tag{4.11}$$

The nonlinear terms in (4.9) are given by[44]

$$\tilde{K}^{n}_{L\gamma} = (\hat{c}_{L\gamma M\alpha} + \Delta c_{2L\gamma M\alpha})\, g^{\mu}_{\alpha,M} + (\hat{e}_{ML\gamma} + \Delta e_{ML\gamma})\, \hat{f}^{\mu}_{,M},$$

$$\tilde{\mathscr{D}}^{n}_{L} = (\hat{e}_{LM\gamma} + \Delta e_{LM\gamma})\, g^{\mu}_{\gamma,M} - (\hat{\varepsilon}_{LM} + \Delta\varepsilon_{LM})\, \hat{f}^{\mu}_{,M}, \tag{4.12}$$

where

$$\hat{c}_{L\gamma M\alpha} = T^{1}_{LM}\delta_{\gamma\delta} + c_{3L\gamma M\alpha AB}E^{1}_{AB} + c_{2L\gamma KM}w_{\alpha,K} + c_{2LKM\alpha}w_{\gamma,K}, \tag{4.13}$$

$$\hat{\varepsilon}_{LM} = b_{LMCD}E^{1}_{CD} - 2\varepsilon_{o}E^{1}_{LM}, \quad \hat{e}_{LM\gamma} = -k_{1LM\gamma BC}E^{1}_{BC} + e_{LMK}w_{\gamma,K}, \tag{4.14}$$

and Δ represents a change in the fundamental material constants due to a change in temperature and is given by $(T - T_0)d/dT$. Equations (4.8) and (4.9) enable the change in frequency of a given linear piezoelectric solution due to a biasing state to be calculated if the linear modal solution and biasing state are known. However, since many of the coefficients in (4.14) are not presently known even for quartz, which is the most important technological material, a very accurate approximate simplified version of (4.9) is employed[45], which is given by

$$H_{\mu} = -\int_{V}\left[(\hat{c}_{L\gamma M\alpha} + \Delta c_{2L\gamma M\alpha})\tilde{g}^{\mu}_{\alpha,M}\tilde{g}^{\mu}_{\gamma,L} - \frac{2e^{2}_{26}}{\varepsilon_{22}}\frac{\Delta e_{26}}{e_{26}}\, g^{(\mu)}_{1,2}\, \frac{g^{(\mu)}_{1}(h)}{h}\right] dV, \tag{4.15}$$

in which the Δe_{26} must be determined from measurements on each orientation of interest[45].

Calculations have been performed for a number of situations using Eq.(4.15) and we present a few of these results here. However, for obvious reasons we do not present either the linear modal solution or the thermoelastic biasing state for any case here, but just present calculated results for two cases which we compare with measurement. Figure 2 shows the calculated temperature coefficient of actual (as opposed to natural[6]) velocity[46] for surface waves on AT-cut quartz substrates as a function of propagation direction at 25°C. The dotted curve shows the average of the calculated values at 0° and 50°C from Ref.36 and the circles are the average of the experimental values at 0° and 50°C, also from Ref.36. It is clear from Fig.2 that the calculations performed in Ref.46 using the proper nonlinearly based formalism, which automatically incorporates skewing effects, are in substantially better agreement with the experimental data of Ref.36 than the calculations performed in Ref.36 using an incomplete linearly based description. Contoured AT-cut quartz crystal resonators are used as high precision oscillators because the pure thickness frequency used in the AT-cut is thermally compensated. The plate is contoured to confine the mode to the vicinity of the center of the plate so that the edges can be supported without damaging the quality factor (Q). An accurate description of[47] the mode shape has been determined. The contouring causes a small change in the orientation of the zero temperature cut, which is very important technologically. Figure 3 shows a comparison[45] of loglog plots of the calculated shift in rotation angle for the zero temperature cut for the fundamental mode for center thicknesses $2h_0 = 0.8258$ mm and $2h_0 = 1.6515$ mm, respectively, which have different electrode sizes, and $2h_0 = 0.3282$ mm, which corresponds to the 5 MHz unelectroded flat plate resonator, and having an electrode diameter of 4 mm, with the known measured design curves for the fundamental mode of the plano-convex resonator. It can be seen from the figure that the calculated results are not straight lines, but curves and, in fact, a different curve for each different center thickness and each electrode size. Although all the calculated curves tend to follow the general trend of the single measured design line, they deviate from it differently for different radii of curvature of the contour. However, the deviation from the straight line is never more than about 4'. Furthermore, the calculated curve for the 5 MHz contoured resonator, which is the most commonly used one, deviates from the measured line by not more than 1.5' over the entire practical range of R.

ACKNOWLEDGEMENTS

The author wishes to thank M.G. Ancona of the Naval Research Laboratory for several helpful and informative discussions. This work was supported in part by the National Science Foundation under Grant No. MEA-8115340.

REFERENCES

1. R.A. Toupin, "The Elastic Dielectric," J. Rational Mech. Anal., 5, 849 (1956).
2. W.F. Brown, Jr., Magnetoelastic Interactions (Springer, Berlin, 1966).
3. R.N. Thurston, "Waves in Solids," in Encyclopedia of Physics, edited by C. Truesdell (Springer, Berlin, 1974), Vol. VIa/4.
4. R.A. Grot, "Relativistic Continuum Physics: Electromagnetic Interactions," in Continuum Physics, edited by A.C. Eringen (Academic, New York, 1976), Vol. III, pp.129-219.
5. G.A. Maugin, "Relativistic Continuum Physics: Micromagnetism," in Continuum Physics, edited by A.C. Eringen (Academic, New York, 1976), Vol.III, pp.221-312.
6. D.F. Nelson, Electric, Optic and Acoustic Interactions in Dielectrics (Wiley, New York, 1979).
7. A.R. Hutson and D.L. White, "Elastic Wave Propagation in Piezoelectric Semiconductors," J. Appl. Phys., 33, 40 (1962).
8. D.L. White, "Amplification of Ultrasonic Waves in Piezoelectric Semiconductors," J. Appl. Phys., 33, 2547 (1962).
9. K.M. Lakin and H.J. Shaw, "Surface Wave Delay Line Amplifiers," IEEE Trans. Microwave Theory Tech. MTT-17, 912 (1969).
10. G.S. Kino and T.M. Reeder, "A Normal Mode Theory for the Rayleigh Wave Amplifier," IEEE Trans. on Electron Devices, ED-18, 909 (1971).
11. P.S. Ramakrishna, "Amplification of Acoustic Surface and Layer Waves," Master's Thesis submitted to McGill University, Montreal, 1971.
12. W. Shockley, Electrons and Holes in Semiconductors (D. Van Nostrand Co., Princeton, 1950).
13. A.C. Smith, J.F. Janak and R.B. Adler, Electronic-Conduction in Solids (McGraw-Hill, New York, 1967).
14. S.M. Sze, Physics of Semiconductor Devices (Wiley, New York, 1969).
15. This is subject to the usual restriction that it must vary slowly compared with the spacing of discrete elements.
16. M.G. Ancona and H.F. Tiersten, "Fully Macroscopic Description of Bounded Semiconductors with an Application to the $Si-SiO_2$ Interface," Phys. Rev. B, 6104 (1980).
17. J. Stratton, Electromagnetic Theory (McGraw-Hill, New York, 1941).
18. J.D. Jackson, Classical Electrodynamics (Wiley, New York, 1975).
19. H.G. de Lorenzi and H.F. Tiersten, "On the Interaction of the Electromagnetic Field with Heat Conducting Deformable Semiconductors," J. Math. Phys., 16, 938 (1975).
20. H.F. Tiersten, "On the Nonlinear Equations of Thermoelectroelasticity," Int. J. Engng. Sci., 9, 587 (1971).
21. B.A. Boley and J.H. Weiner, Theory of Thermal Stresses (Wiley, New York, 1967).
22. S.R. DeGroot and P. Mazur, Nonequilibrium Thermodynamics (North-Holland, Amsterdam, 1963).
23. A.C. Eringen, Nonlinear Theory of Continuous Media (McGraw-Hill, New York, 1962).
24. C. Truesdell and W. Noll, "The Nonlinear Field Theories of Mechanics," in Encyclopedia of Physics, edited by S. Flügge (Springer-Verlag, Berlin 1965), Vol.III/2, Secs.17 and 19.
25. M.F. McCarthy and H.F. Tiersten, "On Integral Forms of the Balance Laws for Deformable Semiconductors," Arch. Rational Mech. Anal., 68, 27 (1978).

26. J.A. Appelbaum and D.R. Hamann, "Surface Potential, Charge Density and Ionization Potential for Si(111) - A Self-Consistent Calculation," Phys. Rev. Lett., 32, 225 (1974).
27. J. Pollmann and S.T. Pantelides, "Scattering Theoretic Approach to the Electronic Structure of Semiconductor Surfaces: The (100) Surface of Tetrahedral Semiconductors and SiO_2," Phys. Rev. B 18, 5524 (1978).
28. Strictly speaking, all semiconductor problems involve boundaries since there are no semiconductors of infinite size. We are concerned here only with problems in which the effects of bounding surfaces on overall response is important.
29. D. Vandorpe, J. Borel, G. Merckel and P. Saintot, "An Accurate Two-Dimensional Numerical Analysis of the MOS Transistor," Solid-State Electron, 15, 547 (1972), Sec.4.
30. P.M. Marcus, "Calculation of the Capacitance of a Semiconductor Surface, with Application to Silicon," IBM Journal, 8, 496 (1964).
31. J. McKenna and N.L. Schryer, "On the Accuracy of the Depletion Layer Approximation for CCD's," Bell Sys. Tech. J., 51, 1471 (1972).
32. S. Zemon and E.N. Conwell, "Effects of Surface States on Surface Wave Amplification," Appl. Phys. Lett., 17, 218 (1970).
33. M.G. Ancona and H.F. Tiersten, "Fully Macroscopic Description of Electrical Conduction in Metal-Insulator-Semiconductor Structures," Phys. Rev. B, 27, 7018. (1983).
34. M.G. Ancona, "Electric Field Induced Effects at the Si-SiO_2 Interface: Theory and Experiment," to be published in the Journal of Applied Physics (1983).
35. E.H. Nicollian and J.R. Brews, MOS Physics and Technology (Wiley, New York, 1982).
36. M.B. Schulz, B.J. Matsinger and M.G. Holland, "Temperature Dependence of Surface Acoustic Wave Velocity on α-Quartz," J. Appl. Phys., 41, 2755 (1970).
37. M.F. Lewis, G. Bell and E. Patterson, "Temperature Dependence of Surface Elastic Wave Delay Lines," J. Appl. Phys., 42, 476 (1971).
38. R. Bechmann, A.D. Ballato and T.J. Lukaszek, "Higher Order Temperature Coefficients of the Elastic Stiffnesses and Compliances of Alpha-Quartz," Proc. IRE 50, 1812 (1962).
39. C. Truesdell and R.A. Toupin, "The Classical Field Theories," in Encyclopedia of Physics, edited by S. Flügge (Springer-Verlag, Berlin, 1961), Vol.III/1, Sec.255.
40. Ref.33, Appendix A.
41. Equation (A12) of Ref.33 has an incorrect sign before the outside bracket of the last term on the right-hand side. As a consequence A^D and D^D have the restrictions in Eq.(3.15) of this work and the signs in Eqs.(2.35) of Ref.33 should be interchanged along with the signs on the last term in (2.39) and (2.40) and β^e and β^h should be defined with minus signs. These errors did not affect any calculations because they were present only in the published version.
42. H.F. Tiersten and C.F. Tsai, "On the Interaction of the Electromagnetic Field with Heat Conducting Deformable Insulators," J. Math. Phys., 13, 361 (1972).
43. J.C. Baumhauer and H.F. Tiersten, "Nonlinear Electroelastic Equations for Small Fields Superposed on a Bias," J. Acoust. Soc. Am., 54, 1017 (1973).
44. H.F. Tiersten, "Perturbation Theory for Linear Electroelastic Equations for Small Fields Superposed on a Bias," J. Acoust. Soc. Am., 64, 832 (1978).
45. D.S. Stevens and H.F. Tiersten, "Temperature Dependence of the Resonant Frequency of Electroded Contoured AT-Cut Quartz Crystal Resonators," J. Appl. Phys., 54, 1704 (1983).
46. B.K. Sinha and H.F. Tiersten, "On the Temperature Dependence of the Velocity of Surface Waves in Quartz," J. Appl. Phys., 51, 4659 (1980).
47. H.F. Tiersten and R.C. Smythe, "An Analysis of Contoured Crystal Resonators Operating in Overtones of Coupled Thickness Shear and Thickness Twist," J. Acoust. Soc. Am., 65, 1455 (1979).

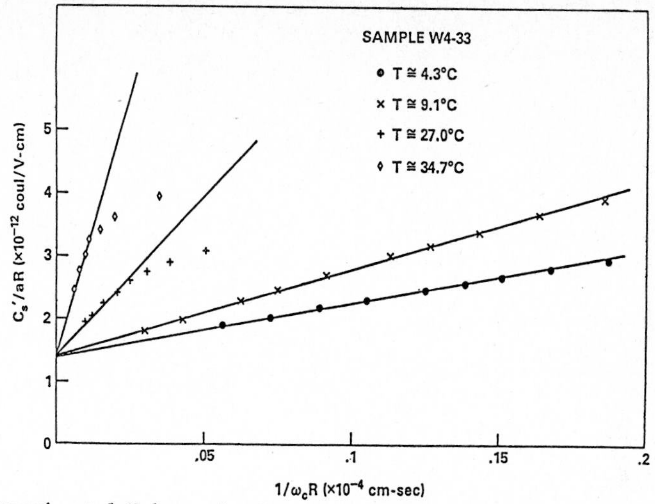

Fig.1 Experimental Values of a Type of Semiconductor Capacitance Versus a Type of Semiconductor Response Frequency in Inversion

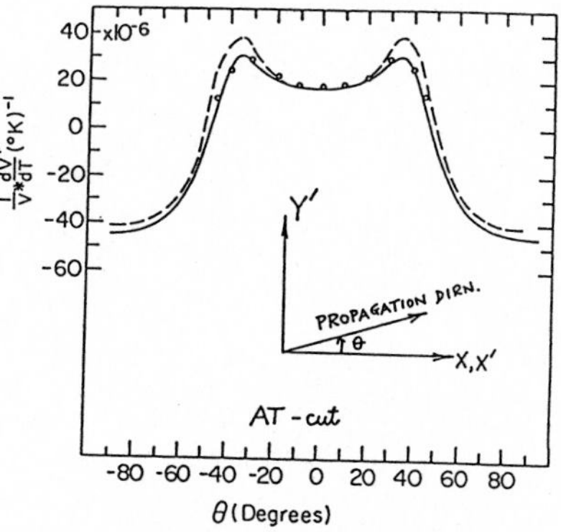

Fig.2 Temperature Coefficients of Actual Velocity for Surface Waves on AT-Cut Quartz as a Function of Propagation Direction Relative to the Digonal Axis at 25°C

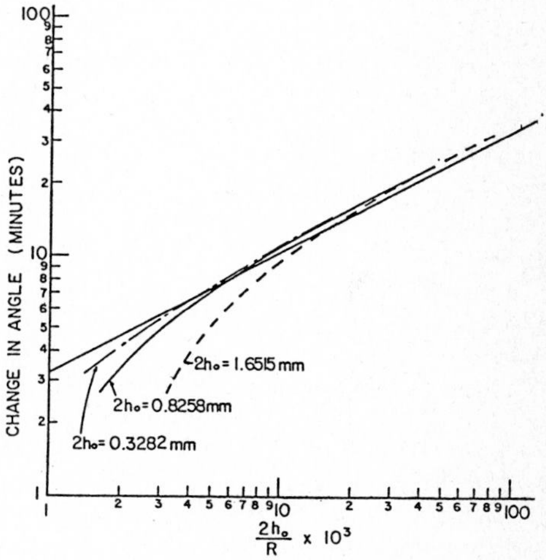

Fig.3 Comparison of the Calculated Change in Rotation Angle $\Delta\theta$ from $\theta = -35^\circ 8.5'$ for the Zero Temperature Cut for the Fundamental Mode of the Contoured Resonator as a Function of the Center Thickness to Radius of Contour Ratio with the Measured Design Curve

The Mechanical Behavior of Electromagnetic Solid Continua
G.A. Maugin (editor)
Elsevier Science Publishers B.V. (North-Holland)

WAVES IN ELASTIC SEMICONDUCTORS IN A BIAS ELECTRIC FIELD

N.DAHER

Université Pierre-et-Marie Curie,Laboratoire de Mécanique Théorique,
Tour 66 , 4 Place Jussieu, 75230 Paris Cedex 05, France.

A nonlinear, rotationally invariant and thermodynamically admissible phenomenological theory of elastic semiconductors is presented. After linearization about a rigid-body solution, a complete set of bulk and surface electroacoustic equations is obtained which provides a general framework for the study of the coupling and excitation of both bulk and surface displacement modes. In particular, we consider two settings of the bias electric field:(i)a longitudinal setting for bulk modes and (ii) an orthogonal setting for bulk and surface modes. The first case is shown to be a generalization of simpler results obtained by various authors while the second case for bulk waves provides useful informations for treating the surface mode which is a generalized Rayleigh problem that, on account of some restrictions, uncouples in a classical Rayleigh problem and a Bleustein-Gulyaev problem on which we focus our attention.

GENERAL NONLINEAR EQUATIONS

In a general manner we use indifferently the direct (intrinsic)dyadic notation or the notation of Cartesian tensors in rectangular co-ordinate systems. The Galilean form of Maxwell's equations is sufficient for the problem we are interested in. The local thermomechanical equations are deduced from the principle of virtual power and the two fundamental principles of thermodynamics [1]. The general continuum equations which govern the elastic piezoelectric semiconductors are as follows:

- continuity equation:

$$\dot{\rho} + \rho \nabla . v = 0 \qquad ,(1)$$

- Euler-Cauchy equations of motion:

$$\rho \bar{v}_i = t_{ij,j} + f_i^{em} \quad \text{in } D_t \quad (i,j=1,2,3) \qquad ,(2)$$

$$T_i + T_i^{em} = t_{ij} n_j \quad \text{on } \partial D_t \qquad ;(3)$$

- Maxwell's equations (that we do not recall)

- Clausius-Duhem inequality:

$$-\rho(\dot{\hat{\phi}} - \hat{\mu}\dot{c} + \eta\dot{\theta}) + \hat{t}_{ij} D_{ij} - P.\overset{*}{\mathcal{E}} - \mathcal{M}.\overset{*}{B} + \mathcal{J}.\hat{\mathcal{E}} - \phi.\nabla\theta \geqq 0 \qquad (4)$$

with

$$t_{ij} = \hat{t}_{ij} + P_i \mathcal{E}_j + \mathcal{M}_i B_j - (P.\mathcal{E} + \mathcal{M}.B)\delta_{ij} \qquad ,(5)$$

$$f^{em} = q_f \mathcal{E} + \frac{1}{c_1}(\mathcal{J} + \overset{*}{P}) \times B + (P.\nabla)\mathcal{E} + (\nabla B).\mathcal{M} \qquad ,(6)$$

$$\underset{\sim}{T}^{em} = \omega_f \langle \underset{\sim}{\mathcal{E}} \rangle + \frac{1}{c_1} \underset{\sim}{\mathcal{K}} \times \langle \underset{\sim}{B} \rangle + (\underset{\sim}{n} . \langle \underset{\sim}{P} \rangle) [\![\underset{\sim}{\mathcal{E}}]\!] + \underset{\sim}{n} (\langle \underset{\sim}{\mathcal{M}} \rangle . [\![\underset{\sim}{B}]\!]) \quad ,(7)$$

$$\underset{\sim}{q} = \theta \underset{\sim}{\phi} + \underset{\sim}{\mathcal{S}} + \lambda_c \hat{\mu} \underset{\sim}{\mathcal{J}} \quad , \quad \underset{\sim}{\mathcal{S}} = c_1 \underset{\sim}{\mathcal{E}} \times \underset{\sim}{\mathcal{H}} \quad ,(8)$$

where $[\![.]\!]$ and $\langle . \rangle$ denote the jump and mean value of their enclosure and $\hat{\psi}$, η and θ are the free energy and entropy densities and the absolute temperature. We assume that the entropy flux deviates from its usual definition by introducing an "extra" entropy flux $\underset{\sim}{k} = -(\underset{\sim}{\mathcal{S}} + \lambda_c \hat{\mu} \underset{\sim}{\mathcal{J}})/\theta$; $\underset{\sim}{\mathcal{J}}$, $\underset{\sim}{\mathcal{E}}, \underset{\sim}{\mathcal{H}}$ and $\underset{\sim}{\mathcal{M}}$ denote the same fields as $\underset{\sim}{J}$, $\underset{\sim}{E}$ $\underset{\sim}{H}$ and $\underset{\sim}{M}$ but in a so-called co-moving frame $R_c(\underline{x},t)$; $\underset{\sim}{\mathcal{S}}$ is Poynting's vector in $R_c(\underline{x},t)$ and

$$\hat{\underset{\sim}{\mathcal{E}}} = \underset{\sim}{\mathcal{E}} - \lambda_c \underset{\sim}{\nabla} \hat{\mu} \quad , \quad \hat{\mu} = \partial \hat{\psi} / \partial c \quad (9)$$

are ,respectively, an effective electromotive intensity and the chemical potential, and

$$\overset{*}{\underset{\sim}{A}} = \dot{\underset{\sim}{A}} - (\underset{\sim}{A}.\underset{\sim}{\nabla})\underset{\sim}{v} + \underset{\sim}{A}(\underset{\sim}{\nabla}.\underset{\sim}{v}) \quad , \quad D_{ij} = v_{(i,j)} \quad .(10)$$

After introducing the conservation of charge in terms of the net concentration in carriers as follows

$$\rho \dot{c} + \lambda_c (\underset{\sim}{\nabla} . \underset{\sim}{\mathcal{J}}) = 0 \quad (11)$$

[with

$$c = (n - p) m_c / \rho \quad , \quad \lambda_c = m_c / q_c \quad , (12)$$

where n , p, m_c and q_c are, respectively, the number of carriers, the number of holes per unit volume in K_t , the mass of a carrier and the individual charge] in the second principle of thermodynamics, we obtain the interesting form of the Clausius-Duhem inequality (4) from which we deduce nonlinear constitutive equations for finitely deformable bodies. For the sake of simplicity only the case of nonmagnetizable solids is considered and we get

$$t_{ij} = \rho [\frac{\partial \tilde{\psi}}{\partial E_{KL}} x_{i,K} x_{j,L} + \frac{\partial \tilde{\psi}}{\partial \mathfrak{E}_K} (\mathfrak{E}_K \delta_{ij} - X_{K,i} x_{j,L} \mathfrak{E}_L)] \quad , (13)$$

$$P_i = - \rho_0 X_{K,i} \frac{\partial \tilde{\psi}}{\partial \mathfrak{E}_K} \quad , \quad \psi = \tilde{\psi}(c, \theta, \underset{\sim}{\mathbb{E}} , \underset{\sim}{\mathfrak{E}}) \quad , (14)$$

where $\mathfrak{E}_K$ is the electric field in the reference configuration and E_{KL} is the Green-Lagrange strain tensor,and

$$(\underset{\sim}{\mathcal{E}} - \lambda_c \underset{\sim}{\nabla} \hat{\mu})_i = J^{-1} x_{i,K} \mathfrak{F}_K (c, \theta , \mathbb{E} , \underset{\sim}{\mathfrak{B}} , \underset{\sim}{\mathfrak{J}}, \underset{\sim}{G}) \quad (15)$$

$$\phi_i = J^{-1} x_{i,K} \varphi_K (c, \theta , \underset{\sim}{\mathbb{E}} , \underset{\sim}{\mathfrak{B}} , \underset{\sim}{\mathfrak{J}}, \underset{\sim}{G}) \quad , (16)$$

which are general statements of Ohm's and Fourier's laws which place in evidence contributions of the net carrier concentration. Here $\mathfrak{B}$, $\underset{\sim}{\mathfrak{J}}$ and $\underset{\sim}{G}$ are the magnetic induction, conduction current and temperature gradient in the reference configuration. Notice that c is one of the arguments in the free energy and that it satisfies a time evolution equation (11) so that c may be considered as a scalar internal variable.(compare [2])

ISOTROPY,INITIAL CONFIGURATION AND SYMMETRY BREAKING

If we consider the case of nonlinear bodies which behave isotropically with respect to the unloaded refernce configuration, we can give representations to all constitutive equations (stress,electric polarization, Ohm's law,heat flux) with the help of representation theorems from the theory of invariants. Although isotropy may appear somewhat restrictive as a material symmetry, we recall that we still are in the framework of a nonlinear theory and that this symmetry is already sufficient to exhibit coupled effects such as the Hall, Ettingshausen, Peltier, Seebeck, Righi-Leduc and Nernst effects as well as elasto- and magneto-resistance [3].Dissipative constitutive equations which are jointly linear in the conduction current and the temperature gradient, while remaining nonlinear in other respects, can also be given.

Most wave propagations take place about an initial configuration K_i which is not as much "ideal" as the refrence configuration K_R. Let us imagine that such a configuration K_i is stationary and a general stationary solution S_o which allows one to pass from K_R to K_i is noted

$$S_o = [\, x_i = \delta_{iK} X_K \,,\, \underset{\sim}{E}_o \,,\, \underset{\sim}{P}_o \,,\, \underset{\sim}{B}_o = \underset{\sim}{H}_o \,,\, \underset{\sim}{J}_o \,,\, {}_oq_f \,,\, \rho_o \,,\, c_o \,,\, \theta_o \,] \qquad .(17)$$

Although the body is free from real deformations in K_i, the electric field $\underset{\sim}{E}_o$ induces internal stresses so that the body of finite extent can be maintained in mechanical equilibrium only if an adequate surface distribution of loads is applied on its boundary. Then we apply a variational method simultaneously to the bulk field equations and the nonlinear constitutive equations with a view to obtaining the bulk equations which govern dynamical perturbations about an initial configuration represented by S_o. A typical feature in this variational procedure which may involve so-called Lagrangian variations (at fixed Lagrangian or material co-ordinates) and Eulerian variations (at fixed Eulerian or current co-ordinates), is that disturbances are shown to take place in a material symmetry which is more involved than the initially selected one because the bias field, which has a vectorial nature,breaks the initial symmetry, so that the initial material coefficients are altered (elasticity coefficients are stiffened) and properties which were not allowed by the natural material symmetry can be allowed by the symmetry induced in the symmetry breaking. Typically, materials which couldn't exhibit piezoelectricity can be made to exhibit such an electromechanical coupling property for the perturbations. Finally, the bulk equations governing perturbations are simplified by neglecting magnetic dragging effects , considering a quasi-electrostatic formulation for electric effects since we are mostly interested in electroacoustics,and considering that the initial electromagnetic fields vary much more slowly over space than the dynamical contributions . This results in a complete set of bulk and surface electroacoustic equations in elastic semiconductors which can be used to study the interaction of acoustic and charge modes which are coupled through piezoelectricity and electric conduction as follows:

. <u>the equations of motion</u>:

$$\rho_o \ddot{\underset{\sim}{u}} = -\, {}_oq_f \underset{\sim}{\nabla}\varphi + C_1 \underset{\sim}{\nabla}(\underset{\sim}{\nabla}.\underset{\sim}{u}) + C_2 \nabla^2 \underset{\sim}{u} + C_3 E_o^2 \underset{\sim}{\nabla}[D(\underset{\sim}{u}.\underset{\sim}{s})] + C_5 E_o^2 D^2 \underset{\sim}{u} \qquad (18)$$

$$+E_o^2 [C_4 \nabla^2(\underset{\sim}{u}.\underset{\sim}{s}) + C_6 D(\underset{\sim}{\nabla}.\underset{\sim}{u}) + C_7 E_o^2 D^2(\underset{\sim}{u}.\underset{\sim}{s})]\underset{\sim}{s} - (f+g - \chi_2 E_o^2) E_o \underset{\sim}{\nabla}(D\varphi)$$

$$-E_o[(f+\chi_1)\nabla^2\varphi + (d+\chi_2)E_o^2 D^2\varphi - ({}_od+\gamma_1)E_o D\bar{c}]\underset{\sim}{s} + (\gamma - {}_od E_o^2)\underset{\sim}{\nabla}\bar{c} + \frac{\rho_o}{\lambda_c} E_o \bar{c}\, \underset{\sim}{s} \; ;$$

. <u>Gauss'equation</u>:

$$\epsilon_1 \nabla^2\varphi + \chi_2 E_o^2 D^2\varphi - (\chi_1 - f)E_o \nabla^2(\underset{\sim}{u}.\underset{\sim}{s}) - (\chi_2 - d)E_o^3 D^2(\underset{\sim}{u}.\underset{\sim}{s}) - {}_od\, E_o D\bar{c}$$

$$-[2\chi_1 + \chi_2 E_o^2 - (f+g)]E_o D(\underset{\sim}{\nabla}.\underset{\sim}{u}) - (\rho_o/\lambda_c)(c_o \underset{\sim}{\nabla}.\underset{\sim}{u} - \bar{c}) = 0 \qquad ;(19)$$

. <u>the equation of conservation of charges</u>:

$$(\rho_o/\lambda_c)\, \dot{\bar{c}} + [\Sigma_{(.)}\underset{\sim}{\nabla}.\underset{\sim}{\nabla}]\varphi - \lambda_c\, {}_o\beta[\Sigma_{(.)}\underset{\sim}{\nabla}.\underset{\sim}{\nabla}]\bar{c} - E_o[\zeta_{(.)}\underset{\sim}{s}.\underset{\sim}{\nabla}]\bar{c} + A_1 E_o D[\Sigma_{(.)}\underset{\sim}{\nabla}.\underset{\sim}{u}]$$

$$+A_1 E_o[\Sigma_{(.)}\underset{\sim}{\nabla}.\underset{\sim}{\nabla}](\underset{\sim}{u}.\underset{\sim}{s}) + A_2 E_o[\Sigma_{(.)}\underset{\sim}{s}.\underset{\sim}{\nabla}](\underset{\sim}{\nabla}.\underset{\sim}{u}) + A_3 E_o^3[\Sigma_{(.)}\underset{\sim}{s}.\underset{\sim}{\nabla}]D(\underset{\sim}{u}.\underset{\sim}{s}) = 0 \qquad .(20)$$

In these equations a superimposed dot denotes partial time differentiation and

$$D = \underset{\sim}{s}.\underset{\sim}{\nabla} \;,\quad \underset{\sim}{s} = \underset{\sim}{E}_o/|\underset{\sim}{E}_o| \;,\quad [T_{(.)}\underset{\sim}{A}.\underset{\sim}{B}] = \sum_{i=1}^{3} T_{(i)}A_iB_i \qquad ,(21)$$

$$\underset{\sim}{\Sigma} = [{}_o\sigma_1 \underset{\sim}{I} + (\xi_1/{}_o\sigma_1)E_o^2 \underset{\sim}{s} \otimes \underset{\sim}{s}]^{-1} \;,\quad \underset{\sim}{\zeta} = (\xi_o/{}_o\sigma_1)\underset{\sim}{\Sigma} \qquad ,(22)$$

$$A_1 = -\left(1 + \frac{{}_o\sigma_2}{2\,{}_o\sigma_1}\right) \;,\quad A_2 = \left(1 - \frac{\xi_3}{{}_o\sigma_1}\right) \;,\quad A_3 = -\frac{1}{{}_o\sigma_1^3}(\xi_1 + \xi_2) \qquad ,(23)$$

where γ_o and γ_1 are related to cross effects between deformation and the concentration in carriers; ${}_od$ is related to cross effects between the electric field and the concentration in carriers ; ${}_o\sigma_1$ and ${}_o\sigma_2$ are resistivity coefficients ; ξ_α are coefficients which account for the dependence of the resistivity on the carrier concentration (ξ_o) and for elasto-resistance (ξ_1,ξ_2,ξ_3).

. <u>stress boundary condition</u>:

$$[(t_1 - \chi_1 E_o^2)\frac{\partial}{\partial n}(\underset{\sim}{u}.\underset{\sim}{n}) + (C_1 - C_2)(\underset{\sim}{\nabla}.\underset{\sim}{u}) + (C_3 - C_4 + f + \chi_1)E_o^2 D(\underset{\sim}{u}.\underset{\sim}{s})]\,\underset{\sim}{n}$$

$$+[(C_4 - f)E_o^2 D(\underset{\sim}{u}.\underset{\sim}{s}) + C_4 E_o^2 \frac{\partial}{\partial n}(\underset{\sim}{u}.\underset{\sim}{s}) - (f + \chi_1)E_o \frac{\partial \varphi}{\partial n} - E_o\bar{\omega}\,]\underset{\sim}{s} + C_2 \frac{\partial}{\partial n}\underset{\sim}{u}$$

$$+[(g - 2\chi_1 - \chi_2 E_o^2)E_o D\varphi + \gamma\,\bar{c}\,]\,\underset{\sim}{n} + (\mu + \chi_1 E_o^2)\underset{\sim}{\nabla}(\underset{\sim}{u}.\underset{\sim}{n}) = \underset{\sim}{0} \qquad ;(24)$$

. <u>electric boundary conditions</u>

$$-[\![\frac{\partial \varphi}{\partial n}]\!] + \chi_1 \frac{\partial \varphi}{\partial n} + fE_o D(\underset{\sim}{u}.\underset{\sim}{n}) + (f - \chi_1)E_o \frac{\partial}{\partial n}(\underset{\sim}{u}.\underset{\sim}{s}) = \bar{\omega} \qquad ,(25)$$

$$[\![\varphi]\!] = 0 \qquad ,(26)$$

$$[\Sigma_{(.)}\underset{\sim}{n}.\underset{\sim}{\nabla}]\varphi + {}_o\sigma_1^{-1}E_o D(\underset{\sim}{u}.\underset{\sim}{n}) - A_1E_o D[\Sigma_{(.)}(\underset{\sim}{u}.\underset{\sim}{n})] + \frac{\partial \bar{\omega}}{\partial t} = 0 \qquad ,(27)$$

where $\partial/\partial n = \underset{\sim}{n}.\underset{\sim}{\nabla}$, $\bar{\omega}$ is the time-varying surface charge density, t_1 is an initial isotropic stress, f and g are electromechanical coupling coefficients, C_α are effective elasticity coefficients, χ_1 and χ_2 are isotropic and anisotropic susceptibility coefficients and μ is the classical Lamé modulus such that $C_2(E_o=0)=\mu$.

COUPLED ELECTROACOUSTIC WAVES

We shall study first the free modes of oscillation for the bulk system (18)through (20), that is, plane harmonic wave solutions in the form

$$\hat{S}^p_{k,\omega} =: (\underset{\sim}{u}, \varphi, \bar{c}) = (\hat{\underset{\sim}{u}}, \hat{\varphi}, \hat{c})\exp[i(\omega t - \underset{\sim}{k}.\underset{\sim}{r})] \qquad ,(28)$$

where ω is a complex angular frequency, $\underset{\sim}{k}$ is the wave vector and $(\hat{\underset{\sim}{u}}, \hat{\varphi}, \hat{c})$ is the set of amplitudes. For a <u>longitudinal</u> setting of the bias electric field

$$\underset{\sim}{k} = k\,\underset{\sim}{s} \qquad ,\quad \hat{\underset{\sim}{u}}.\underset{\sim}{s} = \hat{u}_{//} \qquad ,(29)$$

the transverse elastic modes travel independently of charges,[cf.eqn.(3.11) in Ref.4] while the longitudinal elastic mode is coupled with the charge mode as shown by equation (3.21) in Ref.4.After decomposition of the dispersion relation into real and imaginary parts on account of $\omega = \Omega + i\Gamma$, $k = k_1 + ik_2$, with $\Gamma^2 \ll \Omega^2$ and $k_2^2 \ll k_1^2$, and using perturbation theory, we are left with the results of the form

$$\Omega^2 \simeq \Omega_o^2 + \bar{\Omega}_L^2\,[1 - \mathcal{F}(k_1)] + \nu\,k_2 \qquad (30)$$

and

$$\Gamma \simeq \bar{\Omega}_L (k_2/k_1) - \frac{k_1}{2\bar{\Omega}_L} [\, \xi(k_1) + \nu \,] \qquad (31)$$

with

$$\bar{\Omega}_L^2 = \bar{c}_L^2 k_1^2 \quad , \quad \bar{c}_L^2 = \tilde{c}_L^2 (1+K^2) \quad , \quad \tilde{c}_L^2 = c_L^2 \left[1 + \frac{\bar{t} - 4\bar{\beta} E_o}{\rho_o c_L^2} \right] \quad , \qquad (32)$$

$$K^2 = \frac{(\bar{\beta} - \bar{\chi} E_o)(\bar{\beta} + \chi_1 E_o)}{\rho_o \bar{\epsilon} c_L^2} \quad , \bar{\beta} = (g+2f)E_o + dE_o^3 \quad , \quad c_L^2 = \bar{C}/\rho_o \quad , \qquad (33)$$

where we have set

$$\Omega_o^2 = q_o^2 / \rho_o \bar{\epsilon} \quad , \quad \nu = \frac{\chi_1 + \chi_2}{\rho_o \bar{\epsilon}} E_o q_o \quad ; \qquad (34)$$

see eqn.(3.29) in Ref.4 for the expressions of $\mathcal{F}$ and ξ (cf.[5]).

For an <u>orthogonal</u> setting of the bias electric field

$$\underset{\sim}{k} . \underset{\sim}{s} = \underset{\sim}{k} . \underset{\sim}{E}_o = 0 \qquad . (35)$$

The transverse elastic component polarized orthogonally to the plane spanned by the bias field and the propagation direction whose speed is not altered by the bias field, is uncoupled while all the other field components remain coupled .However, this involved coupling can be characterized by two small parameters arising from (i) cross effects between deformation, electric field and concentration in carriers and (ii)the effect of the ponderomotive force in the presence of an initial charge density, respectively, two other small parameters of a more classical nature being provided by electromechanical couplings and conduction. When the first two of these are vanishingly small, the longitudinal elastic component travels independently of others without attenuation while the second transverse elastic component remains coupled to the electrostatic potential and the charge density. This last mode whose speed is altered by the electromechanical coupling, is rendered slightly dispersive as a result of conduction and the dispersion and attenuation relations are as follows :

$$\Omega_1^2(q) = \bar{\Omega}_{TF}^2 \left[1 - \frac{1+q^2}{T_d^2} \frac{(1-\tilde{\delta}_T)}{(1+q^2)^2 T_d^{-2} + \bar{\Omega}_{TF}^2} \right] \quad , \quad \tilde{\delta}_T = \tilde{\Omega}_{TF}^2 / \bar{\Omega}_{TF}^2 \quad , (36)$$

and

$$\Omega_2(q) = \frac{\bar{\Omega}_{TF}^2}{2\, T_d} \frac{(1-\tilde{\delta}_T)}{(1+q^2)^2 T_d^{-2} + \bar{\Omega}_{TF}^2} \quad , \quad T_d = \tau_d \omega_o \quad , \quad \tau_d = \epsilon_1 / \Sigma_{(1)} \quad , \qquad (37)$$

with

$$\bar{\Omega}_{TF}^2 = \bar{\omega}_{TF}^2 / \omega_o^2 \quad , \quad \bar{\omega}_{TF}^2 = \omega_{TF}^2 (1+K^2) \quad , \quad \omega_{TF}^2 = c_{TF}^2 k^2 \quad , \quad c_{TF}^2 = \frac{C_2 + C_4 E_o^2}{\rho_o} \quad ; \qquad (38)$$

$$K^2 = \frac{(f+\chi_1)(f-\chi_1)}{\rho_o \epsilon_1 c_{TF}^2} E_o^2 \qquad (\omega_o \text{ is a characteristic frequency}) \quad , \qquad (39)$$

$$\tilde{\Omega}_{TF}^2 = \tilde{\omega}_{TF}^2 / \omega_o^2 \quad , \tilde{\omega}_{TF}^2 = \omega_{TF}^2 (1+\alpha_T^2) \quad , \quad \alpha_T^2 = \frac{v_D^2}{c_{TF}^2} \quad , \quad v_D^2 = - \frac{A_1 (f+1+2\chi_1) + f - \chi_1}{\rho_o} E_o^2 . \qquad (40)$$

Having obtained these two preliminary results one can examine , by means of a naïve perturbation procedure, the effects of the first two small parameters mentioned. We see that the longitudinal elastic component remains coupled with the transverse component, the electric potential and the concentration. This will result in a slightly dispersive and attenuated longitudinal mode [4]. In all, the various analytical and asymptotical behaviors obtained for the bulk mode provide precious hints for examining surface wave propagation which, by essence, should be even more complex. Indeed, under the assumption that the semiconductor is initially practically neutral and certain second order effects can be discarded, it is shown that the case corresponding to a transverse bias field set parallel to the limiting plane of the substrate, hence orthogonally to the sagittal plane, splits in two surface wave problems, one of the classical Rayleigh type (elastic component polarized parellely to the sagittal plane) but with mechanical boundary conditions slightly perturbed, and a problem of the Bleustein-Gulyaev type corresponding to horizontally polarized shear waves coupled with the electric potential and charges,which we study in detail only for small conductions. After many approximations and numerical estimates we get for this mode

$$\omega_1^2 \simeq \bar{\omega}_{TF}^2 \left[1 - \frac{h(q_1^2)}{T_d^2} - \alpha K^4\right] \quad , \quad h(q_1^2) = O(1) \tag{41}$$

and

$$\omega_2 \simeq (\Sigma_{(1)}/2\epsilon_1) = \frac{1}{2\tau_d} \quad , \alpha = [(1+\epsilon_1)(1+K^2)]^{-2} \tag{42}$$

and its penetration depth may be approximated by

$$p(\lambda) = \frac{(1+\epsilon_1)(1+K^2)}{2\pi(K^2-T)}\lambda \qquad \text{(compare [6] for } \Sigma_{(1)}=0) \quad ,(43)$$

where λ is the wavelength and $T \ll K^2$. The above-given equations show that small conductions (i) have practically no effects on the penetration depth ,(ii) cause an additional dispersion of the order of ϵ^2 if electromechanical couplings cause a dispersion of order ϵ and (iii) cause an attenuation which, in practice may be considered as proportional to the so-called dielectric relaxation time . In all events, this justifies the neglect of conduction in practical applications insofar as the first two effects are concerned.

REFERENCES

[1] Maugin G.A.,The method of virtual power in continuum mechanics:Application to coupled fields, Acta Mechanica,35(1980) 1-70.

[2] Maugin G.A.,The notion of internal variables in fluid mechanics,Theoretical and Applied Mechanics,pp.671-676(Publ.Bulgarian Acad.Sci.,Sofia, 1981).

[3] Kiréev P., La physique des semiconducteurs (Editions MIR,Moscow, 1975).

[4] Daher N.,Ondes électroacoustiques dans les semiconducteurs piézoélectriques, (Thèse de 3ème Cycle,Université de Paris VI,Paris,1983).

[5] Reik H.G.,Basic Theory of Semiconductor devices, Theory of imperfect crystalline structures (Trieste lectures),pp.409-478(I.A.E.A.Publications,Vienna,1970).

[6] Dieulesaint E. and Royer D.,Ondes élastiques dans les solides,Application au traitement du signal (Masson, Paris, 1974).

The Mechanical Behavior of Electromagnetic Solid Continua
G.A. Maugin (editor)
Elsevier Science Publishers B.V. (North-Holland)

NONLINEAR WAVE PROPAGATION IN ELECTROELASTIC SEMICONDUCTORS

Matthew F. McCarthy

National University of Ireland
University College
Galway, Ireland

A one-dimensional general nonlinear theory of electroelastic semiconductors is used to study the propagation of finite amplitude waves. The propagation and evolutionary behavior of both shock waves and acceleration waves is examined. Modulated simple wave theory is employed to study the propagation of high frequency finite amplitude pulses. The behavior of small amplitude high frequency pulses is examined in detail and the influence of the biasing electric field on the evolutionary behavior of such pulses is discussed in depth.

INTRODUCTION

In this paper the general nonlinear theory of electroelastic semiconductors presented earlier[1] is specialized and the one-dimensional forms of these equations previously studied[2,3] are used to discuss the propagation of finite amplitude waves. Here we show how our earlier results on acceleration waves[2] may be generalized by applying concepts of modulated simple wave theory[4-7] to the study of finite amplitude, one-dimensional pulses in electroelastic semiconductors.

We begin by summarizing the equations which govern the one-dimensional motions of electroelastic semiconductors. The properties of both shock and acceleration waves are examined and, in particular, it is shown that if the dissipative effects of conduction or semiconduction are negligible then the amplitude of an acceleration wave behaves locally as it would if propagating in an elastic dielectric. The properties of such simple waves are then examined in detail and their similarity to acceleration waves is noted. We then apply the general theory of modulated simple waves to the study of finite amplitude pulse propagation. The nonlinear equations which govern the propagation of such pulses are derived and these serve as a basis for the study of small amplitude finite rate pulses. In particular, we show that in such pulses the strain at any point of the pulse may grow or decay depending on the strength of the electric field which acts in the material ahead of the pulse. The behavior of the acceleration in such pulses is also examined in detail and the critical role played by the applied electric field in determining this behavior as well as the possible formation of acoustoelectric domains is discussed.

BASIC EQUATIONS FOR ONE-DIMENSIONAL MOTIONS

In the general nonlinear theory of nonlinear electroelastic semiconductors[1], the one-dimensional, i.e., purely longitudinal, motion of a body may be described by the mapping $y = y(X,t)$ which gives the position y at time t of the lattice point whose reference coordinate is X. In the absence of extrinsic body forces, the motion of the body is governed by the balance laws[2]

$$\rho_o^{-1}\partial_X\Sigma = \alpha_1\partial_X + \alpha_2\partial_X\mathcal{E} + \alpha_3 g = \dot{v}, \quad \partial_X v = \dot{\lambda},$$

$$\partial_X D = \beta_1\partial_X\lambda + \beta_2\partial_X\mathcal{E} = uF + \mu^r(X), \quad \mathcal{E} = FE = \partial_X\Phi,$$

$$\partial_X\mathcal{J} = \gamma_1\partial_X\lambda + \gamma_2\partial_X\mathcal{E} + \gamma_3 g + \gamma_4\partial_X g = u\dot{\lambda} + \dot{u}F, \tag{1}$$

where

$$\lambda = F - 1, \ F = \partial_X y, \ u = \mu^e, \ g = \partial_X u,$$

$$\alpha_1 = \rho_o^{-1}\partial_F\hat{\Sigma}, \ \alpha_2 = \rho_o^{-1}\partial_{\mathcal{E}}\hat{\Sigma}, \ \alpha_3 = \rho_o^{-1}\partial_{\mu^e}\hat{\Sigma}$$

$$\beta_1 = \partial_F\hat{D}, \quad \beta_2 = \partial_{\mathcal{E}}\hat{D},$$

$$\gamma_1 = \partial_F\hat{\mathcal{J}}, \ \gamma_2 = \partial_{\mathcal{E}}\hat{\mathcal{J}}, \ \gamma_3 = \partial_{\mu^e}\hat{\mathcal{J}}, \ \gamma_4 = \partial_g\hat{\mathcal{J}}, \tag{2}$$

with

$$\Sigma = \tau + T - p^e = \hat{\Sigma}(F, \mathcal{E}, \mu^e), \ D = \hat{D}(F, \mathcal{E}), \ \mathcal{J} = \hat{\mathcal{J}}(F, \mathcal{E}, \mu^e, g),$$

$$\tau = \hat{\tau}(F, \mathcal{E}), \ T = 1/2\ \epsilon_o F^{-2}\mathcal{E}^2, \ p^e = \hat{p}^e(\mu^e). \tag{3}$$

In these equations, τ and T represent the usual mechanical and free space Maxwell stresses, respectively; v, E, D and $\mathcal{J}$ represent the velocity of the solid lattice, electric field, electric displacement, electric current, respectively; and μ^e, p^e, ρ_0, Φ, μ^e, μ^r represent the free electronic charge density, the free electronic pressure, mass density of the lattice in the reference configuration, electrical potential, free electronic charge density and residual charge density, respectively, and ϵ_0 is the permittivity of free space. We assume that the response functions $\hat{\Sigma}$ and $\hat{D}$ are of class C^2, while $\mathcal{J}$ is assumed to be a C^1 function. We further assume that the coefficients α_1, β_2 and γ_2 are positive, while the coefficients α_2, β_1, α_3 and γ_4 are assumed to be nonvanishing for all values of their arguments.

If the motion contains a nonmaterial surface of discontinuity whose position at time t is the reference configuration is $X = \tilde{X}(t)$, then Eqs.(1) must be replaced by the relations

$$[\Sigma] = -\rho_o c[v], \quad [v] = -c[\lambda], \ [D] = 0, \ [\Phi] = 0, \ [\mathcal{J}] = c[uF] \tag{4}$$

where $c = c(t) = d\tilde{X}(t)/dt > 0$ is the intrinsic velocity of the surface and $[\psi] = [\psi](t) = \psi^-(t) - \psi^+(t)$, where ψ^+ and ψ^- are the limiting values of $\psi(X,t)$, immediately ahead of and just behind the surface, respectively.

SHOCK WAVES, ACCELERATION WAVES AND SIMPLE WAVES

A motion $y = y(X,t)$ is said to contain a shock wave $X = \tilde{X}(t)$ if $y(.,.)$ is continuous everywhere while F and v and all higher derivatives of these variables suffer jump discontinuities across the propagating surface, but are continuous functions of X and t everywhere else. The continuity properties of the remaining variables may be deduced from (4) and will not be discussed here. The first two of Eqs.(4) imply that the intrinsic velocity of the shock wave is given by the well-known relation

$$c^2 = [\Sigma]/\rho_o[\lambda], \tag{5}$$

so that c is determined by the properties of the material, the conditions prevailing at the wavefront and the strength $[\lambda]$ of the shock wave. In order to study the behavior of the amplitude of the shock wave the kinematical condition of compatibility

$$\frac{\delta}{\delta t}[f] = [\dot{f}] + c[\partial_x f] \tag{6}$$

must be used. The resulting equation for $[\lambda]$ is of the form

$$\frac{\delta}{\delta t}[\lambda] = A\{B - [\partial_x \lambda]\}. \tag{7}$$

Detailed expressions for A and B are given in Ref.3 where the implications of (7) are studied in detail as are the properties of the shock transition. In particular, it is to be noted that, in general, the amplitude of the wave will vary as it traverses the material.

A propagating surface across which $y(X,t)$ and its first derivatives are continuous but at which the second derivatives suffer a jump discontinuity is called an acceleration wave. It has been shown[2] that

$$c^2 = \zeta_1 = \alpha_1 - \alpha_2\beta_1/\beta_2, \quad \zeta_1 = \rho_o^{-1}\partial_F\tilde{\Sigma}(F,D,\mu^e) \tag{8}$$

and, furthermore, if conditions are both uniform and steady ahead of the wave then the amplitude $a(t) = [\dot{v}]$ is given by the formula

$$a(t)/a_c = \theta\,\exp[-\bar{\xi}_o(X-X_1)]/\{1-\theta(1-\exp[-\bar{\xi}_o(X-X_1)])\} \tag{9}$$

where $\theta = a(0)/a_c$, $a_c = \bar{\xi}_o c_o/\beta_o$ is the critical jump in acceleration, $X = c_o t$, $a(0)$ is the value of $a(t)$ when $t = t_1$,

$$\bar{\xi}_o = \{(\alpha_2/\beta_2 + \alpha_3 c_o/\gamma_4)\mu^e + \alpha_3(\gamma_2\beta_1/\beta_2 - \gamma_1)/\gamma_4\}/2c_o^2$$

$$2c_o^3\beta_o = -\zeta_{11} = -\rho_o^{-1}\partial_F^2\tilde{\Sigma}. \tag{10}$$

It is to be noted that, whenever a subscript 0 is attached to a coefficient, this quantity is to be evaluated at the constant values F_o, $\mathcal{D}_o$ and μ_o^e prevailing at the wavefront. It has been observed elsewhere[2] that the sign of $\bar{\xi}_o$ may be positive or negative depending on the condition which prevails ahead of the wave. The expression suggests the definitions $\tau_r = (\bar{\xi}_o c_o)^{-1}$, $L_a = c_o\tau_r$ for the "relaxation time" and "attenuation length" for one-dimensional acceleration waves propagating into a uniform state. We note in particular that in the limit when $|X - X_1| \ll L_a$, so that the magnitude of the attenuation length is very much greater than the distance travelled by the wave, (9) reduces to

$$a(X) = a(0)\{1 + a(0)\zeta_{11}(X - X_1)/2c_o^4\}. \tag{11}$$

The limiting situation which arises when the wavelengths associated with dynamical disturbances are small compared to the magnitude of the attenuation length L_a corresponds to the influence of conduction and semiconduction effects being negligible. The material now behaves locally as an elastic dielectric and its motion is governed by the equations

$$\alpha_1\partial_x\lambda + \alpha_2\partial_x\mathcal{D} = \dot{v}, \quad \partial_x v = \dot{\lambda}, \quad \beta_1\partial_x\lambda + \beta_2\partial_x\mathcal{D} = 0. \tag{12}$$

These equations admit simple wave solutions $\lambda = \lambda(\alpha)$, $v = v(\alpha)$, $\mathcal{D} = \mathcal{D}(\alpha)$ in which the speed of propagation of the wavelet α is determined by (8). Once $c(\alpha)$ has been found, the structure of the wave is determined by the equations

$$v'(\alpha) = -c(\alpha)\lambda'(\alpha), \quad \mathcal{E}'(\alpha) = -(\beta_1/\beta_2)\lambda'(\alpha), \tag{13}$$

where the notation $f'(\alpha) = df(\alpha)/d\alpha$ has been used.

Suppose that λ is prescribed as a function of t at some point X_1. Then $\lambda(X_1,t) = f(t/\tau_p)$, where f will be taken to represent a pulse if $f(y) = 0$ whenever $y<0$ or $y>1$. If $\alpha(X,t)$ is chosen so that $\alpha(X_1,t) = t/\tau_p$ then

$$X - X_1 = c(\alpha)(t - \tau_p\alpha). \tag{14}$$

Formulae for $v(\alpha)$ and $\mathcal{E}(\alpha)$ now follow from (13), and the acceleration on the wavelet α is given by

$$a(X,t) = v'(\alpha)/\{\tau_p + v'(\alpha)\zeta_{11}(\alpha)(x - X_1)/2c(\alpha)^4\}. \tag{15}$$

The similarity of Eqs.(11) and (15) is clear. In particular, we note that (15) holds throughout the wave and reduces to (11) at the wavefront $\alpha = 0$ when $v'(0)$ is replaced by $\tau_p a(0)$.

MODULATED SIMPLE WAVES

We seek to describe the propagation of pulses into the region $X>0$ of an electroelastic conductor, which is initially in a uniform steady state, in terms of the variation in the strain as the pulse passes the point $X=0$. Thus, we seek to solve Eqs.(1) subject to the initial conditions $v=0$, $\lambda=0$, $\mathcal{E}=\mathcal{E}_0$, $u=u_0$ for $t=0$, $X>0$ and the boundary condition

$$\lambda(0,t) = k(t/\tau_p), \quad 0 \le t \le \tau_p. \tag{16}$$

In the limit when $\omega = |\tau_r/\tau_p| >> 1$, Eq.(16) describes a short duration or high frequency pulse. We have seen earlier that when $\xi_0 = 0$ a pulse propagates as a simple wave and this prompts us to describe the propagation of such a pulse by modulated simple wave theory[4-6] when conduction and semiconduction may not be neglected. The simple wave is now modulated by the dissipative mechanisms present in the material and is most conveniently described in terms of the characteristic variable $\alpha(X,t)$ and X. If $t = T(X,\alpha)$ denotes the arrival time of the wavelet α at X and if we write $g(X,t) = \bar{g}(X,\alpha)$

$$\dot{g} = \ell^{-1}\partial_\alpha\bar{g}, \quad \partial_X g = \partial_X\bar{g} - s\ell^{-1}\partial_\alpha\bar{g}, \quad \ell = \partial_\alpha T, \quad s = \partial_X T, \tag{17}$$

where $\ell = \bar{\ell}(X,\alpha)$ is the incremental arrival time and $s = c^{-1} = \bar{s}(X,\alpha)$, the slowness of the wavelet α. It follows from (17) that

$$\partial_X\bar{\ell} = \partial_\alpha\bar{s}. \tag{18}$$

If $\alpha = t/\tau_p$ at $X=0$ then Eqs.(16) and (17) together imply that

$$\bar{\lambda}(0,\alpha) = k(\alpha), \quad \bar{\ell}(0,\alpha) = \tau_p, \quad 0 < \alpha < 1. \tag{19}$$

Equations (1) now assume the forms

$$\partial_\alpha\bar{v} + s(\alpha_1\partial_\alpha\bar{\lambda} + \alpha_2\partial_\alpha\bar{\mathcal{E}} + \alpha_3\partial_\alpha\bar{u}) = \rho_0^{-1}\ell\partial_\alpha\bar{\Sigma}, \quad s\partial_\alpha\bar{v} + \partial_\alpha\bar{v} + \partial_\alpha\bar{\lambda} = \ell\partial_X\bar{v},$$
$$s(\beta_1\partial_\alpha\bar{\lambda} + \beta_2\partial_\alpha\bar{\mathcal{E}}) = \ell(\partial_X\bar{D} - \bar{u}\bar{F} - \mu^r), \quad s\partial_\alpha\Phi = \ell(\mathcal{E} + \partial_X\bar{\Phi}),$$
$$(s\gamma_1 - u)\partial_\alpha\partial_\alpha\bar{\mathcal{E}} + (s\gamma_3 - F)\partial_\alpha\bar{u} + s\gamma_4\partial_\alpha\bar{g} = \ell\partial_X\bar{g}, \tag{20}$$

and it follows from the definitions of v and g that

$$\partial_\alpha \bar{y} = \ell \bar{v}\,, \quad \partial_\alpha \bar{u} = s^{-1} \ell(\partial_X \bar{u} - \bar{g})\,. \tag{21}$$

Equations (18), (20) and (21) describe the motion for any choice of $s = s(X,\alpha)$. When the body is in a steady state the right-hand sides of (20) and (21) vanish identically. Thus, Eqs.(20) and (21) furnish us a set of homogeneous linear equations for $\partial_\alpha \bar{v}$, $\partial_\alpha \lambda$, $\partial_\alpha \mathcal{E}$ and $\partial_\alpha \bar{g}$ at $\alpha = 0$ where $\partial_\alpha \Phi = \partial_\alpha \bar{y} = \partial_\alpha u = 0$. The first three of the homogeneous linear equations which follow from Eqs.(20) have the same form as those which apply throughout the simple wave described earlier. Obviously, any changes in the right-hand sides of (20) and (21) at any point X are induced by the pulse. The modelling of the pulse by a simple wave at points "very close" to the front of the pulse should yield a reasonably accurate description of what is actually happening.

Let us choose $s = c^{-1}$ where c is given by (8). When the second, third and fifth of Eqs.(20) together with the second equation of (21) are used in the first member of (20), we arrive at the nonlinear transport equation

$$\rho_o^{-1} \partial_X \bar{\Sigma} - s^{-1} \partial_X \bar{v} - \alpha_2 \beta_2^{-1} (\partial_X \bar{D} - \bar{u}F - \mu^r) - \alpha_3 \bar{u} - g) = 0\,. \tag{22}$$

The motion of the body is now described by the last four of Eqs.(20) together with Eqs.(21) and (22). Once these equations have been solved, $y'(x,\alpha)$, $\mu^r = u(x,\alpha)$ follow from (21) and $\Phi(X,\alpha)$ follows from $(20)_4$. The function $T(X,\alpha)$ may be obtained by integrating $(17)_4$ and thus all the functions needed to describe the motion of the electroelastic material may be expressed in terms of X and t.

A detailed study of the implications of the foregoing equations for pulses of finite amplitude is beyond the scope of this work and will be presented elsewhere[8].

HIGH FREQUENCY SMALL AMPLITUDE PULSES

Here we extend our earlier results on acceleration waves[2] to cover the propagation of small amplitude finite rate pulses. When dimensionless variables are introduced through the definitions

$$X = L_a X^*,\ y = L_a y^*,\ t = \tau_r t^*,\ \Sigma = \rho_o \bar{c}_o^2 \Sigma^*\,,\ D = -\mu_o^e \bar{c}_o \tau_r D^*,\ u = -\mu_o^e u^*\,,$$
$$\bar{\mu} = \bar{\mu}_o^e \bar{\mu}^*,\ \mathcal{E} = E_o \mathcal{E}^*,\ \Phi = E_o L_a \Phi^*,\ g = -\bar{\mu}_o^e \bar{c}_o g^* \tag{23}$$

and the asterisk is dropped in the resulting dimensionless forms of Eqs.(17), (20) and (21); we find that, with the exception of $(21)_1$, the motion is still governed by these equations which must be solved subject to the initial conditions $\lambda(X,0) = 0$, $\mathcal{E}(X,0) = \mathcal{E}_0$, $\mu^e(X,0) = u_0$ and the boundary conditions

$$\lambda(0,t) = k(\omega t)\,, \quad 0 \le \omega t \le 1\,, \tag{24}$$

where $\omega = L_a/c_0\tau_p$. The dimensionless form of $(21)_1$ follows on setting $\rho_0 = 1$ in the original equation. In the definitions (23) L_a, τ_r and c_0 have the values appropriate to a one-dimensional acceleration wave which is propagating into a region which initially underformed, subject to a uniform field E_0 and in which the mass density and density of free electronic charge have the constant values ρ_0 and $\bar{\mu}_0^e (<0)$, respectively.

We write $k(\xi) = \omega\epsilon f(\xi)$ where $\epsilon > 0$ and $\max|f(\xi)| = 1$, $\xi \mathcal{E}[0,1]$. We assume that $\omega \gg 1$, while $a = \omega^2\epsilon$ is finite so that we are dealing with small amplitude high frequency finite acceleration pulses for which

$$\lambda(0,t) = \frac{a}{\omega} f(\omega,t)\,, \quad 0 \le \omega t < 1\,. \tag{25}$$

Next, we assume that conditions are steady and uniform ahead of the pulse and seek solutions which satisfy the governing equations in the limit as $\omega \to \infty$. We content ourselves with solutions of the form

$$\bar{h} = \bar{h}(X,\alpha) = h_o + \frac{1}{\omega} h^{(1)}(X,\alpha) + \frac{1}{\omega^2} h^{(2)}(X,\alpha) , \tag{26}$$

where $\bar{h}$ represents any one of the variables, $\bar{\lambda}$, $\bar{v}$, $\bar{\mathscr{E}}$, $\bar{u}$, $\bar{g}$, $\bar{\ell}$, $\bar{c}$, $\bar{s}$, $\bar{\Sigma}$, $\bar{g}$, $\bar{D}$ or $\bar{\Phi}$. It follows from (20) and (26) that

$$v^{(1)} = -c_o\lambda^{(1)}, \quad \mathscr{E}^{(1)} = \delta_1\lambda^{(1)}, \quad g^{(1)} = \pi_1\lambda^{(1)}, \quad g^{(1)} = c_o u_o \lambda^{(1)} , \tag{27}$$

where $\delta_1 = -\beta_1^{(o)}/\beta_2^{(o)}$, $\pi_1 = -\{\gamma_1^{(o)} - \gamma_2^{(o)}\beta_1^{(o)}/\beta_2^{(o)} - c_o u_o\}\gamma_4^{(o)}$. Substitution from (26) and (27) into (22) leads to an ordinary linear differential equation for $\lambda^{(1)}(x,\alpha)$ whose solution is

$$\lambda^{(1)}(X,\alpha) = af(\alpha)\exp(-\bar{\xi}_o X) . \tag{28}$$

Since $s = 1/c$, we have

$$s = \frac{1}{c_o} + \frac{a\beta_o}{\omega} f(\alpha)\exp(-\bar{\xi}_o X) , \tag{29}$$

and it follows from (18) and $(17)_4$ that

$$\ell^{(1)} = 1 + a\beta_o f'(\alpha)\{1 - \exp(-\bar{\xi}_o X)\}/\bar{\xi}_o , \tag{30}$$

$$\omega(T - X/c_o) = \alpha + a\beta_o f(\alpha)\{1 - \exp(-\bar{\xi}_o X)\}/\bar{\xi}_o . \tag{31}$$

When (18), $(27)_1$, (28) and (31) are used, the acceleration at the wavelet α follows as

$$a(X,\alpha) = -ac_o f'(\alpha)\exp(-\bar{\xi}_o X)/\{1 + a\beta_o f'(\alpha)[1 - \exp(-\bar{\xi}_o X)]/\bar{\xi}_o\}, \tag{32}$$

and it is to be emphasized that (32) holds throughout the pulse. Equations (28) through (32) have the same form as the corresponding results which arise in the study of high frequency pulse propagation in a viscoelastic material[7]. We now show that the similarity between our results and those given in Ref.7 is purely superficial.

Consider an electroelastic semiconductor for which

$$\mathscr{J} = (\lambda + 1)\{-um\mathscr{E} + D^c g\}, \tag{33}$$

where m and D^c are the mobility and diffusivity, respectively. It has been shown[3] that

$$\bar{\xi}_o = -u_o(\mathscr{E}_o - \mathscr{E}_T)/2c_o^2 , \tag{34}$$

where $\mathscr{E}_T = -c_o/m$ is the threshold field introduced in Ref.9. Since $u_o > 0$, it is clear that $\bar{\xi}_o$ and $(\mathscr{E}_o - \mathscr{E}_T)$ have the same sign. Thus, the possibility of $\bar{\xi}_o < 0$ exists and this is a situation which has no mechanical analogue. If the electroelastic response of the material is linear, i.e., $\bar{\xi}_o = 0$, then all the wavelets will have the same speed of propagation and the strain, velocity acceleration, electric field, stress and conduction current grow or decay as the pulse traverses the material accordingly as $\bar{\xi}_o < 0$ or $\bar{\xi}_o > 0$. The growth which occurs when $\bar{\xi}_o < 0$ is usually associated with the formation of acoustoelectric domains. Finally, when $\beta_o \neq 0$, the strain will grow or decay depending on the sign of $\bar{\xi}_o$. However, the behavior of the acceleration is much more complicated. When $\bar{\xi}_o > 0$, the behavior of the acceleration will be the same as it would be in a viscoelastic material. In particular, if $\text{sgn}(a(0,\alpha)) = \text{sgn}\,\beta_0$, where $a(0,\alpha) = -ac_0 f'(\alpha)$ and $|a(0,\alpha)| > |a_c|$, then the acceleration will become unbounded when the wavelet α has reached the point

$$X = -\bar{\xi}_o^{-1} \ln\{1 - a_c/a(0,\alpha)\}, \tag{35}$$

and this is usually taken to herald the formation of a shock or acoustoelectric domain. If $|a(0,\alpha)| < a_c$, the acceleration will ultimately decay to zero, while if $|a(0,\alpha)| = a_c$ the acceleration at the wavelet α will remain constant as the pulse propagates. On the other hand the case $\bar{\xi}_0 < 0$ has no mechanical analogue. In this case a detailed examination of (32) shows that either $a(X,\alpha) \to a_c$ or else the acceleration becomes unbounded when the wavelet α has reached the point

$$X = |\bar{\xi}_o|^{-1} \ln\{1 + |a_c/a(0,\alpha)|\}. \tag{36}$$

In particular, the acceleration never decays to zero in this case.

REFERENCES

1. H.G. deLorenzi and H.F. Tiersten, "On the Interaction of the Electromagnetic Field with Heat Conducting Deformable Semiconductors," J. Math. Phys., 16, 938 (1975).

2. M.F. McCarthy and H.F. Tiersten, "One-Dimensional Acceleration Waves and Acoustoelectric Domains in Piezoelectric Semiconductors," J. Appl. Phys., 47, 3389 (1976).

3. M.F. McCarthy and H.F. Tiersten, "Shock Waves and Acoustoelectric Domains in Piezoelectric Semiconductors," J. Appl. Phys., 48, 159 (1977).

4. B.R. Seymour and M.P. Mortell, "Nonlinear Geometrical Acoustics," Mechanics Today, 2, 251, Pergamon, Oxford, 1975.

5. D.F. Parker and B.R. Seymour, "Finite Amplitude One-Dimensional Pulses in an Inhomogeneous Granular Material," Arch. Rational Mech. Anal., 72, 265 (1980).

6. E. Varley and T.G. Rogers, "The Propagation of High Frequency Finite Acceleration Pulses and Shocks in Viscoelastic Materials," Peox. Eoy. Aox. A, 296, 498 (1967).

7. B.R. Seymour and E. Varley, "High Frequency, Periodic Disturbances in Dissipative Systems. 1 - Small Amplitude Finite Rate Theory," Proc. Roy. Soc. A, 314, 387 (1970).

8. M.F. McCarthy, "One-Dimensional, Finite Amplitude Pulse Propagation in Electroelastic Semiconductors," forthcoming.

9. A.R. Hutson and D.L. White, "Elastic Wave Propagation in Piezoelectric Semiconductors," J. Appl. Phys., 33, 40 (1962).

The Mechanical Behavior of Electromagnetic Solid Continua
G.A. Maugin (editor)
Elsevier Science Publishers B.V. (North-Holland)

ELECTROTHERMODIFFUSION OF ELECTRONS IN THERMOELASTIC SEMICONDUCTORS

Bogdan Maruszewski

Institute of Technical Mechanics
Technical University of Poznań
ul. Piotrowo 3, 60-965 Poznań
Poland

The coupled electrothermodiffusion theory of charge carriers towards the extrinsic thermoelastic semiconductors is considered. The investigations are based upon the irreversible thermodynamics.

INTRODUCTION

Semiconductors in a macroscopic case appear to be materials with special electrical conductivity; this value causes that the conduction current is comparable with the displacement current. It results from the empirical observations that impurties have the main influence on the conductivity of semiconducting crystals. On the other hand it appears that the electrical current in a semiconductor derives not only from a charge cloud of free electrons but also from the diffusion of charge carriers. Hence the diffusion phenomena in semiconductors can be investigated either as the diffusion of impurities or as the diffusion of charge carriers. In a general case the coupled diffusion theory can be considered as an investigation of influence of the above phenomena on the conductivity or as that of the electrothermoelastic state of the semiconductor /or vice versa/.

This paper aims at investigating the problem of coupled dynamical electrothermodiffusion of charge carriers in an thermoelastic extrinsic semiconductor. All the considerations are based upon the irreversible thermodynamics. Remark that all the indexes „n" and „p" concern electrons and holes respectively and there are no summations over them.

ENERGY BALANCE

We are considering the isotropic, homogeneous, polarizable and extrinsic semiconductor of instantaneous volume V. Omitting the mass of the nonequilibrium electrons and the magnetization of the body we postulate energy balance in the form [1], [2], [3]

$$\frac{d}{dt}\int_V \left[\varrho\left(\frac{v^2}{2} + U\right) + U_e\right] dV = \int_V (\varrho r + f_i v_i)\, dV + \oint_{\partial V}\left[t_{ji} v_i - Q_j - (\underset{\sim}{E} \times \underset{\sim}{H})_j + \mu_n j_{nj} - \mu_p j_{pj} + U_e v_j\right] n_j\, d\partial V. \qquad /2.1/$$

The l.h.s. of /2.1/ represents the time rate of total energy /kinetic, internal and electromagnetic/ enclosed in V. The terms in the r.h.s. denote: heat production by the heat source distribution, rate of work of volume forces $f_i v_i$, of surface forces $\tau_{ji} v_i n_j$, transport of heat $Q_j n_j$, and of electromagnetic energy $(\underset{\sim}{E}\times\underset{\sim}{H})_j n_j$ through the surface into the body / n_j positive outwards/; influx of electromagnetic energy $U_e v_j n_j$ and, finally, diffusion of charge carriers / μ_n, μ_p chemical potentials of electrons and holes, j_{nj}, j_{pj} fluxes of electrons and holes/. τ_{ji} is an electrothermodiffusive stress tensor, yet unknown. ϱ denotes mass density.

After using the Gauss' theorem, the continuity equation

$$\frac{\partial\varrho}{\partial t} + \operatorname{div}(\varrho \underset{\sim}{v}) = 0 \qquad /2.2/$$

and the expression $\frac{d}{dt}\int_V U_e \, dV - \oint_{\partial V} U_e v_n \, d\partial V = \int_V \frac{\partial U_e}{\partial t} dV,$

we obtain the energy balance in the following local form

$$\varrho \frac{d}{dt}\left(\frac{v^2}{2} + U\right) + \frac{\partial U_e}{\partial t} =$$

$$= \varrho r + f_i v_i + \left[\tau_{ji} v_i - Q_j - (\underset{\sim}{E}\times\underset{\sim}{H})_j + \mu_n j_{nj} - \mu_p j_{pj}\right]_{,j}. \qquad /2.3/$$

Taking into account that the processes occurring inside the body are locally isothermal, we take instead of the internal energy

$$U = U(\varepsilon_{ij}, \mathcal{P}_i, n, p, S) \qquad /2.4/$$

the free energy

$$F = U - TS - \mu_n n - \mu_p p, \qquad F = F(\varepsilon_{ij}, \mathcal{P}_i, \mu_n, \mu_p, T), \qquad /2.5/$$

where ε_{ij} denotes elastic strain tensor, $P_i = \varrho \mathcal{P}_i$ - polarization, n and p are the nonequilibrium concentrations of electrons and holes defined as mass densities of the charge, and S - entropy. We also assume the linearity of

$$\varepsilon_{ij} = \frac{1}{2}(u_{i,j} + u_{j,i}), \qquad u_i \text{ -displacement vector.} \qquad /2.6/$$

The electromagnetic phenomena occurring in the considered moving semiconductor are described by Maxwell's equations [4]

$$\nabla\times\underset{\sim}{E} = -\frac{\partial \underset{\sim}{B}}{\partial t}, \qquad \nabla\cdot\underset{\sim}{D} = \varrho(p - p_0 - n + n_0),$$
$$\nabla\times\underset{\sim}{H} = \underset{\sim}{j}' + \frac{\partial \underset{\sim}{D}}{\partial t}, \qquad \nabla\cdot\underset{\sim}{B} = 0, \qquad /2.7/$$

the expressions

$$\underset{\sim}{D} = \varepsilon_0 \underset{\sim}{E} + \underset{\sim}{P}, \qquad \underset{\sim}{B} = \mu_0 \underset{\sim}{H}, \qquad \underset{\sim}{j} = \underset{\sim}{j}_n + \underset{\sim}{j}_p \qquad /2.8/$$

and Poynting's theorem. The total current $\underset{\sim}{j}'$ is given in the form [4]

$$\underset{\sim}{j}' = \underset{\sim}{j} + \varrho(p - p_0 - n + n_0)\underset{\sim}{v} \qquad /2.9/$$

In /2.9/ the term $\nabla\times(\underset{\sim}{P}\times\underset{\sim}{v})$ is omitted because we assume it as very small in comparison with the other terms, then n_0 and p_0 are the equilibrium concentrations of electrons and holes. In addition, the equations /2.7/ and /2.8/ yield the laws of conservation of charge carriers, as follows [5]

$$\rho\frac{\partial n}{\partial t} - \nabla\cdot\dot{j}_n = g_n - r_n\,, \qquad \rho\frac{\partial p}{\partial t} + \nabla\cdot\dot{j}_p = g_p - r_p\,. \qquad /2.10/$$

Here g_n, g_p, r_n, r_p are generation and recombination functions.

Utilization of eqns. /2.2/ - /2.5/ and /2.7/ - /2.10/ allow us to rewrite the energy balance as

$$\begin{aligned}&[\rho\frac{dv_i}{dt} - f_i - \tau_{ji,j} + E_k P_{k,i} + \rho\mu_n n_{,i} + \rho\mu_p p_{,i} - \rho(p-p_0-n+n_0)E_i]v_i + \\ &+ [\sigma_{ij} - \tau_{ji} + (E_k P_k + \rho\mu_n n + \rho\mu_p p)\delta_{ij}]v_{i,j} + \rho\left(\frac{\partial F}{\partial P_i} - E_i\right)\frac{dP_i}{dt} + \\ &+ \rho\left(\frac{\partial F}{\partial \mu_n} + n\right)\frac{d\mu_n}{dt} + \rho\left(\frac{\partial F}{\partial \mu_p} + p\right)\frac{d\mu_p}{dt} + \rho\left(\frac{\partial F}{\partial T} + S\right)\frac{dT}{dt} - j_{ni}(E_i + \mu_{n,i}) - \\ &- j_{pi}(E_i - \mu_{p,i}) + Q_{i,i} + \rho T\frac{dS}{dt} - \Omega = 0\,,\end{aligned} \qquad /2.11/$$

where $\sigma_{ij} = \rho\dfrac{\partial F}{\partial \varepsilon_{ij}}\,, \quad \Omega = \rho r - \mu_n(g_n - r_n) - \mu_p(g_p - r_p)\,.$

ENTROPY BALANCE AND KINETICS

The entropy changes of the system are connected with its exchange with the surroundings and with its positive increase caused by its production. On requiring invariance conditions under superposed rigid body motions of expression /2.11/ and on taking into account that it represents a linear differential form, we obtain the following entopy balance [3]

$$\rho T\frac{dS}{dt} = -Q_{i,i} + j_{ni}(E_i + \mu_{n,i}) + j_{pi}(E_i - \mu_{p,i}) + \Omega\,. \qquad /3.1/$$

The above balance includes as well the free exchange term

$$\rho\frac{dS_e}{dt} = -\left(\frac{Q_i}{T}\right)_{,i} \qquad /3.2/$$

as the term of the production

$$\rho\frac{dS_i}{dt} = -\frac{Q_i}{T^2}T_{,i} + \frac{1}{T}j_{ni}(E_i + \mu_{n,i}) + \frac{1}{T}j_{pi}(E_i - \mu_{p,i}) + \Omega\,. \qquad /3.3/$$

On the other hand, the entropy production can be presented in the form [3]

$$\rho\frac{dS_i}{dt} = L_{ab}K_aK_b\,, \qquad \text{where} \quad J_a = L_{ac}K_c\,. \qquad /3.4/$$

Now, basing upon /3.3/ we get the following matrixes for J_a, K_c and L_{ac}

$$\{J\} = \begin{Bmatrix} \underset{\sim}{Q} \\ \dot{j}_n \\ \dot{j}_p \end{Bmatrix}, \qquad \{\underset{\sim}{K}\} = \begin{Bmatrix} -\frac{1}{T^2}\nabla T \\ \frac{1}{T}(\underset{\sim}{E} + \nabla\mu_n) \\ \frac{1}{T}(\underset{\sim}{E} - \nabla\mu_p) \end{Bmatrix}, \qquad /3.5/$$

$$\{\mathbb{L}\} = \begin{Bmatrix} \varkappa T^2 & \beta_n T^2 & \beta_p T^2 \\ \beta_n T^2 & T\varrho\xi_n n & 0 \\ \beta_p T^2 & 0 & T\varrho\xi_p p \end{Bmatrix} . \qquad /3.6/$$

Finally, we obtain the needed explicit expressions for the fluxes J_a [4]

$$Q_i = -\varkappa T_{,i} + T\beta_n[E_i + (\underset{\sim}{v}\times\underset{\sim}{B})_i + \mu_{n,i}] + T\beta_p[E_i + (\underset{\sim}{v}\times\underset{\sim}{B})_i - \mu_{p,i}]$$

$$j_{ni} = -\beta_n T_{,i} + \varrho\xi_n n[E_i + (\underset{\sim}{v}\times\underset{\sim}{B})_i + \mu_{n,i}]$$

$$j_{pi} = -\beta_p T_{,i} + \varrho\xi_p p[E_i + (\underset{\sim}{v}\times\underset{\sim}{B})_i - \mu_{p,i}]$$

or in more useful form [6]

$$\begin{aligned} Q_i &= -\varkappa T_{,i} + \frac{T\beta_n}{\varrho\xi_n n} j_{ni} + \frac{T\beta_p}{\varrho\xi_p p} j_{pi} \\ j_{ni} &= -\beta_n T_{,i} + \varrho\xi_n n[E_i + (\underset{\sim}{v}\times\underset{\sim}{B})_i + \mu_{n,i}] \\ j_{pi} &= -\beta_p T_{,i} + \varrho\xi_p p[E_i + (\underset{\sim}{v}\times\underset{\sim}{B})_i - \mu_{p,i}] . \end{aligned} \qquad /3.7/$$

In the above mentioned expressions we have assumed that the currents of electrons and holes do not influence each other and we have introduced the following notations: $\varkappa$ - thermal conductivity coefficient, R - thermoelectric constant, β_n, β_p - thermodiffusive constants, ξ_n, ξ_p - mobilities of electrons and holes.

The matrix /3.6/ indicates the symmetry of phenomenological coefficients L_{ac}, which coincide with the Onsager's relations $L_{ac} = L_{ca}$.

Now, taking into account the remarks made at the beginning of this Section and substituting expressions /2.7/, /2.10/, /3.7/ into balance /2.11/, we obtain

- equation of motion

$$\begin{aligned} &\nabla\cdot\mathbb{G} + \underset{\sim}{P}\cdot\nabla\underset{\sim}{E} + \varrho n\nabla\mu_n + \varrho p\nabla\mu_p + \varrho(p - p_0 + n_0 - n)\underset{\sim}{E} + \\ &+ \left(\nabla\times\underset{\sim}{H} - \varepsilon_0\frac{\partial\underset{\sim}{E}}{\partial t} - \frac{\partial\underset{\sim}{P}}{\partial t}\right)\times\underset{\sim}{B} + \underset{\sim}{f} = \varrho\frac{d\underset{\sim}{v}}{dt} , \end{aligned} \qquad /3.8/$$

- equation of heat transfer

$$\begin{aligned} &\varkappa\nabla^2 T - \varrho T\frac{dS}{dt} - \frac{\beta_n T}{\varrho\xi_n n}\nabla\cdot\dot{j}_n - \frac{\beta_n T}{\varrho\xi_n}\dot{j}_n\cdot\nabla\left(\frac{1}{n}\right) - \frac{\beta_p T}{\varrho\xi_p p}\nabla\cdot\dot{j}_p - \\ &- \frac{\beta_p T}{\varrho\xi_p}\dot{j}_p\cdot\nabla\left(\frac{1}{p}\right) = -\frac{\dot{j}_n^2}{\varrho\xi_n n} - \frac{\dot{j}_p^2}{\varrho\xi_p p} - \Omega , \end{aligned} \qquad /3.9/$$

- equations of carrier diffusion

$$\begin{aligned} &\varrho\xi_n n\nabla^2\mu_n + \varrho\xi_n\nabla\cdot(n\underset{\sim}{E}) - \varrho\frac{\partial n}{\partial t} + \varrho\xi_n\nabla n\cdot\nabla\mu_n - \beta_n\nabla^2 T = -(g_n - r_n), \\ &\varrho\xi_p p\nabla^2\mu_p - \varrho\xi_p\nabla\cdot(p\underset{\sim}{E}) - \varrho\frac{\partial p}{\partial t} + \varrho\xi_p\nabla p\cdot\nabla\mu_p + \beta_p\nabla^2 T = -(g_p - r_p), \end{aligned} \qquad /3.10/$$

- electrothermodiffusive stress tensor

$$\mathbb{T}^T = \mathbb{G} + (\underset{\sim}{E}\cdot\underset{\sim}{P} + \varrho\mu_n n + \varrho\mu_p p)\mathbb{I} \qquad /3.11/$$

- and

$$\frac{\partial F}{\partial \mathcal{P}_i} = E_i \,, \quad \frac{\partial F}{\partial \mu_n} = -n \,, \quad \frac{\partial F}{\partial \mu_p} = -p \,, \quad \frac{\partial F}{\partial T} = -S \,.$$

CONSTITUTIVE REALATIONS

Because we will write the basic equations of considered theory in the set of variables $\underset{\sim}{u}$, $\underset{\sim}{E}$, n, p, T, we assume the constitutive relations in the form

$$\begin{aligned} \sigma_{ij} &= \sigma_{ij}(\varepsilon_{ij}, E_i, n, p, T) \\ P_i &= P_i(\varepsilon_{ij}, E_i, n, p, T) \\ \mu_n &= \mu_n(\varepsilon_{ij}, E_i, n, p, T) \\ \mu_p &= \mu_p(\varepsilon_{ij}, E_i, n, p, T) \\ S &= S(\varepsilon_{ij}, E_i, n, p, T). \end{aligned} \qquad /4.1/$$

In the mentioned set of variables the state of the system is described by the following thermodynamical potential [7]

$$G = U - \mathcal{P}_i E_i - TS \,, \quad G = G(\varepsilon_{ij}, E_i, n, p, T). \qquad /4.2/$$

Assuming that in the natural state $P_i = 0$, $\mu_n = 0$, $\mu_p = 0$, $S = 0$, $E_i = 0$, $T = T_o$, $n = n_o$, $p = p_o$, the independence of diffusion fields of electrons and holes to each other and the approaches

$$\Theta = T - T_o \,, \quad \left|\frac{\Theta}{T_o}\right| \ll 1 \,, \quad N = n - n_o \,, \quad \left|\frac{N}{n_o}\right| \ll 1 \,, \quad P = p - p_o \,, \quad \left|\frac{P}{p_o}\right| \ll 1 \,, \qquad /4.3/$$

the constitutive relations take the form

$$\begin{aligned} \sigma_{ij} &= 2\mu\varepsilon_{ij} + (\lambda\varepsilon_{kk} - \gamma_\vartheta\Theta - \gamma_N N - \gamma_p P)\delta_{ij} \\ P_i &= \chi E_i \\ \mu_n &= \frac{\gamma_N}{\varrho}\varepsilon_{kk} + \frac{\delta_N}{n_o} N - \alpha_N \Theta \\ \mu_p &= \frac{\gamma_p}{\varrho}\varepsilon_{kk} + \frac{\delta_p}{p_o} P - \alpha_p \Theta \\ S &= \frac{\gamma_\vartheta}{\varrho}\varepsilon_{kk} + \alpha_N N + \alpha_p P + \frac{c}{T_o}\Theta, \end{aligned} \qquad /4.4/$$

where: λ, μ - Lame's constants, γ_ϑ - thermoelastic constant, γ_N, γ_p - elastodiffusive constants, χ - electrical susceptibility, δ_N, δ_p - diffusive constants, α_N, α_p - thermodiffusive constants, c - specific heat coefficient.

EQUATIONS OF THE THEORY

On substituting eqns. /2.6/, /4.3/, /4.4/ together with eqns./3.8/ - /3.10/, we arrive at the equations of the theory

$$\mu\nabla^2\underset{\sim}{u} + (\lambda+\mu+\gamma_N n_0+\gamma_p p_0)\nabla(\nabla\cdot\underset{\sim}{u}) - (\gamma_\theta+\rho n_0\alpha_N+\rho p_0\alpha_p)\nabla\Theta - (\gamma_N-\rho\delta_N)\nabla N - (\gamma_p-\rho\delta_p)\nabla P +$$
$$+\chi\underset{\sim}{E}\cdot\nabla\underset{\sim}{E} + \rho(P-N)\underset{\sim}{E} + \mu_0\left[\nabla\times\underset{\sim}{H} - (\varepsilon_0+\chi)\frac{\partial\underset{\sim}{E}}{\partial t}\right]\times\underset{\sim}{H} + \underset{\sim}{f} = \rho\ddot{\underset{\sim}{u}}$$

$$\left(\varkappa + \frac{\beta_n^2 T_0}{\rho\xi_n n_0} + \frac{\beta_p^2 T_0}{\rho\xi_p p_0} + \beta_n T_0\alpha_N - \beta_p T_0\alpha_p\right)\nabla^2\Theta - \rho c\dot{\Theta} - T_0\gamma_\theta\nabla\cdot\dot{\underset{\sim}{u}} - \frac{\beta_n T_0\delta_N}{n_0}\nabla^2 N +$$
$$+\frac{\beta_p T_0\delta_p}{p_0}\nabla^2 P - \rho T_0\alpha_N\dot{N} - \rho T_0\alpha_p\dot{P} - \left(\frac{\beta_n T_0\gamma_N}{\rho} - \frac{\beta_p T_0\gamma_p}{\rho}\right)\nabla^2(\nabla\cdot\underset{\sim}{u}) - \frac{\beta_n T_0}{n_0}\nabla\cdot(N\underset{\sim}{E}) -$$
$$-\frac{\beta_p T_0}{p_0}\nabla\cdot(P\underset{\sim}{E}) - (\beta_n T_0+\beta_p T_0)\nabla\cdot\underset{\sim}{E} = -\frac{j_n^2}{\rho\xi_n n_0} - \frac{j_p^2}{\rho\xi_p p_0} - \Omega$$

$$\rho\xi_n\delta_N\nabla^2 N + \rho\xi_n\nabla\cdot(N\underset{\sim}{E}) + \rho\xi_n n_0\nabla\cdot\underset{\sim}{E} - \rho\dot{N} - (\rho\xi_n n_0\alpha_N+\beta_n)\nabla^2\Theta + \qquad /5.1/$$
$$+\xi_n n_0\gamma_N\nabla^2(\nabla\cdot\underset{\sim}{u}) = -(g_n - r_n)$$

$$\rho\xi_p\delta_p\nabla^2 P - \rho\xi_p\nabla\cdot(P\underset{\sim}{E}) - \rho\xi_p p_0\nabla\cdot\underset{\sim}{E} - \rho\dot{P} - (\rho\xi_p p_0\alpha_p - \beta_p)\nabla^2\Theta +$$
$$+\xi_p p_0\gamma_p\nabla^2(\nabla\cdot\underset{\sim}{u}) = -(g_p - r_p)\,.$$

We ought to complete equations /5.1/ with Maxwell's equations /2.7/ which include expressions /3.7/ and /4.4/. For the n - type semiconductors we shall assume for the above mentioned equations that n >> p. We could simplify our equations if we assume the electrical quasi-neutrality of the crystal.

REFERENCES

[1] Parkus, H., Magneto-thermoelasticity /Springer Verlag, Udine, 1972/.
[2] Romano, A., A macroscopic non-linear theory of magnetothermoelastic continua, Arch. Rat.Mech.Anal. 65 /1977/ 1-24.
[3] de Groot, S.R., Mazur, P., Non-equilibrium thermodynamics /North-Holland, Amsterdam, 1962/.
[4] Panofsky, W.K.H, Phillips, M., Classical electricity and magnetism /Addison-Wesley, Cambridge/.
[5] Seeger, K., Semiconductor physics /Springer Verlag, Vien,1973/.
[6] Wiśniewski S., Staniszewski B., Szymanik,R., Thermodynamics of nonequilibrium processes /PWN,Warszawa, 1976/.
[7] Tucker,J.W., Rampton, V.W., Microwave ultrasonics in solid state physics /North-Holland, Amsterdam, 1972/.

Part 4

ELASTIC FERROELECTRICS AND CERAMICS

The Mechanical Behavior of Electromagnetic Solid Continua
G.A. Maugin (editor)
Elsevier Science Publishers B.V. (North-Holland)

HYSTERESIS EFFECTS IN DEFORMABLE CERAMICS

Peter J. Chen

Sandia National Laboratories
Albuquerque, New Mexico 87185
U.S.A.

The responses of ferroelectric ceramics can be quite complex depending on the physical processes to which they are subjected. Their mechanical, electromechanical and dielectric properties depend on domain switching, dipole dynamics and phase transformation. A theory, which describes the responses of these materials, has been formulated, and it is sufficient to characterize various observable phenomena. Specifically, a special case of the theory predicts the nature of the butterfly and hysteresis loops.

INTRODUCTION

The mechanical, electromechanical and dielectric response of ferroelectric ceramics can be quite complex depending on the physical processes to which they are subjected. These responses also depend on domain switching, dipole dynamics and phase transformation which can be caused by external stimuli such as mechanical and electrical loadings and changes in temperature. A theory giving the constitutive relations for the stress and the electric displacement and taking into account the effects of domain switching, dipole dynamics and phase transformation has been formulated. In its present stage of development this theory is sufficient to describe various observable phenomena including the well known hysteresis loops and the effects of dielectric relaxation. Here, I shall demonstrate some of the features of the theory and illustrate some of its predictive capabilities.

BASIC EQUATIONS

To begin with let us restrict our attention to the case of an isothermal theory and neglect the effects of phase transformation. Let the vector $\underset{\sim}{\mu}$ denote the unit cell electric dipole moment. We presume that its magnitude and direction depend on the mechanical strain $\underset{\sim}{S}$ and the external electric field $\underset{\sim}{E}$. We also introduce the vector $\underset{\sim}{N}$, defined by the relation

$$\underset{\sim}{N} = (\Sigma\underset{\sim}{\mu}\cdot\underset{\sim}{j})\underset{\sim}{j} \quad , \tag{1}$$

where $\underset{\sim}{j}$ is a unique unit vector such that if $\Sigma\underset{\sim}{\mu}\cdot\underset{\sim}{j} \neq 0$ then $\Sigma\underset{\sim}{\mu}\cdot\underset{\sim}{j}$ must have maximum value, and where the summation is carried out over each sub-part of a ferroelectric specimen. The direction of $\underset{\sim}{j}$ denotes the direction of polarization, and $\Sigma\underset{\sim}{\mu}\cdot\underset{\sim}{j}$ gives the number of unit cell dipoles effectively aligned in this direction. We also presume that the responses of $\underset{\sim}{\mu}$ to $\underset{\sim}{S}$ and $\underset{\sim}{E}$ can be partitioned into its instantaneous response $\underset{\sim}{\mu}_i$ and its transient response $\underset{\sim}{\mu}_t$ so that

$$\underset{\sim}{\mu} = \underset{\sim}{\mu}_i + \underset{\sim}{\mu}_t \quad . \tag{2}$$

The constitutive relations for the stress $\underset{\sim}{T}$ and the electric displacement $\underset{\sim}{D}$ may, therefore, be given by

$$\underset{\sim}{T} = \tilde{\underset{\sim}{T}}(\underset{\sim}{S},\underset{\sim}{\mu},\underset{\sim}{N}) = \underset{\sim}{T}(\underset{\sim}{S},\underset{\sim}{E},\underset{\sim}{\mu_t},\underset{\sim}{N}) \quad ,$$
$$\underset{\sim}{D} = \tilde{\underset{\sim}{D}}(\underset{\sim}{\mu},\underset{\sim}{N}) = \underset{\sim}{D}(\underset{\sim}{S},\underset{\sim}{E},\underset{\sim}{\mu_t},\underset{\sim}{N}) \quad ; \tag{3}$$

and we have the rate laws

$$\dot{\underset{\sim}{\mu_t}} = \underset{\sim}{f}(\underset{\sim}{S},\underset{\sim}{E},\underset{\sim}{\mu_t},\underset{\sim}{N}) \quad ,$$
$$\dot{\underset{\sim}{N}} = \underset{\sim}{g}(\underset{\sim}{S},\underset{\sim}{E},\underset{\sim}{\mu_t},\underset{\sim}{N}) \quad . \tag{4}$$

The appropriate governing differential equations are, of course, the equation of balance of linear momentum and Gauss' law in the absence of free charge.

A specific representation of the constitutive relations and the rate laws which are relevant to many applications is given by Chen [1]. Here, however, we consider a special case of (3) and (4) which leads to the butterfly and hysteresis loops.

A SPECIAL CASE

The interesting special case which we have in mind corresponds to the physical situation when a ferroelectric specimen is subjected to a slowly varying unidirectional cyclic electric field under stress-free and uniaxial strain conditions. The electric field is of sufficient magnitude to cause domain switching. Since the transient effects of dipole dynamics as described by $(4)_1$ occur on a much faster time scale, we have $\dot{\underset{\sim}{\mu_t}}=\underset{\sim}{0}$. The direction of the electric field is taken as X_3. Therefore,

$$\underset{\sim}{E} = (0,0,E) \quad ,$$
$$\underset{\sim}{N} = (0,0,N) \quad , \tag{5}$$

and the non-zero component of the strain is $S_{33} \equiv S$. It now follows from (3), $(4)_1$ and (5) that the stress component $T_{33} \equiv T$ and the electric displacement component $D_3 \equiv D$ are given by

$$T = T(S,E,N) \quad ,$$
$$D = D(S,E,N) \quad , \tag{6}$$

and the rate law $(4)_2$ becomes

$$\dot{N} = g(E,N) \quad . \tag{7}$$

The effective number of aligned dipoles N is also given by

$$N = N^{\parallel}+N^{\perp} \quad , \tag{8}$$

where $N^{\parallel}$ and $N^{\perp}$ denote, respectively, the effective number of aligned parallel and perpendicular dipoles. In addition, $N^{\parallel}$ and $N^{\perp}$ are given by

$$N^{\parallel} = N_s^{\parallel}+N_n^{\parallel}, \; N^{\perp} = N_s^{\perp}+N_n^{\perp} \quad , \tag{9}$$

where the subscripts s and n denote, respectively, the effective number of aligned permanently switchable and non-permanently switchable parallel or perpendicular dipoles.

We now consider the following specific representation of the constitutive relations (6):

$$T = (C+C'N)S-eNE+h^{\parallel}N^{\parallel}+h^{\perp}N^{\perp} \quad ,$$
$$D = eNS+(\varepsilon+\varepsilon'N)E+kN \quad . \tag{10}$$

In (10), C and ε are the elastic and dielectric constants of the virgin specimen. The term $C'N$ gives the change to C and the term $\varepsilon'N$ gives the change to ε due to domain switching. The coefficient eN with $e>0$ plays the role of the electro-mechanical coupling coefficient. Notice that the values of these terms depend on the effective number of aligned dipoles. The terms $h^{\parallel}N^{\parallel}$ and $h^{\perp}N^{\perp}$ give the additional stress and the term kN gives the additional electric displacement due to domain switching. In particular, we require that

$$\text{sgn}\, h^{\parallel} = -\text{sgn}\, N^{\parallel},\ \text{sgn}\, h^{\perp} = -\text{sgn}\, N^{\perp}\ ,\ k>0\ . \tag{11}$$

We presume that the rate law (7) has the specific representation

$$\dot{N}+\alpha(E)N = \beta(E)\ . \tag{12}$$

This linear differential equation (12) describing the consequences of domain switching under the action of an externally applied electric field seems sufficient and reasonable. However, the dependences of the functions α and β on E are quite complex. First, we note that $1/\alpha(E)$ plays the role of switching time corresponding to each value of the field E. Since switching time must decrease with increasing magnitude of the electric field, we require α to be a positive even function of E which increases monotonically with increasing $|E|$ and $\alpha(0)=0$. Now suppose that a constant applied field has been maintained for a sufficiently long time so that domains have ceased switching. It follows directly from (12) that we have $N_e(E) \equiv\beta(E)/\alpha(E)$. It is clear that $N_e(E)$ gives the total effective number of dipoles which may be aligned by the field E. Hence, we require β/α to be an odd function of E such that $\text{sgn}\{\beta(E)/\alpha(E)\}=\text{sgn}\, E$ and whose absolute value increases monotonically with increasing $|E|$ such that for sufficiently large $|E|$ the values of $|\beta/\alpha|$ are bounded. The properties of the function β may be deduced from those of α and β/α.

As we have remarked, the effective number of aligned dipoles is given by

$$N = N_s^{\parallel}+N_s^{\perp}+N_n^{\parallel}+N_n^{\perp}\ . \tag{13}$$

We may assume that the switching characteristics of these four classes of domains are uncoupled and that we may prescribe separate rate laws for $N_s^{\parallel}$, $N_s^{\perp}$, $N_n^{\parallel}$ and $N_n^{\perp}$. To this end, we define the functions β_s and β_n of E via the relations

$$\beta_s(E) = s\beta(E),\ \beta_n(E) = n\beta(E)\ , \tag{14}$$

where s and n are positive numbers such that $s+n = 1$ and

$$\beta(E) = \beta_s(E)+\beta_n(E)\ . \tag{15}$$

We also introduce the functions N_{se} and N_{ne}, defined by the relations

$$\begin{aligned} N_{se}(E) &= \beta_s(E)/\alpha(E) = sN_e(E),\\ N_{ne}(E) &= \beta_n(E)/\alpha(E) = nN_e(E), \end{aligned} \tag{16}$$

so that

$$N_e(E) = N_{se}(E) + N_{ne}(E). \tag{17}$$

$N_{se}(E)$ and $N_{ne}(E)$ give, respectively, the total effective number of permanently switchable and non-permanently switchable dipoles which may be aligned by the field E.

We now specify the following rate laws for $N_s^{\parallel}$, $N_s^{\perp}$, $N_n^{\parallel}$ and $N_n^{\perp}$:

(i) If at any time t $\text{sgn}\, N_s^{\parallel}(t) = \text{sgn}\, N_{se}(E)$ and $|N_s^{\parallel}(t)|\geq|(1-\delta)N_{se}(E)|$, then

$$\dot{N}_s^{\parallel} = 0 \quad ;$$

otherwise $N_s^{\parallel}$ obeys the rate law

$$\dot{N}_s^{\parallel}+\alpha^{\parallel}(E)N_s^{\parallel} = \alpha^{\parallel}(E)(1-\delta)N_{se}(E) \quad .$$

(ii) If at any time t sgn $N_s^{\perp}(t)$=sgn $N_{se}(E)$ and $|N_s^{\perp}(t)|\geq|\delta N_{se}(E)|$, then

$$\dot{N}_s^{\perp} = 0 \quad ;$$

otherwise $N_s^{\perp}$ obeys the rate law

$$\dot{N}_s^{\perp}+\alpha^{\perp}(E)N_s^{\perp} = \alpha^{\perp}(E)\delta N_{se}(E) \quad .$$

(iii) If at any time t sgn $N_n^{\parallel}(t)$=sgn $N_{ne}(E)$ and $|N_n^{\parallel}(t)|\geq|\frac{1}{3}N_{ne}(E)|$, then

$$\dot{N}_n^{\parallel}+\alpha_n^{\parallel}N_n^{\parallel} = 0 \; ;$$

otherwise $N_n^{\parallel}$ obeys the rate law

$$\dot{N}_n^{\parallel}+q^{\parallel}\alpha^{\parallel}(E)N_n^{\parallel} = \frac{1}{3}q^{\parallel}\alpha^{\parallel}(E)N_{ne}(E) \quad .$$

(iv) If at any time t sgn $N_n^{\perp}(t)$=sgn $N_{ne}(E)$ and $|N_n^{\perp}(t)|\geq|\frac{2}{3}N_{ne}(E)|$, then

$$\dot{N}_n^{\perp}+\alpha_n^{\perp}N^{\perp} = 0 \; ;$$

otherwise $N_n^{\perp}$ obeys the rate law

$$\dot{N}_n^{\perp}+q^{\perp}\alpha^{\perp}(E)N_n^{\perp} = \frac{2}{3}q^{\perp}\alpha^{\perp}(E)N_{ne}(E) \quad .$$

In the preceding equations δ is some number between 0 and 2/3 (for a thoroughly poled specimen $\delta = 0$, and for a virgin specimen $\delta = 2/3$), $q^{\parallel}$ and $q^{\perp}$ are constants, $\alpha^{\parallel}$ and $\alpha^{\perp}$ (also $q^{\parallel}\alpha^{\parallel}$ and $q^{\perp}\alpha^{\perp}$) give the switching times of the parallel and perpendicular domains, and $\alpha_n^{\parallel}$ and $\alpha_n^{\perp}$ give the decay times of the non-permanently switchable parallel and perpendicular domains.

The choice of the values of δ is not as arbitrary as it seems. In the virgin state, 1/3 of the domains have a preferred orientation parallel to a given direction and 2/3 of the domains have a preferred orientation in the plane perpendicular to this direction. The former gives rise to the effectively aligned parallel dipoles, and the latter gives rise to the effectively aligned perpendicular dipoles. After a specimen has been thoroughly poled it is presumed that almost all the domains are parallel domains, except, of course, the non-permanently switchable domains. The consequence of this assumption will be quite obvious when we consider the example.

A thoroughly poled state is that for which

$$T=0, \quad S=S_p, \quad E=0, \quad D=D_p, \quad N=N_s^{\parallel}+N_s^{\perp}=N_p \quad . \tag{18}$$

For convenience, we may normalize N with respect to the poled state, viz.,

$$N_p = \pm 1 \quad . \tag{19}$$

The stress-free condition yields via $(10)_1$ the result

$$S = \frac{eNE-h^{\parallel}N^{\parallel}-h^{\perp}N^{\perp}}{C+C'N} \quad . \tag{20}$$

Substituting (20) into $(10)_2$, we have

$$D = \left(\varepsilon+\varepsilon' N+\frac{e^2N^2}{C+C'N}\right) E - \frac{eN(h^{\parallel}N^{\parallel}+h^{\perp}N^{\perp})}{C+C'N} + kN \quad . \tag{21}$$

The mechanical and dielectric responses of a ferroelectric ceramic specimen during the course of domain switching under the action of an external electric field from the virgin state are, therefore, given by (20) and (21) together with the solutions of the rate laws. The results are valid under isothermal, stress-free and uniaxial strain conditions. In order to obtain the results corresponding to the butterfly and hysteresis loops it is necessary to re-initialize the problem at this juncture. The details of this procedure are given in the papers by Chen and Tucker [2] and Chen and Madsen [3]. They determine the material properties and compare the numerical results to the experimental results concerning the ferroelectric ceramics PZT65/35 and PLZT7/65/35 in a pseudo one dimensional context. These comparisons suggest that the proposed theory does have merit because it predicts the fine details of the butterfly and hysteresis loops. The following example concerning PZT65/35 is taken from the paper by Chen and Tucker [2]. It shows the excellent agreement between theory and experiments.

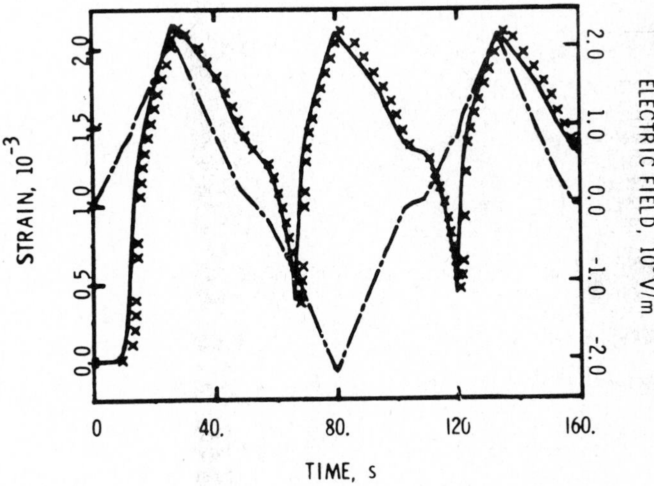

Fig. 1 Comparison of computed strain (solid line) and experimentally measured strain (X) versus time. The associated applied electric field is shown as an interrupted line. The irregularities of the applied field are due to the limitations of the power supply.

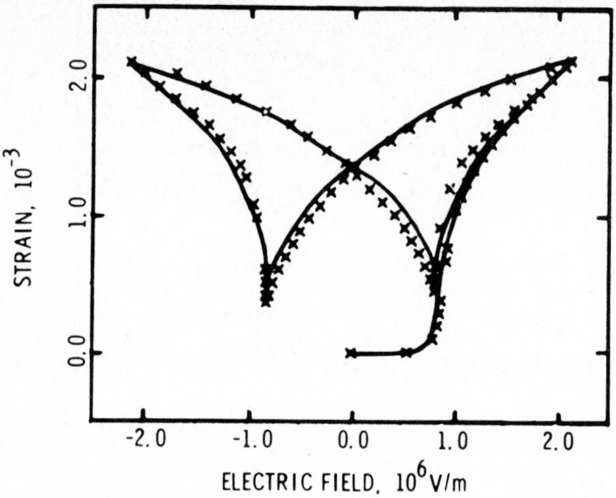

Fig. 2 Cross plot of calculated strain (solid line) and experimentally measured strain (X) versus electric field.

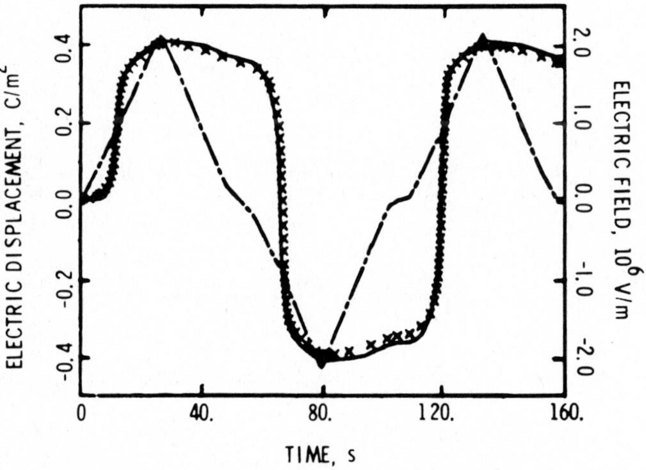

Fig. 3 Comparison of computed electric displacement (solid line) and experimentally measured electric displacement (X) versus time, together with the associated applied electric field (interrupted line).

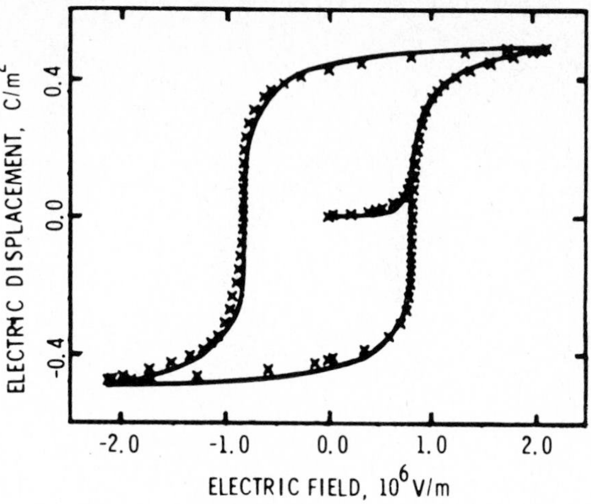

Fig. 4 Cross plot of calculated electric displacement (solid line) and experimentally measured electric displacement (X) versus electric field.

ACKNOWLEDGMENT

This work was sponsored by the U.S. Department of Energy under Contract DE-AC04-76-DP00789 to Sandia National Laboratories, A US Department of Energy facility.

REFERENCES

[1] Chen, P. J., "Three Dimensional Dynamic Electromechanical Constitutive Relations for Ferroelectric Materials," Int. J. Solids Struct., 16, (1980), 1059-1067.

[2] Chen, P. J. and Tucker, T. J., "One Dimensional Polar Mechanical and Dielectric Responses of the Ferroelectric Ceramic PZT65/35 due to Domain Switching," Int. J. Engng. Sci., 19, (1981), 147-158.

[3] Chen, P. J. and Madsen, M. M., "One Dimensional Polar Responses of the Electrooptic Ceramic PLZT7/65/35 due to Domain Switching," Acta Mechanica, 41, (1981), 255-264.

The Mechanical Behavior of Electromagnetic Solid Continua
G.A. Maugin (editor)
Elsevier Science Publishers B.V. (North-Holland)

ELASTIC FERROELECTRIC AND MAGNETIC ASPECTS DESCRIBED BY MICROSCOPIC MODELS

C. CARABATOS, G.E. KUGEL and M.D. FONTANA
Laboratoire de Génie Physique, Faculté des Sciences
Ile du Saulcy, 57045 METZ Cédex, FRANCE.

We give a condensed description of microscopic semiphenomenological lattice and spin dynamical theories, related to "macroscopic" measured data. The illustrative example is $Fe_{1-x}O$.

INTRODUCTION

The problem of the link between microscopic and macroscopic theories is old and still not solved in a complete and satisfactory way. The reasons are various but two of them are predominant : a) the often unsatisfactory knowledge of the interaction potentials in the solids ; b) the big number of degrees of freedom in the microscopic treatment of the problem.

We shall restrict ourselves to the lattice and spin dynamics and relate the theories to experiments such as infrared absorption, elastic constants, phonon and magnon dispersion curves, magnetoelastic constants. We shall illustrate the comparisons in the case of $Fe_{1-x}O$ both in the paramagnetic (lattice dynamics) and the antiferromagnetic (spin dynamics) phases.

II. LATTICE DYNAMICS : EXAMPLE $Fe_{1-x}O$.

First principles and quantum mechanical considerations under several simplifying assumptions, allowed to explain and justify a very commonly used lattice dynamical model known as the "shell Model" (1) (2) (3). The formal resolution of the equations of motion of the nuclei and the electronic clouds (shells) of the ions in an elementary cell of a crystalline solid, leads to a parametrized dynamical matrix $\underset{\sim}{D}$ given by :

$$\underset{\sim}{D} = \underset{\sim}{M}^{-1/2} \{(\underset{\sim}{R} + \underset{\sim}{ZCZ}) - (\underset{\sim}{T} + \underset{\sim}{ZCY}) (\underset{\sim}{S} + \underset{\sim}{K} + \underset{\sim}{YCY})^{-1} (\underset{\sim}{T} + \underset{\sim}{ZCY})^{+}\} \underset{\sim}{M}^{-1/2}$$

where $\underset{\sim}{R}$, $\underset{\sim}{T}$ and $\underset{\sim}{S}$ are short range interactions between nuclei, nucleus one and shell two and between shells respectively. They are submitted to symmetry conditions due to the crystal space group. $\underset{\sim}{K}$ represents the interaction of a nucleus and its own shell. $\underset{\sim}{C}$ is the Coulomb coefficients matrix (long range) ; $\underset{\sim}{Z}$ the charges of the ions and $\underset{\sim}{Y}$ the charges of the shells.

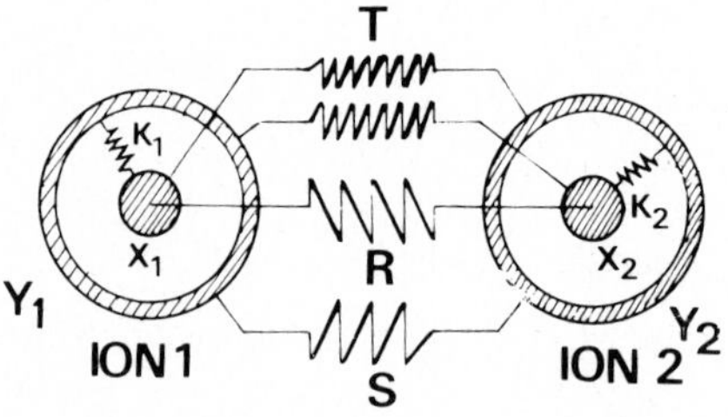

The eigenvalues of $\underset{\sim}{D}$, function of the wavevector $\underset{\sim}{y}$, are the squared frequencies ω^2 of the vibrations and the eigenvectors are the normal modes. If we suppose that the ions are not deformable, $\underset{\sim}{K} \to \infty$ and thus $\underset{\sim}{D} = \underset{\sim}{M}^{-1/2} \{\underset{\sim}{R} + \underset{\sim}{Z}\underset{\sim}{C}\underset{\sim}{Z}\}$. If the Coulomb interactions are rapidly screened by free electrons (case of metals), then $\underset{\sim}{Z} = 0$ and $\underset{\sim}{D} = \underset{\sim}{M}^{-1/2} \underset{\sim}{R}\, \underset{\sim}{M}^{-1/2}$ which is the well known Born von Karman model. (4)

The development of $\underset{\sim}{D}$ up to the second order when the wave vector $\underset{\sim}{y}$ tends to zero and the separation of the macroscopic electric field $\underset{\sim}{E}$, leads to the definition of three algebric expressions : $[\alpha\beta,\gamma\lambda]$, $(\alpha\beta,\gamma\lambda)$ and $(\beta,\alpha\gamma)$ that one can more or less calculate depending on the structure of the crystal and the number of atoms in the elementary cell (4)(5). If the displacements of the nuclei from the equilibrium position is $\underset{\sim}{u}$ then

$$\rho\omega^2 u_\alpha = 4\pi^2 \sum_{\beta\gamma\lambda} \{ [\alpha\beta,\gamma\lambda] + (\alpha\gamma\ \beta\lambda)\}\ y_\gamma y_\lambda u_\beta - 2\pi \sum_{\beta\gamma} [\beta,\alpha\gamma] y_\gamma E_\beta$$

where ρ is the density of the crystal. This last equation compared to the one obtained from the macroscopic theory of elasticity

$$\rho\omega^2 u_\alpha = 4\pi^2 \sum_{\beta\gamma\lambda} C\alpha\gamma,\beta\lambda\ y_\gamma y_\lambda u_\beta + 2\pi i \sum_{\beta\gamma} e_{\beta,\alpha\gamma}\ y_\gamma\ E_\beta$$

implies the following expressions for the elastic constants :

$$C_{\alpha\gamma,\beta\lambda} = [\alpha\beta,\gamma\lambda] + [\beta\gamma,\alpha\lambda] - [\beta\lambda,\alpha\gamma] + (\alpha\gamma,\beta\lambda)$$

The infrared frequencies, dielectric constants and piezoelectric constants can also be derived.

Taking advantage of the group theory prescriptions and the O_h^5 rocksalt symmetry (or nearly) of $Fe_{1-x}O$, one can derive the simplified short range interaction tensors such as $\underset{\sim}{R}$, $\underset{\sim}{S}$ or $\underset{\sim}{T}$. In the case of axially symmetric forces one gets (6)

$$\text{first neighbours}: \begin{vmatrix} A_{12} & 0 & 0 \\ 0 & B_{12} & 0 \\ 0 & 0 & B_{12} \end{vmatrix} \quad \text{and} \quad \text{second neighbours } i = 1 \text{ or } 2: \begin{vmatrix} A_{ii} & 0 & 0 \\ 0 & B_{ii} & 0 \\ 0 & 0 & D_{ii} \end{vmatrix}$$

A, B and D are parameters to be determined by a fitting procedure.
The calculation gives the following expressions for the elastic constants

$$C_{11} = (e^2/Vr_0) \cdot [\tfrac{1}{2}\,(A_{12} + A_{11} + A_{22} + B_{11} + B_{22}) - 2.55604Z'^2]$$

$$C_{12} = (e^2/Vr_0) \cdot [\tfrac{1}{4}\,(-2B_{12} + (A_{11} + A_{22}) - 2(D_{11} + D_{22}) - 3(B_{11} + B_{22})) + 0.11298Z'^2]$$

$$C_{44} = (e^2/Vr_0) \cdot [\tfrac{1}{4}\,(2B_{12} + (A_{11} + A_{22}) + (B_{11} + B_{22}) + 2(D_{11} + D_{22})) + 1.27802Z'^2]$$

Z' is an effective charge (2)(3).

The fitting of the model to the neutron scattering data gives (6)

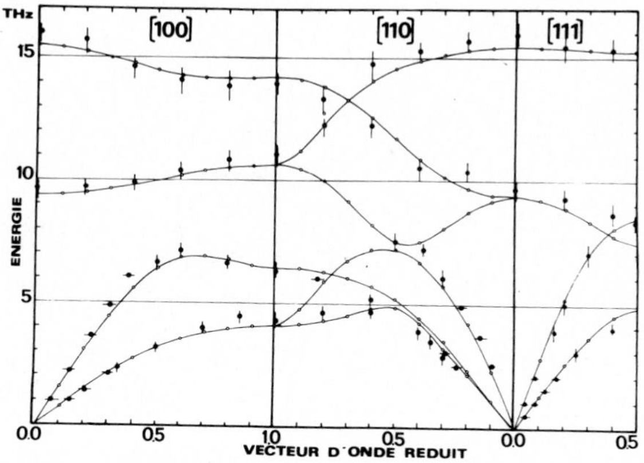

and the following comparison table

Constants	Units	Infrared	Neutron	Calculated
ω_{Lo}	THz	16.0	15.8	15.8
ω_{TO}	THz	8.7	9.6	8.8
ε^{o}	...	32.8	...	31.4
ε^{∞}	...	9.6	...	9.2
C_{11}	10^{12}dyn/cm^2	...	3.5	3.0
C_{12}	"	...	1.5	1.2
C_{44}	"	...	0.6	0.7

We notice that $C_{12} \neq C_{44}$ and conclude that the short range forces are not central (4),(5) and we are able, thanks to the above formulae, to get information concerning the different non central contributions (6),(7).

III. MAGNETIC PROPERTIES : EXAMPLE $Fe_{1-x}O$.

The magnetic behaviour of a system of spins in a perfect crystal is described by a total Hamiltonian $H = H_1 + H_2$; where $H_1 = H_{CF} + H_{SO} + H_{AN} + H_{ME} + H_{MF}$ is the single ion component and $H_2 = H_{EX} - H_{MF}$ is the two ions contribution.
H_{CF} = crystal field, H_{SO} = spin orbit, H_{AN} = magnetic anisotropy, H_{ME} = magnetoelastic, H_{MF} = molecular field, H_{EX} = exchange contributions. The problem is to solve the total Hamiltonian for the particular structure of the substance. Without entering in the details of the mathematical treatment of the (general susceptibility) formalism (8), we simply point out some results for the magnetoelastic effect in $Fe_{1-x}O$. At the phase transition, the elementary cell of the crystal is distorded along [111]. As the lattice distorsion is of magnetoelastic character it can be expressed by (9)

$$H_{ME} = - B_1 (\varepsilon_3 O_2^{\circ} + \sqrt{3}\, \varepsilon_2 O_2^{2}) - B_2(\varepsilon_{xy} P_{xy} + \varepsilon_{yz} P_{yz} + \varepsilon_{zx} P_{zx})$$

where ε_{ij} are the components of the deformation ;

$\varepsilon_3 = \frac{1}{\sqrt{6}} (2\varepsilon_{zz} - \varepsilon_{xx} - \varepsilon_{yy})$, $\varepsilon_2 = \frac{1}{\sqrt{2}} (\varepsilon_{xx} - \varepsilon_{yy})$ and O_2°, O_2°, P_{xy}, P_{yz} and P_{zx}

are the Stevens operators (10). After same calculations one gets

$$H_{ME} = - \frac{B_2^2}{C_{44}} < P_{xy} > . (P_{xy} + P_{yz} + P_{zx})$$

and a distortion constant

$$c = - \frac{B_2^2}{4C_{44}} < P_{xy} > . \frac{3}{2}$$

The fitting of the model Hamiltonian to the neutron scattering data gives (11)

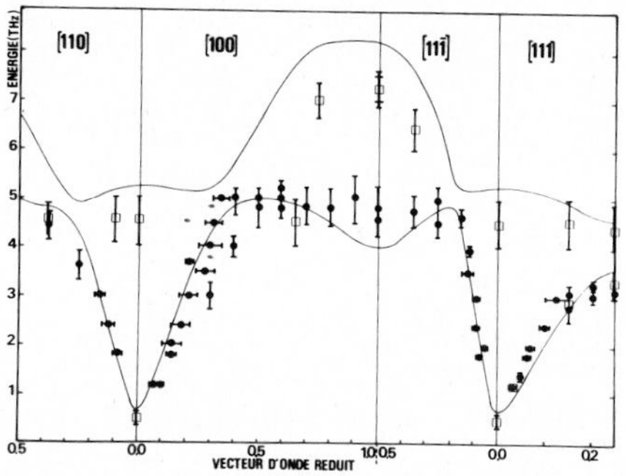

and allows to get reasonable values for the exchange integrals, the magnetic momentum ($4.7\mu_B$) and the spin orbit coupling ($-87 cm^{-1}$). More particularly we get the value $<P_{xy}> = 0.12$ and then

$$B_2 - 2.10^{10} erg/cm^3 = 4.9.10^{13} erg/ion = 2500\ cm^{-1}/ion$$

which gives the importance of the magneto elastic coupling in $Fe_{1-x}O$.

The general susceptibility formalism allows also to study the behaviour of the magnetic properties when the temperature varies, but needs more tedious calculations (12).

All the above considerations illustrate very briefly some possibilities available for the ling between microscopic semi phenomenological theories and macroscopic quantities which are involved in the theories of solid continua.

IV REFERENCES

(1) Dick B.G. and Overhauser A.W., Phys. Rev. <u>112</u> (1958) 90

(2) Schröder U., Solid State Comm. <u>4</u> (1966) 347

(3) Cowley R.A., Proc. Roy. Soc. A268 (1962) 109 and 121

(4) Born M. and Huang K. "Dynamical Theory of Crystal Lattices" Oxford University Press, Oxford 1954

(5) Horton G.K. and Maradudin A.A. "Dynamical Properties of Solids" North Holland, Amsterdam 1974

(6) Kugel G., Carabatos C., Hennion B., Prévot B., Revcolevschi A. and Tocchetti D., Phys. Rev. B16 (1977) 378

(7) Kress W., Reichardt W., Wagner V., Kugel G. and Hennion B. in "Lattice Dynamics" edited by M. Balkanski, Flammarion, Paris 1977

(8) Buyers W.J.L., Holden T.M., Svenson E.C., Cowley R.A. and Hutchings M.T. J. Phys. C4 (1971) 2139

(9) Morin P., Rouchy J. and Du Tremoulet de Lacheisserie, Phys. Rev. B16 (1977) 3182

(10) Stevens K.W.H., Proc. Roy. Soc. A65 (1952) 209

(11) Kugel G., Hennion B. and Carabatos C., Phys. Rev. B18 (1978) 1317

(12) Kugel G., Thesis, University of Metz 1982.

The Mechanical Behavior of Electromagnetic Solid Continua
G.A. Maugin (editor)
Elsevier Science Publishers B.V. (North-Holland)

LATTICE MODEL FOR ELASTIC FERROELECTRICS AND RELATED CONTINUUM THEORIES

Attila Aşkar[+], Joel Pouget[++], Gérard A. Maugin[++]

[+]Boğaziçi Üniversitesi, Matematik Bölümü, Bebek, İstanbul, Turkey
[++]Universite Pierre et Marie Curie, Laboraoire de Mêcanique Theorique associê an CNRS 75230 Paris Cedex 05, France

A lattice model is presented for ferroelectrics of the displacive type for which barium titanate provides a prototype. In its ferroelectric phase a $BaTiO_3$-like crystal is endowed with permanent microscopic dipoles which give rise to an additional rotational degree of freedom. The latter, in turn, yields the ferroelectric soft mode which couples with transverse acoustic vibrations. The macroscopic constants are computed in the long-wave approximation and numerical values are given for $BaTiO_3$ using data from crystallography, elasticity and Raman spectroscopy.

1. INTRODUCTION

Soft modes as predicted by Cochran[1] and Anderson[2], have been studied for numerous materials in their ferroelectric as well as paraelectric phase.[3] In the ferroelectric phase these materials exhibit a strong electromechanical coupling and the latter is notably able to alter the dynamics of both mechanical and dielectric phenomena. The phase transition in such materials governs most of the relevant phenomena and has attracted the interest of many experimentalists. Also important in such media is acoustical wave propagation in reason of their piezoelectric properties. From a continuum mechanical view point, the permanent dipoles form a finite initial state with waves as superimposed perturbation field.

The main phenomenon studied here is the coupling between soft modes and acoustic modes where the lattice theory provides a valuable tool. In addition to the acoustic and optical modes, lattice theories have also been concerned with molecular models accounting for internal motions, in particular, the rotation of groups as rigid bodies.[4-9] At the long wave limit the latter models have coincided with the micropolar theory of elasticity , a continuum theory in which a rigid "microrotation" degree of freedom is added to the usual displacement field. The examination of KNO_3, in this framework provides an example.[6-8] On the other hand, a purely continuum theory has allowed to build phenomenological nonlinear models for electromagneto-mechanical couplings.[10-13] This nonlinear theory is well qualified for building a linearized model about a finite deformation-polarization configuration chosen as a ferroelectric state with a permanent, i.e., spontaneous polarization. These works have concluded in favor of the existence of coupled linearized equations of the piezoelectric type about an initial finite state of polarization. This, in turn, implies a coupling of the resonance type between acoustic and ferroelectric-soft modes, as suggested by experimental studies. The continuum nature of this approach, by providing boundary conditions, has also allowed to exhibit the associated couplings for surface modes.[14,15]

The present work provides a link between a lattice theory and the phenomenological model given by the above-cited continuum theory. This work aims at constructing a dynamical lattice theory for ferroelectric media and then comparing the results of the long-wave limit with those of the continuum theory. The model accounts for the ionic displacements and rotations of the permanent dipoles created by the shift of the ionic group with respect to the center of the structure in ferroelectrics of the displacive type. The vibrations associated with this rigid-body rotational motion lead to a special vibrational mode which is neither acoustical nor optical. This mode presents a cut-off frequency and is identified with a ferroelectric soft mode. The potential instability of this wave motion is indicative of the phase transition. The continuum approximation will then provide the required comparison between the lattice approach and the afore-mentioned continuum theory. In the process the phenomenological constants introduced in the latter are determined in terms of microscopic parameters. Such a study applies particularly well to the case of perovskite crystals of the displacive ferroelectric type for which barium titanate constitutes a good prototype. The elaboration of the lattice model is in effect based on the picture of $BaTiO_3$.

2. THE MODEL

The construction of a lattice model for ferroelectrics is based on an investigation of the crystalline structure of barium titanate - as a prototype of the displacive type. - (Fig. 1). The Ti ions are shifted by an amount δ ($\delta \ll a$ = unit cell size) with respect to the center of the unit cell. There consequently exists a permanent microscopic polarization $P_0 = \delta Q$, where Q is the ionic charge of the Ti ion ($Q = 4e$ with e being the electronic charge).

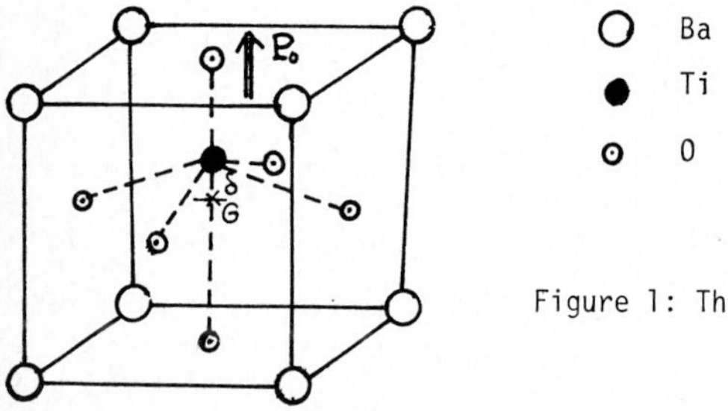

Figure 1: The unit cell for $BaTiO_3$

In the unit cell, the TiO_3 group forms a "solid" ion whose motion can be described by (i) a longitudinal displacement, (ii) a transverse displacement, and (iii) a rigid-body rotation associated with the orientation of the polarization attached to the group. This schematization is justified by the fact that the internal vibration modes of the TiO_3 group occur at much higher frequencies than those related to the motion of the O atoms with respect to the Ti ions. Similarly, the modes of vibration of Ba atoms are at much higher frequencies than those of TiO_3 groups.[16] Consequently, we are primarily interested in the coupling between the rotational mode of the TiO_3 group and the transverse acoustic mode. A more complicated model accounting for a diatomic structure is discussed in a forthcoming paper.[17]

Following along the line of the Born-Von Karman model, we schematize the ionic structure by a point, which stands for the whole of TiO_3. This point is equipped with the mass of this group and the (neutral) charge of the ($Ba + TiO_3$) structure. In addition, microscopic polarization $\vec{P}_0$ is attributed to the lattice point and, in general, $\vec{P}_0$ makes an angle Θ_0 with the axis of the structure (cf. Figure 2). The polarization is subjected to small rotations which are related to the rotation of the TiO_3 group. Let I be the corresponding moment of inertia. $\vec{X}_n$ denotes the position of the n-th particle at rest; $\vec{x}_n$ is its position after motion in such a way that the displacement $\vec{u}_n = (u_n, v_n)$ is defined as usual as $\vec{u}_n = \vec{x}_n - \vec{X}_n$. The rotational motion of the dipole $\vec{P}_0$ at x_n superimposes itself on this motion. Let Θ_n denote the perturbation angle about Θ_0.

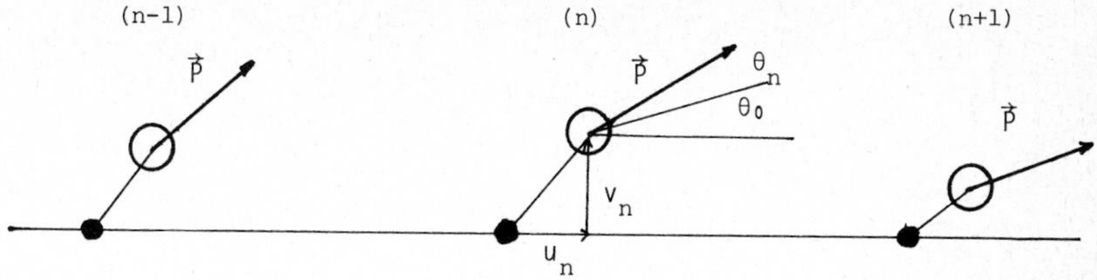

Figure 2: The notation. Lattice points n,n+1: displacement components u_n, v_n; dipole orientation: actual θ_n, initial θ_0.

The forces acting on the n-th particle result from the interaction with neighboring particles. These effects can be represented by springs with elongation coefficient $k_\parallel$ and bending coefficient $k_\perp$ for transverse motion of the chain of particles. The quadratic energy associated with the displacement motion can be written as

$$W_{def} = \frac{1}{2}\Sigma\, k_\parallel[(u_{n-1} - u_n)^2 + (u_{n+1} - u_n)^2] + k_\perp[(v_{n-1})^2 + (v_{n+1} - v_n)^2] \qquad (1)$$

The coefficients $k_\parallel$ and $k_\perp$ in fact represent the global effects of the electrostatic attractions and repulsions between the atoms. In addition to this, we must also account for the electrostatic interaction between neighboring dipoles. For two dipoles (a) and (b) - see Figure 3 - this electrostatic energy reads

$$W_{elec} \quad - \vec{P}.\vec{E}$$

where $\vec{E}$ is the electric field produced by the dipole (a) at the site of (b).

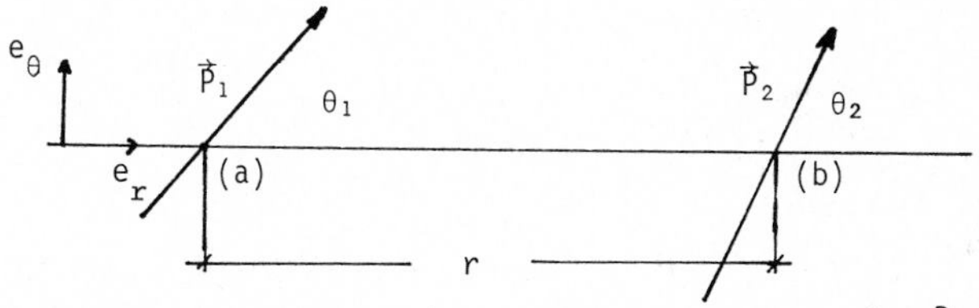

Figure 3: Interaction of two dipoles. Electric field at b: $\vec{E} = \frac{P_1}{4\pi\varepsilon_0 r^3}(\cos\theta_1, \sin\theta_1)$. Interaction energy: $W_{elec} = \frac{P_1 P_2}{4\pi\varepsilon_0 r^3}\cos(\theta_1 - \theta_2)$.

The energy for the particles n and n-1 for the motion defined by the perturbations $(u_n, u_{n-1}, v_n, v_{n-1}, \theta_n, \theta_{n-1})$ superimposed on the permanent polarization $\vec{P}_0$ is:

$$W_n^{n-1} = - \frac{P_0^2}{4\pi\varepsilon_0 a^3}\Bigg\{ 1 + \cos^2\theta_0 - 3(1+\cos\theta_0)\left(\frac{u_n - u_{n-1}}{a}\right)$$

$$- \sin\theta_0 \cos\theta_0(\theta_{n-1} + \theta_n - 2\phi_{n-1}) + 6(1+\cos^2\theta_0)\left(\frac{u_n - u_{n-1}}{a}\right)^2$$

$$- \frac{3}{2}(1+\cos\theta_0)\left(\frac{v_n - v_{n-1}}{a}\right) - \frac{1}{2}(\theta_n - \theta_{n-1})^2 \qquad (2)$$

$$- \frac{1}{2} (\cos^2\theta_0)(\theta_{n-1}-\phi_{n-1})^2 - \frac{1}{2} (\cos^2\theta_0)(\theta_n-\phi_{n-1})^2 + (\sin^2\theta_0)$$

$$(\theta_{n-1}-\phi_{n-1})(\theta_n-\phi_{n-1}) + 3 \sin\theta_0 \cos\theta_0 (\frac{u_n-u_{n-1}}{a})(\theta_{n-1}+\theta_n-2\phi_{n-1})\Big\}$$

$$+O(\theta_n^3, \phi_n^3, n_n^3, v_n^3) \qquad ; \qquad \phi_n \equiv (v_n-v_{n-1})/a$$

where ε_0 is the vacuum dielectric constant. Similarly, the kinetic energy for the combined longitudinal, transverse and rotational motions is given by

$$T = \frac{1}{2} \Sigma\, m(\dot{u}_n^2+\dot{v}_n^2) + I\dot{\theta}_n^2 \tag{3}$$

where m is the mass of the structure.

3. EQUATIONS OF MOTION

For coherence between the model and its applicability, we consider that $\vec{P}_0$ is aligned with the chain; i.e., $\theta_0 = 0$. The Euler-Lagrange equations associated with $L = T - (W_{def} + W_{elec})$ yield the equations of motion as

$$\begin{aligned} \hat{k}_{\parallel}(u_{n+1}-2u_n+v_{n-1}) &= m\ddot{u}_n \\ \hat{k}_{\perp}(v_{n+1}-2v_n+v_{n-1}) -\varepsilon(\theta_{n+1}-\theta_{n-1}) &= m\ddot{v}_n \\ \varepsilon a(\theta_{n+1}-4\theta_n+\theta_{n-1}) + \varepsilon(v_{n+1}-v_{n-1}) &= I\ddot{\theta}_n \end{aligned} \tag{4}$$

Here we have set

$$\hat{k}_{\parallel} = k_{\parallel} + 24\varepsilon/a \qquad \hat{k}_{\perp} = k_{\perp} + 8\varepsilon/a \qquad \varepsilon = P_0^2/4\pi\varepsilon_0 a^4 \tag{5}$$

The last two of eqs. (4) are coupled (coupling between transverse-displacement and dipole-rotation motions), while the longitudinal motion has the same uncoupled equation as for a monoatomic chain. However, $\hat{k}_{\parallel}$ and $\hat{k}_{\perp}$ are effective spring and bending rigidities since they contain contributions due to the microscopic polarization. This is to be compared with the perturbation stiffening of the elasticity coefficients by an initial polarization density in the linearized version of the nonlinear continuum theory of deformable ferroelectrics.[12,13] On a close look at the electrostatic energy in Eq. (2) we also remark that terms of the type $\theta_n - (v_n - v_{n-1})/a$ correspond to the difference between the "microrotation" θ_n and the "macrorotation" $(v_n - v_{n-1})/a$ (compare theories of micropolar continua in Refs. 5 and 9).

For a comparison with piezoelectric ferroelectrics, we express the rotations in terms of polarizations. For a rigid dipole rotated by an infinitesimal angle θ_n in the general setting of Figure 2, the microscopic polarization is given by

$$\vec{P} = P_0(\cos(\theta_0+\theta_n), \sin(\theta_0+\theta_n)) \simeq \vec{P}_0 + P_0\,(-\sin\theta_0, \cos\theta_0)\theta_n \tag{6}$$

In the case $\theta_0 = 0$, this allows one to define a perturbation polarization about the uniform polarization as

$$\vec{p}_n = (0, P_0\theta_n) \equiv (0, p_n) \tag{7}$$

By substituting $\theta_n = p_n/P_0$ in the lattice equations (4) and rearranging coefficients in order to give them the proper physical dimension, we obtain

$$\begin{aligned} \hat{k}_{\parallel}(u_{n+1}-2u_n+u_{n-1}) &= m\ddot{u}_n \\ \hat{k}_{\perp}(v_{n+1}-2v_n+v_{n-1}) - (\varepsilon/P_0)(p_{n+1}-p_{n-1}) &= m\ddot{v}_n \\ (\varepsilon a/P_0^2)(p_{n+1}-4p_n+p_{n-1}) + (\varepsilon/P_0)(v_{n+1}-v_{n-1}) &= (I/P_0^2)\ddot{p}_n \end{aligned} \tag{8}$$

Here (I/P_0) is the "microscopic polarization inertia".

This same reasoning can be extended to the diatomic case, straightforwardly in a conceptual sense but with lots more algebra. The dispersion curve obtained through such an extension is presented for $BaTiO_3$ in Figure 4.

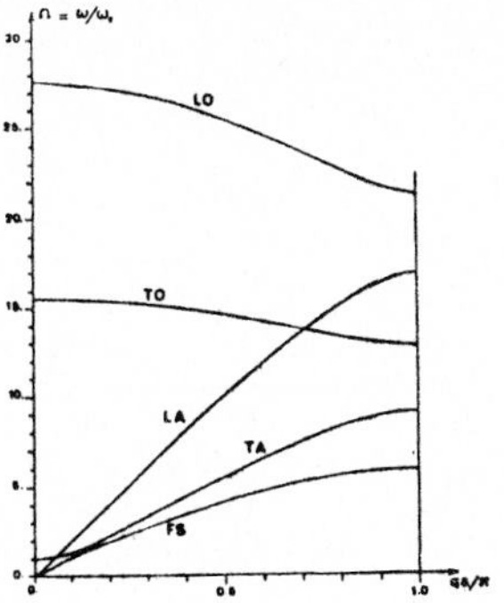

Figure 4: Dispersion relation for $BaTiO_3$. L: longitudinal; T: transversal; A: acoustic; O: optical.

The parameters of the models are obtained from crystal structure, elasticity coefficients and Raman scattering for the cut-off frequency for the soft mode.[17]

4. THE CONTINUUM APPROXIMATION

Towards establishing a comparison with the field equations obtained within the continuum mechanics framework for micropolar media (cf. Ref 7) and piezoelectric ferroelectrics (cf. Refs. 12,13), we consider the long wavelength approximation for which $qa \ll 1$. For that purpose, following along classical arguments, we expand displacements u_n, v_n and rotations θ_n or polarizations p_n in Taylor series such as e.g.

$$u_{n\underline{+}1} = u_n \pm a u_x + \frac{1}{2} a^2 u_{xx} \pm \ldots \tag{9}$$

The system (8) yields a system of partial differential equations for the continuous variables u, v and p_n as

$$\begin{aligned} (\hat{k}/a)\, u_{xx} &= \rho u_{tt} \\ (\hat{k}/a)\, v_{xx} - (2\varepsilon/{}_0Pa^2)\, p_x &= \rho v_{tt} \\ (\varepsilon/{}_0P^2)\, p_{xx} - (2\varepsilon/{}_0P^2a^2)\, p + (2\varepsilon/{}_0Pa^2)\, v_x &= d p_{tt} \end{aligned} \tag{10}$$

Here ${}_0P$ is the macroscopic polarization per unit volume (${}_0P = P_0/a^3$), so that p is the macroscopic-polarization perturbation. The "inertia" associated with the polarization "motion" is defined by $d = I/{}_0P^2$. On the other hand, the three-dimensional equations of the continuum theory of piezoelectric ferroelectrics (cf. Refs. 29, 30) specialized to the one-dimensional motion considered in the present study,

$$\begin{aligned} (C_1+C_2+2{}_0P^2(C_3+C_4+C_5+2{}_0P^2C_4))\, u_{xx} &= \rho u_{tt} \\ (C_2+{}_0P^2(C_4+C_5-C_8))\, v_{xx}+2\zeta\,{}_0PC_8\pi_x &= \rho v_{tt} \\ (\overline{C}_{14}+{}_0P^2\overline{C}_{17})\pi_{xx}-2\zeta C_4\pi-2\,{}_0PC_8v_x &= d\pi_{tt} \end{aligned} \tag{11}$$

where π now is the macroscopic polarization per unit mass and the C_i's and $\overline{C}_i$'s are phenomenological coefficients. In view of the approximation in eq. (10), macroscopic coefficients of a continuous piezoelectric ferroelectric medium can be identified with combinations of microscopic parameters of the crystalline structure. In eq. (10) the coefficient $(2\varepsilon/{}_0P^2a^2)$ is the piezoelectric coefficient of the continuum theory while the term involving the coefficient $(\varepsilon/_0P^2)$ is of the same type as the term resulting from the inclusion of polarization gradients in the continuum theory (in Refs. 11, 12) whose purpose was to account for the ferroelectric ordering. Remark that in the continuum theory the piezoelectric constant C_8 and the dielectric susceptibility C_4 are a priori independent of one another, while in the continuum approximation to the lattice theory the corresponding coefficients are related since we can write $(2\varepsilon a/_0P^2) = a^3(2\varepsilon/_0P\ a^2)/_0P$. The same remark holds true for the polarization-gradient coefficient.

5. DISCUSSION

The goal of this work was, on the one hand, to present a simple lattice model for the dynamics of ferroelectric crystals of the displacive type and, on the other, to establish a physical basis for two phenomenological theories, the micropolar elasticity of crystals and the nonlinear piezoelectricity theory of solids with permanent dipoles. The latter goal is achieved by considering the long-wave limit of the lattice equations and the identification of the macroscopic coefficients in terms of the lattice parameters allows one to calculate the numerical values of the former for specific crystals. The results for the material coefficients are presented for the case of the well-studied barium titanate in a forthcoming paper[17] which accounts also for the diatomic nature of the crystal with Ba and (TiO_3) as lattice points. Furthermore, the onset of instability for the soft optical mode provides the criterion for the phase transition. This can be incorporated through the temperature dependence of the material coefficients, in particular in $\hat{k}_\perp$.

REFERENCES:

[1] Cochran, W., Advan.Phys., 9, (1960) 387.
[2] Anderson, P.W., in Fizika Dielektrikov, edited by G.I. Skamavi (Akad.Nauk SSSR, Fizicheskii Inst. Lebedeva, Moscow, 1960).
[3] Scott, J.F., Rev. Modern Phys. 46 (1974) 83.
[4] Ludwig, W., Recent Developments in Lattice Theory (Springer-Verlag, Berlin, 1967).
[5] Askar, A. and Cakmak, A.S., Int. J. Engng. Sci., 6 (1968) 583.
[6] Askar, A., Int. J. Engng. Sci., 10 (1972) 293.
[7] Askar, A., J. Phys. Chem. Solids, 34 (1973) 1901.
[8] Fischer-Hjalmars, I., Int. J. Engng. Sci., 19 (1981) 1765.
[9] Eringen, A.C., in: Fracture: A treatise, edited by H. Liebowitz, p. 621, Vol. II (Academic Press, New York, 1968).
[10] Maugin, G.A. and Eringen, A.C., J. Mécanique, 16 (1977) 101.
[11] Maugin, G.A., Acta Mechanica, 35 (1980 1-70.
[12] Maugin, G.A. and Pouget, J., J. Acoust. Soc. Amer., 68 (1980) 575.
[13] Pouget, J. and Maugin, G.A., J. Acoust. Soc. Amer., 68 (1980) 588.
[14] Pouget, J. and Maugin, G.A., J. Acoust. Soc. Amer., 69 (1981) 1304.
[15] Pouget, J. and Maugin, G.A., J. Acoust. Soc. Amer., 69 (1981) 1319.
[16] Barret, H.H., in: Physical Acoustics, edited by W.P. Mason, Vol. VI, pp 65-108 (Academic Press, New York, 1970).
[17] Pouget, J, Askar, A., and Maugin, G.A., Diatomic Lattice Model for Elastic Ferroelectrics: Couple Wave Modes and Long Wave Approximation, preprint,1983 .

The Mechanical Behavior of Electromagnetic Solid Continua
G.A. Maugin (editor)
Elsevier Science Publishers B.V. (North-Holland)

SHOCK WAVES IN DEFORMABLE FERROELECTRIC MATERIALS

Bernard Collet

Laboratoire de Mécanique Théorique associé au C.N.R.S.
Université Pierre et Marie Curie
Tour 66, 4 place Jussieu, 75230 Paris Cedex 05
France

The behavior of one-dimensional shock waves in deformable ferroelectric materials which are non-conductor of heat and in which internal state variables describe the mechanical and electrical relaxation processes, is analyzed in the case of quasi-electrostatics. The differential equation governing the amplitude of wave and criteria concerning the polarization per unit mass and temperature changes across the shock are deduced.

INTRODUCTION

We analyze the behavior of one-dimensional shock waves propagating in nonlinear ferroelectric materials which are non-conductor of heat and in which internal state variables describe the mechanical and electrical relaxation processes [1-2]. We review the kinematics and the relevant field equations for this class of materials. We introduce the constitutive assumptions and note the restrictions imposed by the rational thermodynamics. In the proposed modelization, we use the recent notion of electromagnetic internal state variable [3-6]. The constitutive equations obtained are well adapted to the response of ferroelectric ceramics subjected to mechanical or electrical disturbances. We note the quasi-exclusive use of these materials in devices of electromechanical energy conversion under shock compression. After having derived equations for the propagation velocity and the amplitude of the shock wave as a function of the jumps in certain variables across, and the state of the material immediately ahead of, the wave front, we examine the properties of the shock transition. In particular, we find that the classical results of shock transition can be generalized to the present situation. It is found that the evolutionary behavior of the amplitude of the shock depends on the relative magnitudes of the jump in strain gradient and the quantity λ_c, called the critical jump in strain gradient. We examine the properties of the polarization per unit mass and temperature during shock transition. Finally, we specialize the analysis to the case of weak shocks.

FIELD EQUATIONS AND CONSTITUTIVE ASSUMPTIONS

The one-dimensional generalized motion of deformable ferroelectrics which do not conduct heat and are placed in a quasi-electrostatic background, is described by $x = \chi(X,t)$ and $\pi = \chi_\pi(X,t)$ giving the spatial position x and the polarization per unit mass at time t of the particle which occupied the position X in the reference configuration K_o with mass density ρ_o. We associate with this generalized motion the following quantities: $F = \partial_X \chi(X,t) > 0$, the deformation gradient; $T(X,t)$, the mechanical stress; $T^{em}(X,t)$, the Maxwell stress; $\overline{T}^{em}(X,t)$, the "effective" Maxwell stress; $b(X,t)$, the body force density of non electromagnetic origin; ${}^L E(X,t)$, the local electric field; $E(X,t)$ the actual electric field; $\overline{E}^{em}(X,t)$, the associated em electric field; $e(X,t)$, the internal energy per unit mass; $\overline{e}^{em}(X,t)$, the "effective" ponderomotive electromagnetic energy per unit mass; $r(X,t)$, the heat supply density; $\theta(X,t)>0$, the thermodynamical temperature; $\eta(X,t)$, the entropy per unit mass; $\overline{E}(X,t)$, the reference electric field and $D(X,t)$ the electric displacement. In addition, we define the energy per unit mass $\Gamma(X,t)$, the "effective" stress $\Sigma(X,t)$ and the "Γ"

local electric field ${}^{L}E_r(X,t)$ by the relations

$$\Gamma = e + \bar{e}^{em}, \qquad \Sigma = T + \bar{T}^{em}, \qquad {}^{L}E_r = {}^{L}E + \bar{E}^{em}, \tag{1}$$

where

$$\bar{e}^{em} = \frac{\rho_o}{2} F^{-1}\pi^2, \qquad \bar{T}^{em} = (T^{em} - \frac{1}{2} D^2) = \rho_o \partial_F \bar{e}^{em},$$
$$\bar{E}^{em} = -\partial_\pi \bar{e}^{em}, \qquad \bar{E} = FE, \qquad D = F^{-1}(\bar{E} + \rho_o \pi). \tag{2}$$

Ferroelectrics are quite common dielectrics materials with an ordered state which possess the essential property to exhibit a spontaneous electric polarization below a certain phase-transition temperature θ_c [7]. The local field equations and the associated jump conditions† across the shock wave, which propagates in deformable ferroelectrics, can be written as follows [8-9]:

(i) Balance of momentum

$$\partial_X \Sigma + \rho_o b = \rho_o \ddot{x}, \qquad [\Sigma] = -\rho_o U[\dot{x}], \qquad [\partial_X \Sigma] = \rho_o[\ddot{x}], \tag{3}$$

(ii) Balance of electric fields

$${}^{L}E_r + D = 0, \qquad [{}^{L}E_r + D] = 0, \tag{4}$$

(iii) Balance of energy (first principle of thermodynamics)

$$\rho_o \dot{\Gamma} = \Sigma \dot{F} - \rho_o {}^{L}E_r \dot{\pi} + \rho_o r,$$
$$-\rho_o U[\Gamma + \frac{FD^2}{2\rho} - \pi D + \frac{1}{2}\dot{x}^2] = [(\Sigma + \frac{1}{2}D^2)\dot{x}] - [\Phi \dot{D}], \tag{5}$$
$$\rho_o[\dot{\Gamma}] = [\Sigma \dot{F}] - \rho_o[{}^{L}E_r \dot{\pi}],$$

(iv) Entropy inequality (second principle of thermodynamics)

$$\theta\dot{\eta} \geqslant r, \qquad [\eta] \geqslant 0, \qquad [\theta\dot{\eta}] \geqslant 0, \tag{6}$$

(v) The quasi-electrostatic Maxwell equations (in the absence of free charges)

$$\bar{E} = -\partial_X \Phi, \qquad [\Phi] = 0, \qquad [\bar{E}] = -[\partial_X \Phi],$$
$$\partial_X D = 0, \qquad [D] = 0, \qquad [\partial_X D] = 0, \tag{7}$$

where $U > 0$ is the intrinsic velocity of the shock and $[f] = f^- - f^+$ denotes the jump of a quantity f across the wave, f^- and f^+ are respectively, the limiting values of f immediately behind and just in front of the shock wave.

According to the working hypotheses of internal-variable theory, we assume that the responses of the material are characterized by:

$$\Gamma = \tilde{\Gamma}(g(X,t)), \quad \Sigma = \tilde{\Sigma}(g(X,t)), \quad {}^{L}E_r = {}^{L}\tilde{E}_r(g(X,t)), \quad \theta = \tilde{\theta}(g(X,t)),$$
$$g(X,t) = (F, \pi, \eta, F^{int}, \pi^{int}), \tag{8}$$
$$\dot{F}^{int} = \tilde{A}(g(X,t)), \quad \dot{\pi}^{int} = \tilde{B}(g(X,t)), \quad F^{int}(t_o) = F_o^{int}, \quad \pi^{int}(t_o) = \pi_o^{int},$$

where F^{int} and π^{int} are anelastic and electric internal-variables (with the same physical dimension as F and π respectively). It is well known that the entropy inequality $(6)_1$ can only hold for every thermodynamic process if the responses functions (8) satisfy certain restrictions. These are:

† The jump conditions are given in the absence of b and r.

$$\Sigma = \rho_o \partial_F \tilde{\Gamma}, \qquad {}^L E_r = - \partial_\pi \tilde{\Gamma}, \qquad \theta = \partial_\eta \tilde{\Gamma},$$
$$\Sigma^{int} \dot{F}^{int} + E^{int} \dot{\pi}^{int} \geq 0, \tag{9}$$

where Σ^{int} and E^{int} are defined by

$$\Sigma^{int} = - \partial_{F^{int}} \tilde{\Gamma}, \qquad E^{int} = - \partial_{\pi^{int}} \tilde{\Gamma}. \tag{10}$$

It is simple matter to show from constitutive eqns $(9)_{1-3}$ and definitions (10) that $(5)_1$ in the absence of heat supply density, reduces to

$$\theta \dot{\eta} = \Sigma^{int} \dot{F}^{int} + E^{int} \dot{\pi}^{int}. \tag{11}$$

We suppose there exists a scalar function R(X,t) or pseudo-potential [1]

$$R = \overline{R}(\overline{g}(X,t)), \qquad \overline{g}(X,t) = (F, \pi, \eta, \Sigma^{int}, E^{int}), \tag{12}$$

a convex function in its arguments Σ^{int} and E^{int} and such that

$$\tilde{A} = \partial_{\Sigma^{int}} \overline{R}, \qquad \tilde{B} = \partial_{E^{int}} \overline{R}. \tag{13}$$

From (8) it is clear that (13) is the "classical" form of evolution equations of the theory of deformable ferroelectrics with internal state variables (or hidden variables) [5]. If the "relaxation potential" R is at most a quadratic positive definite function with respect Σ^{int} and E^{int}, the dissipation inequality $(9)_4$ is satisfied. Futhermore if $\tilde{\Gamma}$ is quadratic in F^{int} and π^{int} then (13) is a linear relatioship between $(\dot{F}^{int}, \dot{\pi}^{int})$ and (F^{int}, π^{int}) from which it is possible to obtain realistic physical relaxation laws.
We now assume that the function $\tilde{\Gamma}$ is of class C^3; thus by $(9)_{1-3}$, Σ, ${}^L E_r$ and θ are of class C^2. For later use we define the following quantities:

$$a_{1,2,3,4,5} = \partial_{F,\pi,\eta,F^{int},\pi^{int}} \tilde{\Sigma}, \qquad a_1 > 0, \qquad a_2 < 0,$$
$$b_2 = \partial_\pi {}^L\tilde{E}_r < 0, \qquad b_{3,4,5} = - \partial_{\eta,F^{int},\pi^{int}} {}^L\tilde{E}_r,$$
$$G = a_3 + b_3 I, \qquad H = a_1 + (a_2/\rho_o) I, \qquad I = a_2/b_2, \tag{14}$$
$$J = I^- + I^+ - 2 \frac{[\pi]}{[F]}, \qquad K = a_4 + b_4 I, \qquad L = a_5 + b_5 I,$$
$$H_1 = \partial_F H, \qquad H_2 = \partial_\pi H, \qquad G_1 = \partial_F G, \qquad K_1 = \partial_F K, \qquad L_1 = \partial_F L.$$

ON THE BEHAVIOR OF SHOCK WAVES

The generalized motion is said to contain a shock wave if:
S.1 the functions χ, F^{int}, π^{int} and Φ are continuous everywhere,
S.2 the functions $\dot{x}$, F, χ_π, η, $\dot{F}^{int}$, $\partial_X F^{int}$, $\dot{\pi}^{int}$, $\partial_X \pi^{int}$, $\dot{\Phi}$, $\partial_X \Phi$ and all higher order derivatives may suffer jump discontinuities across the shock wave but are continuous everywhere else.
The method of derivation of the expressions which govern the properties of shock waves are well-know [10-14]. Hence rather than repeating some lengtly intermediate calculations, we present the main results and examine their consequences.

(i) The intrinsic velocity of the shock wave and the generalization of the familiar Hugoniot relation† to the case of non heat conducting deformable ferroelectrics are respectively given by

$$\rho_o U^2 = [\Sigma]/[F],$$
$$\rho_o [\Gamma] - \frac{1}{2} (\Sigma^+ + \Sigma^-)[F] + \rho_o {}^L E_r^+ [\pi] = 0. \tag{15}$$

† Note that this relation takes an other remarkable form if the definitions (1) are used.

(ii) The amplitude of a shock wave propagating in nonlinear deformable ferroelectric materials with internal state variables which do not conduct heat obeys the equation ($\delta/\delta t$ denotes the displacement derivative)

$$\frac{\delta}{\delta t}[F] = \frac{U(1-\xi)(2\tau-1)}{(3\xi+1)\tau-(3\xi-1)}\{\lambda_c - [\partial_X F]\}, \tag{16}$$

where

$$\begin{aligned}
\lambda_c = -\frac{1}{H^-(1-\xi)(2\tau-1)} & \Big\{\{3H^-(1-\xi)+(\tau-2)[H]-\frac{3}{2}\frac{\overset{+}{a}_2}{\rho_o}J-(\tau-\frac{1}{2})\frac{\overset{+}{a}_2}{\rho_o}[I]\}\frac{\dot{F}^+}{U} \\
& +3\{H^-(1-\xi)+(\tau-1)[H]\}(\partial_X F)^+ \\
& +\{\frac{3}{2}\overset{+}{b}_2 J+(\tau-\frac{1}{2})\overset{+}{b}_2[I]\}\frac{\dot{\pi}^+}{U} \\
& +3\{\frac{1}{2}(G^-+G^+)+(\tau-\frac{1}{2})[G]-\rho_o\frac{[\theta]}{[F]}\}(\partial_X\eta)^+ \\
& +\{3K^+ +(\tau+1)[K]-(\frac{3}{2}\overset{+}{b}_4 J+(\tau-\frac{1}{2})\overset{+}{b}_4[I]) \\
& +(3G^+ +(\tau+1)[G]-(\frac{3}{2}\overset{+}{b}_3 J+(\tau-\frac{1}{2})\overset{+}{b}_3[I]))\frac{\Sigma^{int+}}{\theta^+} \\
& +\frac{\rho_o(1+\tau)}{\tau}(\frac{[\Sigma^{int}]}{[F]}-\frac{\Sigma^{int+}}{\theta^+}\frac{[\theta]}{[F]})\}\frac{\dot{F}^{int+}}{U} \\
& +3\{\frac{1}{2}(K^-+K^+)+(\tau-\frac{1}{2})[K]+\rho_o\frac{[\Sigma^{int}]}{[F]}\}(\partial_X F^{int})^+ \\
& +\{3L^+ +(\tau+1)[L]-(\frac{3}{2}\overset{+}{b}_5 J+(\tau-\frac{1}{2})\overset{+}{b}_5[I]) \\
& +(3G^+ +(\tau+1)[G]-(\frac{3}{2}\overset{+}{b}_3 J+(\tau-\frac{1}{2})\overset{+}{b}_3[I]))\frac{E^{int+}}{\theta^+} \\
& +\frac{\rho_o(1+\tau)}{\tau}(\frac{[E^{int}]}{[F]}-\frac{E^{int+}}{\theta^+}\frac{[\theta]}{[F]})\}\frac{\dot{\pi}^{int+}}{U} \\
& +3\{\frac{1}{2}(L^-+L^+)+(\tau-\frac{1}{2})[L]+\rho_o\frac{[E^{int}]}{[F]}\}(\partial_X\pi^{int})^+ \\
& +\{(2\tau-1)((K^-+\frac{\rho_o}{\tau}\frac{\Sigma^{int-}}{[F]})[\partial_X F^{int}]+(L^-+\frac{\rho_o}{\tau}\frac{E^{int-}}{[F]})[\partial_X\pi^{int}])\}\Big\}
\end{aligned} \tag{17}$$

and

$$\xi = \rho_o U^2/H^-, \qquad \tau = \rho_o\theta^-/G^-[F]. \tag{18}$$

Clearly, equation (16) is extremely complicated and we can only hope to deduce useful informations from it by adopting additional assumptions concerning the properties of the material and the nature of the shock wave under consideration.

(iii) Let us turn our attention to the properties of shock transition. In view of our assumption $b_2 < 0$, it follows that ${}^L\tilde{E}_r$ is invertible, i.e., there exists a function

$$\pi = \hat{\pi}(h(X,t)), \qquad h(X,t) = (F, {}^LE_r, \eta, F^{int}, \pi^{int}). \tag{19}$$

Thus, in particular, we can define the functions $\hat{\Gamma}$ and $\hat{\Sigma}$ and determine the derivatives of $\hat{\Sigma}$ with respect to the deformation gradient and the entropy per unit mass and assume these derivatives satisfy the conditions

$$\Gamma = \hat{\Gamma}(h(X,t)), \qquad \Sigma = \hat{\Sigma}(h(X,t)), \qquad \partial_F\hat{\Sigma} = H > 0, \qquad \partial_\eta\hat{\Sigma} = G < 0. \tag{20}$$

It follows from $(20)_{2,4}$ that there exists a fonction $\hat{\hat{\eta}}$ such that

$$\eta = \hat{\hat{\eta}}(\hat{h}(X,t)), \qquad \hat{h} = (F, {}^{L}E_{r}, \Sigma, F^{int}, \pi^{int}). \tag{21}$$

If, at given time, we now fix the values of F^{+}, π^{+}, η^{+}, F^{int+} and π^{int+} just ahead of the shock wave, and in view of $(8)_3$ the value of ${}^{L}E_{r}^{+}$, then the thermodynamic state immediately behind the shock wave is determined once F^{-} and Σ^{-} are known. Obviously, F^{-} and Σ^{-} are not arbitrary but they must satisfy the Hugoniot relation $(15)_2$. As in usual shock wave studies, we assume that this relation can be represented by the Hugoniot $\Sigma^{-} = \Sigma_{H}(F^{-})$ curve. The function Σ_{H} also depends on F^{+}, ${}^{L}E_{r}^{+}$, η^{+}, F^{int+} and π^{int+} or, alternatively, on F^{+}, π^{+}, η^{+}, F^{int+} and π^{int+}.
With the preceding assumption, we can prove the following:
If in addition to $(20)_3$ we assume that

$$\partial_{F}\hat{\Sigma}(h(X,t)) < 0 \quad \text{for all } h(X,t),\ F < 1 \tag{22}$$

then

(a) The wave is compressive, i.e., $[F] < 0$;
(b) The intrinsic speed of the shock is supersonic with respect to the material ahead of the shock and subsonic with respect to the material behind the shock, i.e.,

$$H^{+} < \rho_{o} U^{2} < H^{-}, \tag{23}$$

(c) Along the whole Hugoniot curve the entropy increases with decreasing deformation gradient.
The functions introduced in (19) and (21) and our assumptions concerning the Hugoniot relation (15) imply that there exist functions η_{H} and π_{H} such that

$$\begin{aligned}[][\eta] &= \eta_{H}([F], F^{+}, \pi^{+}, \eta^{+}, F^{int+}, \pi^{int+}),\\ [\pi] &= \pi_{H}([F], F^{+}, \pi^{+}, \eta^{+}, F^{int+}, \pi^{int+}).\end{aligned} \tag{24}$$

While we do not, in general, know the explicit forms of the functions η_{H} and π_{H}, we can determine explicit expressions for their derivatives with respect to any variables on which they depend; we obtain

$$\begin{aligned}
\frac{\partial\eta_{H}}{\partial[F]} &= \frac{H^{-}(1-\xi)}{G^{-}(2\tau-1)},\\
\frac{\partial\eta_{H}}{\partial F^{+}} &= \{2 - \frac{[H]}{H^{-}(1-\xi)} - \frac{a_{2}^{+}J}{\rho_{o}H^{-}(1-\xi)}\}\frac{\partial\eta_{H}}{\partial[F]},\\
\frac{\partial\eta_{H}}{\partial\pi^{+}} &= \frac{b_{2}^{+}J}{H^{-}(1-\xi)}\frac{\partial\eta_{H}}{\partial[F]},\\
\frac{\partial\eta_{H}}{\partial\eta^{+}} &= \frac{1}{H^{-}(1-\xi)}\{G^{-} + G^{+} - b_{3}^{+}J - 2\rho_{o}\frac{[\theta]}{[F]}\}\frac{\partial\eta_{H}}{\partial[F]},\\
\frac{\partial\eta_{H}}{\partial F^{int+}} &= \frac{1}{H^{-}(1-\xi)}\{K^{-} + K^{+} - b_{4}^{+}J + 2\rho_{o}\frac{[\Sigma^{int}]}{[F]}\}\frac{\partial\eta_{H}}{\partial[F]},\\
\frac{\partial\eta_{H}}{\partial\pi^{int+}} &= \frac{1}{H^{-}(1-\xi)}\{L^{-} + L^{+} - b_{5}^{+}J + 2\rho_{o}\frac{[E^{int}]}{[F]}\}\frac{\partial\eta_{H}}{\partial[F]}
\end{aligned} \tag{25}$$

and

$$\frac{\partial\pi_{H}}{\partial[F]} = \frac{a_{2}^{-}}{\rho_{o}b_{2}^{-}} + \frac{b_{3}^{-}}{b_{2}^{-}}\frac{\partial\eta_{H}}{\partial[F]}, \qquad \frac{\partial\pi_{H}}{\partial F^{+}} = \frac{[a_{2}]}{\rho_{o}b_{2}^{-}} + \frac{b_{3}^{-}}{b_{2}^{-}}\frac{\partial\eta_{H}}{\partial F^{+}},$$

$$\frac{\partial \pi_H}{\partial \pi^+} = -\frac{[b_2]}{b_2^-} + \frac{b_3^-}{b_2^-}\frac{\partial \eta_H}{\partial \pi^+}, \qquad \frac{\partial \pi_H}{\partial \eta^+} = \frac{[b_3]}{b_2^-} + \frac{b_3^-}{b_2^-}\frac{\partial \eta_H}{\partial \eta^+}, \tag{26}$$

$$\frac{\partial \pi_H}{\partial F^{int+}} = \frac{[b_4]}{b_2^-} + \frac{b_3^-}{b_2^-}\frac{\partial \eta_H}{\partial F^{int+}}, \qquad \frac{\partial \pi_H}{\partial \pi^{int+}} = \frac{[b_5]}{b_5^-} + \frac{b_3^-}{b_2^-}\frac{\partial \eta_H}{\partial \pi^{int+}}.$$

Henceforth let us consider a compressive shock entering a material which is initially in compression, i.e., $[F] < 0$, $F^+ < 1$. A straightforward use of the preceding results allows one to obtain the following criteria:

$$[\partial_X F] \lesseqgtr \lambda_c \Longleftrightarrow \frac{\delta}{\delta t} \,\|[F]\| \lesseqgtr 0. \tag{27}$$

In accord with the traditional nomenclature, we call λ_c the critical jump in the gradient of the deformation gradient, or strain gradient, for shock waves in deformable ferroelectric materials. On the other hand, considerable simplification of (17) is obtained if we consider a shock wave propagating in a region which is a steady and uniform state ahead of the wave; the expression (17) then reduces to

$$\lambda_c = - N/H^-(1 - \xi), \tag{28}$$

where

$$N = \{(K^- + \frac{\rho_o}{\tau}\frac{\Sigma^{int-}}{[F]})(\partial_X F^{int})^- + (L^- + \frac{\rho_o}{\tau}\frac{E^{int-}}{[F]})(\partial_X \pi^{int})^-\}.$$

It follows from (27) and (28) that the amplitude of the shock wave will remain constant as the wave traverses the material if

$$(\partial_X F)^- + N/H^-(1 - \xi) = 0. \tag{29}$$

(iv) Let us now we examine the properties of the polarization per unit mass and temperature during a shock transition. In order to examine the properties of the jump in temperature across the shock, we assume that $\tilde{\theta}(F,\pi,..,F^{int},\pi^{int})$ is invertible and introduce the specific heat c supposed strictly positive for all $g(X,t)$ and define by the relation

$$c = \tilde{c}(g(X,t)) = \tilde{\theta}(g(X,t))\{\partial_\eta \tilde{\theta}(g(X,t))\}^{-1}. \tag{30}$$

Differentiating the jump of the temperature across the wave with respect to $[F]$, we find

$$\frac{\partial[\theta]}{\partial[F]} = \frac{G^-}{\rho_o} + \{\frac{\theta^-}{c^-} + \frac{(b_3^-)^2}{b_2^-}\}\frac{\partial \eta_H}{\partial[F]}. \tag{31}$$

It follows that we obtain, for a compressive shock with $F^+ < 1$, the following:

(a) If $(\theta^-/c^-) > -(b_3^-)^2/b_2^- \geqslant 0$, then $\frac{\partial[\theta]}{\partial[F]} < 0$, (32)

(b) If $(\theta^-/c^-) < -(b_3^-)^2/b_2^-$, then

$$\frac{\partial[\theta]}{\partial[F]} < 0 \quad \text{whenever} \quad -\frac{G^-}{\rho_o} > \{(\theta^-/c^-) + (b_3^-)^2/b_2^-\}\frac{\partial \eta_H}{\partial[F]},$$

or (33)

$$\frac{\partial[\theta]}{\partial[F]} > 0 \quad \text{whenever} \quad -\frac{G^-}{\rho_o} < \{(\theta^-/c^-) + (b_3^-)^2/b_2^-\}\frac{\partial \eta_H}{\partial[F]}.$$

Further, for a compressive shock wave with $F^+ < 1$, it is possible to derive the following:

(a) If $b_3^- \geqslant 0$, then $\dfrac{\partial[\pi]}{\partial[F]} > 0$, (34)

(b) If $b_3^- < 0$, then

$$\frac{\partial[\pi]}{\partial[F]} > 0 \quad \text{whenever} \quad -(a_2^-/\rho_o) > b_3^- \frac{\partial\eta_H}{\partial[F]} ,$$

or (35)

$$\frac{\partial[\pi]}{\partial[F]} < 0 \quad \text{whenever} \quad -(a_2^-/\rho_o) < b_3^- \frac{\partial\eta_H}{\partial[F]} .$$

(v) We report some asymptotic results for weak shock waves, i.e., we consider the case of a shock wave of infinitesimal amplitude ($\|[F]\|$) << 1. We give only asymptotic expressions for $[\pi]$, $[\eta]$, $[\Sigma]$, $[\theta]$, $[\Sigma^{int}]$, $[E^{int}]$ and U

$$\begin{aligned}
[\pi] &= \frac{1}{\rho_o} \{I^+[F] + \frac{1}{2} \frac{H_2^+}{b_2^+}[F]^2\} + o([F]^2), \\
[\eta] &= \frac{1}{12\rho_o\theta^+}\{H_1^+[F]^3\} + o([F]^3), \\
[\Sigma] &= H^+[F] + \frac{1}{2} H_1^+[F]^2 + o([F]^2), \\
[\theta] &= \frac{1}{\rho_o} \{G^+[F] + \frac{1}{2} G_1^+[F]^2\} + o([F]^2), \qquad (36)\\
[\Sigma^{int}] &= -\frac{1}{\rho_o} \{K^+[F] + \frac{1}{2} K_1^+[F]^2\} + o([F]^2), \\
[E^{int}] &= -\frac{1}{\rho_o} \{L^+[F] + \frac{1}{2} L_1[F]^2\} + o([F]^2), \\
U^2 &= U_o^2 + \frac{1}{2\rho_o} H_1^+[F] + o([F]), \quad U_o^2 = H^+/\rho_o .
\end{aligned}$$

We remark that the presence of the polarization per unit mass and the internal state variables does not modify the conclusion that the jump in entropy per unit mass across a weak shock is of order three in the jump in the deformation gradient. This result is not surprising since the material is considered as non conductor of heat and the internal state variables are assumed continuous everywhere.

REFERENCES

[1] SIDOROFF, F., Variables Internes en Viscoélasticité et Plasticité, Thèse de Doctorat D'Etat Es Sciences Mathématiques, Univ.Paris VI, mimeographed, Paris. (1976).
[2] DANIEL, V.V., Dielectric Relaxation (Academic Press,New-York,1967).
[3] KLUITENBERG, G.A., Physica.87A (1977) 302-330; 109 (1981) 91-122.
[4] MAUGIN, G.A., J. Mécanique. 18 (1979) 541-563.
[5] MAUGIN, G.A., Arch. Mech. 33 (1981) 927-935.
[6] CHEN, P.J. and PEERCY.P.S., Acta Mechanica. 31 (1979) 231-241.
[7] JONA, F. and SHIRANE, G., Ferroelectric Crystals (Mac Millan,New-York,1962).
[8] MAUGIN, G.A., Acta Mechanica. 35 (1980) 1-70.
[9] COLLET, B., Int. J. Engng Sci. 20 (1982) 1145-1160.
[10] CHEN, P.J., Growth and Decay of Waves in Solids,in: C.Truesdell.(Ed), Handbuch der Physik, Band VI-3 (Springer,Berlin,1973).
[11] CHEN, P.J. and McCARTHY, M.F., Int. J. Solids Structures. 10 (1974) 1229-1242.
[12] CHEN, P.J., McCARTHY, M.F. and O'LEARY, T.R., Arch. Rat. Mech. Anal. 62 (1976) 189-207.
[13] McCARTHY, M.F. and TIERSTEN, H.F., J. Appl. Phys. 48 (1977) 159-166.
[14] KOSINSKI, W., Arch. Mech. 27 (1975) 445-458.

Part 5

PIEZOELECTRIC POWDERS

The Mechanical Behavior of Electromagnetic Solid Continua
G.A. Maugin (editor)
Elsevier Science Publishers B.V. (North-Holland)

ELECTROACOUSTIC ECHOES IN PIEZOELECTRIC POWDERS

Georgii A. Smolensky and Nikolai K. Yushin

A.F. Ioffe Physico-Technical Institute
of the Academy of Sciences of the USSR
194021 Leningrad K21
USSR

The electroacoustic echoes are of two general types - dynamic and memory echoes. The dynamic echoes are found to be a consequence of the anharmonicity connected with fourth-order elastic constants. Various hypotheses proposed to describe the memory echoes include either a physical rotation of individual particles or the formation of a static internal polarization and strain of the particles.

INTRODUCTION

Electroacoustic echoes represent the coherent pulse reemision by a piezoelectric sample at certain times after application to the sample of rf pulses at times $t = 0, \tau$, and T. It is schematicaly shown in Figure 1. The electroacoustic echoes in piezoelectrics are one of a wide class of phenomena which are named the echoes, and observed in a lot of nonlinear system [1].

This report gives a review of experimental and theoretical studies of the electroacoustic echoes. On the basis of present experiments it is concluded that two-pulse echoes in piezoelectric powders are due to an anharmonic oscillator interaction while the echoes in single crystals result from a parametric electroacoustic interaction.

Two models describing three-pulse memory echoes are discussed. One is associated with a physical rotation of particles. The second describes occurence of the echoes via inner nonlinear processes resulting in a long-lived acoustic hologram.

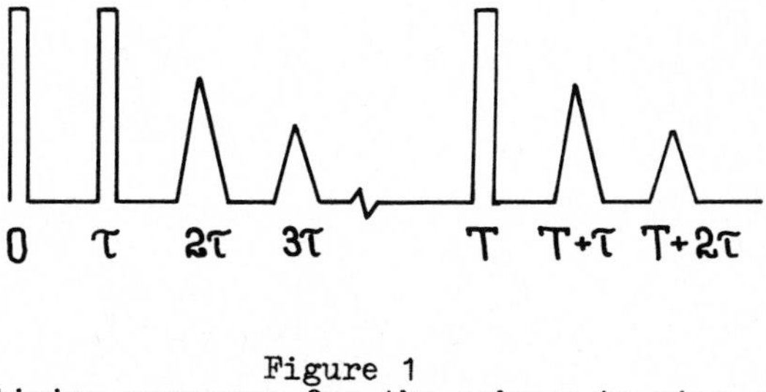

Figure 1
Schematic of timing sequence for the echoes in piezoelectric powder

TWO-PULSE DYNAMIC ECHOES

Experimental results

The two-pulse electroacoustic echoes were discovered in 1970 [2, 3]. Since then the echoes were studied in a large number of piezoelectrics [1].

Radio and microwave spectroscopy as well as ultrasonics equipments are used for observation of the electroacoustic echoes. The method of two-pulse echo observation is as follows. Two rf pulses with amplitudes A_1 and A_2 and widths Δ_1 and Δ_2 are applied to the powder sample at times $t = 0$ and τ. Usually, the pulse width ranges between o.1 and 20 μs. In experiment the pulse repetition rate is to be less than $1/\tau$ to avoid an interference jamming of the echo amplitude and a heating of the sample . The pulse amplitudes are adjustable up to a few kilovolts.

The preparation of powder samples consists of fine grounding of single crystals and separating in size by sieving. Then the powder is washed and dried. In most cases the powder is sealed in a glass holder. The sample is located in a capacitor gap. In first experiments it was placed in an inductance space [2, 3]. A typical sample consists of about 10^6 particles. It should be noted that the echoes in powders involving in principle a cooperative effect can be observed only in a large number of particles [4].

Already first experiments have shown an acoustic nature of the echoes [2]. Indeed, the echoes are observed in piezoelectrics, and the frequency dependence of the two-pulse echo amplitude A_{2e} has a low frequency cutoff corresponding to the fundamental acoustic resonance of the particles [5]. It is shown in Figure 2.

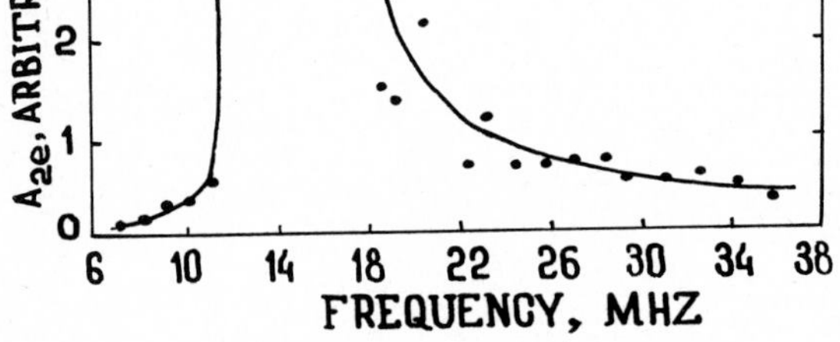

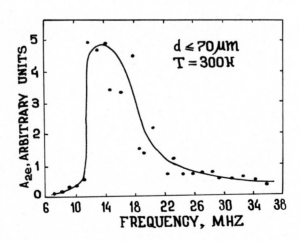

Figure 2
Frequency dependence of two-pulse echo amplitude A_{2e} in $Bi_{12}GeO_{20}$ powder [5]

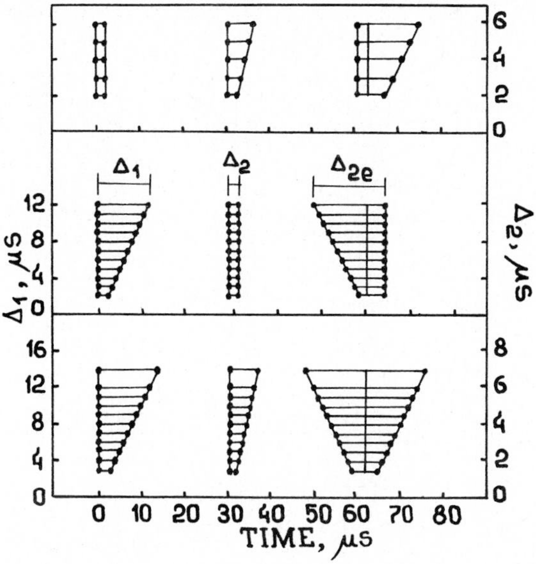

Figure 3
Two-pulse echo width versus first and second pulses widths in SbSI [6]

The two-pulse echo shape and particularly its width Δ_{2e} are defined by shape and width of the applied pulses. For small pulse amplitudes the relationship $\Delta_{2e} = \Delta_1 + \Delta_2$ is valid as illustrated in Figure 3. The echo amplitude A_{2e} depends on the applied pulse amplitudes A_1 and A_2 as third power of E when $A_1 = A_2 = E_i$ (see Figure 4). The width of the applied pulses affects the echo amplitude [4]. The behavior of A_{2e} as a function of pulse separation τ is represented in Figures 5 and 6. Figure 6 shows the dependence of A_{2e} on τ at different values of amplitude and frequency of the applied pulses. In this experiment [7], observation of echo signals at very small τ was difficult because of the blocking of the detector after the rf pulse.

Theoretical models

The two-pulse echoes in piezoelectrics were theoretically studied in Refs. 8-11. The description of echoes on the basis of phase conjugation is proposed by Korpel [1]. For delta-function pulses applied at times t = 0 and τ, $A_1 = \delta(t)$ and $A_2 = \delta(t-\tau)$, the Fourier transformation is $\Phi(\omega) \propto 1 + \exp(-i\omega\tau)$. For nonlinear interaction such as $A_{2e} \propto A_1 A_2^2$ we obtain the spectrum

$$\Phi_{2e} \propto \Phi^*(\omega)\Phi^2(\omega),$$

which includes also the term $\exp(-i2\omega\tau)$. Fourier transforming the latter yields

$$\int d\omega \exp\left[i\omega(t-2\tau)\right] = \delta(t-2\tau) = A_{2e}.$$

This is an echo occuring at time $t = 2\tau$. Below we omit an explanation of the echo occurence at $t = 2\tau$ as being obvious for a nonlinearity of the $A_1 A_2^2$ type.

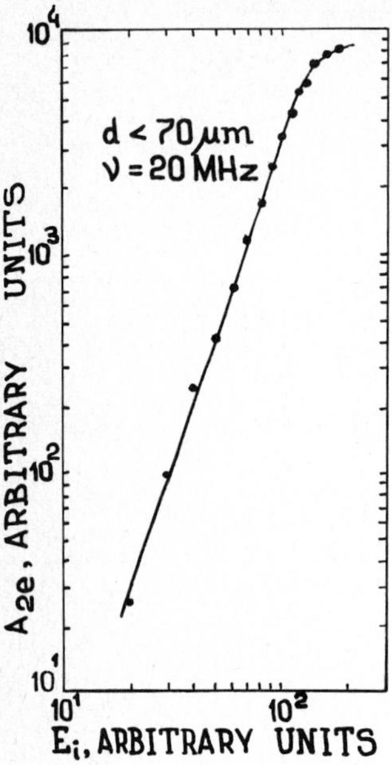

Figure 4
Two-pulse echo amplitude A_{2e} versus amplitudes of the applied pulses [7]

Two models should be noted of the advanced theories. One describes the echo formation in powders via parametric interaction between elastic oscillations excited by the first pulse and an electric field of the second pulse. The parametric electroacoustic interaction and the echo formation according to this mechanism have been studied in some detail on single crystals in Refs. 11-13. For piezoelectric powders, this model has been considered in Ref. 8. The treatment is based on using the equations of motion of individual particles, the elastic displacement of a particle, u_i^a, being represented as an expansion in series of oscillation modes of the particle, $u_{i\lambda}^a$,

$$u_i^a = \sum_\lambda C_{a\lambda} u_{i\lambda}^a$$

(here, a is the particle number). Taking into account nonlinear electroacoustic terms the coefficients $C_{a\lambda}$ are obtained from

$$\frac{d^2 C_{a\lambda}}{dt^2} + \omega_\lambda^2 C_{a\lambda} = Q\mathcal{E} + \sum_{\beta\mu} (\mathcal{E} M^{(1)}_{a\lambda,\beta\mu} + \mathcal{E}^2 M^{(2)}_{a\lambda,\beta\mu}) C_{\beta\mu}, \quad (1)$$

where ω_λ is the frequency of λ mode, Q defines piezoelectric excitation by the electric field $\mathcal{E}$. The many of high-order electroacoustic constants contribute to the coefficients $M^{(1)}$ and $M^{(2)}$.

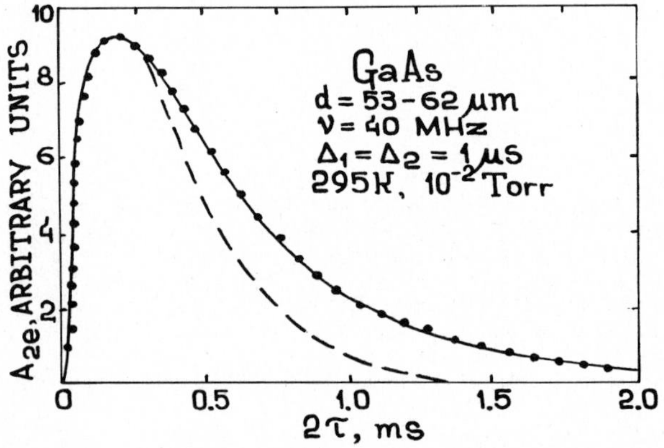

Figure 5
Dependence of two-pulse echo amplitude A_{2e} on twice the pulse separation 2τ in GaAs powder [4]

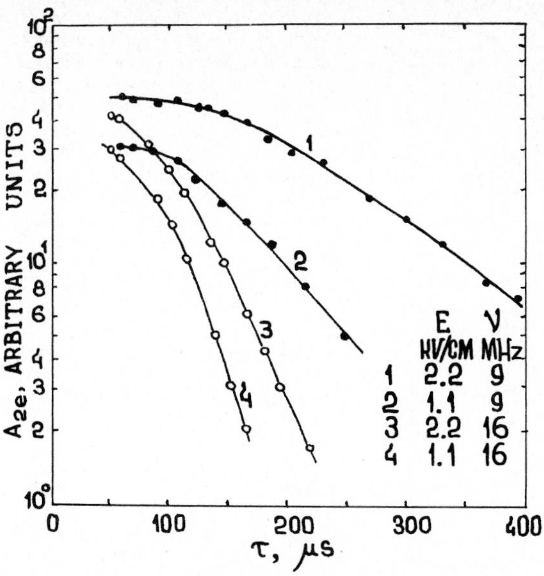

Figure 6
Two-pulse echo amplitude A_{2e} versus pulse separation in $Bi_{12}GeO_{20}$ powder [7]

The electric current measured in the echo experiment, $J = dP/dt$ (where P is the mean dipole moment of the sample), is found to be

$$J = -in(\omega)B_p \mathcal{E}_1^* \mathcal{E}_2^2 f(t)\exp(-i\omega t) + c.c., \qquad (2)$$

where B_p is the nonlinear coefficient connected with $M^{(1)}$ and $M^{(2)}$, $n(\omega)$ is the spectral density of the sample, $\mathcal{E}_1$ and $\mathcal{E}_2$ are the electric field of the first and second pulses, respectively, and $f(t)$ is the echo shape (according to Ref. 11). The parametric model gives the echo occurence at $t = 2\tau$, the echo amplitude A_{2e} being varied as

$$A_{2e} \propto A_1 A_2^2 \exp(-\Gamma\tau).$$

Here Γ is the coefficient of elastic oscillation damping. The echo width Δ_{2e} is equal to $\Delta_1 + 2\Delta_2$.

The second model is based on an anharmonic oscillator interaction in the powder sample [4, 9, and 14]. In this model, the elastic oscillations excited by an electric field of the rf pulses behave anharmonically. The echo occurence is assumed to result from a cumulative phase reversal due to the anharmonic coupling. The procedure of consideration is similar to that of the parametric model, but the elastic nonlinearity has been taken into account. This account lead to the following equation for $C_{a\lambda}$ instead of Eq. (1)

$$\frac{d^2C_{a\lambda}}{dt^2} + \omega_\lambda^2 C_{a\lambda} = Q\mathcal{E} + \sum_{\mu\nu} N^{(1)}_{\mu\nu} C_{a\mu}C_{a\nu} + \sum_{\mu\nu\varepsilon} N^{(2)}_{\mu\nu\varepsilon} C_{a\mu}C_{a\nu}C_{a\varepsilon},$$

where the symbols are the same as in Eqs. (1) and (2), and $N^{(1)}$ and $N^{(2)}$ are the nonlinear coefficients connected with the third-order, C_3, and fourth-order, C_4, elastic constants, respectively.

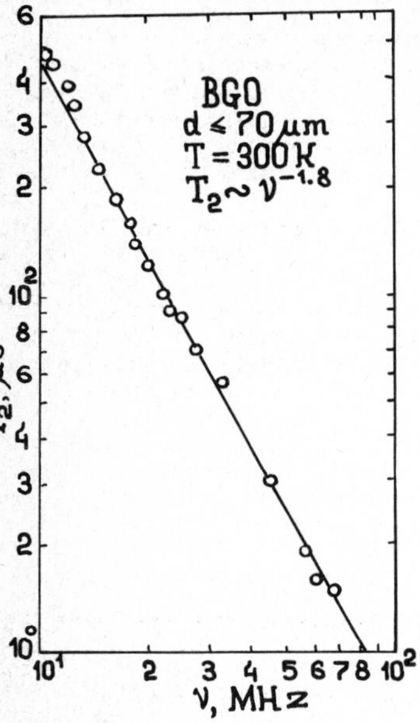

Figure 7
Relaxation time T_2 versus frequency ν in $Bi_{12}GeO_{20}$ powder at room temperature [15]

The term including $N^{(2)}$ is responsible for the echo occurence at $t = 2\tau$. The elastic oscillations excited by the first and second pulses interact, via the anharmonic coupling, between themselves during the time interval $\tau < t < 2\tau$. This causes the echo amplitude to initially increase with τ, reach a maximum, and finally decrease for large τ. The echo amplitude A_{2e} varies as

$$A_{2e} \propto A_1 A_2^2 \exp(-\Gamma\tau)\left[1 - \exp(-\Gamma\tau)\right].$$

Comparing the two models discussed above shows that the difference between them mainly consist in the dependence of A_{2e} on τ, the analysis of a quantity relationship being difficult due to the lack of experimental data on the nonlinear coefficients. The observed dependences of A_{2e} on τ (see Figures 5 and 6) are in agreement with the anharmonic oscillator model. As far as other experimental results are concerned they are in accordance with both the models.

It is important to note that for large τ A_{2e} decreases as $\exp(-\Gamma\tau)$ and a comparison with the echo decay in single crystals, $\exp(-2\tau/T_2)$, gives the decay constant $T_2 = 2/\Gamma$ as a relaxation time of the two-pulse echoes. The time T_2 depends on frequency (see Figure 7) and amplitude (Figure 6) of the applied rf pulses.

THREE-PULSE MEMORY ECHOES

Experimental results

Three-pulse echoes may occur in any system of anharmonic oscillators, but if the time of the third pulse application, T, does not exceed the life-time of the excited oscillators. The long-lived storage echoes in piezoelectric powders first reported in Refs. 16 and 17 consist in that the echoes at $t = T + k\tau$ $(k = 1,2,3,...)$ are observed at room temperatures, T exceeding days and weeks.

The three-pulse echo amplitude A_{3e} decreases with increasing τ as

$$A_{3e} \propto \exp(-2\tau/T_2 - T/T_1)$$

(see Figure 8). However, at large T, the amplitude A_{3e} tends to some constant level rather than zero. Thus for large enough T the echo amplitude does not practically depend on T, i.e. there are the "static" or memory echoes independet of T and the dynamic echoes varied with T and τ. The relaxation time of the dynamic three-pulse echoes, T_1, is close to T_2.

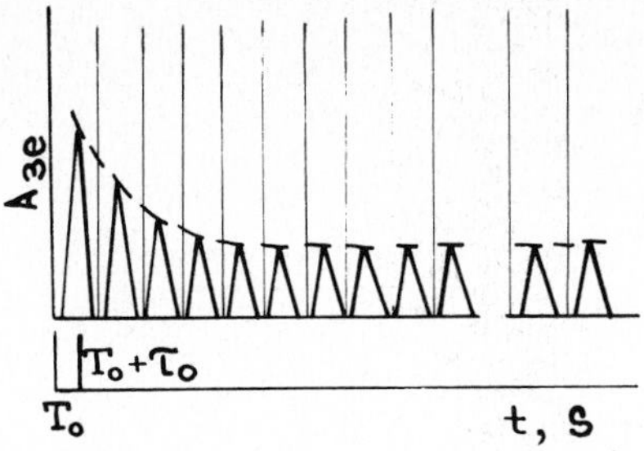

Figure 8
Dependence of three-pulse echo amplitude A_{3e} on T [16]

In the case of incoherent rf pulses a single application of the first two pulses and repeating the third pulse lead the memory echo amplitude to increase more than one order of magnitude. The investigation was carried out at equal frequencies $\nu_1=\nu_2$ of the first and second pulses. The memory echo signal at $t = T + \tau$ measured in this regime has the following features.

1. The echo amplitude A_{3e} is varied as

$$A_{3e} \propto A_1 A_2 A_3$$

(see Figure 9) and depends on the pulse widths, a relative maximum of A_{3e} taking place when $\Delta_1 \simeq \Delta_2$ [7].

2. The echo width Δ_{3e} is equal to the sum of the pulse widths [16].

3. Mixing the particles of the sample causes a decrease in the echo amplitude [16], but the initial amplitude partly restores after a slight shaking [18].

4. The echo amplitude decrease as the sample is rotated by angle γ

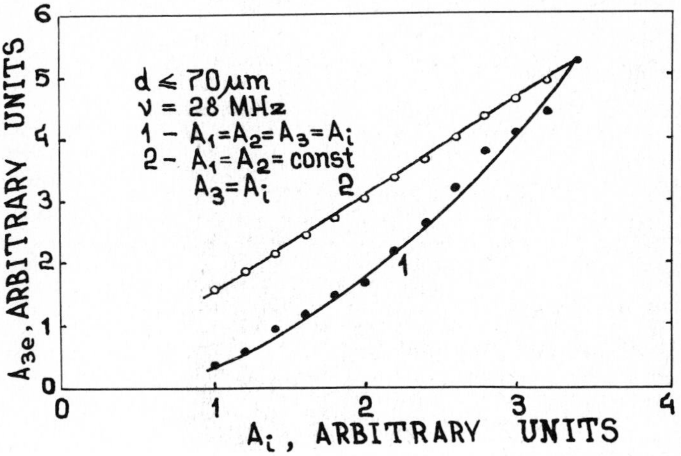

Figure 9
Three-pulse echo amplitude A_{3e} versus pulse amplitude E with $A_1= A_2= A_3= A_i$ (bottom curve) and $A_3= A_i$ and $A_1= A_2=$ const (top curve) [7]

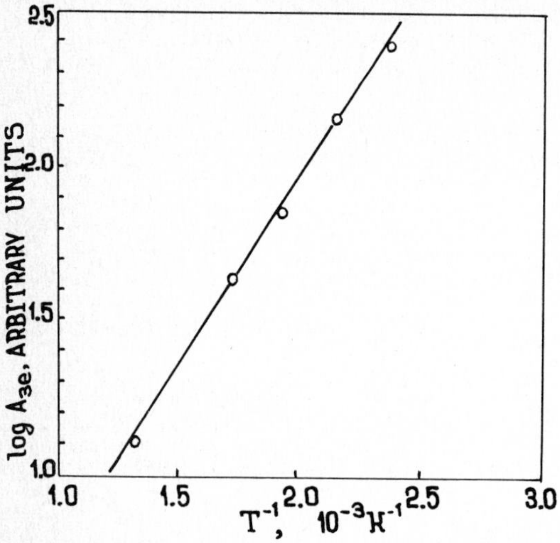

Figure 10
Three-pulse echo amplitude A_{3e} measured at room temperature versus the inverse temperature of the powder annealing [24]

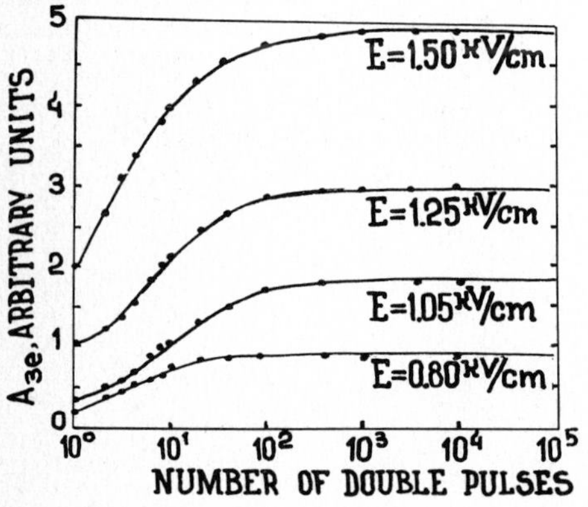

Figure 11
Three-pulse echo amplitude A_{3e} versus number of recording pairs of pulses, N, when $A_1 = A_2 = A_3 = E$ [24]

and almost vanishes at $\gamma = \pm 90°$. Rotation through an angle 180° results in the same echo amplitude as at $\gamma = 0°$ [19-21].

5. The memory echoes persist, although reduce in amplitude after sieving the powder [22, 20].

6. The same sample may contain a lot of the storages produced at various frequencies [23] and/or at different pulse separations τ [7].

7. The echo amplitude decreases as the third pulse amplitude and width exceed the amplitude and width of the first two pulses [7].

8. The memory echoes can be destroyed by heating the sample to a high temperature (see Figure 10).

In the case of coherent rf pulses the echoes have the following features.

1. The echo amplitude increases several times of magnitude at first with increasing number of the repeated pair of the pulses, N, and then attains a saturation at $N \simeq 5 \cdot 10^4$ (see Figure 11).

2. The echo phase has a 180° shift relative to phase of the third rf pulse [24].

Theoretical description of the long-lived storage

There are two basic models for the memory echoes. One involves the physical rotation of individual particles. Second involves the formation of a static internal deformation of the individual particles.

The rotation model of the memory echoes has been proposed in Refs. 25-26. According to this model the first rf pulse excites elastic oscillations which induce, via piezoelectric effect, the dipole moment p(t). During the action of the second rf pulse the electric field E(t) exerts a torque on the particles, [pxE]. An individual particle is rotated through an angle θ due to the torque, and the spatial orientation of the dipole moments takes place. The third rf pulse applied after the acoustic oscillations are damped reexcites the oscillations of the oriented particles. At the time $t = T + \tau$, the phase of new oscillations coincide with the initial oscillation phase. The calculations [19, 21] have shown that the echo amplitude A_{3e} has to depend on angles as $A_{3e} \propto \cos 2\theta_0 \cos(2\gamma + 2\theta_0)$ where γ is the angle of the sample rotation and θ_0 is the angle between the piezoelectric axis of a particle and the electric field direction of the first two pulses. The experimental dependence of A_{3e} on γ is $A_{3e} \propto \cos^2\gamma$ [19-21]. This fact and persistence of the echoes after sieving [22, 20] are not be explained by the rotation model.

The second model is based on analysis of elastic oscillations of individual particles having line dislocations which are displaced under static mechanical stress [27]. The mean static stress occurs in a particle via the elastic anharmonicity. The dislocations move to new positions when σ exceed a certain threshold stress σ_0. The action of two first pulses results in a creation of a certain inhomogeneous distribution of the dislocations remaining for a long time. The induced residual stress σ_r depends on the phase shift of the two pulses, i.e. the σ_r includes a term proportional to $\cos(\omega_a\tau)$ (where ω_a is a resonance frequency of the particle). Thus after the two pulses, the distribution of dislocations keeps a memory of the rf pulse phase shift. The stress σ_r changes the resonance frequency of the particle. This frequency change can be described as a burn a hole in the previously uniform frequency spectrum of the particles. The third rf pulse is applied to the sample with the new spectrum. The dipole moment of the sample, P, determined by this spectrum $Z(\omega_a)$ is

$$P(t) = \int d\omega_a Z(\omega_a) p ,$$

where p is the dipole moment of a particle. Averaging P(t) over the sample gives the nonzero P(t) at times $t = T + k\tau$, i.e. the echoes occur. The echo amplitude A_{3e} is $A_{3e} \propto S_3 \omega \tau \exp(-\Gamma\tau)\sigma_0/C$, where S_3 is the oscillation amplitude just after the third pulse application. According to this model the echoes have to vanish after annealing the sample as it was observed in experiments (see Figure 10). The echoes persist after sieving the powder as it was found in Refs. 22 and 20. The storage is erased under irradiation by a strong single pulse applied at $\tau < t < T$ since this pulse creates a new residual dislocation structure disconnected with that produced by the first two pulses.

ACKNOLEDGMENT

The authors would like to acknoledge helpful discussions with S.N. Popov.

REFERENCES

[1] Korpel, A. and Chatterjee, M., Proc. IEEE 69 (1981) 1539-1556.

[2] Popov, S.N. and Krainik, N.N., Fiz. Tverd. Tela 12 (1970) 3022-3027.

[3] Kessel, A.R. et al., Fiz. Tverd. Tela 12 (1970) 3070-3072.

[4] Fossheim, K. et al., Phys. Rev. B17 (1978) 964-997.

[5] Krainik, N.N. et al., Fiz. Tverd. Tela 17 (1975) 2462-2464.

[6] Smolensky, G.A. et al., Izv. Akad. Nauk SSSR, Ser. Fiz. 39 (1975) 965-969.

[7] Smolensky, G.A. et al., Zh. Eksp. Teor. Fiz. 72 (1977) 1427-1438.

[8] Laikhtman, B.D., Fiz. Tverd. Tela 17 (1975) 3278-3283.

[9] Laikhtman, B.D., Fiz. Tverd. Tela 18 (1976) 612-614.

[10] Kessel, A.R. et al., Fiz. Tverd. Tela 18 (1976) 826-831.

[11] Fedders, P.A. and Lu, E.Y.C., Appl. Phys. Lett. 23 (1973) 502-503.

[12] Billmann, A. et al., J. Phys. (Paris) 34 (1973) 453-470.

[13] Yushin, N.K. et al., Fiz. Tverd. Tela 16 (1974) 2789-2791.

[14] Gould, R.W., Phys. Lett. 19 (1965) 477-478.

[15] Smolensky, G.A. et al., Ferroelectrics 14 (1976) 571-573.

[16] Popov, S.N. et al., Pis'ma Zh. Eksp. Teor. Fiz. 21 (1975) 543-546.

[17] Asadullin, Ya.Ya. et al., Pis'ma Zh. Eksp. Teor. Fiz. 22 (1975) 285-288.

[18] Asadullin, Ya.Ya. et al., Pis'ma Zh. Tekh. Fiz. 3 (1977) 298-301.

[19] Cheeke, D. et al., Sol. St. Comm. 25 (1978) 289-291.

[20] Billmann, A. et al., J. Phys. (Paris) 39 (1978) L407-410.

[21] Ettinger, H. et al., Can. J. Phys. 56 (1978) 865-871.

[22] Berezov, V.M. and Romanov, V.S., Pis'ma Zh. Eksp. Teor. Fiz. 25 (1977) 165-167.

[23] Sawatzky, G.A. and Huizinga, S., Appl. Phys. Lett. 28 (1976) 476-478.

[24] Smolensky, G.A. et al., Fiz. Tverd. Tela 19 (1977) 1968-1972.

[25] Melcher, R.L. and Shiren, N.S., Phys. Rev. Lett. 36 (1976) 888-891.

[26] Chaban, A.A., Pis'ma Zh. Eksp. Teor. Fiz. 23 (1976) 389-391.

[27] Kosevich, A.M. and Bogoboyashchii, V.V., Fiz. Tverd. Tela 24 (1982) 3110-3119.

The Mechanical Behavior of Electromagnetic Solid Continua
G.A. Maugin (editor)
Elsevier Science Publishers B.V. (North-Holland)

ELECTRO-ACOUSTIC ECHOES IN PIEZOELECTRIC POWDERS

Joël Pouget

Laboratoire de Mécanique Théorique associé au C.N.R.S.
Université Pierre et Marie Curie
Tour 66, 4 Place Jussieu, 75230 PARIS Cedex 05
France

Based on an oriented continuum model, a continuum approach for powders made of piezoelectric grains is applied to the study of electroacoustic echoes in resonant piezoelectric powders. The nonlinear equations obtained through this approach govern the piezoelectric oscillations and the micro-gyration of each grain and lead to the echo formation. The evaluations of the dynamical echo at 2τ and the memory echo at $T+\tau$ are presented. A memory phenomenon lying in the reorientation of grains due to a ponderomotive couple inside the powder sample is placed in evidence. The physical interpretations are given. Lastly, we give an application to electroacoustic devices : convolution by means of electroacoustic echoes.

INTRODUCTION

Phonon echo phenomena have been experimentably identified by Solid-State physicists. Indeed, if electromagnetic pulses (of radio-frequency) are applied to a body at appropriately chosen lapses of time, after a certain latency period, the body radiates a pulse called echo. This a matter of nonlinear interaction phenomena between fields rather than phenomena of refexion such as usual echoes in acoustics. This type of echoes takes place in all sorts of media where nonlinear interactions are involved (for instance : spin echoes, cyclotron echoes in plasmas, echoes in superconductors, phenon echoes in ionic crystals, halographic-polarization echoes in simple-crystal piezoelectric semiconductors, electro-acoustic echoes in piezoelectric crystal, etc, etc..).

Here we are interested in a kind of echoes that we can classify in the above mentioned category, but their behavior is more peculiar. They are electroacoustic echoes in piezoelectric resonant powders. In this case the phenomenon exhibits a memory-effect [1, 2. 3]. Schematically, an electroacoustic-echo experiment on resonant piezoelectric powders consists in applying a radio-frequency pulse (ω_o) to powder sample at time $t = 0$; a non-coherent signal, which rapidly decays, is radiated by the sample. After a second pulse at $t = \tau$ and a lapse of time the sample radiates signal echoes in a coherent manner with the same frequency (ω_o) at the times 2τ , 3τ , ... with an amplitude which decays exponentially (FIG.1). After a third pulse at time $t = T$, the sample emits coherent signals at the times $T+\tau$, $T+2\tau$, ... with exponentially decreasing amplitude as well. A new aspect of the echo phenomenon becomes evident in the mechanical resonant case. Indeed, whereas T has the same order of magnitude as τ in the dynamical-echo case in piezoelectric powders [4] , or in piezoelectric crystals [5] , here $T >> \tau$ and it can reach a very long time (from several minutes to hours and even days), which exceeds the "lifetime" of any reasonable dynamical process. Only a static process can explain such a phenomenon. This means that the first two pulses -the "writing sequence"- are used for "freezing" some information inside the sample, and the third pulse -the "read-out sequence"- allows us to read the memory. A few models have been proposed to explain this curious phenomenon. A first model was elaborated ; it involves the formation of a static internal

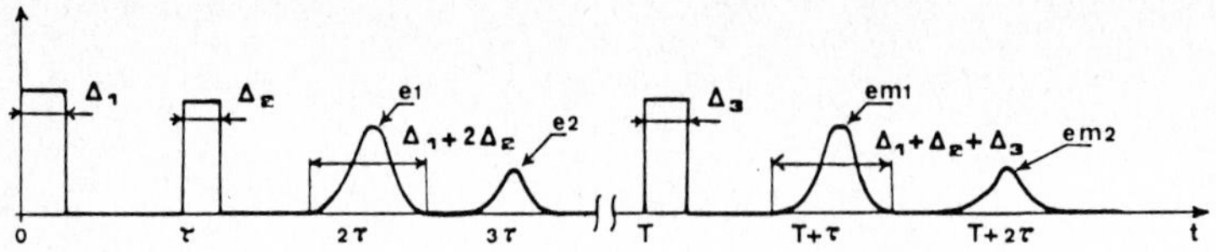

Fig.1 - Writing sequence : pulses at t=0 and t=τ, echoes at times 2τ,3τ, etc..., read out sequence : pulse at t=T, memory echoes at times T+τ , T+2τ , etc

polarization or deformation of the individual particles as a consequence of applied pulses [2] . A "reorientation model" or "torque-rotation model [3] was also proposed. This model involves the physical rotation of individual particules due to the torque exerted by the applied pulses on the oscillating particles excited by the same pulses. In this mechanism the phase of the oscillations at the beginning of the second pulse is "stored" as the angle through which the particle has rotated [6, 7]

In short, physically, the origin of the long lifetime storage can be explained by the grain reorientation caused by the mechanical torque due to the electric field. In general, echo phenomena are recognized an originating from the resonator non-linearity and the collective mode behavior [8] . A powder sample contains a large number of piezoelectric grains. Each grain is endowed with a resonant oscillation mode and a piezoelectric axis. These modes are randomly distributed in a certain frequence range. A simple radio-frequency pulse excites those grains of which the resonant frequencies are in the spectral region distributed around the carrier frequency (ω_o). In the inhomogeneously distributed resonators, a two-pulse sequence induces the oscillations satisfying a condition such that the oscillations excited by the first pulse become coherent with that excited by the second pulse. The condition is determined by the pulse separation

The aim of the present study is to model the echo phenomena in the spirit of the "torque-orientation" model but on the basis of a fully continuous approach. We remark that the phenomena occuring in this type of medium have a characteristic length (here the wavelength of vibrations of grains) of the order of magnitude of the inhomogeneities in the medium (the size of grains). It seems reasonable to model a piezoelectric powder by a continuum approach leaning on the microstructure-continuum theory. In a general manner, if we consider the material points of a continuum as small deformable bodies, at each point of this continuum there will exist a "microstrain tensor" [9, 10] . Thereby, the medium is subjected to deformation at two different scales : the "macroscopic deformation" at the usual scale of continuum mechanics and the "microscopic deformation" at the scale of inhomogeneities schematized by the material point. Here, we consider a microstructural description in which a set of vectors, so-called directors, is attached to each material point. These directors can deform, i.e. they can rotate and stretch [9, 11] . Besides the mechanical properties, there exist electromagnetic interactions at both micro- and macro- levels. We, therefore, expect to have interactions between electromagnetic fields and the micro-deformation, some other electromagnetic interactions being transmitted to the usual continuum, i.e., the mass centre of grains. The microstructure itself exhibits dielectric properties. That is, an electric polarization is defined at any point in the microstructure.This study is devised in the spirit of modern continuum mechanics along with the necessary ingredients from electrodynamics.

THE MODEL

A good enough description of piezoelectric powders could be given by a continuum approach based on a microstructure-continuum theory. In this approach each grain constitutes a microstructure and is considered as a small deformable piezoelectric crystalline body which possesses a piezoelectric axis. This means that the electromechanical couplings take place along a priviliged direction $\underset{\sim}{d}$ associated with the grain. Then, the applied electric fields will induce a deformation of the grain in the $\underset{\sim}{d}$ - direction and, reciprocally, only a deformation of the grain in this direction will give rise to an electric polarization. In other words, the piezoelectric effect in a grain occurs along the direction $\underset{\sim}{d}$ attached to the grain. The motion of the mass center of grains can be identified, in some sense, with the motion of the classical continuum, it is described by the deformation mapping $\underset{\sim}{x} = \underset{\sim}{\mathcal{X}}(\underset{\sim}{X},t)$. In addition, at the macroscopic-scale level, to account for the distribution of solid grains in a granular body, a kinematical variable ν, called the volume distribution of grains, is introduced as a continuum function defined throughout the body [16, 17]. This function is such that $o < \nu(\underset{\sim}{X},t) < 1$. If γ is the mass density of the grain, then we obviously have $\rho = \gamma\nu$ which is the mass density of the global continuum. Insofar as the microscopic scale is concerned, the micro-motion (deformation of grains themselves) is well described in terms of a director field $\underset{\sim}{d}(\underset{\sim}{X},t)$ (FIG.2). The micro-motion of a grain consists in i) a micro-gyration about its mass center and ii) a micro-stretch along the direction $\underset{\sim}{d}$; this micro-stretch, in fact, corresponds to eigen-modes of oscillations of a grain along the direction $\underset{\sim}{d}$ [18]. In order to account for the dielectric properties of grains a volume distribution of electric polarization $\underset{\sim}{P}(\underset{\sim}{X},t)$, defined at the mass center of the grain, is introduced. In summary, the electrokinematical state of the continuum built from dielectric grains is described, at each material point $\underset{\sim}{X}$, by the set

$$\mathcal{E}(\underset{\sim}{X},t) = \{ \underset{\sim}{\mathcal{X}}(\underset{\sim}{X},t), \quad \nu(\underset{\sim}{X},t), \ \underset{\sim}{d}(\underset{\sim}{X},t), \ \underset{\sim}{P}(\underset{\sim}{X},t)\}. \qquad (1)$$

We associate with the motion of the grain some kinetical quantities, among them we quote a tensor of moment of inertia, a moment of momentum and a moment of mass acceleration [9, 10, 18]. In order to construct the equations which govern the motion of a continuum made of dielectric grains we employ the method of virtual power [13, 14]. In this method the primary notion is that of virtual velocity field here adequately generalized so as to account for velocities of quantities which are not "motions" in the classical sense of the term (the polarization, for instance) [12, 13, 14]. In the framework of the first gradient theory with respect to the macro -and micro- motion and a zero gradient theory with respect to the polarization we obtain a set of nonlinear local field equations which govern a con-

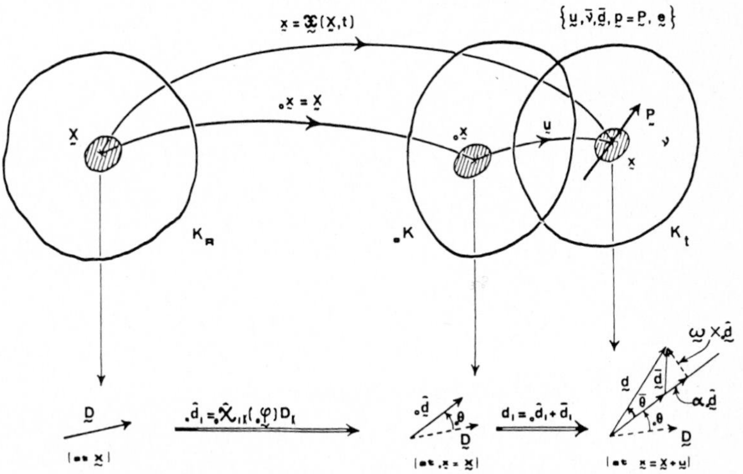

Fig.2 - Generalized motion of the continuum and local change in the director field.

tinuum of dielectric grains. Here, we do not give these equations, but we refer the reader to Ref. [18] . Then, in order to deduce the constitutive laws we develop the thermodynamical concepts (principles of virtual power for real velocity field, the first and second principles of thermodynamics [15]) for media made of piezoelectric grains. We specialize the study to the case of materials made of elastic dielectric grains with the possible occurence of some simple thermodynamically irreversible processes (heat conduction, viscosity, orientational relaxation of grains, etc...). In agreement with the usual modelling of granular media [16,17] the global behavior of the granular medium will be one of a compressible viscous fluid while the behavior of grains, themselves, will be one of an elastic dielectric medium. We assume that the global behavior of the medium or fluid behavior and the behavior of grains are entirely-uncoupled, the dissipative processes are entirely uncoupled as well. Here, we do not write the constitutive equations thus obtained (see Ref. [18]).

PERTURBED FIELD EQUATIONS

In order to apply the present theory to signal processing are need to obtain the field equations which govern fields about an initial configuration $_\circ K$ which is typical of a powder of piezoelectric grains in a virgin state (no electrical signal applied). The static solution $_\circ S$ corresponding to this configuration $_\circ K$ can therefore be written as

$$_\circ S = \{ {}_\circ x_i = \delta_{ik} X_k ,\ \nu = {}_\circ\nu ,\ {}_\circ \underset{\sim}{d} ,\ {}_\circ \underset{\sim}{E} = \underset{\sim}{0} ,\ {}_\circ \underset{\sim}{P} = \underset{\sim}{0} \} , \quad (2)$$

and such that the initial configuration is free of strains, mechanical loods and electric field, but (i) there, of course, exists an initial volume distribution function $_\circ\nu$, with $_\circ(\underset{\sim}{\nabla}\nu) = \underset{\sim}{0}$ and (ii) the piezoelectric axes of grains, represented by $\underset{\sim}{d}$, are randomly oriented so that $\underset{\sim}{d}({}_\circ K) = {}_\circ\hat{\underset{\sim}{d}}(\underset{\sim}{x})$. The medium therefore is nonhomogeneous from this standpoint in $_\circ K$. The material constants which result from the perturbation process are affected by the initial director field $_\circ\hat{\underset{\sim}{d}}$. We perform a perturbation [18] of nonlinear equations. Then we specialize the equations thus obtained to the case where heat conduction and initial pressure are discarded, we assume that the influence of the displacement gradient is small as compared to those of other parameters. The useful perturbed equations take on the following form [18, 19]:

$$\rho_\circ \ddot{u}_i = -{}_\circ\nu^2(c_1+2c_2)\eta_{,j}\,\eta_{,ji} - ({}_\circ\nu^2 c_1)\eta_{,jj}\eta_{,i} + (a_1-b)\eta_{,i} + (\lambda+\mu)\dot{u}_{j,ji} + \mu\,\dot{u}_{i,jj} + t^{INT}_{ij,j} , \quad (3)$$

with

$$\underset{\sim}{t}^{INT} = -\rho_\circ(A\alpha + \beta\,\dot{\alpha} + \varepsilon_1\, {}_\circ\hat{\underset{\sim}{d}}\cdot\underset{\sim}{\pi})\,{}_\circ\hat{\underset{\sim}{d}}\otimes{}_\circ\hat{\underset{\sim}{d}} - \rho_\circ\beta_1\,{}_\circ\hat{\underset{\sim}{d}}\otimes(\hat{\underset{\sim}{\nu}}\times{}_\circ\hat{\underset{\sim}{d}}) , \quad (4)$$

$$\Delta^2\eta = a\,\eta + c\,\underset{\sim}{\nabla}\eta\cdot\underset{\sim}{\nabla}\eta , \quad (5)$$

$$\mathscr{I}\ddot{\alpha} + \beta\dot{\alpha} + A\alpha = -\varepsilon_1\,{}_\circ\hat{\underset{\sim}{d}}\cdot\hat{\underset{\sim}{\pi}} , \quad (6)$$

$$\rho_\circ\,\underset{\sim}{j}\,\dot{\hat{\underset{\sim}{\nu}}} + \rho_\circ\beta_1 \left|\, {}_\circ\hat{\underset{\sim}{d}}\times(\hat{\underset{\sim}{\nu}}\times{}_\circ\hat{\underset{\sim}{d}})\,\right| = \rho_\circ\varepsilon_1\,\hat{\underset{\sim}{\chi}}(-\phi)(\hat{\underset{\sim}{d}}\times\hat{\underset{\sim}{\pi}})\,\alpha , \quad (7)$$

$$\chi^{-1}\underset{\sim}{\pi} + \rho_\circ\,\varepsilon_1\,\alpha\,{}_\circ\hat{\underset{\sim}{d}} = {}_\circ\underset{\sim}{f} + \underset{\sim}{f} , \quad (8)$$

where we have set :

$$\underset{\sim}{j} = \mathscr{I}(\underset{\sim}{1} - {}_\circ\hat{\underset{\sim}{d}}\otimes{}_\circ\hat{\underset{\sim}{d}}) , \quad (9)$$

$$\eta = \nu / {}_\circ\nu ,\quad \underset{\sim}{\pi} = \hat{\underset{\sim}{\chi}}(-\underset{\sim}{\phi})\underset{\sim}{p} ,\quad \underset{\sim}{f} = \hat{\underset{\sim}{\chi}}(-\underset{\sim}{\phi})\underset{\sim}{e} ,\quad {}_\circ\underset{\sim}{f} = \hat{\underset{\sim}{\chi}}(-\underset{\sim}{\phi})\,{}_\circ\underset{\sim}{e} . \quad (10)$$

$\underset{\sim}{u}$ is the infinitesimal displacement field, $\hat{\underset{\sim}{\nu}}$ is the micro-gyration velocity, α is the micro-stretch, $\underset{\sim}{\phi}$ is the rigid-body rotation of the grain $\hat{\underset{\sim}{\chi}}$ is the orthogonal tensor associated with the rigid-body rotation as in micropolar media and $\underset{\sim}{j}$ is the tensor of moment of inertia. The expressions (10) mean that $\underset{\sim}{\pi}$, $\underset{\sim}{f}$ and ${}_\circ\underset{\sim}{f}$ are the polarization, electric field and applied electric field expressed in coordinates which rotate with the grain. Then the constants c_1, c_2, a_1, b,

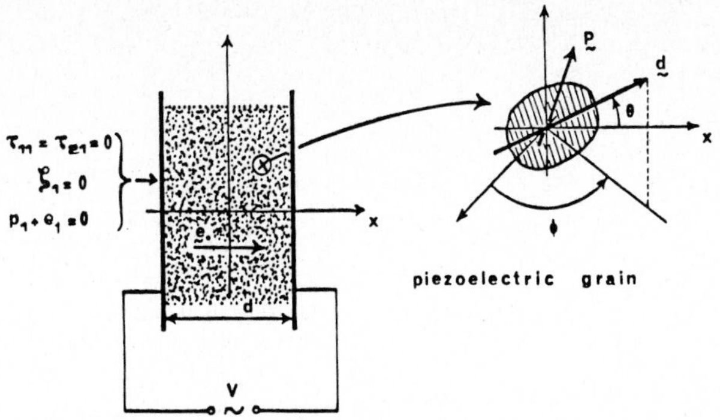

Fig.3 - Schematical device between the two plates of a capacitor and geometry of the echo problem.

λ , μ , A, β , ε_1 , c and β_1 are some material coefficients. Equation (3) is the equation of the motion of the mass center of the grain, (4) is the continuum-director field interaction tensor, eqn. (5) is the balance equation associated with the volume distribution of grain, eqn. (6) governs the micro-stretch motion (or piezoelectric vibrations), eqn. (7) governs the micro-gyration of a grain about its mass center and eqn. (8) is the balance equation of electric fields. χ^{-1} is the electric susceptibility, ε_1 is the "micro-piezoelectricity" coefficient, β and β_1 are the micro-viscosity constants connected with the micro-stretch and micro-gyration motion, respectively. We can also write the associated boundary conditions [18,19].

We consider a sample of piezoelectric powder placed between the two broad parallel infinite plane electrodes of a capacitor (FIG.3) so that the applied electric field is quite uniform thoughout the sample. Therefore, we have a one-dimensional problem and all fields depend only on time and on x in the direction of the width of the capacitor. We now write down the equations useful in the echo problem and which govern the vibrations of grain. After change of time scale ($t = \omega_\circ t$) and introducing obvious nondimentional notations, we obtain :

$$\ddot{\alpha} + 2\Gamma\dot{\alpha} + \Omega^2\alpha = -\ \varepsilon \cos({}_\circ\theta + \theta)\hat{e}_\circ(t) \quad , \tag{11}$$

$$\ddot{\theta} + \gamma\dot{\theta} = \varepsilon\alpha\hat{e}_\circ(t) \sin({}_\circ\theta + \theta). \tag{12}$$

On account of the definitions (10) , the polarization $\underset{\sim}{p}$ can then be written as

$$\hat{p}_1 = \hat{e}_\circ + \alpha \cos({}_\circ\theta + \theta) \ , \qquad \hat{p}_2 = \alpha \sin({}_\circ\theta + \theta) \quad , \tag{13}$$

where we have set

$$\Omega^2 = \left[A + \varepsilon/\mathscr{I}\right]\mathscr{I}\omega_\circ^2 \ , \qquad 2\Gamma = \beta/\mathscr{I}\omega_\circ \ , \qquad \gamma = \beta_1/\mathscr{I}\omega_\circ \ ,$$
$$\varepsilon = \rho_\circ\varepsilon_1^2/\omega_\circ^2\mathscr{I}(1+\chi^{-1}) \ , \qquad \hat{\underset{\sim}{p}} = \underset{\sim}{p}(1+\chi^{-1})/\rho_\circ\varepsilon_1 \ , \qquad \hat{e}_\circ = e_\circ/\rho_\circ\varepsilon_1 . \tag{14}$$

Here, Γ is the damping factor associated with micro-stretch vibration of grains, Ω is the natural frequency of the grain, γ is a damping factor connected with the micro-gyration of grains (rotatory vibration) and ε is a small parameter introduced by the electromechanical coupling of grains. We assume that only the natural frequency Ω as well as the initial orientation ${}_\circ\theta$ depend on the position of the mass center of the grain within the sample. In fact, we suppose that there exists a continuous random distribution of natural frequencies and initial orientations throughout the sample of powder. Equations (11)-(12) are coupled and nonlinear.

ECHO PHENOMENON

With a view to studying the echo phenomena as described in the introduction we consider an applied electric field $\hat{e}_\circ(t)$ consisting of three pulses with the amplitudes $\mathscr{E}_1$, $\mathscr{E}_2$ and $\mathscr{E}_3$ and the widths Δ_1, Δ_2 and Δ_3 , respectively, (see FIG.1).

The initial conditions are (sample at rest) : $\alpha(0) = \dot{\alpha}(0) = \theta(0) = 0$ and $\dot{\theta}(0) = {}_o\theta$. During the pulses eqns. (11) and (12) are nonlinear. Solutions in terms of a small perturbation ε are considered :

$$\alpha = \alpha_o + \varepsilon\alpha_1 + \varepsilon^2\alpha_2 + \ldots\ldots\ldots ,$$
$$\theta = {}_o\theta + \varepsilon\theta_1 + \varepsilon^2\theta_2 + \ldots\ldots\ldots . \tag{15}$$

The α solution will be $O(\varepsilon)$ whereas $(\theta - {}_o\theta)$ will be $O(\varepsilon^2)$. The response of an individual grain is given by the polarization $\hat{p}$ eqn. (13) . We make the choice to compute the response in the direction of the applied electric field, and then consider the $\hat{p}_1$-component. On expanding $\hat{p}_1$, we obtain thus

$$\hat{p}_1 = \hat{e}_o(t) + \varepsilon\alpha_1\cos(2{}_o\theta) + \varepsilon^3(\alpha_3\cos(2{}_o\theta) - \alpha_1\theta_2\sin(2{}_o\theta)) + O(\varepsilon^5). \tag{16}$$

For the first dynamical echo and the memory echo we need the expressions only up to first order for the α-solution and up to second order for the θ-solution. Accordingly, in the expression (16) of $\hat{p}_1$, only the nonlinear term $\varepsilon^3\alpha_1\theta_2\sin(2{}_o\theta)$ is of interest because it is the one that mostly contributes to the first echo. Insofar as the second echo is concerned, we must consider the nonlinear terms of order ε^5.

The global response of the sample is obtained by summing up all the elementary responses of all grains along the applied-field direction. In agreement with the continuum approach, the global response is obtained through on integrale representation taken over the frequency distribution $R(\omega)$ $(\omega = \Omega - 1)$ and the solid angle $\sin{}_o\theta d{}_o\theta d\Phi/4\pi$, where Φ is the azimuthal angle in the plane normal to the applied-field direction. Assuming a uniform distribution of grain orientation, the global response of the sample can be written as

$$Q(t) = \langle\hat{p}_1\rangle_{\omega,{}_o\theta} = \frac{1}{4\pi}\int_o^{2\pi} d\Phi \int_o^{\pi} \sin{}_o\theta d{}_o\theta \int_{-\infty}^{+\infty} \hat{p}_1(t,{}_o\theta,\omega)R(\omega)d\omega , \tag{17}$$

where $\hat{p}_1(t,{}_o\theta,\omega)$ is given by (16). Note, nevertheless, that the relative frequency ω and the initial orientation ${}_o\theta$ depend on the spatial variable x in a random manner.

(a) For the two-pulse echo, on substituting from eqn(16) into eqn(17), the following expression is obtained for the coherent response generated by the simple after a two-pulse sequence [19] :

$$Q(t) = -\frac{23}{126}\varepsilon^3\mathcal{E}_1\mathcal{E}_2^2\Delta_1\Delta_2^2\exp\left[i(t-2\tau)\right]I(t) , \tag{18}$$

$$I(t) = \int_{-\infty}^{+\infty}\left\{\frac{\Delta_2}{2}\frac{G(\xi_2^*)}{\zeta_2^*} - \frac{i}{\gamma}\left[1-\exp(-\gamma(t-\tau-\Delta_2))\right]\exp(-\gamma\Delta_2)\right.$$
$$\left.\times G(\zeta_2^*)\right\} G^*(\xi_1)G(\xi_2)\exp(i\omega(t-2\tau))R(\omega)d\omega \tag{19}$$

where we have defined

$$G(\xi_j) = \exp(-i\xi_j)(\sin\xi_j)/\xi_j ,$$
$$\xi_j = (\omega + it)\Delta_j/2 , \quad \zeta_j = \left[\omega + i(\Gamma-\gamma)\right]\Delta_j/2 , \quad (j = 1,2,3) , \tag{20}$$

(ξ^* denotes the conjugate of ξ). The integral factor $I(t)$ thus defined gives the shape of the two-pulse echo at $t = 2\tau$ and shows that this response is nonzero only in the neighborhood of 2τ (FIG.1). After a rather longthly computation we can obtain the following results [19] : (i) the width of the two-pulse echo is $\Delta_1 + 2\Delta_2$ **(ii)** the signal amplitude is proportional to the electromechanical coupling constant ε^3 and the shape factor of grains $(1/\langle\rho_o\mathcal{I}\rangle\omega_o^2)^3$, where $\langle\ \rangle$ is an average of the inertia moment of grains ; (iii) the signal amplitude depends on the applied-field amplitudes $\mathcal{E}_1$ and $\mathcal{E}_2$ as $\mathcal{E}_1\mathcal{E}_2^2$; (iV) the maximum of the echo amplitude is reached at time $t = 2\tau-(\Delta_1-2\Delta_2)/2$. The behavior of the amplitude Q^e with the pulse separation τ is of the form $f(\tau)\exp(-2\Gamma\tau)$; the function $f(\tau)$ depends on the pulse widths Δ_1 and Δ_2 and on dampling constants Γ and γ connected with the electromechanical vibration and the rotation of grains, respectively ; $f(\tau)$ is such that $f(0) \sim 0$. However, for $\tau \to \infty$, we have $Q^{e}(2\tau) \simeq \exp(-2\Gamma\tau)$ and we recover the classical dependence of the amplitude echo on τ that exists in the classical theory of dynamical-polarization echoes [4] . In the present model, the formation of the two-pulse echo can be favorably compared to the few experimental results for various crystalline powders [4, 20, 21] .

(b) Memory echo . - After a long enough time $t = T$, the sample returns to rest and all the dynamical effects connected with the natural oscillations α of grains vanish. But the initial orientation now is modified, so that the new initial orientation of grains is $\theta = {}_{\circ}\theta + \theta_{l}({}_{\circ}\theta, \omega, \tau)$, where θ_{l} is the small perturbation in θ obtained for a very long lapse of time. The global response of the sample can be computed in the same manner as for the two-pulse echo from the expression (17). After computation we obtain [19] :

$$Q^{m}(t) = -\frac{23}{126}\, \varepsilon^{3}\, \mathcal{E}_{1} \mathcal{E}_{2} \mathcal{E}_{3}\, \Delta_{1}\Delta_{2}\Delta_{3} \exp\left[\ (i-\Gamma)(t-T-\tau) - 2\Gamma\tau\ \right]\ J(t)\ , \tag{21}$$

$$J(t) = \int_{-\infty}^{+\infty} \left\{ \frac{\Delta_2}{2} \frac{G(\xi_2^*)}{\zeta_2^*} - \frac{i}{\gamma} \exp(-\gamma\Delta_2) G(\zeta_2^*) \right\} G(\xi_1^*) G(\xi_3) \times R(\omega) \exp(i\omega(t-T-\tau))\, d\omega. \tag{22}$$

The latter integral factor gives the shape of the memory echo and this factor is essentially nonzero in a neighbourhood of $T + \tau$ (FIG.1). We then obtain the following results : (i) the width of the memory echo is $\Delta_1+\Delta_2+\Delta_3$ and its shape depends essentially on the integral factor $J(t)$ (eqn(22)) ; (ii) the maximum of the memory echo amplitude is reached at the time $t = T + \tau - (\Delta_1-\Delta_2-\Delta_3)/2$; (iii) this amplitude is proportional to the product of pulse amplitudes as $\mathcal{E}_1 \mathcal{E}_2 \mathcal{E}_3$; (iv) the behavior of the amplitude Q^{m} is proportional to the factor $\exp(-2\Gamma\tau)$; it also depends nontrivially on the pulse widths Δ_1, Δ_2 and Δ_3 as well as on the damping constants Γ and γ [19]

CONCLUSION AND APPLICATION

In this paper a continuum approach to media made of piezoelectric grains has been successfully applied to the study of electroacoustic echoes in resonant piezoelectric powders [18, 19] . Several points concerning physical interpretations can be emphasized. The first pulse excites the piezoelectric oscillations at the pulse frequency and the grain oscillates in rotation about its mass center. After the first pulse all the grains oscillate at their natural frequency and there follows from this a dephasing and a concellation of the oscillations. At the time τ of the second pulse, the phase of the oscillations of grains is randomly distributed inside the sample and the piezoelectric and rotational oscillations depend, therefore, on the phase of the grain. Then, the second pulse excites the grains from phases reached at the time τ . The second pulse causes the "time-inversion" in order that the phase evolution of the oscillations between τ and 2τ be opposite to that between 0 and τ and all the oscillations become coherent at the time 2τ . Another effect of the second pulse is to freeze in the phases reached by the oscillations of grains at $t = \tau$ through the angle θ_{l}. Then, inside the sample of powder, there exists a "network" of phases reached by the oscillations at $t = \tau$. Therefore, a third pulse, even a long time after the first two ones, stimulates the grains from the phases reached at τ and the grains rotate about the new angle ${}_{\circ}\theta + \theta_{l}$. It clearly follows that the "torque-orientation model" places a memory phenomena is evidence. This effect lies in the micropolar contribution in the continuum description. It is possible to show that, at the scale of the global medium, the global behavior of the sample is effected by the reorientation of grains [19] . Nevertheless, this continuum approach to the torque-orientation model does not elucidate all questions. Indeed, the contribution to the echoes would not be limited only to particle orientation inside the sample, but might also be caused by defect motion phenomena inside each particle [20,22,23,24] . In the memorisation would, therefore, be localized in internal processes for each grain. The present model bases on a continuum approach allows us to get rid of inevitable hypotheses introduced in the elementary models [3,25,26] .

The important technological development of electroacoustic devices using the piezoelectricity of materials forecasts a promissing use of echo phenomena in powders [27]. The electroacoustic echoes can be used to perform operations of convolution [28]. If we resolve the equations (11) and (12) without specifying the shape of first two pulses, we can show that the global response of sample (17) after two-pulse sequence takes on the following form [28] .

$$Q^e(t) = -\frac{23\pi}{252\gamma}\, i\varepsilon^3\{e^{(i-\Gamma)t-2\tau\Gamma}\left[S_1(t)-e^{-\gamma(t+\tau)}S_2(t)\right]\}*\delta(t-2\tau) \tag{23}$$

$$\text{with} \quad \begin{aligned} S_1(t) &= \mathcal{E}_2(\Gamma,t)*\mathcal{E}_1^*(\Gamma,-t)*\mathcal{E}_2(-\Gamma,t) \quad , \\ S_2(t) &= \mathcal{E}_2(\Gamma,t)*\mathcal{E}_1^*(\Gamma,-t)*\mathcal{E}_2(\gamma-\Gamma,t) \quad , \end{aligned} \tag{24}$$

where we have defined

$$\mathcal{E}_j(\Gamma,t) = \mathcal{E}_j(t)e^{\Gamma t} \quad , \quad (j=1,2). \tag{25}$$

Then, $\mathcal{E}_1(t)$ and $\mathcal{E}_2(t)$ are the shapes of the first and second pulse, respectively, δ denote the Dirac impulsion function and $*$ stands for the convolution product. We have thus placed in evidence that the shape of the signal echo can be considered as the convolution of the shapes of the applied pulses. The same can be done for the memory echo.

REFERENCES

[1] Frenois Ch., Joffrin J. and Levelut A.- J. Physique-Lettres, 35, (1974), L221-L223.

[2] Popov S.N., Krainik N.N. and Smolenskii G.A.- J.E.T.P. Letters, 21, (1975), 253-254.

[3] Melcher R.L. and Shiren N.S.- Phys. Rev. Letters, 36, (1976), 888-891.

[4] Fossheim K., Kajimura K.Kazyaka T.G., Melcher R.L. and Shiren N.S. -Phys. Rev. B, 17, (1978), 964-998.

[5] Billman A. Frenois Ch., Joffrin J., Levelut A. and Zioliewicz.- J. Physique, 34, (1973), 453-470.

[6] Chaban A.A. - JETP Letters, 23, (1976), 350-352.

[7] Chaban A.A. - Soviet Phys. - Acoustic, 26, (1978), 533-535.

[8] Herrmann G.F., Kaplan D.E. and Hill R.M.- Phys. Rev., 181, (1969), 829-841.

[9] Eringen A.C.- Int.J.Engng.Sci., 2, (1964), 189-203.

[10] Eringen A.C.-Foundation of Micropolar Thermoelasticity, C.I.S.M., Lecture Note, (Udine, Italy, 1970).

[11] Stojanovic R.-Mechanics of Polar Continua : Theory and Applications, C.I.S.M., Lecture Note, (Udine, Italy, 1969).

[12] Maugin G.A. and Pouget J.- J. Acoust. Soc. Amer., 68, (1980), 575-587.

[13] Germain P.- SIAM Appl. Math., 25, (1973), 556-575.

[14] Maugin G.A.- Acta Mechanica, 35, (1980), 1-70.

[15] Maugin G.A. and Eringen A.C.- J. de Mécanique, 16, (1977), 101-147.

[16] Goodman M.A. and Cowin S.C.- J.Fluid. Mech., 45, (1971), 321-339.

[17] Goodman M.A. and Cowin S.C.- Arch.Rat.Mech.Anal., 44, (1971), 249-266.

[18] Pouget J. and Maugin G.A. - J.Acoust.Soc.Amer., 71, (1983).

[19] Pouget J. and Maugin G.A. - J.Acoust.Soc.Amer., 71, (1983).

[20] Popov N.S., Krainik N.N. and Smolenskii G.A. Soviet Phys.- J.E.T.P., 42, (1975), 494-496.

[21] Smolenskii G.A., Popov S.N., Krainik N.N., Laikhtman B.D. and Tarakanov E.A.- Soviet Phys.- J.E.T.P., 45, (1977), 749-753.

[22] Berezov V.M. and Romanov V.S.-Soviet Phys.- JETP Letters, 25, (1977), 151-153.

[23] Billmann A., Frenois Ch., Guillot G. and Levelut A.- J. Physique-Lettres, 39, (1978), L407-L410.

[24] Kessel A.R.- Ferroelectrics, 22, (1978), 759-761.

[25] Ettinger H., Lahkani A.A. and Cheeke D.- Canad.J.Phys., 56, (1978), 865-877.

[26] Pouget J.- Mech.Res.Commun., 9, (1980), 59-65.

[27] Smolenskii G.A. and Krainik N.N.- Ferroelectrics, 24, (1980), 247-253.

[28] Pouget J.- C.R.Acad.Sc.- Paris. T.295, Série II, (1982), 845-848.

Part 6

MODELS AND PROBLEMS OF ELECTROELASTICITY

The Mechanical Behavior of Electromagnetic Solid Continua
G.A. Maugin (editor)
Elsevier Science Publishers B.V. (North-Holland)

POINT CHARGE, INFRA-RED DISPERSION AND CONDUCTION IN NONLOCAL PIEZOELECTRICITY

A. Cemal Eringen

Princeton University
Princeton, NJ 08544
U.S.A.

After an outline of the nonlocal theory of electromagnetic elastic solids which I developed in several previous papers [1]-[3], discussions are given on several applications which include point charge, defect, space dispersions of optical modes, infrared dispersion and lattice vibrations, anomalous skin effect and superconductivity.

INTRODUCTION

According to Maxwell's electromagnetic theory, plane waves are non-dispersive. Therefore, the refractive index in an isotropic non-dissipative medium is constant. Yet experiments show that the refractive index depends on frequency and wave length. At transition frequencies to exciton states in a crystal, the dependence of exciton energy on the wave vector cannot be neglected. Even for static fields of stationary electrons, we have the Debye screening that results from nonlocal interactions. In plasma physics, at low temperatures, the effect of the spatial dispersion on electromagnetic (E-M) properties of metals is considerable. Interactions of optical modes with lattice vibrations cause infrared dispersion. Natural optical activity in non-centrosymmetric crystals is caused by the distant electric fields. Anomalous skin effect arises in a conductor when a highly non-uniform field is established.

All of the above-mentioned phenomena and many more are the result of nonlocality, i.e. the effect of E-M fields and strains, in the neighborhood of a reference point, on the state of that point. Usually, these phenomena are treated either by means of atomic theory of electrons or some heuristic modifications of the classical (local) continuum theory.

Raison d'être of the present paper is two-fold:

(i) To present a brief outline of the newly developed nonlocal E-M theory of deformable solids, and

(ii) To discuss some of the above-mentioned physical Phenomena by means of nonlocal theory.

The five problems that I have selected should be adequate to indicate the power and the potential of nonlocal theory. Space limitation does not allow me to give a full account on the subject. Interested readers should consult References [1]-[3].

2. BALANCE LAWS

Recently I developed a nonlocal continuum theory of electromagnetic-elastic solids [1]-[3]. For inert bodies and negligible body force and internal energy residuals, the balance laws of the nonlocal theory are identical to those of the classical (local) theory. Thus, we have the usual Maxwell's equations and the mechanical balance laws supplemented by the E-M body forces and energy.

A. Maxwell's Equations, expressed in Lorentz-Heaviside Units are:

$$(2.1)\qquad \nabla\cdot\mathbf{D} = q\,, \qquad\qquad (2.2)\qquad \nabla\times\mathbf{E} + \frac{1}{c}\frac{\partial\mathbf{B}}{\partial t} = \mathbf{0}$$

$$(2.3)\qquad \nabla\cdot\mathbf{B} = 0\,, \qquad\qquad (2.4)\qquad \nabla\times\mathbf{H} - \frac{1}{c}\frac{\partial\mathbf{D}}{\partial t} = \frac{1}{c}\mathbf{J}$$

$$(2.5)\qquad \frac{\partial q}{\partial t} + \nabla\cdot\mathbf{J} = 0$$

where $\mathbf{D}$, $\mathbf{E}$, $\mathbf{B}$, $\mathbf{H}$ and $\mathbf{J}$ are, respectively, the electric displacement vector, electric field vector, magnetic induction vector, magnetic field vector and total current vector, c is the speed of light in vacuum.

B. Mechanical Balance Laws

$$(2.6)\qquad \rho_0/\rho = \det x_{k,K}\,,$$

$$(2.7)\qquad t_{k\ell,k} + \rho(f_\ell - \dot{v}_\ell) + {}_Mf_\ell = 0$$

$$(2.8)\qquad t_{k\ell} + P_k\mathcal{E}_\ell + \mathcal{M}_kB_\ell = t_{\ell k} + P_\ell\mathcal{E}_k + \mathcal{M}_\ell B_k \equiv {}_Et_{k\ell}$$

$$(2.9)\qquad \rho\dot{\varepsilon} - t_{k\ell}v_{\ell,k} - q_{k,k} - \rho h - \rho\boldsymbol{\mathcal{E}}\cdot(\dot{\overline{\mathbf{P}/\rho}}) + \boldsymbol{\mathcal{M}}\cdot\dot{\mathbf{B}} - \boldsymbol{\mathcal{J}}\cdot\boldsymbol{\mathcal{E}} = 0$$

where ρ_0 is the mass density in the reference state V and ρ, $t_{k\ell}$, f_ℓ, v_ℓ, ε, q_k and h are, respectively, the mass density, stress tensor, body force density, velocity vector, internal energy density, heat vector and heat source in the deformed state $\mathcal{V}$. $\boldsymbol{\mathcal{E}}$, $\boldsymbol{\mathcal{M}}$ and $\boldsymbol{\mathcal{J}}$ are the electric, magnetization, and current vectors in the proper (co-moving) frame as defined by

$$(2.10)\qquad \boldsymbol{\mathcal{E}} = \mathbf{E} + \frac{1}{c}\mathbf{v}\times\mathbf{B}\,, \qquad \boldsymbol{\mathcal{M}} = \mathbf{M} + \frac{1}{c}\mathbf{v}\times\mathbf{P}\,, \qquad \boldsymbol{\mathcal{J}} = \mathbf{J} - q\mathbf{v}$$

Here, $\mathbf{M}$ and $\mathbf{P}$ are respectively, the magnetization and polarization vectors in the fixed frame so that

$$(2.11)\qquad \mathbf{D} = \mathbf{E} + \mathbf{P}\,, \qquad \mathbf{B} = \mathbf{H} + \mathbf{M}$$

The E-M body force is given by [4], [5].

$$(2.12)\qquad \underset{\sim}{M}^f = q\underset{\sim}{E} + \frac{1}{c}\underset{\sim}{J}\times\underset{\sim}{B} + (\nabla\underset{\sim}{E})\cdot\underset{\sim}{P} + (\nabla\underset{\sim}{B})\cdot\underset{\sim}{M} + \frac{1}{c}[(\underset{\sim}{P}\times\underset{\sim}{B})v_k]_{,k} + \frac{1}{c}\frac{\partial}{\partial t}(\underset{\sim}{P}\times\underset{\sim}{B}) .$$

For the boundary conditions, see (Ref. 5).

3. CONSTITUTIVE EQUATIONS

Constitutive equations of nonlocal E-M elastic solids are developed by assuming that the free energy Ψ, η, ${}_E T_{KL}$, Q_K, Π_K, M_K and J_K are <u>functionals</u> of C_{KL}, θ, $\theta_{,K}$, E_K and B_K, where $\theta > 0$ is the absolute temperature and

$$(3.1)\qquad {}_E T_{KL} = \frac{\rho_0}{\rho} X_{K,k} X_{L,\ell}\, {}_E t_{k\ell} , \qquad \Psi = \varepsilon - \theta\eta - \rho_0^{-1}\,\Pi_K E_K ,$$

$$(3.2)\qquad \{Q_K, \Pi_K, M_K J_K\} = \frac{\rho_0}{\rho} X_{K,k}\{q_k, P_k, M_k, J_k\} ,$$

$$C_{KL} = x_{k,K} x_{k,L} , \quad \{\theta_{,K}, E_K, B_K\} = x_{k,K}\{\theta_{,k}, E_k, B_k\}$$

are material fields. Constitutive equation for Ψ reads

$$(3.3)\qquad \Psi(\underset{\sim}{X},t) = F[C'_{KL}, \theta', \theta'_{,K}, E'_K, B'_K]$$

where a prime indicates value at <u>all</u> points $\underset{\sim}{X}'$ of the body, at <u>all</u> past times, e.g.

$$(3.4)\qquad C'_{KL} \equiv C_{KL}(\underset{\sim}{X}', t-\tau') , \qquad \underset{\sim}{X}' \varepsilon V, \qquad 0 \le \tau' < \infty .$$

The response functionals (such as F) are assumed to have a suitable Hilbert space topology and obey the second law of thermodynamics, in the global form,

$$(3.5)\qquad \int_V \frac{1}{\theta}[-\rho_0(\dot{\Psi} + \eta\dot{\theta}) + \frac{1}{2}\,{}_E T_{KL}\,\dot{C}_{KL} + \frac{1}{\theta} Q_K \theta_{,K} - \Pi_K \dot{E}_K - M_K \dot{B}_K + J_K E_K]\, dV \ge 0$$

It follows that [3],

(3.6) $$\eta = -\frac{\partial\Psi}{\partial\theta}\,, \quad {}_E T_{KL} = 2\rho_0 \frac{\partial\Psi}{\partial C_{KL}}\,, \quad \frac{\partial\Psi}{\partial\theta_{,K}} = 0\,,$$

$$\Pi_K = -\rho_0 \frac{\partial\Psi}{\partial \mathcal{E}_K}\,, \quad M_K = -\rho_0 \frac{\partial\Psi}{\partial B_K}\,,$$

(3.7) $$\int_V \left(\frac{1}{\theta^2} Q_K\, \theta_{,K} + \frac{1}{\theta} J_K \mathcal{E}_K - \frac{\rho_0}{\theta}\delta\Psi\right) dV \geq 0$$

where $\delta\Psi$ is the Fréchet derivative at points $\underset{\sim}{X}' \neq \underset{\sim}{X}$ and $\tau' \neq 0$. Spatial constitutive equations follow from (3.1) and (3.2). For heat and current separate constitutive equations are written subject to the restrictions of (3.7). The general nonlinear theory so constructed can be used to write approximate theories. For these, see Refs. [2] and [3]. Below, we give a set of linear constitutive equations for elastic dielectrics that do not possess memory of past motions and E-M fields. For conduction, see Sections 8 and 9.

4. LINEAR ELASTIC DIELECTRICS

Most dielectrics are non-magnetizable and non-conductors so that the effect of magnetization and conductions are negligible. The relevant linear constitutive equations are of the form

(4.1) $$E^{t}{}_{k\ell} = \int_V (\sigma_{k\ell mn}\, e'_{\ell mn} - e_{mk\ell} E'_m)\, dv'\,,$$

(4.2) $$D_k = \int_V (e_{k\ell m} e'_{\ell m} + \varepsilon_{k\ell} E'_\ell)\, dv'$$

where $\sigma_{k\ell mn}$, $e_{mk\ell}$ and $\varepsilon_{k\ell}$ functions of $\underset{\sim}{x}'-\underset{\sim}{x}$ and a prime indicates dependence on $\underset{\sim}{x}'$, e.g., $e'_{mn} \equiv e_{mn}(\underset{\sim}{x}',t)$. These moduli depend on an internal characteristic length a which may be taken as the lattice paramter, or granular distance depending on the constitution of the solids. When $a \to 0$, they must become a Dirac Delta measure reducing (4.1) and (4.2) to their classical (local) forms (c.f. Ref. [5]).

For a solid of infinite extent, the Fourier transforms of (4.1) and (4.2) are useful

(4.3) $$E^{\bar{t}}{}_{k\ell} = \bar{\sigma}_{k\ell mn}(\underset{\sim}{\xi})\, \bar{e}'_{mn}(\underset{\sim}{\xi},t) - \bar{e}_{mk\ell}(\underset{\sim}{\xi})\, \bar{E}'_m(\underset{\sim}{\xi},t)\,,$$

(4.4) $$\bar{D}_k = \bar{e}_{k\ell m}(\underset{\sim}{\xi})\, \bar{e}'_{\ell m}(\underset{\sim}{\xi},t) + \bar{\varepsilon}_{k\ell}(\underset{\sim}{\xi})\, \bar{E}'_{\ell}(\underset{\sim}{\xi},t)$$

where a superposed bar indicates the Fourier transform with respect to $\underset{\sim}{x}$ and $\underset{\sim}{\xi}$ is the wave vector. For isotropic solids, we have

(4.5) $$\bar{\varepsilon}_{k\ell} = (\delta_{k\ell} - \xi^{-2}\, \xi_k\xi_\ell)\, \varepsilon_T(\xi^2) + \xi^{-2}\, \xi_k\xi_\ell\, \varepsilon_L(\xi^2) ,$$

(4.6) $$\bar{e}_{k\ell m} = \xi^{-1}(\gamma_1\xi_k\delta_{\ell m} + \gamma_2\xi_\ell\delta_{km} + \gamma_2\xi_m\delta_{k\ell}) + \gamma_3\, \xi^{-3}\, \xi_k\xi_\ell\xi_m ,$$

(4.7) $$\begin{aligned}\bar{\sigma}_{k\ell mn} &= \bar{\lambda}\, \delta_{k\ell}\delta_{mn} + \bar{\mu}(\delta_{km}\delta_{\ell n} + \delta_{kn}\delta_{\ell m}) + \lambda_1\, \xi^{-2}(\xi_m\xi_n\delta_{k\ell} + \xi_k\xi_\ell\, \delta_{mn}) \\ &\quad + \lambda_2\, \xi^{-2}(\xi_k\xi_m\delta_{\ell n} + \xi_k\xi_n\delta_{\ell m} + \xi_\ell\xi_m\, \delta_{kn} + \xi_\ell\xi_n\delta_{km}) \\ &\quad + \lambda_3\, \xi^{-4}\, \xi_k\xi_\ell\xi_m\xi_n\end{aligned}$$

where ε_T and ε_L are the transverse and longitudinal dielectric moduli, γ_α, $\bar{\lambda}$, $\bar{\mu}$ and $\bar{\lambda}_\alpha$ are functions of $\xi^2 = \xi\cdot\xi$. The special case of rigid dielectrics ($\bar{e}_{k\ell m} = \bar{\sigma}_{k\ell mn} = 0$) was reviewed in Ref. [6].

5. POINT CHARGE

Consider a point charge $q = e\,\delta(\underset{\sim}{x})$ located at $\underset{\sim}{x} = \underset{\sim}{0}$ in an elastic dielectric of infinite extent. Basic equations to be solved are

(5.1) $$\nabla\cdot\underset{\sim}{D} = q , \qquad \nabla\times\underset{\sim}{E} = \underset{\sim}{0} , \qquad E^t{}_{k\ell,k} = 0$$

Combining the Fourier transforms of (5.1) with (4.3) to (4.7), we obtain

(5.2) $$\begin{aligned}\gamma\xi^{-1}\, \underset{\sim}{\xi}\cdot\underset{\sim}{u} - \varepsilon_L\, \bar{\phi} &= -\,\bar{q}\, \xi^{-2} \\ \alpha\underset{\sim}{\xi}\, \underset{\sim}{\xi}\cdot\underset{\sim}{u} + \beta\xi^2\, \underset{\sim}{u} + \gamma\underset{\sim}{\xi}\, \xi\bar{\phi} &= 0\end{aligned}$$

where ϕ is the electric potential and

(5.3) $$\gamma(\xi^2) \equiv \gamma_1 + 2\gamma_2 + \gamma_3 , \quad \alpha(\xi^2) = \bar{\lambda} + \bar{\mu} + 2\lambda_1 + 3\lambda_2 + \lambda_3 ,$$
$$\beta(\xi^2) = \bar{\mu} + \bar{\lambda}_2$$

From (5.2) we determine

(5.4) $$\bar{\phi} = \bar{q}\,\xi^{-2}[\varepsilon_L + \gamma^2(\alpha+\beta)^{-1}]^{-1} , \quad \bar{\underset{\sim}{u}} = -\gamma(\alpha+\beta)^{-1}\,\underset{\sim}{\xi}\,\xi^{-1}\,\bar{\phi} .$$

In the case of rigid dielectrics $\gamma = 0$ and if we select $\varepsilon_L/\varepsilon_0 = 1 + (r_s^2\xi^2)^{-1}$ where r_s is constant, we obtain

(5.5) $$\phi(\underset{\sim}{x}) = (e/4\pi\,\varepsilon_0 r)\exp(-r/r_s) , \qquad r \equiv |\underset{\sim}{x}|$$

The fact that such a potential corresponds to the Debye screening of the field, is well-known.

For elastic dielectrics, we take $\gamma/(\alpha+\beta) = i\,b(\xi^2)/\xi$, then integrations of (5.4) gives

(5.6) $$\phi = \frac{e}{4\pi\varepsilon_0 r}\,e^{-r/r_m} , \qquad u_r = \frac{be}{4\pi\varepsilon_0}\,\frac{r_m}{r}\,(1 + \frac{r_m}{r})\,e^{-r/r_m} + \frac{C}{r^2}$$

where $r_m^{-2} = r_s^{-2} - \dfrac{b^2(\alpha+\beta)}{\varepsilon_0} = \text{const.}$ This result shows that the Debye screening radius will be different for elastic dielectrics. In fact, for $r_m^{-2} = 0$ the Debye radius may be very large. Whether this is possible or not depends on the material moduli and their signs, The displacement field u_r given by $(5.6)_2$ shows that the point change in a defect decreases the strength of singularity by an amount $b\,e r_m^2/\varepsilon_0$ over the case of defect with no charge, since the volume change is now $\delta v = (b\,e\,r_m^2/\varepsilon_0) + 4\pi\,C$.

6. INFRA-RED DISPERSION AND LATTICE VIBRATIONS

Optical modes in the infrared regions can be affected by lattice vibrations. Ignoring the magnetization, conductions and body forces, we have the relevant equations

(6.1) $$\underset{\sim}{\xi}\cdot\bar{\underset{\sim}{D}} = 0 ,$$ (6.2) $$\underset{\sim}{\xi}\times\bar{\underset{\sim}{E}} + \frac{\omega}{c}\bar{\underset{\sim}{H}} = \underset{\sim}{0} ,$$

(6.3) $$\underset{\sim}{\xi}\cdot\underset{\sim}{H} = 0 ,$$ (6.4) $$\underset{\sim}{\xi}\times\bar{\underset{\sim}{H}} - \frac{\omega}{c}\bar{\underset{\sim}{D}} = \underset{\sim}{0}$$

(6.5) $$i\,\xi_k\,\bar{E}^t_{k\ell} - \rho\,\omega^2\,\bar{u}_\ell = 0$$

These equations when combined with (4.3) to (4.7) for isotropic solids lead to the following solutions:

(6.6) $\rho \bar{\varepsilon}_L = \gamma^2/[c_1^2 - (\omega/\xi)^2]$, $(\bar{\underset{\sim}{E}}\cdot\underset{\sim}{\xi} \neq 0 , \quad \bar{\underset{\sim}{u}}\cdot\underset{\sim}{\xi} \neq 0)$

(6.7) $c_1^2(\xi) \equiv (\bar{\lambda} + 2\bar{\mu} + 2\lambda_1 + 4\lambda_2 + \lambda_3)/\rho , \quad \gamma(\xi) \equiv \gamma_1 + 2\gamma_2 + \gamma_3$

If $\bar{\underset{\sim}{E}}\cdot\underset{\sim}{\xi} = 0$, then for $\gamma \neq 0$ $\bar{\underset{\sim}{u}}\cdot\xi = 0$ and we have transverse waves only. In this case we obtain

(6.8) $\bar{D}_k = \varepsilon \bar{E}_k ; \quad \varepsilon \equiv \varepsilon_T - \rho^{-1} \gamma^2\xi^2[(\bar{\mu} + \lambda_2)(\xi^2/\rho) - \omega^2]^{-1}$

comparing this result with the classical treatment of photon-phonon interaction[7]

(6.9) $\varepsilon_\infty = \varepsilon_T , \quad \varepsilon_0 - \varepsilon_\infty = - \gamma^2/(\bar{\mu} + \lambda_2) , \quad \omega_0^2 = (\bar{\mu} + \lambda_2)(\xi^2/\rho)$

In classical treatment ε_∞ , ε_0 and ω_0^2 are considered constants. Here they depend on ξ^2 so that the space dispersion is fully accounted.

A second set of solutions of the above system is obtained for non-zero $\underset{\sim}{\bar{H}}$ and $\underset{\sim}{\xi} \times \underset{\sim}{\bar{u}}$ leading to the dispersion relations

(6.10) $\omega^4 - [c^2\rho + (\bar{\mu}+\lambda_2)\varepsilon_T - \gamma_2^2] (\xi^2\omega^2/\rho\varepsilon_T) + (\bar{\mu}+\lambda_2)(c^2\xi^4/\rho\varepsilon_T) = 0$

This is the dispersion relation for optical modes that possess nonvanishing transverse components $\underset{\sim}{\xi} \times \bar{\underset{\sim}{u}}$ and $\underset{\sim}{\xi} \times \bar{\underset{\sim}{E}}$. It is valid when $\underset{\sim}{\xi}\cdot\bar{\underset{\sim}{E}} = \underset{\sim}{\xi}\cdot\bar{\underset{\sim}{u}} = 0$. If we employ expressions (6.9), equation (6.10) can be expressed in the form

(6.11) $\xi^2c^2/\omega^2 = \varepsilon_\infty + (\varepsilon_0 - \varepsilon_\infty) \omega_0^2(\omega_0^2 - \omega^2)^{-1}$

which is a well-known result for constant ε_0, ε_∞ and ω_0 .

7. NATURAL OPTICAL ACTIVITY

The dependence of the dielectric modulus $\bar{\varepsilon}_{k\ell}$ on the wave-vector $\underset{\sim}{\xi}$ give rise to optical activity. From the total free energy consideration

(7.1) $\int_V D_kE_k \, dv = \int_V\int_V \varepsilon_{k\ell}(\underset{\sim}{x}-\underset{\sim}{x}') \, E_k(\underset{\sim}{x}) \, E_\ell(\underset{\sim}{x}') \, dv(\underset{\sim}{x}) \, dv(\underset{\sim}{x}')$

it is clear that $\varepsilon_{k\ell}(\underset{\sim}{x}-\underset{\sim}{x}') = \varepsilon_{\ell k}(\underset{\sim}{x}'-\underset{\sim}{x})$. Consequently, $\bar{\varepsilon}_{k\ell}(\underset{\sim}{\xi}) = \bar{\varepsilon}_{\ell k}(-\underset{\sim}{\xi})$. If we expand $\bar{\varepsilon}$ into power series about $\underset{\sim}{\xi} = \underset{\sim}{0}$, retain the first two terms and note that $\underset{\sim}{\bar{\varepsilon}}^* = - \underset{\sim}{\bar{\varepsilon}}$, then we will have

(7.2) $\varepsilon_{k\ell}(\underset{\sim}{\xi}) = \varepsilon_{k\ell}^0 - i \, \gamma_{k\ell m}\xi_m$

with the symmetry regulation

(7.3) $\varepsilon_{k\ell}^0 = \varepsilon_{\ell k}^0 , \quad \gamma_{k\ell m} = - \gamma_{\ell km} = \gamma_{k\ell m}^*$

where an asterisk denotes the complex conjugate. This approximation is equivalent to

(7.4) $$D_k = \varepsilon^0_{k\ell} E_\ell + \gamma_{k\ell m} (\partial E_\ell / \partial x_m)$$

which is identical to the classical expression [8] leading to natural optical activity.

8. ANOMALOUS SKIN EFFECT

Anomalous skin effect arises when a highly non-uniform field is established in a conductor. For the half-space $x_3 \geq 0$, in the absence of the displacement current and magnetization, the Fourier transforms of Maxwell's equations reduce to

(8.1) $$\underset{\sim}{\xi} \times \bar{\underset{\sim}{E}} + \frac{\omega}{c} \bar{\underset{\sim}{H}} = \underset{\sim}{0}, \qquad \underset{\sim}{\xi} \times \underset{\sim}{H} = \frac{i}{c} \bar{\underset{\sim}{J}}$$

where $\bar{\underset{\sim}{J}}$ is given by the constitutive equation [3],

(8.2) $$\bar{J}_k = \bar{\sigma}_{k\ell} \bar{E}_\ell ; \qquad \bar{\sigma}_{k\ell} = \sigma_0 \delta_{k\ell} + \sigma_1 \xi^{-2} \xi_k \xi_\ell$$

for isotropic solids. Here σ_0 and σ_1 depend on $\xi^2 \equiv \underset{\sim}{\xi} \cdot \underset{\sim}{\xi}$ and If E-M fields are independent of x_2 and E-field is in x_1-direction, these equations give

(8.3) $$i \omega c^{-2} \bar{\sigma}_{11}(\xi_3^2, \omega) - \xi_3^2 = 0$$

which shows that the penetration depth depends on the frequency. This result can be used to determine the conductivity σ_{11} by comparing it with atomic calculations.

The differential equation for the electric field in the space domain, is given by

(8.4) $$\frac{d^2 E_1}{dx_3^2} + \frac{i\omega}{c^2} \int_{-\infty}^{\infty} \sigma_{11}(x_3 - x_3', \omega) E_1(x_3') dx_3' = 0$$

This equation is identical to that obtained by Reuter and Sondheimer [9] for the case of specular reflection. In this case, σ_{11} is given by

(8.5) $$\sigma_{11}(x) = \sigma_0 \{ Ei_1[(1 - i\omega\tau)|x|] - Ei_3[(1 - i\omega\tau)|x|]$$

where σ_0, τ are constants $Ei_n(x)$ is the exponential function.

(8.6) $$Ei_n(x) = \int_1^\infty s^{-n} e^{-sx} ds$$

9. SUPERCONDUCTIVITY

Below a critical temperature ranging from less than 1°K to 18°K, a large number of metals and alloys are superconductors. At this stage, there is no resistance to the electric field inside of the metal. Here we show that the superconductivity is included in nonlocal theory.

Two surviving Maxwell's equations have the form

(9.1) $$\nabla \times \underset{\sim}{H} = \underset{\sim}{J}/c\,, \qquad \nabla \cdot \underset{\sim}{H} = 0$$

The constitutive equations for $\underset{\sim}{J}$ is given by [3],

(9.2) $$\underset{\sim}{J} = \int_V \sigma_0(R)\, \underset{\sim}{R} \times \underset{\sim}{H}(\underset{\sim}{x}')\, dv(\underset{\sim}{x}')$$

where $\underset{\sim}{R} = \underset{\sim}{x}' - \underset{\sim}{x}$ and $R \equiv (\underset{\sim}{R} \cdot \underset{\sim}{R})^{\frac{1}{2}}$. Boundary conditions are:

(9.3) $$\underset{\sim}{H} \cdot \underset{\sim}{n} = 0\,, \qquad [\underset{\sim}{H} \times \underset{\sim}{n}] = \underset{\sim}{0} \qquad \text{on} \quad \partial V$$

where [ ] indicate jump across ∂V. If we take the curl of (9.1) and use (9.2) and (9.3), we obtain

(9.4) $$\nabla \times \nabla \times \underset{\sim}{H} + \frac{1}{c} \int_V \gamma(R)\, \underset{\sim}{H}(\underset{\sim}{x}')\, dv(\underset{\sim}{x}') = \underset{\sim}{0}$$

where $\gamma(R) = \nabla \cdot (\sigma_0 \underset{\sim}{R})$. This equation reduces to London's equation in the classical limit when $\gamma(R)$ becomes proportional to a Dirac-delta measure.

If we introduce vector potential $\underset{\sim}{A}$ by

(9.5) $$\underset{\sim}{H} = \nabla \times \underset{\sim}{A}\,, \qquad \nabla \cdot \underset{\sim}{A} = 0$$

and assume $\sigma_0 = (C/2R^2)\exp(-R/\xi_0)$ where C and ξ_0 are constant, then (9.2) leads to Pippard's [10] form for $\underset{\sim}{J}$:

(9.6) $$\underset{\sim}{J} = C \int_V (\underset{\sim}{A} \cdot \underset{\sim}{R})\, R^{-4}\, \underset{\sim}{R} \exp(-R/\xi_0)\, dv(\underset{\sim}{x}')$$

This latter assumption is justified on the basis of interatomic attenuation of the nonlocality.

REFERENCES

[1] A. Cemal Eringen, "Theory of Nonlocal Electromagnetic Elastic Solids," J. Math. Phys., 14, 733-740, 1973.

[2] A. Cemal Eringen, "Theory of Nonlocal Piezoelectricity," to appear in J. Math. Phys.

[3] A. Cemal Eringen, "Electrodynamics of Memory-Dependent Nonlocal Elastic Continua," submitted for publication.

[4] G.A. Maugin and A. Cemal Eringen, "On the Equations of the Electrodynamics of Deformable Bodies of Finite Extent," Journal de Mécanique 16, 101-147, 1977.

[5] A. Cemal Eringen, "Mechanics of Continua," Second Edition, Krieger, 1980.

[6] A.A. Rukhadze and V.P. Silin, "Electrodynamics of Media with Spatial Dispersion," Soviet Physics, Uspekhi, 4, 459-484, 1961.

[7] Max Born and Kun Huang, Dynamical Theory of Lattices, Oxford U. Press, Section II.8, 1954.

[8] L.D. Landau and E.M. Lifshitz, Electrodynamics of Continuous Media, Pergamon Press, 1960, p. 338.

[9] G.E.H. Reuter and E. H. Sondheimer, Proc. Roy. Soc., A., 195, 336, 1948.

[10] A.B. Pippard, The Dynamics of Conduction Electrons, Gordon and Breach, 1965.

The Mechanical Behavior of Electromagnetic Solid Continua
G.A. Maugin (editor)
Elsevier Science Publishers B.V. (North-Holland)

ELECTROELASTIC EFFECTS IN CELL MEMBRANES

Joseph M. Crowley

Applied Electrostatics Research Laboratory
Department of Electrical Engineering
University of Illinois
Urbana, Illinois 61801
U.S.A.

Biological cells are surrounded by a thin membrane consisting primarily of insulating lipids supporting an electric field which usually exceeds 25 MV/m, and may become much larger under some conditions. At these high field strengths, the electric force can cause appreciable deformation of the relatively compliant membrane, and may lead to rupture. An elastic model of the membrane has been used to study such instabilities, and has predicted breakdown voltages which agree quite well with those reported in experiments on biological cells and rapidly stressed artificial membranes.

INTRODUCTION

The membrane which surrounds a biological cell is one of its most important constituents. All chemicals which enter or leave the cell must pass through the membrane, which exercises a great deal of selectivity in this gate-keeping function, whether dealing with large protein molecules or small ions. Even the very similar Na^+ and K^+ ions are maintained at different concentrations within the cell, often by orders of magnitude. As a result of this ionic selection, the membrane is always subject to an electric potential difference.

The electric potential across a membrane is not just an unimportant byproduct of cell biochemistry; indeed, it plays a key roll in many cell functions. The best known of these is the propagation of nerve impulses along the axon, which has long been considered as an electrochemical wave. Conversely, the external application of a voltage to the membrane will affect the cell function or even its structural integrity. For these reasons, basic biophysical studies of cells and their membranes often involve electrical stimulation.

Many biomedical instruments also involve the application of voltage to a cell. Typical of these is the Coulter counter, which measures the concentration of red blood and other cells through the decrease in current flow between electrodes caused by the presence of a cell. Most of the voltage drop associated with the presence of the cell appears across the cell membrane, so passage through a Coulter counter can often affect the cell function. In fact, one of the major limitations of this device is the rupture of the membrane which occurs if the voltage drop is much greater than one volt (Zimmermann et al. 1974).

Failure of the cell membrane under electrical stress is not always a liability if the voltage is applied as a very short pulse. The membrane breaks down (or at least becomes more permeable) for a time, but then heals itself (Kinoshita and Tsong 1977, Benz et al. 1979). This temporary breakdown, called electroporation, furnishes an opportunity to open a cell, insert some chemicals which would normally be rejected by the membrane, and then return the cell to its original milieu. The most exciting application of this technique is the transfer of DNA into a foreign cell in genetic engineering (Auer 1976, Neumann et al. 1982).

Some time ago it was suggested that the electrical breakdown of a membrane could be explained as the failure of an elastic material under compression caused by electrostatic forces (Crowley 1973). This model assumed that the membrane could in fact be treated as elastic, and that it would be soft enough to deform under the electric pressures associated with potential differences on the order of one volt. Since then, the model has been put to the test for a variety of natural and artificial membranes. The object of the present paper is to review this model, and to assess its validity in the light of these tests.

ELECTROELASTIC MODELS OF CELL MEMBRANES

Electrically, the membrane consists of an electrically insulating layer surrounded by good conductors as shown in Figure 1.

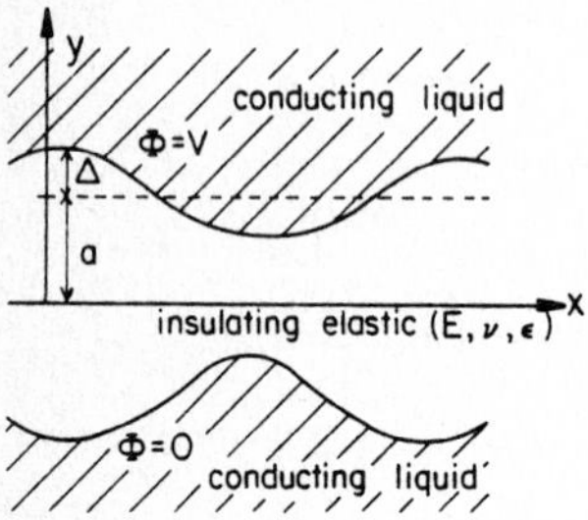

Figure 1. Elastic model of a membrane

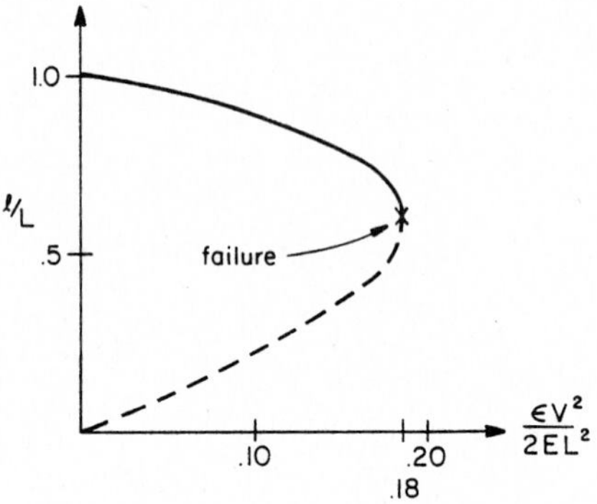

Figure 2. Membrane compression due to voltage

There is a large imbalance between the ions inside and outside a living cell membrane which leads to an appreciable voltage, in the range 0.1 to 1.0 V. The entire voltage drop appears across the 4 nm thickness of the membrane, so the electric field there,

$$V/\ell \simeq 25 \text{ to } 250 \text{ MV/m} \tag{1}$$

is enormous by macroscopic standards. (Sparks form in air at 3 MV/m.)

The elastic properties are represented by a Young's modulus, Y. The compression arising from the electric field across the membrane is given by (Crowley 1973)

$$(\ell/L)^2 \ln(\ell/L) = -\frac{\varepsilon V^2}{2YL^2} \tag{2}$$

which for small deflections

$$\delta\ell = \ell - L << L \tag{3}$$

takes the form

$$\frac{\delta\ell}{L} \simeq \frac{\varepsilon V^2}{2YL^2} \tag{4}$$

A plot of membrane thickness against electric pressure is shown in Figure 2.

An interesting aspect of the voltage dependence is the infinite compression (or failure) which occurs at a finite value of voltage, given by

$$\frac{\varepsilon V^2}{2YL^2} \simeq 0.18 \tag{5}$$

This is a manifestation of an electro-elastic instability, since no stable equilibrium is possible at higher voltages. This instability occurs because the voltage is held constant while the opposite sides of the membrane approach each other, which causes the electric field and electric pressure to increase, further compressing the membrane. Above the critical voltage the membrane is no longer able to resist the additional compression, and it fails. Similar breakdown mechanisms also occur in soft solids such as plastics (O'Dwyer 1973).

The one-dimensional elastic model based on Hooke's Law offers a clear picture of the nature of the electroelastic failure of the membrane, but it seems a poor choice to describe the breakdown of a membrane with significant extension in the directions perpendicular to the compression. To meet this objection, the electroelastic instability was re-examined using a three-dimensional isotropic elastic model (Crowley 1973). The stability condition for the three-dimensional model is

$$\frac{\varepsilon V^2}{2YL^2} \simeq .17 \text{ to } .18 \tag{6}$$

depending on the value of ν. This is essentially the same condition derived for the one-dimensional model.

EXPERIMENTAL CONFIRMATION

There are two principal predictions of the electroelastic model: The membrane should exhibit compression initially proportional to the square of the voltage, and it should fail in compression above a critical voltage. Because of the exceedingly small thickness of a biological membrane, it is extremely difficult to measure changes in thickness directly, so early confirmation of the model was sought in measurements of the electric capacitance of the membrane. For a thin insulating membrane, the capacitance per unit area,

$$C = \frac{\varepsilon}{\ell} \tag{7}$$

is fairly large because ℓ is so small. Changes in thickness should be easily measured as the membrane is compressed, and these changes are, in fact, proportional to V^2 for low voltages, as shown in Figure 3 which is taken from Alvarez and Latorre (1978).

This figure gives results for a particular synthetic bimolecular lipid membrane (BLM) composed of the lipid glycerol monooleate, but it is typical of the voltage dependence obtained from measurements on most membranes. Since all of the properties of the membranes in equation 6 are known with the exception of the Young's modulus, results like Figure 3 can be used as an experimental determination of the Young's modulus of the membrane. For the membrane shown in that figure, as an example, these measurements correspond to a relatively high value of $Y \simeq 140$ MPa. Similar voltage dependence has been obtained by many others (Wobschall 1982, White 1974, Carius 1976). Thus it seems clear that the membrane can be treated as an elastic material compressed by electric forces.

The second major prediction of the model is failue of the membrane at a critical voltage which depends on the elastic properties of the membrane. At the time the model was proposed, data for elastic properties and breakdown voltages for BLM's were not available for the same preparations, but estimates based on individual reported values were in approximate agreement with the theory. Since then, experiments have been carried out on a number of natural and artificial membranes. A selection of results from these experiments is shown in Figure 4.

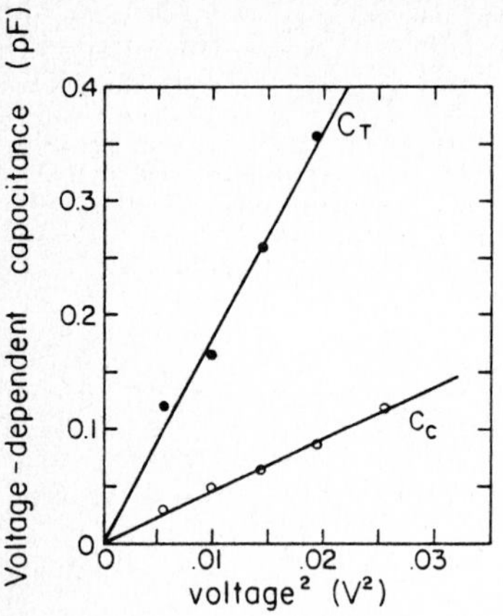

Figure 3. Capacitance variation with voltage for a stiff BLM

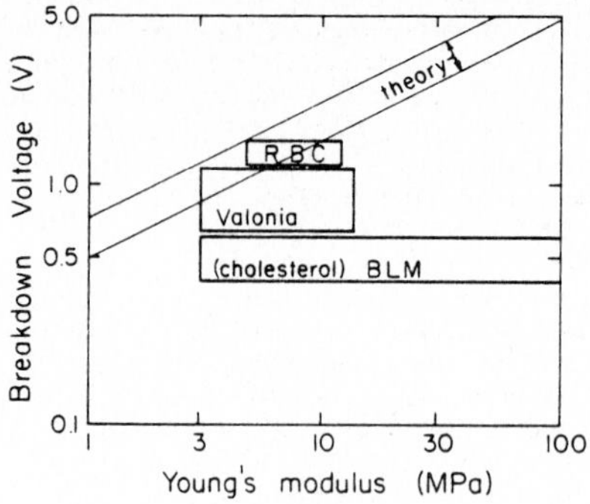

Figure 4. Breakdown predictions based on Young's modulus

From the range of measured values, represented by the boxes, it is clear that neither the Young's modulus nor the breakdown voltage can be predicted with great accuracy. Even so, the correlation between the two seems to justify the electroelastic model, especially for the biological cells (red blood cells and the alga Valonia). The synthetic membranes appear not to satisfy the electroelastic breakdown, especially for the high values of Young's modulus which have been obtained with very pure preparations.

DIFFICULTIES AND ALTERNATE MODELS

A closer look at the experimental evidence described above led to the suspicion that the homogeneous, istropic elastic model might be inconsistant with some of the phenomena observed in membranes. The first objections (Requena et al. 1975, Benz and Janko 1976) concerned the determination of Young's modulus by capacitance measurements. This method assumed that only compression of the lipid molecules contributed to capacitance changes. An alternate mechanism, however, relies on the observation that the synthetic membranes (BLM's) are not uniform across their surface. Rather, they contain numerous small microlenses of excess solvent, which are considerably thicker than the membrane itself. When a voltage is applied across the membrane, these microlenses grow as more solvent is squeezed from the bilayer structure. In this picture, the membrane thins not by elastic compression but by fluid flow through the membrane interior. The flow continues until the pressures in the lenses and the bilayer are equalized.

This flow is very slow because the membrane is so thin, and it may take seconds or minutes until it reaches equilibrium. When the voltage is applied for a shorter time (or at a higher frequency), these flows do not occur, and the membrane responds according to the elastic model. Measurements of the capacitance change for short times (Benz 1976, Carius 1976) suggest that the Young's modulus should be > 10 MPa, instead of < 1 MPa which is inferred from the DC measurements. According to the failure criterion, this would raise the breakdown voltage by a factor of exceeding $\sqrt{10}$. Similar conclusions can be drawn from measurements on BLM's made without solvent (Alvarez and Latorre 1978). Unlike the solvent BLM's, these preparations exhibit a capacitance much less dependent on applied voltage, indicating a stiffer membrane. As in the high frequency measurements on solvent BLM's, the apparent Young's modulus is much larger than initial estimates.

A second objection concerned the simplicity of the elastic model for the membrane. Even if a continuum model based on elasticity is valid for a material which is two molecules thick, it seems unlikely that a single elastic constant such as Young's modulus will be sufficient to describe the deformation and stability. In particular, the anisotropy of the bilayer structure seems to require an anisotropic elastic model. Such models have been proposed (Evans and Simon 1975a, b, Crowley 1976), but they have not proved particularly useful to date. The main problem with the use of more complex elastic models is the difficulty in determining the appropriate elastic constants for a small, inaccessible structure like a cell membrane.

An alternate approach to the anisotropy of the membrane is a microscopic model of the molecular arrangement (Crowley 1977, Abidor 1978, Sugár 1978, 1979). The most elaborate of these models (Sugár 1979) is based on the statistical mechanical description of the interactions between various parts of the modecules, such as the head groups and hydrocarbon tails, as well as the effects of elastic deformation and electric compression. This model also allows for phase transistions which alter the modecular arrangement. Since this is a thermodynamic model, it can express the electromechanical properties in terms of other thermodynamic coefficients related to thermal expansion or heats of reaction, quantities which are more easily measured. This model predicts large variations in Young's modulus with temperature (.5 to 9000 MPa), which seem to correlate with observed changes in membrane behavior. It also predicts the electrocompressive failure of the bilayer at voltages on the order of those observed.

VALIDITY OF THE ELECTROELASTIC MODEL

The preceding sections summarized a great deal of partially conflicting evidence on the usefulness of the electroelastic model of membranes. It is now time to take stock of the current situation, and to discuss where the model may justifiably be used. It will be helfpul to keep in mind that there are two types of membranes under consideration--the synthetic laboratory model membrane (BLM) and the membrane of a biological cell. Each has its own peculiar sources of variation. The BLM, for instance, is made from known chemical constituents, but it usually contains numerous microlenses of solvent scattered across the molecular bilayer. These films are inhomogeneous even on a microscopic scale, and the solvent is free to move between bilayer and lens. This process is much softer and much slower than elastic deformation of the membrane, since it involves fluid flow through small spaces. This picture of the membrane leads to thinning under applied voltage, but not to immediate breakdown, because after all the solvent is expelled the stiff lipid structures will support the electrocompression. Electroelastic breakdown can still occur, but the Young's modulus will be much larger than the apparent Young's modulus derived from low frequency measurements. When using high frequencies ($>$ 10kHz) or short pulses ($<$ 100μs), this problem does not arise, since the solvent does not have time to flow into the microlenses. Under these conditions, the breakdown voltage agrees with the apparent Young's modulus, confirming the model for use in the BLMs.

A similar conclusion can be drawn for explanations of breakdown as statistical defects in the membrane. Estimates of the time required for random pores to grow in response to an applied electric field suggest a delay of approximately 0.1 ms (Shchipunov and Drachev 1982). Such mechanisms may be important at low frequencies, but for fast pulses and high frequencies, they can have no effect. Thus the electroelastic effect is always present in BLM's, but it may compete with the microlens effect or statistical pore formation at low frequencies. For high frequencies or short pulses, it appears to be the only possible breakdown mechanism.

The second type of membrane is that surrounding a living cell. Unlike the BLM, this membrane may be composed of a large number of complex molecules, although the primary constituent is also a lipid. It is simpler in one respect, however;

it does not contain microlenses of excess solvent. Thus the principal source of difficulty with the synthetic BLM's is removed. As a result, there have been no reported experiments on biological cells which contradict the electroelastic model. Of course, experiments on living cells do not yield the precision possible with BLM's, so this conclusion may be modified as experimental techniques improve. Nonetheless, the electroelastic model is clearly the best available at the prsent time to explain the interrelations among compression, transport and breakdown under applied voltages.

REFERENCES

[1] Abidor, I. G. et al., "Electrical breakthrough of bilayer lipid membranes," Doklady Ak. Nauk SSSR 240, pp. 733-736 (1978).

[2] Alvarez, O. and R. Latorre, "Voltage-dependent capacitance in lipid bilayers made from monolayers," Biophys. J. 21, pp. 1-17 (1978).

[3] Auer, D. et al., "Electric breakdown of the red blood cell membrane and uptake of SV 40 DNA and mammalian cell RNA," Naturwissenshaften 63, p. 391 (1976).

[4] Benz, R. et al., "Reversible electrical breakdown of lipid bilayer membranes: A charge pulse relaxation study," J. Membrane Biol. 48, pp. 181-204 (1979).

[5] Carius, W., "Voltage dependence of bilayer membrane capacitance," J. Colloid Interface Sci. 57, pp. 301-307 (1976).

[6] Crowley, J. M., "Electrical breakdown of bimolecular lipid membranes as an electromechanical instability," Biophys. J. 13, pp. 711-725 (1973).

[7] Crowley, J. M., "On mechanics of bilayer membranes," J. Colloid Interface Sci. 56, pp. 186-187 (1976).

[8] Crowley, J. M., "The buckling of a composite membrane under electric stress," Intern. Symp. of Quantum Biology, Sanibel, Florida (1977).

[9] Evans, E. and S. Simon, "Mechanics of bilayer membranes," J. Colloid Interface Sci. 51, pp. 266-271 (1975a).

[10] Evans, E. A. and S. Simon, "Mechanics of electrocompression of lipid bilayer membranes," Biophys J. 15, pp. 850-852 (1975b).

[11] Kinoshita, K., and T. Y. Tsong, "Voltage-induced pore formation and hemolysis of human erythrocytes," Biochem. Biophys. Acta 471, pp. 227-242, (1977).

[12] Mueller, P. et al., "Formation and properties of bimolecular lipid membranes," Prog. Surface Memb. Sci. 1, pp. 379-393, (1964).

[13] Neumann, E. et al., "Gene transfer into mouse lyoma cells by electroporation in high electric fields," EMBO J. 1, pp. 841-845 (1982).

[14] O'Dwyer, J. J., The Theory of Electrical Conduction and Breakdown in Solid State Dielectrics, Clarendon Press, Oxford, pp. 286-289 (1973).

[15] Requena, J. et al., "Lenses and the compression of black lipid membranes by an electric field," Biophys. J. 15, pp. 77-81 (1975).

[16] Shchipunov, Yu. A. and G. Yu. Drachev, "An investigation of electrical breakdown of bimolecular lipid membranes," Biochem. Biohysica Acta 691, pp. 353-358 (1982).

[17] Sugar, I. P., et al., "Electric field induced defect-forming mechanisms in lipid bilayers," Acta Biochem. Biophys. Acad. Scie. Hung. 13, pp. 193-200 (1978).

[18] Sugar, I. P., "A theory of the electric field-induced phase transistion of phospholipid bilayers," Biochem. Biophysica Acta 556, pp. 72-85 (1979).

[19] White, S. H., "Comments on 'Electrical Breakdown of Biomolecular Lipid Membranes as an electromechanical Instability,'" Biophysical J. 14, pp. 155-159 (1974).

[20] Wobschall, D., "Voltage dependence of bilayer membrane capacitance," J. Colloid Interface Sci. 40, pp. 417-423 (1972).

[21] Zimmerman, U. et al., "Dielectric breakdown of cell membranes," Biophys. J. 14, pp. 881-899 (1974).

[22] Zimmermann, U. et al., "Effects of external electrical fields on cell membranes," Bioelectrochem. Bioenerg. 3, pp. 58-83 (1976).

The Mechanical Behavior of Electromagnetic Solid Continua
G.A. Maugin (editor)
Elsevier Science Publishers B.V. (North-Holland)

SONIC STUDIES OF THE MECHANICAL AND ELECTRICAL PROPERTIES OF A MEMORY ALLOY

R.S.Geng, A.H.Kafhagy , A.B.Rajab and R.W.B.Stephens

Physics Department,Chelsea College
University of London,Pulton Place,London SW6 5PR
United Kingdom.

A comprehensive series of measurements have been made of the martensitic transformation of a copper-based alloy:the experiments are designed to investigate possible correlation between its mechanical and electrical behaviour when the specimen is subjected to mechanical and/or thermal cyclings. Experimental techniques involving acoustic emission (AE),photoacoustic spectroscopy (PAS),internal friction and electrical resistivity are employed. Such procedures for example enable the critical stress,which brings about a phase change in the alloy system at a constant temperature,to be determined by both electrical and acoustic emission measurements, and the phase transition owing to the change of temperature to be determined accurately by the change of internal friction or, correspondingly, of sound velocity. The thermal diffusivity is determined by means of the photoacoustic technique.The ultrasonic measurements have also proved useful in evaluating the effect of heat-treatment on the properties of the alloy.

INTRODUCTION

Since the discovery of the shape-memory alloy in the 1930s /1/ much attention has been paid to the fabrication and application of alloys which exhibit the shape-memory effect (S.M.E). This refers to the phenomenon in which an alloy, plastically deformed at a certain temperature, will completely, or almost, recover its initial shape on being raised to a higher temperature. The Ni-Ti based memory alloy, Nitinol, has a good memory effect and mechanical properties, but its high cost has restricted its wide application. It was only following the development of a new family of shape-memory alloys based on copper, zinc and aluminium, which are much cheaper than the Nitinol and easier to fabricate, that the large scale of application of the phenomenon became possible. This new alloy is termed Betalloy (Beta + alloy), since it possesses the β-phase at high temperatures. Betalloy has generally a large grain size, ranging from 100μm up to 1mm or even more, and its mechanical properties are relatively poor compared to Nitinol, e.g. it possesses a short fatigue life. Fig.1 shows a micrograph of Betalloy, indicating an average grain size of about 250 μm.
The controversy on the mechanism of the S.M.E. has been prolonged, but it is generally accepted that the effect is due mainly to the martensitic - austenitic transformation /2/. Means of investigating the shape-memory alloy covers a wide range of experimental techniques: X-ray diffraction, neutron diffraction, calorimetry, resistivity measurements and ultrasonic inspection, etc. Since a phase transition in solids is accompanied by changes in many of the material properties, measurements of any sensitive property around the phase transition will, in principle, provide a means of investigating the transition. Efforts have been made to increase the memory effect and to improve mechanical properties of the Betalloy by changing its composition and/or by a proper heat-treatment.
A combination of ultrasonic techniques are in use in the authors' laboratory for material investigations. It has been found that ultrasonic tests can supply a simple and effective means of evaluating the S.M.E. because it gives not only the changes of some macroscopic parameters, but also involves the

changes of physical parameters closely relating to the microstructure of the alloy.

THE EFFECT OF HEAT-TREATMENT ON THE PROPERTIES OF BETALLOY

Since the martensitic transformation is responsible for the S.M.E., the transition temperature and the fraction of martensite under various conditions are therefore of a primary concern. It was generally assumed that martensitic transformation is athermal and involves no thermal activated transfer of atoms at the interphase boundary. The phase transition temperature depends only on the composition of the alloy and the fraction of martensite on temperature as long as a long term of time period is considered. It means that the "intermediate state" of the material has no effect on the fraction of martensite it finally comprises. Hence, the effect of high temperature annealing (betalizing) on the S.M.E. has received less investigation. However, many experiments have shown that there would be no significant fraction of martensite and subsequent S.M.E. without this betalizing and proper quenching process being followed. This is possibly because betalizing can eliminate the defects created by mechanical process and can relieve external and internal stresses. These stresses may change the relative stability of the phases existing in the alloy, leading to various sequences of phase transition. An example is given here.

Two types of Betalloy of different compositions have been tested. They were designated BM40 and BP50, respectively. The phase transition temperatures (Ms) are -40°C and +50°C respectively if they were annealed at 550°C for two hours and then rapidly quenched in ice-cold water. For the as-received specimens of the two types, however, the longitudinal sound velocity (c) and damping coefficient (α) exhibited no significant difference. After being annealed at 550°C for two hours and then quenched in water of 15°C, a significant difference in c and α has been observed since both c and α changed in the opposite direction for the two types of material, respectively. For example, for the as-received specimens, c_{BP50} = 4690 m/s and c_{BM40} = 4700 m/s, but after the heat-treatment, these were respectively 4650 and 4760 m/s. One of such changes is shown in Fig.1(a) and (b) for the BP50. The damping coefficient is so high, for the BP50, after the heat-treatment that a specimen of shorter length has to be employed in order to observe the reflected echoes. This proves that a distinct phase state is produced after being betalized and quenched.

A pronounced difference has also been observed from the AE measurements. The movements of the twinning boundaries of the Betalloy have been regarded as responsible for the large amount of AE energy release under relatively low stresses/3/, and a detailed description of this work and its significance in evaluating the alloy can be found in /3/ and /4/. An important conclusion is that the AE is very sensitive to the heat-treatment of the alloy and that a rather slow quenching process (e.g. slow furnace-cooling) may eventually lead to an insignificant fraction of the martensite being produced and therefore the disappearance of shape-memory effect. This may be seen from Fig.3(a) and (b), in which both BP50X and BP50F were annealed at 550°C for two hours, but were then the specimens ice-water quenched and furnace cooled to room temperature, respectively. The ordinary methods of stress-strain measurements are obviously not so sensitive as acoustic emission.

CHANGES OF RESISTIVITY WITH STRESS AND ITS COMPARISON WITH THE STRESS-INDUCED TRANSITION

For BM40 alloy, it was reported /3/ that a reversible acoustic energy release process accompanies the stress-induced phase transition and that the critical stress to produce this reversible emission is a function of the test temperature. This critical stress therefore has been defined as the phase transition initiation stress (σ_T) and an approximate formula for σ_T and T (Kelvin scale) is expressed as

$$\sigma_T = 1.25(T^2 - M_s^2)^{\frac{1}{2}} \quad N/mm^2 , \qquad (1)$$

where $M_s \sim 240K$.
It was also found that for a fixed stress, the acoustic emission energy is dependent on temperature and that the AE energy - T curves can be divided into two "wings" with different slopes and, furthermore, the intercept point of the two parts corresponds to the phase transition temperature, for that specific stress/3/. For a stress of 250 N/mm^2, the interception point corresponds to 28°C.
The electrical resistivity change of BM40 seems to follow the same abrupt pattern. Fig.4(a) shows the resistivity (ρ) change versus the applied external stress. The abrupt change of ρ occurs also at a magnitude of stress which corresponds to the phase transition initiation stress at that temperature.(The change of length dimensions under tension may be neglected cf. actual resistance change.) However, for BP50, when there is no phase transition involved, the resistivity does not show any abrupt change around the same value of stress.

THE INTERNAL FRICTION OF BETALLOY DURING THERMAL CYCLINGS

The study of a material by using ultrasonic waves, except for the above-mentioned AE measurements, is in essence to investigate the change of two parameters: ultrasound velocity and attenuation. In metals, attenuation can be attributed to the internal friction, elastic hysteresis and grain scattering. Many experiments have been made of the change of internal friction during thermal cycling of alloys during martensitic transformation/5,6/, however, the extension of internal friction study to higher frequencies (KHz) is only of recent development /7/. By doing so, an immediate advantage is the elimination of the effect of the change rate of temperature on measured internal friction. This is because the internal friction (Q^{-1}) can be split to two terms as /8/

$$Q^{-1} = \alpha_1 \dot{T}/f + Q^{-1}(T, \varepsilon) \qquad (2)$$

where f is the frequency, ε the strain, $\dot{T}$ temperature change rate and α_1 a proportional coefficient. When f is in the range of KHz, Eqn.(2) can be simplified to

$$Q^{-1} = Q^{-1}(T, \varepsilon) \qquad (3)$$

By adopting the system developed by Bell et al /7/, the effect of ε (strain amplitude in a pendulum system) can also be minimised since now the external agency is in general not large enough to induce any non-linear vibration. Therefore, the experimental results seem more consistent during consecutive tests.
Experimental results are shown in Fig.5, from Fig.5(a) it can be seen that a very sharp increase in internal friction for the BM40 appears at -50°C (for cooling) and -30°C (for heating), respectively. The peak of internal friction is in each case accompanied by a sharp decrease of the resonant frequency (which is proportional to the sound velocity), corresponding to the softening of the material. These two peaks are evidently correspond to the M_s and A_s, respectively. The relatively high internal friction in martensitic phase is

in agreement with results described in section 2 for the observation on damping coefficient. This high attenuation is due to the movement of twin boundary (i.e. the movement of the interfaces existing between the martensite plates). As described previously, martensitic transition involves the movement of a large number of atoms in a very short period, hence it is accompanied by a sharp peak in internal friction. This coincides also with the fact that a peak acoustic emission energy release occurs during this transition. The disappearance of the internal friction peak for the BP50, see Fig.5(b), is possibly due to the high damping in the martensitic phase, which may significantly reduce the amplitude of the exciting signal. However, the sharp decrease of the resonant frequency still gives an indication that the phase transition temperature for the BP50 is $45^{\circ}C(M_s)$ or $60^{\circ}C(A_s)$. Unlike the observation in /9/, we found that these curves are reproducible during consecutive thermal cyclings and, therefore, the time effect is not obviously observed. The internal friction measurement provides an easy and convenient means of evaluating the S.M.E. and corresponding phase transition temperatures of a material.

As mentioned above, the sharp decrease of the resonant frequency (or sound velocity) at the transition temperature represents the weakening of the alloy over the transition region. This effect has to be taken into account when designing a device or a structure involving the use of the Betalloy, even though the working temperature is well above the phase transition temperature. For instance, when under an external stress which is large enough to induce the phase transition at that temperature, because then the structure is further weakened owing to the "phase-transition-softening"; in such circumstances a severe damage may possibly be imposed on the structure.

PHOTOACOUSTIC MEASUREMENTS

In the photoacoustic measurements the test material is irradiated with modulated electromagnetic radiation which is initially absorbed in the medium and subsequently leads to a vibration transmitted into the surrounding gas and detected by a microphone. Two forms of the photoacoustic measuring system are shown in Fig.6 and the essential component is a lock-in amplifier to eliminate random noise problems.

In one type of measurement it is the phase angle of the received signal with respect to the incident radiation which is measured and in the other its amplitude. Figs.7 and 8 show these measurements for the Betalloy, respectively. From the two measurements the thermal diffusivity and optical absorption of the specimens may be evaluated. The obtained values of the thermal diffusivity (α^*) and optical absorption coefficient (β), for the Betalloy are as follows:

$$\alpha^*_{BM40} = 8.13\times10^{-2}\ cm^2/sec., \qquad \alpha^*_{BP50} = 8.85\times10^{-2}\ cm^2/sec.;$$

and

$$\beta_{BM40} = 13.3\ cm^{-1}, \qquad \beta_{BP50} = 13.9\ cm^{-1}.$$

DISCUSSION OF RESULTS

Despite the limitation imposed by measurements being restricted to room temperature a few general observations may be made. The reduced velocity in the martensitic phase would imply a smaller elasticity and thus a more flexible structure. This in turn implies a greater mobility for dislocation and defect motion i.e. an increase in internal friction and in thermal diffusivity, which would appear to be reconcilable with the experimental data given in Table I. It is also interesting to note that the thermal damping factor (i.e. internal friction) is given by

$$Q^{-1} = 2f\alpha^*(E^{\phi}/E^{\theta} - 1)/c^2 \qquad (4)$$

where f is the frequency, E^{ϕ} and E^{θ} are respectively the adiabatic and isothermal elasticities and c is the acoustic velocity. In words it implies that the thermal diffusivity α^* is proportional to the internal friction which is borne out by the experimental results. A possibly closer correlation of mechanical and electrical properties would involve measurement extended to low temperatures.

REFERENCES

1. A.B.Greninger and V.G.Mooradian, Trans. Met. Soc. AIME 128,(1938), 337.
2. C.W.Wayman and K.Shimizu, Met. Sci. J. 6, (1972) 175.
3. R.S.Geng, W.G.B.Britton and R.W.B.Stephens, J. Acoust. Soc. Am., 73(4), April 1983.
4. R.S.Geng, W.G.B.Britton and R.W.B.Stephens, Fortschritte der Akustik - FASE/DAGA'82.
5. B.Tirbonod and W.Benoit, in "Internal friction and ultrasonic attenuation in solids", edited by C.C.Smith. Pergamon Press, 1980.
6. S.Koshimizu, M.Mondino and W.Benoit, Ibid.
7. J.F.Bell and J.M.Pelmore, J. of Physics E, Scientific Instruments, 1977 Vol.10.
8. S.Koshimizu and W.Benoit, Jour. de Physique, Tom 43, Dec. 1982.
9. J.Van Humbeeck and L.Delaey, Ibid.
10. M.J.Adam and G.F.Kirkbright, Analyst, April 1977, Vol 102, pp.281-292.
11. A.Rosencwaig, Photoacoustics and Photoacoustic spectroscopy, J.Wiley&Son.

Table I. Summary of Experimental Data for the Betalloy

	Martensitic phase	Austenitic phase
Sound velocity (c): m/sec	4650	4760
Internal friction (Q^{-1})	3×10^{-3}	2×10^{-3}
Thermal diffusivity (α^*): cm^2/sec.	8.85×10^{-2}	8.13×10^{-2}
Optical absorption (β): 1/cm	13.9	13.3
Resistivity (ρ): $\mu\Omega$·cm (without external stress)	41.2	28.2

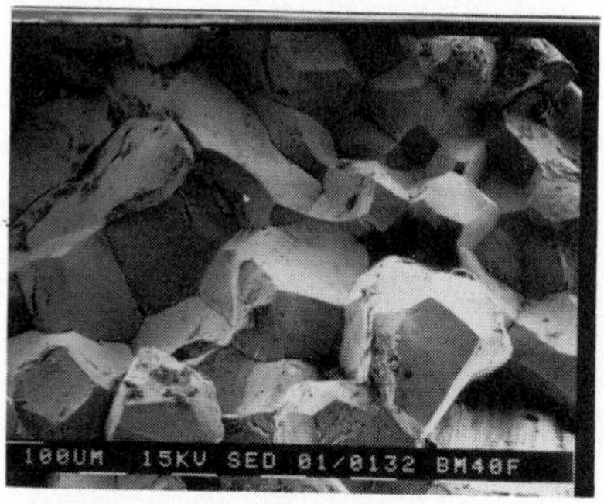

Fig.1. Micrograph of BM40, showing the average grain size of 0.25 mm.

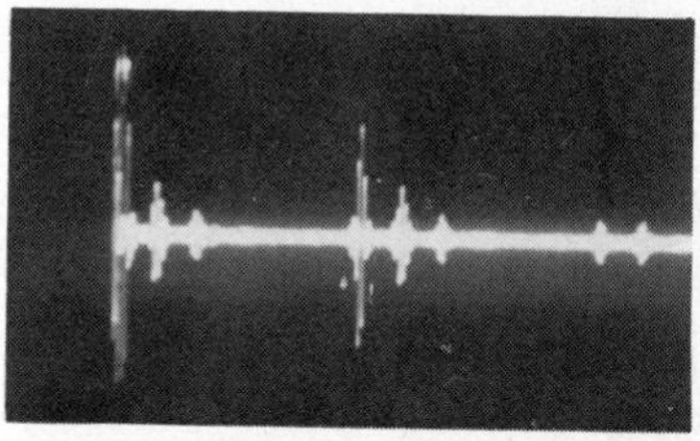

Fig.2(a). Before heat-treatment. l(specimen length) = 75 mm

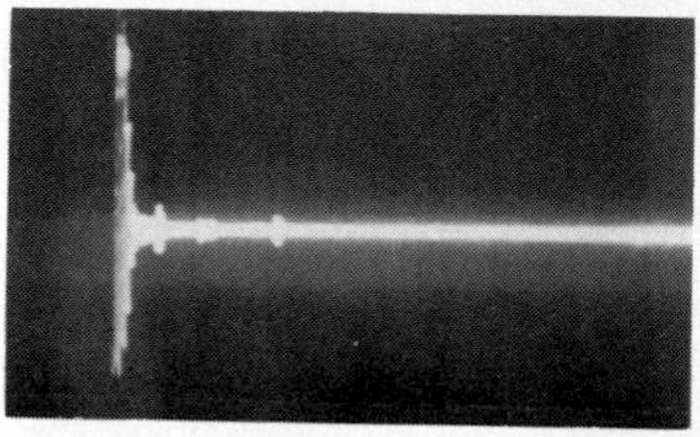

Fig.2(b). After heat-treatment. l = 50 mm.

Fig.2. Comparison of sound velocity and damping of the BP50 before and after a heat-treatment. X: 10 μs/div. Y: 0.5 v/div.

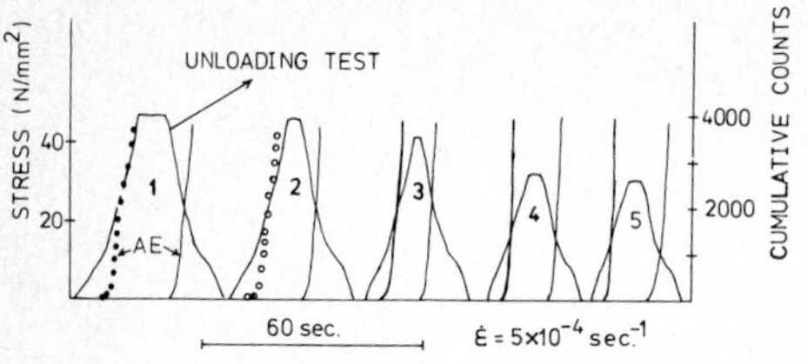

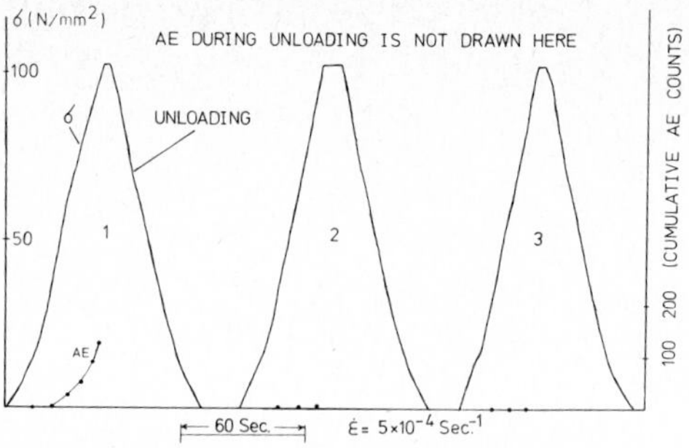

Fig.3(a).AE of BP50X during cycling test

Fig.3(b).AE of BP50F during cycling test →

Fig.3 Adifferent AE pattern for the BP50 specimen when the quenching rate is extremely different.

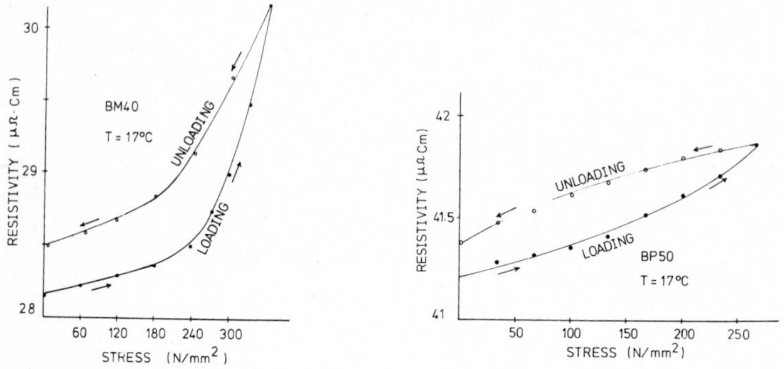

(a). BM40 (b). BP50

Fig.4. Resistivity versus stress at room temperature

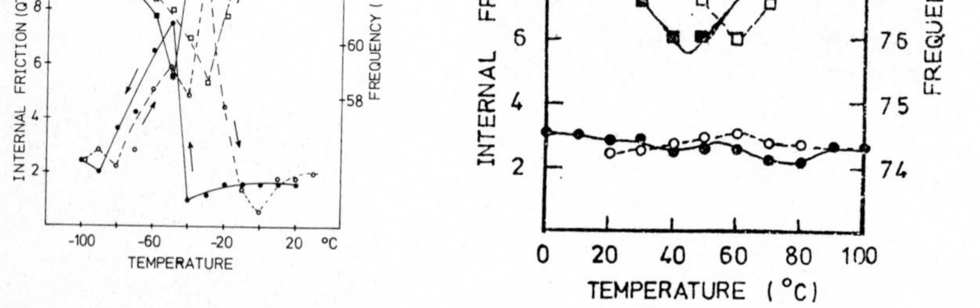

(a). BM40 (b). BP50

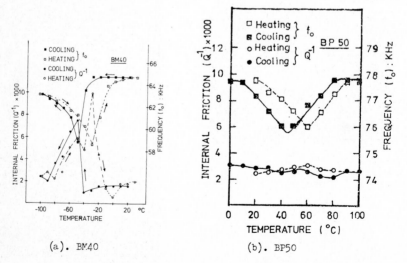

Fig.5. Internal friction and resonant frequency of the Betalloy versus temperature

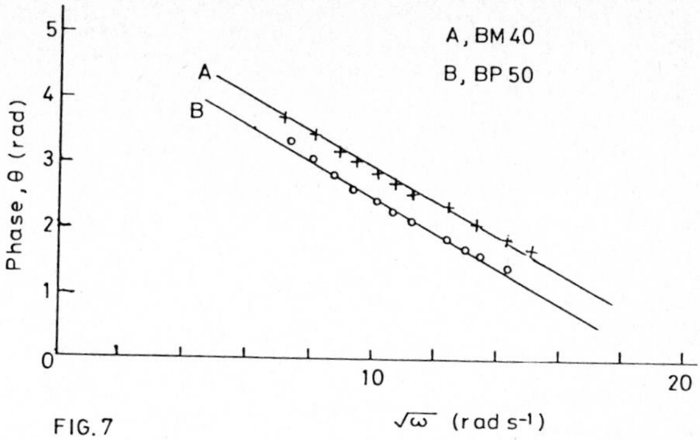

FIG.7

Phase angle, θ versus $\sqrt{\omega}$ for BM40 and BP50.

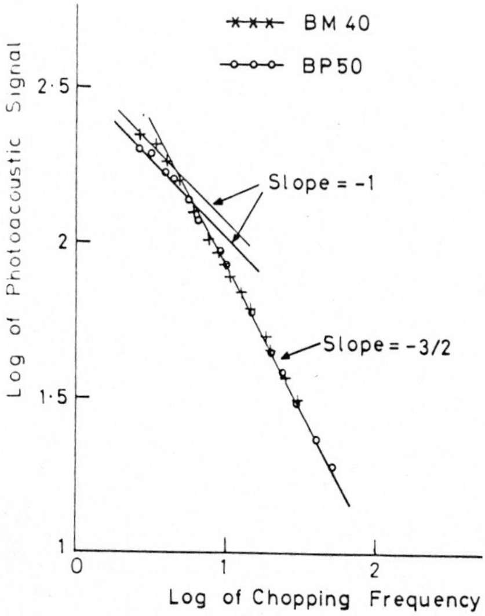

FIG.8

A log-log plot of PA signal versus chopping frequency for BM40 and BP50.

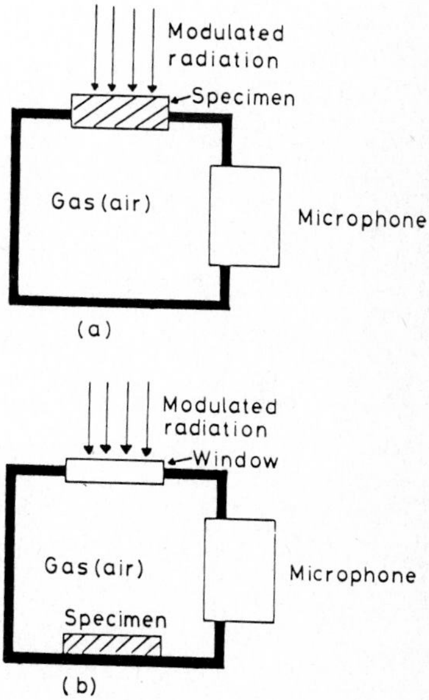

FIG.6 Simple diagram of the photoacoustic cell, (a) rear-surface illumination (b) front-surface illumination.

The Mechanical Behavior of Electromagnetic Solid Continua
G.A. Maugin (editor)
Elsevier Science Publishers B.V. (North-Holland)

PROPAGATION OF ACCELERATION WAVES IN GENERALIZED THERMOELASTIC DIELECTRICS[1]

Sadik Dost[2], Marcelo Epstein[3] and Sedat Gözde[4]

Department of Mechanical Engineering
The University of Calgary
Calgary, Alberta, Canada T2N 1N4

The propagation of acceleration waves of arbitrary form in deformable thermoelastic dielectrics with polarization effect is investigated. Implicit equations for wave speeds corresponding to nonzero amplitudes of acceleration waves are obtained by computing the jumps of the balance equations on the singular surface. For isotropic materials, assuming that the material is at rest and in thermal equilibrium ahead of the wave-front, the wave speeds for longitudinal and transverse waves are obtained in explicit form and the conditions of existence for real wave speeds are investigated.

INTRODUCTION

The theory of singular surfaces was developed by Hadamard [1], and it has been applied successfully to the analysis of acceleration waves. The propagation of acceleration waves in generalized thermoelastic materials was studied by McCarthy [2] and later by Dost and Suhubi [3]. The thermodynamic influence on the propagation and the variation of the amplitude of waves was examined in [4]. The general behaviour of 1-dimensional acceleration waves was investigated for elastic dielectrics with memory in [5] and with polarization effects in [6]. Acceleration waves in elastic dielectrics were also examined in [7].

The purpose of this paper is to study the propagation conditions of "3-dimensional" acceleration waves in generalized thermoelastic dielectrics with polarization gradient. First, the basic equations of thermoelastic dielectrics obtained in [8] and [9] are presented in the deformed state. In the next section, an acceleration wave is defined as a second order propagation surface of discontinuity on which the position vector, the polarization vector, Maxwell potential and temperature gradient and their first order derivatives with respect to time and space coordinates are continuous while the second order derivatives may suffer jumps but are continuous elsewhere. By computing the jumps of the balance equations on the wavefront, implicit equations for wave speeds corresponding to non-zero amplitudes of the wave are obtained. It is found that the jumps of the second order derivatives of Maxwell potential vanish on the wave front. Assuming a homogeneous deformation, thermal equilibrium and no polarization field in the medium ahead of the wave front, and further considering principal waves, the longitudinal and transverse wave speeds are obtained in explicit forms and the conditions of existence for real wave speeds are investigated. It is found that squared wave speeds are the roots of a cubic equation and, under certain conditions, there could be at most four real wave speeds. Finally, the effects of heat and polarization on the existence of real wave speeds are discussed.

BASIC EQUATIONS

The basic equations of thermoelastic dielectrics are given below [8] - [11] in the deformed state[5]:

Field equations:

$$t_{k\ell,k} + \rho f_\ell = \rho \dot{v}_\ell \quad , \quad E_{k\ell,k} + E_\ell = \tilde{E}_\ell \quad , \quad C + d_{k,k} = 0 \tag{1}$$

$$\rho \dot{\varepsilon} + \dot{U} = (t_{k\ell} - U\delta_{k\ell})v_{\ell,k} + \bar{C}\dot{\phi} + d_k (\dot{\phi})_{,k} + E_{k\ell} (\dot{p}_k)_{,\ell} + \tilde{E}_k \dot{p}_k + q_{k,k} + \rho h$$

Constitutive Equations:

$$t_{k\ell} = \frac{\rho}{\rho_o}\frac{\partial \Sigma}{\partial x_{\ell,K}} x_{k,K} + U\delta_{k\ell} - \frac{\partial U}{\partial \phi_{,\ell}} \phi_{,k} \quad , \quad q_k = \frac{\rho}{\rho_o}\kappa \frac{\partial \Sigma}{\partial \theta_{,K}} x_{k,K} \quad ,$$

$$E_{k\ell} = \frac{\rho}{\rho_o}\frac{\partial \Sigma}{\partial p_{\ell,K}} x_{k,K} \quad , \quad \tilde{E}_\ell = \frac{\rho}{\rho_o}\frac{\partial \Sigma}{\partial p_\ell} + \frac{\partial U}{\partial p_\ell} \quad , \quad d_k = \frac{\partial U}{\partial \phi_{,k}} \quad , \quad \kappa = \Psi\left(\frac{\partial \Psi}{\partial \dot{\theta}}\right)^{-1} \quad , \tag{2}$$

$$\Sigma = \rho_o \varepsilon - \rho_o \eta \Psi = \Sigma(x_{k,K}, p_{k,K}, p_k, \theta_{,K}, \dot{\theta}, \theta) \quad , \quad U = U(\phi_{,k}, p_k)$$

where x_k and X_K $(k,K = 1,2,3)$ are spatial and material cartesian coordinates respectively, the notations (,) and (.) denote partial and material differentiation with respect to coordinates and time; $t_{k\ell}$ and $E_{k\ell}$ are the Cauchy stress and electric tensors; f_k and E_k are body force per unit mass and electric field per unit volume, v_ℓ and $\tilde{E}_\ell$ are velocity and electric vectors; ρ, Σ, U and C denotes the actual density, energy density of deformation, electrical energy per unit volume and charge density per unit volume while ρ_o, p_k, ϕ and d_k represent the density in the reference state, the polarization vector, Maxwell potential and electrical displacement vector; ε, η and q_k denote the internal energy density, entropy density and heat flux vector; Ψ is the thermodynamic temperature which reduces to the absolute temperature θ for $\dot{\theta} = 0$; h is the heat source per unit mass per unit time.

For isotropic materials, the local Cauchy stress and electrical tensors are obtained from $(2)_1$ and $(2)_3$ as follow

$$_L t_{k\ell} = \frac{\rho}{\rho_o}\left[2 \frac{\partial \Sigma}{\partial c^{-1}_{\ell s}} c^{-1}_{ks} + \frac{\partial \Sigma}{\partial \pi^{-1}_{\ell s}} \pi^{-1}_{ks} - \frac{\partial \Sigma}{\partial p_k} p_\ell\right] \quad , \quad E_{k\ell} = \frac{\rho}{\rho_o}\frac{\partial \Sigma}{\partial \pi^{-1}_{s\ell}} c^{-1}_{ks} \tag{3}$$

where $c^{-1}_{k\ell} = x_{k,K}\, x_{\ell,K}$ and $\pi^{-1}_{k\ell} = x_{k,K}\, p_{\ell,K}$ are respectively Finger's deformation and the polarization tensors.

PROPAGATION OF ACCELERATION WAVES

We define an acceleration wave in a generalized thermoelastic dielectric as a propagating surface of discontinuity γ on which x_k, p_k, ϕ, θ and their first order derivatives with respect to time and X_K are continuous but second and higher order derivatives suffer jumps. On denoting the jump of a quantity by $[\ ]$, we write on γ

$$[x_{k,KL}] = a_k\, x_{m,K}\, x_{n,L}\, n_m n_n, \quad [\dot{x}_{k,K}] = -Va_k\, x_{\ell,K}\, n_\ell, \quad [\ddot{x}_k] = V^2 a_k \tag{4}$$

where $\underset{\sim}{n}$ is the unit normal to γ, V the speed of propagation and $a_k = c_n^{-2}[x_{k,KL}] N_K N_L$ the amplitude of acceleration jump where $\underset{\sim}{N}$ denotes the unit normal to the material image of γ and $c_n^2 = \underset{\sim}{c}^{-1}\, \underset{\sim}{n} \cdot \underset{\sim}{n}$. The jump conditions for p_k, ϕ and θ follow similarly where $\underset{\sim}{a}$ is to be replaced by $b_k = c_n^{-2}[p_{k,KL}] N_K N_L$, $\omega = c_n^{-2}[\phi_{,KL}] N_K N_L$ and $w = c_n^{-2}[\theta_{,KL}] N_K N_L$, which may be called as the amplitudes of polarization, of Maxwell potential, and of the thermal jumps, respectively.

The jumps of Eqs. $(1)_{1-2}$ and $(1)_4$ on γ, after some manipulations, take the following forms,

$$Q_{\ell n}\, a_n + P_{\ell n}\, b_n + S_\ell\, w = \rho V^2 a_\ell \quad , \quad P_{\ell n}\, a_n + R_{\ell n}\, b_n + T_\ell\, w = 0$$

$$S_n\, a_n + T_n\, b_n + Zw = 0 \tag{5}$$

where the coefficients $\underset{\sim}{Q}$, $\underset{\sim}{P}$, $\underset{\sim}{R}$, $\underset{\sim}{S}$, $\underset{\sim}{T}$ and $\underset{\sim}{Z}$ are given in Eqs. (A.1) for isotropic materials. The jump of Eq. $(1)_3$ yields that the second and higher order derivatives of Maxwell potential are continuous on the wave front. These results were also deduced in [5] and [6] in the case of one-dimensional waves. For a non-trivial solution for the amplitudes $\underset{\sim}{a}$, $\underset{\sim}{b}$ and w, the determinant of the coefficients in Eqs. (5) must be zero, which will yield the speeds of propagation. Since the determinant contains the speed V in the form of V^2, the characteristic equation is a polynomial of eight degree in V. It is, in addition, an implicit function of V. Therefore, at this stage, it seems not as easy to determine the wave speeds explicitly. Even in the case of isotropic materials, it is still an implicit function of V. Therefore, for the sake of simplicity, we shall consider the case of homogeneous deformation (i.e. $\underset{\sim}{c}^{-1}$ = const., $\underset{\sim}{\pi}^{-1} = 0$ and $\underset{\sim}{p}$ = const.) and further assume that the material is in thermal equilibrium (i.e. $\dot{\theta} = 0$ and $\mathrm{grad}\theta = \underset{\sim}{0}$) ahead of the acceleration wave.

Eliminating $\underset{\sim}{b}$ and w from Eqs. (5) and introducing $\underset{\sim}{a} = a\underset{\sim}{\nu}$, where $\underset{\sim}{\nu}$ is a unit vector, we obtain

$$(\Phi - \rho V^2 - \xi)(\gamma - \alpha V^2) + \delta\beta V^2 = 0 \tag{6}$$

where Φ, ξ, γ, α, δ and β are given in Eqs. (A.2). Eq. (6) yields a cubic equation in V^2 and it can be solved for V^2 if the directions of $\underset{\sim}{\nu}$ are known.

PRINCIPAL WAVES

If the direction of propagation coincides with one of the principal directions of the present deformation field then the wave is called a principal wave. In isotropic materials a principal wave is either longitudinal or transverse. We denote the principal stretches by λ_α (α = 1,2,3) and a unit vector in the principal direction by $\underset{\sim}{n}$. We also assume that the medium is in thermal equilibrium ahead of the wave front. If $\underset{\sim}{n}^\alpha$ is one of the principal directions then

$$\underset{\sim}{c}^{-1}\, \underset{\sim}{n}^\alpha = \lambda_\alpha\, \underset{\sim}{n}^\alpha \qquad \text{(no sum on } \alpha) \tag{7}$$

For simplicity, we will also assume that the polarization vector $\underset{\sim}{p}$ can be along the principal directions. If $\underset{\sim}{\sigma}$ is a unit vector along $\underset{\sim}{p}$, then $\underset{\sim}{p} = p_o\underset{\sim}{\sigma}$

(i) *Longitudinal Waves*

In this case, $\underset{\sim}{\nu} = \underset{\sim}{n}^1$; and we also assume $\underset{\sim}{\sigma}//\underset{\sim}{\nu}$, then using equation (7) we obtain the coefficients in Eq. (6) as follows,

$$\begin{aligned}
&\Phi_L = Q_1 + Q_2 + \lambda_1^2 Q_3 + \lambda_1^4 Q_4 + p_o^2 Q_5 + p_o Q_6 \ , \quad R_L = R_1 + R_2 + R_3 \\
&P_L = P_1 + P_2 + \lambda_1^2 P_3 + \lambda_1^4 P_4 + p_o P_5 \ , \quad \gamma_L = h_{16} \ , \quad \alpha_L = \frac{\rho}{\kappa}\frac{\partial\varepsilon}{\partial\dot{\theta}} \\
&\delta_L = \lambda_1^2 \left(\frac{\partial h_7}{\partial\dot{\theta}} + \frac{\partial h_8}{\partial\dot{\theta}}\lambda_1^2 + \frac{\partial h_9}{\partial\dot{\theta}}\lambda_1^4\right), \quad \beta_L = \lambda_1^2 \left(\frac{\partial h_1}{\partial\dot{\theta}} + \frac{\partial h_{13}}{\partial\dot{\theta}}\lambda_1^2 + \frac{\partial h_{15}}{\partial\dot{\theta}}\lambda_1^4\right) \\
&\xi_L = [P - (\gamma - \alpha V^2)^{-1}\delta^2 V^2][P - (\alpha - \alpha V^2)^{-1}\beta^2 V^2][R - (\gamma - \alpha V^2)^{-1}\delta\beta V^2]^{-1}
\end{aligned} \tag{8}$$

where $Q_1, \ldots, Q_6$, $P_1, \ldots, P_5$, R_1, R_2 and R_3 are given explicitly in Eqs. (A.3). On the other hand, if $\underset{\sim}{\sigma} \perp \underset{\sim}{\nu}$ then the terms containing p_o and R_3 in Φ, P and R in Eqs. (8) will vanish.

(ii) *Transverse Waves*

Here, $\nu = \underset{\sim}{n}^2$ or $\underset{\sim}{n}^3$. In this case, the coefficients do not depend on the direction chosen for $\underset{\sim}{\sigma}$. Then

$$\begin{aligned}
&\Phi_T = Q_2 + \lambda_2^2 Q_3 + \lambda_2^4 Q_4 \ , \qquad P_T = P_2 + \lambda_2^2 P_3 + \lambda_2^4 P_4, \\
&\xi_T = \frac{1}{R_1}(P_2 + \lambda_2^2 P_3 + \lambda_2^4 P_4)^2 \ , \quad \beta_T = \delta_T = 0, \ \gamma_T = h_{16}, \ \alpha_T = \frac{\rho}{\kappa}\frac{\partial\varepsilon}{\partial\dot{\theta}}
\end{aligned} \tag{9}$$

Now, substitution of the above coefficients in Eqs. (8) and (9) into Eq. (6) yields the following cubic equation in V^2,

$$V^6 + \mu_1 V^4 + \mu_2 V^2 + \mu_3 = 0 \tag{10}$$

where the coefficients μ_1, μ_2 and μ_3 are listed in Eqs. (A.5) for the longitudinal and transverse waves, respectively. Introducing the following parameters for Eq. (10),

$$r = \mu_1 - \frac{2}{3}\mu_1^2 + \mu_2 \quad , \quad q = \frac{2}{27}\mu_1^3 - \frac{1}{3}\mu_1\mu_2 + \mu_3 \tag{11}$$

the roots of Eq. (10) can be obtained explicitly. These roots will be real or imaginary depending on the relationship between r and q. Investigation of these relations shows that the number of real wave speeds could be at most four when $q^2/4 + r^3/27 < 0$. Hence, at least two of the wave speeds will be imaginary. This fact rules out the apparent possibility of six real waves propagating with different wave speeds in a thermoelastic dielectric. These conditions for real wave speeds are examined in detail in [12].

In the absence of heat effects, in which case the material is an elastic dielectric under the influence of a constant polarization vector with the magnitude p_0, the squared wave speeds reduce to the following simpler expressions.

(i) *Longitudinal Waves*

$$\rho V_L^2 = \Phi_L - R_L^{-1} P_L^2 \quad \text{when} \quad \underset{\sim}{\sigma} // \underset{\sim}{n}^1 \tag{12}$$

and

$$V_L^2 = (Q_1 + Q_2 + \lambda_1^2 Q_3 + \lambda_1^4 Q_4) - \frac{(P_1 + P_2 + \lambda_1^2 P_3 + \lambda_1^4 P_4)^2}{R_1 + R_2} \quad \text{when} \quad \underset{\sim}{\sigma} \perp \underset{\sim}{n}^1. \tag{13}$$

(ii) *Transverse Waves*

$$\rho V_T^2 = \Phi_T - R_1^{-1} P_T^2 \tag{14}$$

The first terms on the right hand sides of Eqs. (12-14) represent the well-known wave speeds for elastic materials. The second term, on the other hand, characterizes the polarization effects on the wave speeds. For instance, in Eq. (12) the second term will be negative if the denominator $R_L < 0$, and positive if $R_L > 0$. Therefore, depending on the sign of R_L, the longitudinal wave speeds will be increased or lowered. Similar observations can be made for other cases given in Eqs. (13) and (14). This shows that the presence of a polarization field will alter the wave speeds in the medium. It is also possible that the denominators of the second terms cause the R.H.S. of the above expressions to be negative in which case there will be no real waves in the medium. Hence, the presence of a polarization field might also hinder the development of a real acceleration wave in the material. It is noteworthy that the results given in Eqs. (12-14) for wave speeds are in complete agreement with those obtained in [6] in the case of one-dimensional acceleration waves.

In the absence of polarization effects, while heat effects are retained, the squared wave speeds reduce to simple expressions and are the same as those given in [3].

---

1 The results presented here were obtained in the course of research sponsored by the Natural Science and Engineering Council of Canada, Grant No. A-1628.

2,3 Associate Professor

4 Research Assistant

5 Cartesian coordinates are used throughout the paper.

REFERENCES

1. HADAMARD, J., "Lecons sur la Propagation des Ondes et les Equations de l'Hydrodynamique", (Herman, Paris 1903).
2. McCARTHY, M.F., "Wave Propagation in Generalized Thermoelasticity", Int. J. Engg. Sci., 10, (1973) 593-602.
3. DOST, S. and SUHUBI, E.S., "Propagation of Acceleration Waves in Generalized Thermoelastic Solids", Letters Appl. Engg. Sci., Vol. 3, (1975) 71-79.
4. McCARTHY, M.F., "Thermodynamic Influences on the Propagation Waves in Electro-elastic Materials", Int. J. Engg. Sci., 11, (1973) 1301-1316.
5. CHEN, P.J., McCARTHY, M.F. and O'PEARY, T.R., "One-Dimensional Shock and Acceleration Waves in Deformable Dielectric Materials with Memory", Arch. Rat. Mech. Anal., 62, (1976) 189.
6. COLLET, B. "One-Dimensional Acceleration Waves in Deformable Dielectrics with Polarization Gradients", Int. J. Engg. Sci., 19, (1981) 389-407.
7. DOST, S., "Acceleration Waves in Elastic Dielectrics with Polarization Effects", Int. J. Engg. Sci., (in print).
8. CHOWDHURY, K.L., EPSTEIN, M. and GLOCKNER, P.G., "On the Thermodynamics of Non-Linear Elastic Dielectrics", Int. J. Non-Linear Mech. Vol. 13, (1979) 311-322.
9. DOST, S., "On the Generalized Thermoelastic Dielectrics", J. Thermal Stress 4, No. 1, (1981).
10. CHOWDHURY, K.L. and GLOCKNER, P.G., "Representations for Isotropic Thermo-elastic Dielectrics", Int. J. Non-Linear Mech. vol. 15, (1980) 263-269.
11. DOST, S. "Constitutive Theory of Generalized Thermoelastic Dielectrics", Proc. CAN CAM 83, (1983) 253.
12. DOST, S., EPSTEIN, M. and GÖZDE, S., " Propagation of Acceleration Waves in Generalized Thermoelastic Dielectrics", Department of Mechanical Engineering Report No. 263, University of Calgary, (June 1983).

APPENDIX

The terms appearing in Eqs. (5) are defined as follows [12],

$$Q_{\ell n} = \left(\frac{\partial_L t_{k\ell}}{\partial c_{nb}^{-1}} + \frac{\partial_L t_{k\ell}}{\partial c_{bn}^{-1}}\right) c_{bm}^{-1} n_k n_m + \frac{\partial_L t_{k\ell}}{\partial \pi_{nb}^{-1}} \pi^{-1}_{mb} n_k n_m - \frac{\partial_L t_{k\ell}}{\partial p_n} p_m n_k n_m$$

$$P_{\ell n} = \frac{\partial_L t_{k\ell}}{\partial \pi_{bn}^{-1}} c_{bm}^{-1} n_k n_m \ , \quad R_{\ell n} = \frac{\partial E_{k\ell}}{\partial \pi_{bn}^{-1}} c_{bm}^{-1} n_k n_m$$

$$S_\ell = \frac{\partial_L t_{k\ell}}{\partial \theta_{,m}} n_m n_k - \frac{\partial_L t_{k\ell}}{\partial \dot{\theta}} V n_k \ , \quad T_\ell = \frac{\partial E_{k\ell}}{\partial \theta_{,m}} n_m n_k - \frac{\partial E_{k\ell}}{\partial \dot{\theta}} V n_k \tag{A.1}$$

$$Z = \frac{1}{\kappa}\left\{\frac{\partial q_m}{\partial \theta_{,\ell}} n_m n_\ell - [2\frac{\partial q_m}{\partial \dot{\theta}} - \frac{1}{\kappa} q_m (\frac{\partial \kappa}{\partial \dot{\theta}} + 1)] n_m V - \rho \frac{\partial \varepsilon}{\partial \dot{\theta}} V^2\right\}$$

The terms appearing in Eq. (6) are

$$\Phi = Q_{k\ell} \nu_k \nu_\ell \ , \quad \delta = \frac{\partial t_{k\ell}}{\partial \dot{\theta}} n_k \nu_\ell \ , \quad \alpha = \frac{\rho}{\kappa}\frac{\partial \varepsilon}{\partial \theta} \ , \quad \gamma = \frac{1}{\kappa}\frac{\partial q_k}{\partial \theta_{,m}} \nu_k \nu_\ell$$

$$\xi = (P_{kn} - Z^{-1} S_k B_n)(R_{\ell n} - Z^{-1} T_\ell B_n)^{-1} (P_{\ell m} - Z^{-1} T_\ell A_m) \nu_k \nu_m \tag{A.2}$$

$$A_m = -\frac{\partial E_{km}}{\partial \dot{\theta}} V n_k \ , \quad B_n = -\frac{\partial t_{kn}}{\partial \dot{\theta}} V n_k \ , \quad \beta = \frac{\partial E_{k\ell}}{\partial \dot{\theta}} n_k \nu_\ell \ , \quad Z = h_{16} - \alpha V^2$$

Definitions of the terms appearing in Eq. (8) are as follows

$$Q_1 = \lambda_1^2[h_7 + 2\lambda_1^2 h_8 + 3\lambda_1^4 h_9 + 2\lambda_1^2 (\phi^{(7)} + \lambda_1^2 \phi^{(8)} + \lambda_1^4 \phi^{(9)})]$$

$$Q_2 = \lambda_1^2(h_7 + \lambda_1^2 h_8 + \lambda_1^4 h_9) + h_{10} p_o^2 , \quad Q_3 = \lambda_1^2 (h_8 + \lambda_1^2 h_9)$$

$$Q_4 = \lambda_1^2 h_9, \quad Q_5 = \frac{\partial h_{10}}{\partial I_{10}} p_o^2 , \quad Q_6 = h_{10} p_o \tag{A.3}$$

$$P_1 = \lambda_1^4 (\psi^{(7)} + \lambda_1^2 \psi^{(8)} + \lambda_1^4 \psi^{(9)}) , \quad P_2 = \lambda_1^2 (h_1 + \lambda_1^2 h_{13} + \lambda_1^4 h_{15})$$

$$P_3 = \lambda_1^2 (h_{13} + \lambda_1^2 h_{15}) , \quad P_4 = \lambda_1^2 h_{15}, \quad P_5 = -\lambda_1^2 (h_{11} + \psi^{(10)})p_o$$

$$R_1 = \lambda_1^2 h_4 , \quad R_2 = \lambda_1^4 (h_2 + 2\lambda_1^2 h_4 + \psi^{(1)} + \lambda_1^2 \psi^{(13)} + \lambda_1^4 \psi^{(15)})$$

$$R_3 = p_o^2 (2h_{12} + \lambda_1^4 \psi^{(11)})$$

where the expressions for $\phi^{(a)}$ and $\psi^{(a)}$ are as follows

$$\phi^{(a)} = \frac{\partial h_a}{\partial I_7} + 2\lambda_1^2 \frac{\partial h_a}{\partial I_8} + 3\lambda_1^4 \frac{\partial h_a}{\partial I_9} , \quad \psi^{(a)} = \frac{\partial h_a}{\partial I_1} + \lambda_1^2 \frac{\partial h_a}{\partial I_{13}} + \lambda_1^4 \frac{\partial h_a}{\partial I_{15}} , \tag{A.4}$$

and the coefficients h_i are given by [11]

$$h_i = \frac{\rho}{\rho_o} \frac{\partial \Sigma}{\partial I_i} , \quad i = 1, \ldots, 18$$

where I_i the eighteen invariants of $\underset{\sim}{c}^{-1}$, $\underset{\sim}{\pi}^{-1}$, $\underset{\sim}{p}$, and gradθ [10].

The coefficients in eq. (10) are as follow

(a) *Longitudinal Waves*

$$\mu_1 = -[\rho\alpha(\alpha R + \delta\beta)]^{-1} [\alpha^2(\Phi R - P^2) + \alpha\beta\delta(\Phi - R) - \alpha P(\delta^2 + \beta^2) + 2\alpha\gamma\rho R + \gamma\rho\delta\beta - 2\delta^2\beta^2] \tag{A.5}$$

$$\mu_2 = [\rho\alpha(\alpha R + \delta\beta)]^{-1}[2\alpha\gamma(\Phi R - P^2) + \gamma\delta\beta(\Phi - R) - \gamma P(\delta^2 + \beta^2) + \rho\gamma^2 R]$$

$$\mu_3 = -[\rho\alpha(\alpha R + \delta\beta)]^{-1}[\gamma^2(\Phi R - P^2)],$$

(b) *Transverse Waves*

In this case μ_1, μ_2, μ_3 will be obtained from Eqs. (A.5) replacing R by R_1 and substituting $\delta = \beta = 0$.

The Mechanical Behavior of Electromagnetic Solid Continua
G.A. Maugin (editor)
Elsevier Science Publishers B.V. (North-Holland)

A VARIATIONAL PRINCIPLE FOR ELASTIC DIELECTRICS

Angelo Morro

Department of Biophysical and Electronic Engineering
University of Genova
Genova, Italy

As a first step towards a comprehensive variational formulation of electrodynamics in deformable continua, the behaviour of an elastic dielectric in the quasi-static approximation is investigated. By having recourse to the theory pertaining to the inverse problem of the calculus of variations, the most general constitutive equations for the stress and the electric displacement are derived which allow the model to admit a variational formulation. Then the explicit form of the Lagrangian density is determined; consistently with the adoption of the material description, the Lagrangian density ascribes to the dielectric the structure of a particle-like system.

INTRODUCTION

The wide literature on the subject bears evidence of the existence of a great many formulations for the electrodynamics in deformable media. Although none of these formulations has gained a general credit as yet, the four theories reviewed by Pao (1978) are in fact the favourite ones in applications. The coexistence of many theories for such a subject should come as no surprise inasmuch as the electromagnetic field inside matter is expressed in terms of field variables which cannot be directly measured and then a selection through crucial experiments is not feasible. This circumstance motivates the attempts to set up theories of electrodynamics in deformable media on the basis of general methods and principles such as the principle of virtual power and the principles of continuum thermodynamics (Maugin (1980)).

Among the general principles or guidelines to be followed in constructing a theory one may well consider the feature that the theory admits a variational formulation, this being important for both conceptual and operative aspects. Now, there are essentially two methods for elaborating a variational theory. First, one may guess at the form of the Lagrangian and then to judge the plausibility of Euler-Lagrange equations as balance or evolution equations for the system. The guess hinges usually on analogies or, rather, on operative rules such as that of Lax and Nelson (1976) according to which the Lagrangian of a deformable body in the presence of the electromagnetic field consists of the sum of three Lagrangians describing the field, the field-matter interaction, and the matter. Second, one may start from balance equations and constitutive equations in a rather general form. Then the compatibility conditions are placed which enable a system of differential equations to arise from a variational formulation, these conditions being delivered by the theory pertaining to the inverse problem of the calculus of variations (Bampi and Morro (1982)). The compatibility conditions lead often to severe restrictions on the constitutive properties; it is just the form of the restricted constitutive equations which allow us to judge the plausibility of the theory under construction. As the last step, the Lagrangian is determined through a standard procedure.

So as to give new insights into the general framework of the electrodynamics of deformable media, while avoiding any controversy about the right balance equations, in this lecture the second method is employed in connection with the elastic dielectric in the quasi-static approximation. As a result it is shown that a wide class of constitutive equations allows the existence of a Lagrangian for the dielectric.

ESSENTIALS OF THE ELASTIC DIELECTRIC IN THE QUASI-STATIC APPROXIMATION

The dielectric is supposed to consist of particles labelled by their position $\mathbf{X}$ in a suitable reference configuration; so $\mathbf{x}(\mathbf{X},t)$ is the position of the particle $\mathbf{X}$ at time t, $\dot{\mathbf{x}} = \partial\mathbf{x}(\mathbf{X},t)/\partial t$ is the velocity, $\mathbf{F} = \mathrm{Grad}\,\mathbf{x}(\mathbf{X},t)$ while $\mathrm{Grad} = \partial/\partial\mathbf{X}$. Owing to the conservation of mass, the mass density ρ is determined through the motion by

$$\rho J = \rho_0(\mathbf{X})$$

where $J = \det\mathbf{F}$ and ρ_0 is the reference mass density. As we are dealing with the material description, it is convenient to look at the reference measures $\mathcal{E}$ and $\mathfrak{D}$ of the electric field $\mathbf{E}$ and the electric displacement $\mathbf{D}$ (McCarthy and Tiersten (1978)), namely

$$\mathcal{E} = \mathbf{F}\,\mathbf{E} \qquad \mathfrak{D} = J\,\mathbf{F}^{-1}\mathbf{D} \quad ,$$

the magnetic quantities $\mathcal{H}$, $\mathfrak{B}$ being irrelevant in the quasi-static approximation. Analogously, the stress of the body is conveniently described via the first Piola -Kirchhoff stress tensor $\mathbf{S}$. These fields enter the pertinent balance equations in the form

$$\begin{aligned} \rho_0\ddot{\mathbf{x}} &= \mathrm{Div}\,\mathbf{S} + \rho_0\mathbf{b} \\ \mathrm{Div}\,\mathfrak{D} &= \mu_0 , \\ \mathrm{Curl}\,\mathcal{E} &= \mathbf{0} , \end{aligned} \tag{1}$$

$\mathbf{b}$ being the body force and μ_0 the charge density. Owing to $(1)_3$ the electric field $\mathcal{E}$ is potential, namely there exists a scalar function ϕ such that

$$\mathcal{E} = -\,\mathrm{Grad}\phi \ . \tag{2}$$

So as to characterise the nature of the elastic dielectric we adopt the constitutive equations

$$\begin{aligned} \mathbf{S} &= \mathbf{S}(\mathbf{F}, \mathcal{E}, \mathbf{X}, t) , \\ \mathfrak{D} &= \mathfrak{D}(\mathbf{F}, \mathcal{E}, \mathbf{X}, t) , \end{aligned}$$

and regard $\mathbf{b}$, ρ_0, and μ_0 as assigned functions of $\mathbf{X}$ and t. Then, in view of $(1)_{1,2}$ and (2), we are led to account for the behaviour of the dielectric via the system of second order differential equations

$$\begin{aligned} \rho_0\ddot{\mathbf{x}} - \mathrm{Div}\,\mathbf{S}(\mathrm{Grad}\,\mathbf{x}, \mathrm{Grad}\phi, \mathbf{X}, t) - \rho_0\mathbf{b}(\mathbf{X},t) &= \mathbf{0} , \\ \mathrm{Div}\,\mathfrak{D}(\mathrm{Grad}\,\mathbf{x}, \mathrm{Grad}\phi, \mathbf{X}, t) - \mu_0(\mathbf{X},t) &= 0 , \end{aligned} \tag{3}$$

in the unknown functions $\mathbf{x}(\mathbf{X},t)$, $\phi(\mathbf{X},t)$.

THE LAGRANGIAN DENSITY FOR THE ELASTIC DIELECTRIC

Henceforth we use Cartesian tensor notation, the convention that capital indices refer to reference coordinates and lower case indices to present coordinates of material points, and the convention that round (square) brackets, enclosing a pair of indices, denote symmetrization (skew-symmetrization). The system (3) may be written in a more explicit form as

$$\begin{aligned} \chi_i &:= \rho_0 \ddot{x}_i - \frac{\partial S_{i(M}}{\partial x_{p,Q)}} x_{p,QM} - \frac{\partial S_{i(M}}{\partial \phi_{,Q)}} \phi_{,QM} - \frac{\partial S_{iM}}{\partial X_M} - \rho_0 b_i = 0 , \\ \Phi &:= \frac{\partial \mathfrak{D}_{(M}}{\partial x_{p,Q)}} x_{p,QM} + \frac{\partial \mathfrak{D}_{(M}}{\partial \phi_{,Q)}} \phi_{,QM} + \frac{\partial \mathfrak{D}_M}{\partial X_M} - \mu_0 = 0 , \end{aligned} \tag{4}$$

where latin indices run over 1,2,3 and a comma followed by a letter denotes differentiation with respect to the corresponding coordinate. According to the general theory (Bampi and Morro (1982)) the system (4) admits a variational formulation if and only if

$$\frac{\partial \chi_i}{\partial x_{p,QM}} = \frac{\partial \chi_p}{\partial x_{i,QM}} , \tag{5}$$

$$\frac{\partial \chi_i}{\partial \phi_{,QM}} = \frac{\partial \Phi}{\partial x_{i,QM}} , \tag{6}$$

$$\frac{\partial \chi_i}{\partial x_{p,Q}} = - \frac{\partial \chi_p}{\partial x_{i,Q}} + 2 \left(\frac{\partial \chi_p}{\partial x_{i,QM}}\right)_{,M} , \tag{7}$$

$$\frac{\partial \chi_i}{\partial \phi_{,Q}} = - \frac{\partial \Phi}{\partial x_{i,Q}} + 2 \left(\frac{\partial \Phi}{\partial x_{i,QM}}\right)_{,M} , \tag{8}$$

$$\frac{\partial \Phi}{\partial \phi_{,Q}} = - \frac{\partial \Phi}{\partial \phi_{,Q}} + 2 \left(\frac{\partial \Phi}{\partial \phi_{,QM}}\right)_{,M} , \tag{9}$$

$$\frac{\partial \chi_i}{\partial x_p} = \frac{\partial \chi_p}{\partial x_i} - \left[\frac{\partial \chi_p}{\partial x_{i,Q}} - \left(\frac{\partial \chi_p}{\partial x_{i,QM}}\right)_{,M} \right]_{,Q} , \tag{10}$$

$$\frac{\partial \chi_i}{\partial \phi} = \frac{\partial \Phi}{\partial x_i} - \left[\frac{\partial \Phi}{\partial x_{i,Q}} - \left(\frac{\partial \Phi}{\partial x_{i,QM}}\right)_{,M} \right]_{,Q} ; \tag{11}$$

other relations are omitted because they result in trivial identities. The analysis of the conditions (5)-(11) is simplified by observing that the identity

$$\frac{\partial}{\partial u_{\Gamma,Q}} h_{M,M} = \left(\frac{\partial h_M}{\partial u_{\Gamma,Q}}\right)_{,M} \tag{12}$$

holds for any vector function **h** which depends on $u_{\Gamma,Q}$ ($u_\Gamma = x_i , \phi$) and on higher order derivatives.

Observe first that

$$\frac{\partial \chi_i}{\partial x_{p,QM}} = - \frac{\partial S_{i(M}}{\partial x_{p,Q)}}$$

and that, in view of (12),

$$\frac{\partial \chi_i}{\partial x_{p,Q}} = - \frac{\partial}{\partial x_{p,Q}} S_{iM,M} = - \left(\frac{\partial S_{iM}}{\partial x_{p,Q}}\right)_{,M} .$$

Then we have

$$\frac{\partial \chi_i}{\partial x_{p,Q}} - \left(\frac{\partial \chi_i}{\partial x_{p,QM}}\right)_{,M} = \left(\frac{\partial S_{i[Q}}{\partial x_{p,M]}}\right)_{,M} .$$

Accordingly, (5) and (7) yield

$$\frac{\partial S_{i(M}}{\partial x_{p,Q)}} - \frac{\partial S_{p(M}}{\partial x_{i,Q)}} = 0 , \tag{13}$$

$$\left(\frac{\partial S_{i[M}}{\partial x_{p,Q]}} + \frac{\partial S_{p[Q}}{\partial x_{i,Q]}}\right)_{,M} = 0 . \tag{14}$$

Upon integration, (14) provides

$$\frac{\partial S_{i[M}}{\partial x_{p,Q]}} + \frac{\partial S_{p[M}}{\partial x_{i,Q]}} = 2\varepsilon_{MQR}\,\psi_{ip,R} \tag{15}$$

where $\psi_{ip} = \psi_{(ip)}$ is an arbitrary tensor function of $\mathbf{X}$, t and ε_{MQR} is the Levi-Civita symbol. Addition of (13) and (15) gives

$$\frac{\partial S_{iM}}{\partial x_{p,Q}} - \frac{\partial S_{pQ}}{\partial x_{i,M}} = 2\varepsilon_{MQR}\,\psi_{ip,R}\,.$$

This implies the existence of a scalar function $\Sigma = \Sigma(\mathrm{Grad}\,\mathbf{x}, \mathrm{Grad}\,\phi, \mathbf{X}, t)$ such that

$$S_{iM} = \frac{\partial \Sigma}{\partial x_{i,M}} + \varepsilon_{MQR}\,\psi_{ip,R}\,x_{p,Q}\,. \tag{16}$$

As a consequence, (10) is immediately seen to be an identity.

Following along an analogous procedure we find that (9) implies

$$\mathfrak{D}_M = \frac{\partial \Lambda}{\partial \phi_{,M}} + \varepsilon_{MQR}\,\Xi_{,R}\,\phi_{,Q} \tag{17}$$

where $\Lambda = \Lambda(\mathrm{Grad}\,\mathbf{x}, \mathrm{Grad}\,\phi, \mathbf{X}, t)$ and $\Xi = \Xi(\mathbf{X},t)$ are arbitrary scalar functions.

As the last step we have to examine the mixed conditions (6), (8), and (11). Straightforward calculations lead to the relationship

$$\Sigma + \Lambda = \sigma + \lambda + \varepsilon_{MQR}\,\Theta_{i,R}\,x_{i,M}\,\phi_{,Q} \tag{18}$$

where $\sigma = \sigma(\mathrm{Grad}\,\phi, \mathbf{X}, t)$, $\lambda = \lambda(\mathrm{Grad}\,\mathbf{x}, \mathbf{X}, t)$, $\Theta_i = \Theta_i(\mathbf{X},t)$ are arbitrary functions. In connection with σ and λ it is of interest to point out that the transformation $\Sigma \to \Sigma + \sigma$, $\Lambda \to \Lambda + \lambda$, leaving $\mathbf{S}$ and $\mathfrak{D}$ unaffected, may be viewed as a gauge transformation. So, letting $\sigma = 0$, $\lambda = 0$ amounts to choosing a particular gauge.

Since the functions (16) and (17), along with the restriction (18), make equations (4) meet the potentialness conditions (5)-(11) we are now in a position to determine the corresponding Lagrangian density. To this end we follow the standard procedure (Bampi and Morro (1983)). Upon choosing the gauge $\sigma = 0$, $\lambda = 0$ we have

$$\begin{aligned} S_{iM} &= \frac{\partial \Sigma}{\partial x_{i,M}} + \varepsilon_{MQR}\,\psi_{ip,R}\,x_{p,Q} \\ \mathfrak{D}_M &= -\frac{\partial \Sigma}{\partial \phi_{,M}} + \varepsilon_{MQR}(\Theta_{i,R}\,x_{i,Q} + \Xi_{,R}\,\phi_{,Q})\,. \end{aligned} \tag{19}$$

Hence it follows that

$$S_{iM,M} = \Big(\frac{\partial \Sigma}{\partial x_{i,M}}\Big)_{,M}\,, \qquad \mathfrak{D}_{M,M} = -\Big(\frac{\partial \Sigma}{\partial \phi_{,M}}\Big)_{,M}\,.$$

Now, the addition to a Lagrangian density of material divergences and of material time derivatives leaves the Euler-Lagrange equations unchanged. Accordingly, the Lagrangian density $L(\mathbf{x},\phi)$ is given by

$$\begin{aligned} L(\mathbf{x},\phi) = {} & \tfrac{1}{2}\rho_0\,\dot{x}_i\dot{x}_i + \rho_0\,b_i\,x_i + \mu_0\,\phi \\ & + \int_0^1 \Big[(x_i - \hat{x}_i)\Big(\frac{\partial \Sigma}{\partial x_{i,M}}\Big)_{,M}(\tilde{\mathbf{x}},\tilde{\phi}) + \phi\Big(\frac{\partial \Sigma}{\partial \phi_{,M}}\Big)_{,M}(\tilde{\mathbf{x}},\tilde{\phi})\Big]\,d\xi \end{aligned}$$

where $\hat{\mathbf{x}}$ is a given function of $\mathbf{X}$ and $\tilde{\mathbf{x}} = \hat{\mathbf{x}} + \xi(\mathbf{x} - \hat{\mathbf{x}})$, $\xi \in [0,1]$. Because

$$\frac{\partial \Sigma}{\partial \xi} = (x_i - \hat{x}_i)_{,M}\frac{\partial \Sigma}{\partial x_{i,M}} + \phi_{,M}\frac{\partial \Sigma}{\partial \phi_{,M}}$$

on disregarding an inessential material divergence we arrive at the sought Lagrangian density

$$L(\mathbf{x},\phi) = \tfrac{1}{2}\rho_0\,\dot{\mathbf{x}}\cdot\dot{\mathbf{x}} - \Sigma(\mathrm{Grad}\,\mathbf{x}, \mathrm{Grad}\,\phi) + \rho_0\,\mathbf{b}\cdot\mathbf{x} + \mu_0\,\phi \tag{20}$$

the explicit dependence on $\mathbf{X}$, t being understood. According to the Lagrangian density (20) the function Σ plays the role of (total) potential energy. This feature is hardly surprising although usually, in continuum mechanics, the prescription

L = T - V only works for conservative particle-like systems. Indeed, the structure of a particle-like system occurs just because the use of Lagrangian coordinates preserves the similarity with a system of discrete particles.

REMARKS

At least in the case of the quasi-static approximation for elastic dielectrics, the second law of thermodynamics implies that **S** and $\mathfrak{D}$ are given by a potential function, ε say, such that (Grot (1976))

$$S_{iM} = \rho_0 \frac{\partial \varepsilon}{\partial x_{i,M}} \qquad \mathfrak{D}_M = -\rho_0 \frac{\partial \varepsilon}{\partial \phi_{,M}} \; . \tag{21}$$

It is apparent that eqs. (19) are less restrictive than (21) in that they allow **S** and $\mathfrak{D}$ to be affected by the additive terms

$$S^*_{iM} = \varepsilon_{MQR}\, \psi_{ip,R}\, x_{p,Q} \qquad \mathfrak{D}^*_M = \varepsilon_{MQR}(\Theta_{i,R}\, x_{i,Q} + \Xi_{,R}\, \phi_{,Q}) \; .$$

These terms seem to have not yet been exhibited in any variational or thermodynamic approach. This should come as no surprise as long as **S*** and $\mathfrak{D}^*$ do not enter the balance equations nor do they enter the expression (19) for the Lagrangian density; the physical relevance of **S*** and $\mathfrak{D}^*$ is obviously related to boundary conditions.

Should we look at **b** as a function of the position **x**, an inspection of the compatibility conditions shows that, owing to (10), **b** must satisfy

$$\frac{\partial b_i}{\partial x_p} = \frac{\partial b_p}{\partial x_i} \tag{22}$$

namely **b** must be irrotational. In such a case there exists a scalar function U(**x**) such that **b** = $-\partial U/\partial \mathbf{x}$ and hence $\rho_0 \mathbf{b}\cdot\mathbf{x}$ in (19) is replaced with $-\rho_0 U(\mathbf{x})$. That (22) is a sufficient condition for **b** to allow a variational formulation was already known at least in the framework of elasticity (Suhubi (1975)); here we have proved that (22) is also a necessary condition as far as elastic dielectrics are concerned.

REFERENCES

[1] Bampi, F. and Morro, A., The inverse problem of the calculus of variations applied to continuum physics, J. Math. Phys. 23 (1982) 2312-2321.

[2] Bampi, F. and Morro, A., A Lagrangian for the dynamics of elastic dielectrics, Int. J. Non-Linear Mech., to appear.

[3] Grot, R. A., Relativistic continuum physics: electromagnetic interactions, in: Eringen, A. C. (ed.), Continuum Physics III (Academic, New York, 1976).

[4] Lax, M. and Nelson, D. F., Electrodynamics of elastic pyroelectrics, Phys. Rev. B13 (1976) 1759-1769.

[5] McCarthy, M. F. and Tiersten, H. F., On integral forms of the balance laws for deformable semiconductors, Arch. Rational Mech. Anal. 68 (1978) 27-36.

[6] Maugin, G. A., The method of virtual power in continuum mechanics: application to coupled fields, Acta Mech. 35 (1980) 1-70.

[7] Pao, Y.-H., Electromagnetic forces in deformable continua, in: Nemat-Nasser, S. (ed.), Mechanics Today 4 (Pergamon, New York, 1978).

[8] Suhubi, E. S., Thermoelastic solids, in: Eringen, A. C. (ed.), Continuum Physics II (Academic, New York, 1975).

The Mechanical Behavior of Electromagnetic Solid Continua
G.A. Maugin (editor)
Elsevier Science Publishers B.V. (North-Holland)

SOME DYNAMICAL PROBLEMS OF ELASTIC DIELECTRICS WITH POLARIZATION GRADIENT

Jerzy Paweł Nowacki

Institute of Fundamental Technological Research
Polish Academy of Sciences
Warsaw, Poland

The paper is devoted to the investigation of the propagation of waves in dielectric with polarization gradient. First, we consider the waves in a layer, assuming that the solution depends on the variables x_1, x_2 and t only. The equations of piezoelectricity with this functional dependance lead to two independent systems. Next we investigate the torsional waves in a infinite rod with circular cross-section. Using Boggio theorem we obtain the analitic form of solution. In both cases the nature of the vibration is determined by the corresponding characteristic equation.

I. INTRODUCTION

Elastic dielectric materials, in which occurs the linear piezoelectric effect have become important in modern technology. The classical linear theory of piezoelectricity created by Voigt [1] and developed by Toupin [2] is concerned with the interaction between mechanical strain and the electric field and the electric polarization. This theory was extended by Mindlin [3] by assuming the dependance of the stored energy function also on the polarization gradient. This extension accomodates observed phenomena, such as electromechanical interaction in symmetric and non-symmetric materials and capacitance phenomena of thin dielectric films. The equations proposed by Mindlin are derivable as a long wave limit of the equation of the modern dynamical theory of crystal lattice of electronicaly polarizable atoms [4]. It also include a surface energy of deformation and polarization, which is absent from the previous theories but which has been observed in the laboratory and calculated from atomic considerations. [5,6].

In recent years this theory has received considerable attention from researches in the field of continuum mechanics. A number of papers have been devoted to the dynamical problems. For example, by including magnetic effects, plane monochromatic waves in quartz have been investigated in [7]. Glockner and his coworkers considered the propagation of waves in thermoelastic centrosymmetric isotropic dielectric [8,9,10]. The Love's waves and the wave reflection and refraction in elastic dielectric has been investigated in [11,12].

In the present paper we will consider two problems. The propagation of waves in homogenous dielectric plate bounded by two parallel planes and the propagation of torsional waves in infinite cylinder with circular cross-section. Oscilation of an elastic plate, the surface of which is free of stresses were investigated first by Rayleigh and is described in detail in [13]. In the problems of wave propagation in layered media a special case having some practical interest is that in which boundaries are cylinders: the cylindrical rod in a vacuum being the simplest example. Three types of vibration in cylindrical rods are observed: longitudinal, lateral and torsional. Our considerations are confined to the torsional vibration only.

The system of basic linear equations for a elastic dielectric including polarization gradient effects occupying the domain V and bounded by the surface S are given by

i The field equations

$$T_{ij,i} + pf_j = p\ddot{u}_j; \quad T_{ij} = T_{ji} \tag{1.1}$$

$$\epsilon_{ij,i} + {}_LE_j + E_j^{MS} = -E_j^o \tag{1.2}$$

$$-\epsilon_o \phi_{,ii} + P_{i,i} = -p_c \quad \text{in} V \tag{1.3}$$

$$\phi_{,ii} = 0 \quad \text{in } V^* \tag{1.4}$$

ii The boundary conditions

$$T_{ij}n_i = k_j; \quad \epsilon_{ij}n_i = S_j; \tag{1.5}$$

$$n_i[P_i - \epsilon_o \| \phi_{,i} \|] = \beta. \tag{1.6}$$

iii The constitutive relations

$$T_{ij} = d_{12}\delta_{ij}P_{n,n} + d_{44}(P_{j,i} + P_{i,j}) + c_{12}\delta_{ij}S_{kk} + 2c_{44}S_{ij} \tag{1.7}$$

$$-{}_LE_k = aP_k \tag{1.8}$$

$$\epsilon_{ij} = b_{12}\delta_{ij}P_{n,n} + b_{44}(P_{i,j} + P_{j,i}) + d_{12}\delta_{ij}S_{kk} + 2d_{44}S_{ij} + b_{77}(P_{j,i} - P_{i,j}) + b_o\delta_{ij} \tag{1.9}$$

iv The kinematic relations

$$S_{ij} = \frac{1}{2}(u_{i,j} + u_{j,i}) \tag{1.10}$$

$$E_i^{MS} = -\phi_{,i} \tag{1.11}$$

where T_{ij}, ϵ_{ij} and S_{ij} designate the components of the stress tensor, and strain tensor, respectively; u_i, P_i, ${}_LE_i$, E_i^{MS}, f_i, n_i are components of the displacement vector, polarization vector, local electric vector, Maxwell self-field vector, body force vector and the exterior unit normal vector, respectively; ϕ, $\| \phi_{,i} \|$, p_c, denote Maxwell potential, jump in $\phi_{,i}$ across S and charge density, respectively; k_i, S_i and β are surface traction, surface electric force and surface charge respectively, while V^* stands for outer vacuum and ϵ_o its permittivity; a, $b_{\alpha\beta}$, $c_{\alpha\beta}$, $d_{\alpha\beta}$, are material constants.

Substituting eqns (1.7) - (1.11) into (1.1) - (1.4) yields Navier's equations of dielectrics as

$$c_{44}\nabla^2\bar{u} + (c_{12} + c_{22})\bar{\nabla}\bar{\nabla} \cdot \bar{u} + d_{44}\nabla^2\bar{P} + (d_{12} + d_{44})\bar{\nabla}\bar{\nabla}\bar{P} + p\bar{f} = p\ddot{\bar{u}} \tag{1.12}$$

$$d_{44}\nabla^2\bar{u} + (d_{12} + d_{44})\bar{\nabla}\bar{\nabla}\cdot\bar{u} + (b_{44} + b_{77})\nabla^2\bar{P} + \qquad (1.13)$$

$$+ (b_{12} + b_{44} - b_{77})\bar{\nabla}\bar{\nabla}\cdot\bar{P} + a\bar{P} - \bar{\nabla}\phi + \bar{E}^o = 0$$

$$\bar{\nabla}\cdot\bar{P} - \epsilon_o\nabla^2\phi = -p_c \qquad (1.14)$$

II. Propagation of waves in the plate

Let us consider a wave motion in a homogenous dielectric layer of thickness 2h and of infinite in-plane dimension. We assume that the edges of layer $x_1 = \pm h$ are free of stresses. The following conditions should be satisfied on these edges

$$T_{11} = T_{12} = 0\,,\ \epsilon_{11} = \epsilon_{12} = 0$$

$$P_1 - \epsilon_o \,\|\partial_1\phi\| = 0 \qquad (2.1)$$

and

$$T_{13} = 0,\ \epsilon_{13} = 0 \qquad \text{for } x = \pm h$$

Assume that all effects depend only on the variables x_1, x_2 and t. With this functional dependance for $\bar{u}$, $\bar{P}$ and ϕ, eqns (1.12)-(1.14) lead to the following two independent systems

$$c_{44}\nabla_1^2 u_1 + (c_{12} + c_{44})\partial_1(\partial_1 u_1 + \partial_2 u_2) + d_{44}\nabla_1^2 P_1 + (d_{12} + d_{44}) \qquad (2.2)$$
$$\times\partial_1(\partial_1 P_1 + \partial_2 P_2) = \rho\ddot{u}_1$$

$$c_{44}\nabla_1^2 u_2 + (c_{12} + c_{44})\partial_2(\partial_1 u_1 + \partial_2 u_2) + d_{44}\nabla_1^2 P_2 + (d_{12} + d_{44}) \qquad (2.3)$$
$$\times\partial_2(\partial_1 P_1 + \partial_2 P_2) = \rho\ddot{u}_2$$

$$d_{44}\nabla_1^2 u_1 + (d_{12} + d_{44})\partial_1(\partial_1 u_1 + \partial_2 u_2) + (b_{44} + b_{77})\nabla_1^2 P_1 + \qquad (2.4)$$
$$(b_{12} + b_{44} - b_{77})\partial_1(\partial_1 P_1 + \partial_2 P_2) - aP_1 - \partial_1\phi = 0$$

$$d_{44}\nabla_1^2 u_2 + (d_{12} + d_{44})\partial_2(\partial_1 u_1 + \partial_2 u_2) + (b_{44} + b_{77})\nabla_1^2 P_2 + \qquad (2.5)$$
$$+ (b_{12} + b_{44} - b_{77})\partial_2(\partial_1 P_1 + \partial_2 P_2) - aP_2 - \partial_2\phi = 0$$

$$\bar{\nabla}\cdot\bar{P} - \epsilon_o\nabla_1^2\phi = 0 \qquad (2.6)$$

and

$$c_{44}\nabla_1^2 u_3 + d_{44}\nabla_1^2 P_3 = \rho\ddot{u}_3 \qquad (2.7)$$

$$d_{44}\nabla_1^2 u_3 + (b_{44} + b_{77})\nabla_1^2 P_3 - aP_3 = 0 \qquad (2.8)$$

In these equations, it was assumed that the body forces, $\bar{f}$, the electric body force, $\bar{E}^o$ and the charge density p_c, are zero. The system (2.2)-(2.6) corresponds to the following form for the stress tensor, $\bar{T}$ and electric tensor, $\bar{\bar{\epsilon}}$

$$\bar{\bar{T}} \Rightarrow \begin{vmatrix} T_{11} & T_{12} & 0 \\ T_{21} & T_{22} & 0 \\ 0 & 0 & T_{33} \end{vmatrix}, \qquad \bar{\bar{\epsilon}} \Rightarrow \begin{vmatrix} \epsilon_{11} & \epsilon_{12} & 0 \\ \epsilon_{21} & \epsilon_{22} & 0 \\ 0 & 0 & \epsilon_{33} \end{vmatrix}$$

while eqns (2.7) and (2.8) are associated with

$$\bar{\bar{T}} => \begin{vmatrix} 0 & 0 & T_{13} \\ 0 & 0 & T_{23} \\ T_{31} & T_{32} & 0 \end{vmatrix}, \qquad \bar{\bar{\epsilon}} => \begin{vmatrix} 0 & 0 & \epsilon_{13} \\ 0 & 0 & \epsilon_{23} \\ \epsilon_{31} & \epsilon_{32} & 0 \end{vmatrix}$$

The system (2.2)-(2.6), in which (u_1, u_2, P_1, P_2 and ϕ) are unknown fields, is rather complex. Thus, to find the solution, we introduce the functions ψ, χ, Γ and Λ connected to the displacement and polarization fields by relations

$$u_1 = \partial_1\psi + \partial_2\Gamma, \qquad u_2 = \partial_2\psi - \partial_1\Gamma \tag{2.9}$$
$$P_1 = \partial_1\chi + \partial_2\Lambda, \qquad P_2 = \partial_2\chi - \partial_1\Lambda$$

Substituting eqn (3.2) into eqns (2.19)-(2.24), one obtains

$$c_{11}\nabla_1^2\psi + d_{11}\nabla_1^2\chi - p\ddot{\psi} = 0 \tag{2.10}$$
$$d_{11}\nabla_1^2\psi + b_{11}\nabla_1^2\chi - a\chi - \phi = 0 \tag{2.11}$$

$$\nabla_1^2\chi - \epsilon_o\nabla_1^2\phi = 0 \tag{2.12}$$

and

$$c_{44}\nabla_1^2\Gamma + d_{44}\nabla_1^2\Lambda - p\ddot{\Gamma} = 0 \tag{2.13}$$
$$d_{44}\nabla_1^2\Gamma + (b_{44} + b_{77})\nabla_1^2\Lambda - a\Lambda = 0 \tag{2.14}$$

where

$$x_{11} = x_{12} + 2x_{44}; \quad x = b,c,d$$

After some algebra, eqns (2.10)-(2.12) can be recast in the form

$$\nabla_1^2[l_1^2\nabla_1^2\nabla_1^2 - \nabla_1^2(1 + \frac{b_{11}}{\hat{a}c_1^2}\partial_t^2) + \frac{1}{c_1^2}\partial_t^2](\psi, \chi, \phi) = 0 \tag{2.15}$$
$$[l_2^2\nabla_1^2\nabla_1^2 - \nabla_1^2(1 + \frac{\hat{b}_{44}}{ac_2^2}\partial_t^2) + \frac{1}{c_2^2}\partial_t^2](\Lambda, \Gamma) = 0 \tag{2.16}$$

where

$$\hat{a} = a + \epsilon_o^{-1}, \quad \hat{b}_{44} = b_{44} + b_{77}, \quad c_1^2 = \frac{c_{11}}{\varrho}, \quad c_2^2 = \frac{c_{44}}{\varrho}$$
$$l_1^2 = \frac{c_{11}b_{11} - d_{11}^2}{\hat{a}\,c_{11}} > 0, \quad l_2^2 = \frac{c_{44}\hat{b}_{44} - d_{44}^2}{a\,c_{44}} > 0$$

We confine ourselves to monochromatic waves propoagate along the x_2 axis. Thus, if we put

$$(\psi, \chi, \phi, \Lambda, \Gamma)(x_1, x_2, t) = (\psi^*, \chi^*, \phi^*, \Lambda^*, \Gamma^*)(x_1)\exp[ikx_2 - iwt] \tag{2.17}$$

then eqns (2.15)-(2.16) takes the form:

$$(\partial_1^2 - k^2)[l_1^2(\partial_1^2 - k^2)^2 - (\partial_1^2 - k^2)(1 - \frac{b_{11}w^2}{\hat{a}c_1}) - \frac{w^2}{c_1}](\psi^*, \chi^*, \phi^*) = 0 \tag{2.18}$$

$$[l_2^2(\partial_1^2 - k^2)^2 - (\partial_1^2 - k^2)(1 - \frac{\hat{b}_{44}w^2}{a\,c_2^2}) - \frac{w^2}{c_2^2}](\Lambda^*, \Gamma^*) = 0 \tag{2.19}$$

Solving eqns (2.18) - (2.19), we assume

$$\psi^* = Ae^{\beta x_1}, \quad \chi^* = \bar{A}e^{\beta x_1}, \quad \phi^* = \bar{\bar{A}}e^{\beta x_1}, \quad \Lambda^* = Be^{\gamma x_1}, \quad \Gamma^* = \bar{B}e^{\gamma x_1} \tag{2.20}$$

Substituting (2.20) into (2.18)-(2.19) we obtain the equations:

$$(\beta^2 - k^2)[l_1^2(\beta^2 - k^2)^2 + (\eta_1 - 1)(\beta^2 - k^2) - \sigma_1^2] = 0$$
$$l_2^2(\gamma^2 - k^2)^2 + (\eta_2 - 1)(\gamma^2 - k^2) - \sigma_2^2 = 0 \qquad (2.21)$$

where

$$\eta_1 = \frac{b_{11}w^2}{\hat{a}c_1^2}, \quad \eta_2 = \frac{\hat{b}_{44}w^2}{ac_2^2}, \quad \sigma_1 = \frac{w}{c_1}, \quad \sigma_2 = \frac{w}{c_2}$$

The solution of (2.21) has the form

$$\beta_{1,3} = \pm\left[k^2 + \frac{1}{l_1^2}\left\{(1 - \eta_1) + [(1-\eta_1)^2 + 4\sigma_1^2 l_1^2]^{1/2}\right\}\right]^{1/2}$$
$$\gamma_{1,3} = \pm\left[k^2 + \frac{1}{l_2^2}\left\{(1 - \eta_2) + [(1-\eta_2)^2 + 4\sigma_2^2 l_2^2]^{1/2}\right\}\right]^{1/2}$$
$$\beta_{5,6} = \pm k \qquad \beta_{2,4} = \pm i\bar{\beta}_{2,4}, \quad \gamma_{2,4} = \pm i\bar{\gamma}_{2,4}$$

where

$$\bar{\beta}_{2,4} = \pm\left[-k^2 - \frac{1}{2l_1^2}\left\{(1 - \eta_1) - [(1-\eta_1)^2 + 4\sigma_1^2 l_1^2]^{1/2}\right\}\right]^{1/2}$$
$$\bar{\gamma}_{2,4} = \pm\left[-k^2 - \frac{1}{2l_2^2}\left\{(1 - \eta_2) - [(1-\eta_2)^2 + 4\sigma_2^2 l_2^2]^{1/2}\right\}\right]^{1/2}$$

Let us observe, that β_1 and γ_1 are real and positive. Moreover, if we assume that

$$2k^2 l_\alpha^2 < \eta_\alpha - 1 + [(1 - \eta_\alpha)^2 + 4\sigma_\alpha^2 l_\alpha^2]^{1/2} \qquad \alpha = 1,2$$

then $\bar{\beta}_2$ and $\bar{\gamma}_2$ are real and positive.
Therefore, the final solution of the eqns (2.18)-(2.19) is given by:

$$\begin{aligned}
\psi^* &= A_1 \operatorname{ch}\beta_1 x_1 + A_2 \operatorname{sh}\beta_1 x_1 + A_3\cos\bar{\beta}_2 x_1 + A_4 \sin\bar{\beta}_2 x_1 + A_5 \operatorname{ch} kx_1 + A_6 \operatorname{sh} kx_1 \\
\chi^* &= \bar{A}_1 \operatorname{ch}\beta_1 x_1 + \bar{A}_2 \operatorname{sh}\beta_1 x_1 + \bar{A}_3\cos\bar{\beta}_2 x_1 + \bar{A}_4 \sin\bar{\beta}_2 x_1 + \bar{A}_5 \operatorname{ch} kx_1 + \bar{A}_6 \operatorname{sh} kx_1 \\
\phi^* &= \bar{\bar{A}}_1 \operatorname{ch}\beta_1 x_1 + \bar{\bar{A}}_2 \operatorname{sh}\beta_1 x_1 + \bar{\bar{A}}_3\cos\bar{\beta}_2 x_1 + \bar{\bar{A}}_4 \sin\bar{\beta}_2 x_1 + \bar{\bar{A}}_5 \operatorname{ch} kx_1 + \bar{\bar{A}}_6 \operatorname{sh} kx_1 \\
\Lambda^* &= B_1 \operatorname{ch}\gamma_1 x_1 + B_2 \operatorname{sh}\gamma_1 x_1 + B_3 \cos\bar{\gamma}_2 x_1 + B_4 \sin\bar{\gamma}_2 x_1 \qquad (2.22) \\
\Gamma^* &= \bar{B}_1 \operatorname{ch}\gamma_1 x_1 + \bar{B}_2 \operatorname{sh}\gamma_1 x_1 + \bar{B}_3 \cos\bar{\gamma}_2 x_1 + \bar{B}_4 \sin\bar{\gamma}_2 x_1
\end{aligned}$$

In view of the coupling of the equations (2.10)-(2.12) we obtain

$$\bar{A}_\alpha = \varkappa_1 A_\alpha, \quad \bar{\bar{A}}_\alpha = \epsilon_0^{-1}\bar{A}_\alpha, \quad \bar{B}_\alpha = \tau_1 B_\alpha \qquad \alpha = 1,2$$
$$\bar{A}_\delta = \varkappa_2 A_\delta, \quad \bar{\bar{A}}_\delta = \epsilon_0^{-1}\bar{A}_\delta, \quad \bar{B}_\delta = \tau_2 B_\delta \qquad \delta = 3,4$$
$$A_5 = A_6 = \bar{A}_5 = \bar{A}_6 = 0$$

where

$$\varkappa_\alpha = -\frac{c_{11}}{d_{11}} \frac{\beta_\alpha^2 - k^2 + \sigma_1^2}{\beta_\alpha^2 - k^2} \qquad \tau_\alpha = -\frac{c_{44}}{d_{44}} \frac{\gamma_\alpha^2 - k^2 + \sigma_2^2}{\gamma_\alpha^2 - k^2}$$

Similary as in classical elastokinetics, the general problem of propagation of waves may be reduced to the solution of two simple problems, i.e. to the consideration of the symmetric and anty-

symmetric vibrations only. These vibrations are characterized by the symmetry of displacement u_2 and polarization P_2 with respect to the plane x = 0. The displacement u_1 and the polarization P_1 are antisymmetric with respect to the same plane.
In this case the eqns (2.22) reduce to

$$
\begin{aligned}
\psi^* &= A_1 \mathrm{ch}\beta_1 x_1 + A_3 \cos\bar{\beta}_2 x_1 \\
\chi^* &= \varkappa_1 A_1 \mathrm{ch}\beta_1 x_1 + \varkappa_2 A_3 \cos\bar{\beta}_2 x_1 \\
\phi^* &= \epsilon_0^{-1}(\varkappa_1 A_1 \mathrm{ch}\beta_1 x_1 + \varkappa_2 A_3 \cos\bar{\beta}_2 x_1) + \bar{\bar{A}}_5 \mathrm{ch} k x_1 \\
\Lambda^* &= B_2 \mathrm{sh}\,\gamma_1 x_1 + B_4 \sin\bar{\gamma}_2 x_1 \\
\Gamma^* &= \tau_1 B_2 \mathrm{sh}\,\gamma_1 x_1 + \tau_2 B_4 \sin\bar{\gamma}_2 x_1
\end{aligned}
\tag{2.23}
$$

The boundary conditions $(2.1)_1$ will be expressed by the functions $\psi, \chi, \phi, \Lambda$ and Γ. We get

$$
\begin{aligned}
&T_{11}\big|_{x_1=h} = 2c_{44}(\partial_1^2\psi + \partial_1\partial_2\Lambda) + c_{12}\nabla_1^2\psi + 2d_{44}(\partial_1^2\chi + \partial_1\partial_2\Gamma) + \\
&+ d_{12}\nabla^2\chi\big|_{x_1=h} = 0 \\
&T_{12}\big|_{x_1=h} = c_{44}[\partial_1\partial_2\psi + (\partial_2^2 - \partial_1^2)\Lambda] + d_{44}[2\partial_1\partial_2\chi + \\
&+(\partial_2^2 - \partial_1^2)\Gamma]\big|_{x_1=h} = 0 \\
&\epsilon_{11}\big|_{x_1=h} = 2d_{44}(\partial_1^2\psi + \partial_1\partial_2\Lambda) + d_{12}\nabla_1^2\psi + 2b_{44}(\partial_1^2\chi + \\
&+\partial_1\partial_2\Gamma) + b_{12}\nabla^2\chi\big|_{x_1=h} = 0 \\
&\epsilon_{12}\big|_{x_1=h} = d_{44}[2\partial_1\partial_2\psi + (\partial_2^2 - \partial_1^2)]\Lambda + b_{44}[2\partial_1\partial_2\chi + \\
&+(\partial_2^2 - \partial_1^2)\Gamma]\big|_{x_1=h} = 0 \\
&-\epsilon_0|\partial_1\phi^+ - \partial_1\phi^-| + \partial_1\chi + \partial_2\Gamma\big|_{x_1=h} = 0
\end{aligned}
\tag{2.24}
$$

For vacuum we have

$$
\nabla^2\phi^+ = 0 \qquad x \in V^* \qquad |x_1| \gg h \tag{2.25}
$$

The solution of (2.25) has the form:

$$
\phi^+(x_1, x_2, t) = A_0\,\phi^{*+}(x_1)\exp[-k(x_1 - h) - iwt] \tag{2.26}
$$

and the electric potential ϕ must be continous in the plane $x_1 = h$, i.e.

$$
(\phi^+ - \phi^-)\big|_{x_1=h} = 0 \tag{2.27}
$$

The conditions (2.24) and (2.27) lead to a system of equations for six integration constants $A_1, A_3, \bar{A}_5, B_1, B_2$, and A_0. Making equal to zero the determinant of this system, we arrive at the characteristic equation. From this equation one can obtain an infinite number of roots k. To each of these roots there corresponds a definite form of vibrations.
Let us now consider the harmonic waves, propagating along the x_2-axis, described by the system of eqns (2.7)-(2.8). After some manipulation we get the equation identical as to its structure with equ. (2.19)

$$\left[l_2^2(\partial_1^2 - k^2) - (\partial_1^2 - k^2)(1 - \eta_2) - \sigma_2^2\right](u_3^*, P_3^*) = 0 \quad (2.28)$$

The solution of (2.28) has the form:

$$\begin{aligned} u_3^*(x_1) &= C_1 \mathrm{ch}\,\gamma_1 x_1 + C_2 \mathrm{sh}\,\gamma_1 x_1 + C_3 \cos\bar{\gamma}_2 x_1 + C_4 \sin\bar{\gamma}_2 x_1 \\ P_3^*(x_1) &= \tau_1 C_1 \mathrm{ch}\,\gamma_1 x_1 + \tau_1 C_2 \mathrm{sh}\,\gamma_1 x_1 + \tau_2 C_3 \cos\bar{\gamma}_2 x_1 + \tau_2 C_4 \sin\bar{\gamma}_2 x_1 \end{aligned} \quad (2.29)$$

Integration constants in above equation are bound by the boundary conditions $(2.1)_2$ which expressed by u_3 and P_3 take the form:

$$\begin{aligned} T_{13}\Big|_{x_1=h} &= c_{44}u_{3,1} + d_{44}P_{3,1}\Big|_{x_1=h} = 0 \\ \mathcal{E}_{13}\Big|_{x_1=h} &= d_{44}u_{3,1} + \hat{b}_{44}P_{3,1}\Big|_{x_1=h} = 0 \end{aligned} \quad (2.30)$$

If we consider the symmetric and antysymmetric vibrations separetly we obtain the following characteristic equations

$$\begin{aligned} \cos\bar{\gamma}_2 h &= 0 \\ \sin\bar{\gamma}_2 h &= 0 \end{aligned} \quad (2.31)$$

The transcendental equations (2.31) permit to determine the phase velocity $c = \frac{\omega}{k}$ of the propagation of wave.
The expressions (2.29) represent dispersion waves, γ_1 and γ_2 being dependent on ω.

III. Propagation of waves in cylinder

Let us assume a torsional wave propagating in an infinite cylinder with circular cross-section, along the z-axis with constant phase velocity c. Our considerations are conducted within the system of cylindrical coordinates (r, θ, z). We assume

$$\underline{u} = (0, u_\theta, 0), \quad \underline{P} = (0, P_\theta, 0)$$

and u_θ, P_θ are independent of angle θ
Under these assumption the system of the following two equations is what remains from the system of equation of the problem (1.12)-(1.14) within the system of cylindrical coordinates:

$$\begin{aligned} c_{44}\left(\nabla^2 u_\theta - \frac{u_\theta}{r^2}\right) + d_{44}\left(\nabla^2 P_\theta - \frac{P_\theta}{r^2}\right) - \rho\ddot{u}_\theta &= 0 \\ d_{44}\left(\nabla^2 u_\theta - \frac{u_\theta}{r^2}\right) + \hat{b}_{44}\left(\nabla^2 P_\theta - \frac{P_\theta}{r^2}\right) - aP_\theta &= 0 \end{aligned} \quad (3.1)$$

where

$$\nabla^2 \equiv \frac{\partial^2}{\partial r^2} + \frac{1}{r}\frac{\partial}{\partial r} + \frac{\partial^2}{\partial z^2}$$

The field of displacement u_θ and polarization P_θ induces the following state of stresses

$$\bar{\bar{T}} \Rightarrow \begin{vmatrix} 0 & T_{r\theta} & 0 \\ T_{\theta r} & 0 & T_{\theta z} \\ 0 & T_{z\theta} & 0 \end{vmatrix} \qquad \bar{\bar{\mathcal{E}}} \Rightarrow \begin{vmatrix} 0 & \mathcal{E}_{r\theta} & 0 \\ \mathcal{E}_{\theta r} & 0 & \mathcal{E}_{\theta z} \\ 0 & \mathcal{E}_{z\theta} & 0 \end{vmatrix}$$

Let us consider the monochromatic waves only

$u_\theta = u_\theta^*(r)\exp(ikz - iwt)$, $P_\Theta = P_\Theta^*(r)\exp(ikz - iwt)$
Then eqns (3.1) take the form

$$c_{44}\left(\frac{\partial^2}{\partial r^2} + \frac{1}{r}\frac{\partial}{\partial r} - k^2 - \frac{1}{r^2} + \frac{\varrho\omega^2}{c_{44}}\right)u_\theta^* + d_{44}\left(\frac{\partial^2}{\partial r^2} + \frac{1}{r}\frac{\partial}{\partial r} - k^2 - \frac{1}{r^2}\right)P_\theta^* = 0$$

$$d_{44}\left(\frac{\partial^2}{\partial r^2} + \frac{1}{r}\frac{\partial}{\partial r} - k^2 - \frac{1}{r^2}\right)u_\theta^* + \hat{b}_{44}\left(\frac{\partial^2}{\partial r^2} + \frac{1}{r^2}\frac{\partial}{\partial r} - k^2 - \frac{1}{r^2} - \frac{\hat{a}}{\hat{b}_{44}}\right)P_\theta^* = 0 \tag{3.2}$$

After some algebra eqns (3.2) can be recast to

$$(D + \mu_1^2)(D + \mu_2^2)(u_\theta^*, P_\theta^*) = 0 \tag{3.3}$$

where
$$D = \frac{\partial}{\partial r^2} + \frac{1}{r}\frac{\partial}{\partial r} - k^2 - \frac{1}{r^2}$$

$$\mu_{1,2} = \frac{1}{2l_2^2}\left\{\eta_2 - 1 \pm \left[(\eta_2 - 1)^2 + 4\sigma_2^2 l_2^2\right]^{1/2}\right\}$$

Eqs. (3.3) is equivalent to $(D - \gamma_1^2)(D + \bar{\gamma}_2^2)(u_\theta^*, P_\theta^*) = 0$ (3.3')

Using the Boggios theorem [14] we can write the solution of eq. (3.3) in the form:

$$u_\theta^* = u_\theta^{*'} + u_\theta^{*''} \qquad P_\theta^* = P_\theta^{*'} + P_\theta^{*''} \tag{3.4}$$

where $u_\theta^{*'}$, $u_\theta^{*''}$, $P_\theta^{*'}$, $P_\theta^{*''}$ satisfy the equations:

$$(D - \gamma_1^2)(u_\theta^{*'}, P_\theta^{*'}) = 0, \quad (D + \bar{\gamma}_2^2)(u_\theta^{*''}, P_\theta^{*''}) = 0 \tag{3.5}$$

The soultion of eqns (3.5) has the following form:

$$\ddot{u}_\theta^{*'} = A_1 I_1(\gamma_1 r) + \bar{A}_1 K_1(\gamma_1 r) \qquad u_\theta^{*''} = B_1 J_1(\bar{\gamma}_2 r) + \bar{B}_1 Y_1(\bar{\gamma}_2 r)$$

$$P_\theta^{*'} = A_2 I_1(\gamma_1 r) + \bar{A}_2 K_1(\gamma_1 r) \qquad P_\theta^{*''} = B_2 J_1(\bar{\gamma}_2 r) + \bar{B}_2 Y_1(\bar{\gamma}_2 r) \tag{3.6}$$

The $J_1(\beta r)$ and $Y_1(\beta r)$ are the Bessel function of first and second kind, respectively, while $I_1(\beta r)$ and $K_1(\beta r)$ represents the modified Bessel functions of the first and second kind, respectively. We will neglect the part of solution with $Y_1(\beta r)$ and $K_1(\beta r)$ because $K_1, Y_1(\beta r) \xrightarrow[r \to 0]{} \infty$ Now, if we substitute (3.6) into (3.2) we obtain the relations between A_1, B_1 and A_2, B_2.

$$A_2 = \tau_1 A_1 \qquad B_2 = \tau_2 A_2$$

In order to get the unique solution of the problem thus formulated we need also the boundary conditions. We assume, that the boundary surface of the cylinder is free of stresses, i.e. the following conditions have to be fulfilled

$$T_{r\theta} = 0, \ \epsilon_{r\theta} = 0 \quad \text{for } r = r_o \tag{3.7}$$

where r_o is a radius of cross-section of cylinder. These conditions expressed in displacement and polarization takes the form:

$$c_{44}\left(\frac{\partial u_\theta^*}{\partial r}-\frac{u_\theta^*}{r}\right)+d_{44}\left(\frac{\partial P_\theta^*}{\partial r}-\frac{P_\theta^*}{r}\right)\Bigg|_{r=r_0}=0$$

$$d_{44}\left(\frac{\partial u_\theta^*}{\partial r}-\frac{u_\theta^*}{r}\right)+b_{44}\left(\frac{\partial P_\theta^*}{\partial r}-\frac{P_\theta^*}{r}\right)\Bigg|_{r=r_0}=0 \qquad (3.8)$$

The conditions (3.8) lead to the system of equation for two integration constants which will have a non-zero solution if the transcendental equation

$$\left[k^4+\frac{1}{l_2^2}(1-\eta_2)k^2-\frac{\sigma_2^2}{l_2^2}\right]^{\frac{1}{2}}\left\{a\left(\eta_2^2+4\sigma_2^2 l_2^2+1\right)^{\frac{1}{2}} I_2(\gamma_1 r_o)\, x\right.$$
$$x\ J_2(\bar\gamma_2 r_o)+b_{77}\left[(\sigma_2^2-\mu_2^2)I_2(\gamma_1 r_o)J_1'(\bar\gamma_2 r_o)-(\sigma_2^2-\mu_1^2)\,x\right.$$
$$\left.\left. x\ J_2(\bar\gamma_2 r_o)\, I_1'(\gamma_1 r_o)\right]\right\}=0 \qquad (3.9)$$

is satisfied.

Equation (3.9) is complicated and in its general form unsuitable for discussion. In a particular case, when the wave is long as compared with the radius of the cylinder we get-expanding the Bessel function into a series and retaining but first two terms of the expanded expression- the first approximation in the form

$$\left(k^2-\frac{\omega^2}{c_2^2}\right)\left[l_2^2k^4+(1-\eta_2)k^2-\frac{\omega^2}{c_2^2}\right]^{\frac{1}{2}}=0 \qquad (3.10)$$

The torsional wave undergoes dispersion which is due - as to seen from (3.10) - to the fact that the phase velocity depends on the frequency.

For wave short as compared with the radius of the cylinder the characteristc equation, after asymptotic transformation - reduces to the equation

$$(\mu_1^2-k^2)(\mu_2^2-k^2)=0$$

This waves are also dispersed.

REFERENCES

1. W.Voigt, Lehrbuch der Kristallphysik, Teubner, Leipzig,1910.

2. R.A. Toupin, The Elastic Dielectric, J.Ration.Mech.Anal., vol. 5, pp.849-915,1956.

3. R.D.Mindlin, Polarization Gradient in Elastic Dielectric, Int. J.Solids Struct., vol.4, pp.637-642,1968.

4. W.Cochran, Phonons in Perfect Lattices, Plenum Press, New York (1966)

5. M.P.Tosi, Solid State Physics, Vol. 16. Academic Press, New York (1964)

6. G.C.Benson and K.S.Yun, The Solid-Gas Interface Edited by E.A. Flood . Dekker, New York(1967).

7. R.D.Mindlin and R.A.Toupin, Int.J.Solids Structures 7, 1219 (1971)

8. K.L.Chowdhury and P.G.Glockner, On Thermoelastic Dielectrics, Int. J.Solids Struct. 13, 1173-1182 (1977)

9. J.P.Nowacki and P.G.Glockner, Some Dynamical Problems of thermo-elastic dielectrics, Int. J.Solids Structures 15, 183(1979).

10. J.P.Nowacki and P.G.Glockner, Propagation of waves in the interior of a thermoelastic dielectric half-space, Int. J.Engng Sci. Vol. 19, pp.603-613

11. K.Majorkowska - Knapp. The Love's Waves in Elastic Isotropic Solid Dielectrics, Bull.Acad.Polon.Sci.Ser.Sci. Techn. vol.28,No.11-12, (1980).

12. K.Majorkowska - Knapp. Wave Reflection and Refraction in Solid Elastic Dielectrics , Bull Acad.Polon.Sci.Ser.Sci.Techn. vol. 29,No.7-8 (1981).

13. W.M.Ewing, W.S.Jardetzkiy, F.Press, Elastic Waves in Layered Media, Mc Graw-Hill, New York-Toronto-London, 1957

14. T.Boggio,Sull integrazione di alcuna equationi linerari alle derivate parziale, Ann.Mat., ser,III, vol.8, p.181 (1903).

The Mechanical Behavior of Electromagnetic Solid Continua
G.A. Maugin (editor)
Elsevier Science Publishers B.V. (North-Holland)

A GAUGE THEORETICAL APPROACH TO MACROSCOPIC ELECTRODYNAMICS

Isaak A. Kunin

Department of Mechanical Engineering
University of Houston
Houston, Texas 77004
U.S.A.

Gauge field theories give the most powerful method for describing interactions between fundamental physical fields. Recently the gauge approach was extended to certain domains of classical physics, in particular, to defects in solids. In this paper, interaction between electromagnetic field and deformable bodies is treated from gauge theoretical point of view.

INTRODUCTION

There is a well developed phenomenological theory for interactions of macroscopic electromagnetic fields with deformable bodies (see e.g., [1,2]). The main assumption of the theory is that electromagnetic field interacts with a body through the Lorentz forces acting on free or bound charges and currents in the body. This assumption being combined with principles of continuum mechanics, leads to electrodynamics of deformable bodies.

On the other hand, it is well known that a microscopic electromagnetic field is the simplest example of so called gauge fields. For fundamental fields in particle physics, gauge theories proved to be the most powerful method to establish interactions between them in a non ad hoc way (see, e.g., [3,4]). The first idea of "gauging" was introduced by H. Weyl (1918) in an attempt to construct a unified theory of gravitational and electromagnetic interaction. It was immediately pointed out by A. Einstein that Weyl's theory was untenable by physical reasons. Later, after quantum mechanics was established, F. London and H. Weyl showed that microscopic electrodynamics was indeed a U(1) gauge theory describing interaction between electromagnetic field and quantum mechanical electron field.

One of the important consequences of the gauge theory for electrodynamics is a conclusion that the electromagnetic field $F = E + iB$ does not contain all the physical information. Some of the information is contained in the vector-potential A, though the latter is defined up to gauge transformations. As a result Bohm-Aharonov type effects are possible (their existence is characteristic for any gauge theory). Interaction between electromagnetic field and a charged particle reduces to the Lorentz force in the classical limit and the Bohm-Aharonov effect vanishes.

The next crucial step in development of gauge theories was made in the pioneering work of C. Yang and R. Mills (1954) who extended the isospin symmetry of nuclear interactions to SU(2) gauge theory. Later on many other outstanding results were obtained based on the principle of gauge invariance. Among them was a unified gauge theory of electroweak interactions based on the gauge group U(1) x SU(2). In modern field theories, the principle of gauge invariance is considered as one of the basic physical principles.

Gauge theories were also extended to certain domains of classical physics. First it was shown that general relativity is a gauge theory. Recently, gauge theories of liquid crystals, amorphous solids, dislocations and other defects in solids were developed [5].

A question now arises. Is it also possible to interpret macroscopic electrodynamics as a gauge theory? There are arguments pro and con. Let us consider interaction between electromagnetic field and free or bound classical charges and currents. In this case, interaction is described by the Lorentz force. This means that all physical information about electromagnetic field is contained in E and B at points of interaction. The vector-potential A is a pure auxiliary quantity. Aharonov-Bohm type effects are impossible in this case, and consequently, the field F = E + iB is not a gauge field.

Now let us consider electromagnetic interaction with deformable dielectrics or ferromagnetics. The corresponding microscopic theory should be quantum mechanical. It is not clear in this case whether collective motions result in some Aharonov-Bohm type effects. If such effects are possible in principle, it means that physical information is contained not only in F but also in A. In this case, the macroscopic electromagnetic field can be a gauge field in the scope of the corresponding phenomenological model. In the present paper we consider briefly such a model while omitting technical details.

MICROSCOPIC ELECTRODYNAMICS AS A GAUGE THEORY

First let us consider a standard gauge approach to interaction between the electron field (matter field) and electromagnetic field (gauge field). The starting Lagrangian for a free Dirac particle of mass m is

$$L_o = \bar{\psi}(i\not{\partial} - m)\psi , \tag{1}$$

where ψ is a four component spinor wave function, and $\not{\partial} = \gamma^\mu\partial_\mu$ where γ^μ are Dirac matrices. The Lagrangian L_o is invariant with respect to the following phase transformation (global gauge transformation):

$$\psi(x) \to \psi'(x) = e^{-i\alpha}\psi(x) , \tag{2}$$

where α is an arbitrary constant phase angle. It is said that the Lagrangian L_o of the matter field ψ is invariant with respect to the gauge group G_o of global transformations, where the group G_o in this case is U(1).

The first step of "gauging" is substituting the one-dimensional group G_o with an infinite-dimensional group G of local transformations

$$\psi(x) \to \psi'(x) = e^{-i\alpha(x)}\psi(x) , \tag{3}$$

where $\alpha(x)$ is now an arbitrary space and time dependent function. The main requirement is that the matter field Lagrangian L_o should be changed in such a way that the new Lagrangian is invariant with respect to G. The gauge theory gives a recipe for restoring the symmetry of the Lagrangian. One introduces a compensating (or gauge) field $A_\mu(x)$ with the law of transformation

$$A_\mu(x) \to A_\mu(x) + \partial_\mu\alpha(x) , \tag{4}$$

under action of G, and substitutes the derivatives ∂_μ in L_o by covariant derivatives

$$D_\mu = \partial_\mu + iA_\mu(x) \quad . \tag{5}$$

This "minimal replacement" $L_o(\partial\psi) \to L_o(D\psi)$ provides us with a G-invariant (gauge invariant) Lagrangian, while A_μ is interpreted as vector-potential for electromagnetic field.

The second recipe indicates how to construct a G-invariant Lagrangian L_{em} for A_μ ("minimal coupling"). The tensor

$$F = dA \quad \text{or} \quad F_{\lambda\mu} = \partial_\lambda A_\mu - \partial_\mu A_\lambda \tag{6}$$

is invariant under transformations (4). The invariant Lagrangian L_{em} is taken in the form

$$L_{em} = -\frac{1}{4} F_{\lambda\mu} F^{\lambda\mu} \quad , \tag{7}$$

while F is interpreted as an electromagnetic field.

To the total Lagrangian $L = L_o(D\psi) + L_{em}$ there corresponds the Euler equations

$$D\psi + im\psi = 0, \; \partial_\nu F^{\mu\nu} = j^\mu \; , \tag{8}$$

with the conserved current $j^\mu(x)$ being given by

$$j^\mu = \bar{\psi}\gamma^\mu\psi \quad (\partial_\mu j^\mu = 0) \quad . \tag{9}$$

The equations (8) and (6) are the Dirac-Maxwell equations coupling the matter field ψ and the compensating field A.

The field A is often written in the form qA where q is a coupling constant proportional to the electron charge e.

THE GAUGE COUPLING OF ELASTIC AND ELECTROMAGNETIC FIELDS

Now we shall try to couple deformations of dielectrics and ferromagnetics with macroscopic electromagnetic field using analogous gauge procedure with necessary corrections due to the phenomenological character of the model.

Although we consider the non-relativistic case, to make formulas transparent we use four-dimensional notations $x = (\vec{x},t)$ for a point in space-time and $u = (\vec{u},0)$ for displacement.

Assuming for simplicity that deformations are small and external body forces are absent we can write the elasticity Lagrangian in a compact four-dimensional form

$$L_o = \frac{1}{2} \, \partial u \, C \, \partial u \tag{10}$$

where C is a block matrix including elastic constants and the mass density.

With every point of the material capable of interacting with electromagnetic field there should be associated a charge space. The main assumption is that, as motivated earlier, the charge space does not reduce in general to classical free or bound charges and currents only. Collective motions lead to the appearance of additional hidden microparameters of the state. Simplest model corresponds to the assumption that the space of microparameters is one-dimensional and boils down to a phase factor, the phase defined up to an arbitrary constant. In the absence of electromagnetic field the phase is constant, hence is not observable.

This means formally that in the absence of electromagnetic field displacement is described by a complex vector field

$$\psi(x) = u(x)e^{i\theta} \quad , \quad \theta = \text{const} \tag{11}$$

with an additional condition of the invariance of the Lagrangian with respect to a phase shift by a constant: $\theta \to \theta - \alpha$ (global gauge transformation). For physical equivalence of the model with the initial elasticity model, the Lagrangian is written in the form

$$L_o = (\partial\psi)^* C(\partial\psi) \quad . \tag{12}$$

This Lagrangian is considered now as a starting matter field model. To introduce interaction with electromagnetic fields we provide the first step of gauging

$$L_o(\partial\psi) \to L_o(D\psi) \quad , \tag{13}$$

where

$$D_\mu = \partial_\mu + i\eta A_\mu \quad , \tag{14}$$

in terms of the vector-potential A_μ and the coupling constant η. If A is dimensionless, then η is a characteristic wave number. If A has the dimension of an electromagnetic vector-potential, then

$$\eta = \frac{1}{B_o \lambda_o^2} = \frac{1}{\Phi_o} \tag{15}$$

where λ_o, B_o, Φ_o are characteristic wavelength, magnetic field and magnetic flux. It is convenient to put temporarily $\eta = 1$.

The next step is to construct a total Lagrangian invariant under local gauge transformations

$$\psi(x) \to e^{-i\alpha(x)}\psi(x), \quad A(x) \to A(x) + \partial\alpha(x) \quad . \tag{16}$$

We choose this Lagrangian in the form

$$L = L_o + L_{em} + L_1 \quad . \tag{17}$$

The first term $L_o = L_o(D\psi)$ is gauge invariant because of (12) and the transformation property of $D\psi$

$$(D\psi)(x) \to e^{-i\alpha(x)}(D\psi)(x) . \tag{18}$$

The second term L_{em} is a pure electromagnetic Lagrangian. We assumed earlier the existence of hidden internal microparameters. To be consistent with this assumption, it is natural to represent the total electromagnetic field as a sume of two components

$$F = dA \quad \text{and} \quad F' = dA' \quad . \tag{19}$$

The vector-potential A is responsible mainly for gauge interaction with the hidden parameters and transforms in accordance with (16), i.e., its law of transformation is coupled to that of ψ. The vector-potential A' has nothing to do with local gauge transformations (16). It gives no contribution to the covariant derivative $D = \partial + iA$ and is defined up to transformations $A' \to A' + \partial\phi$ where ϕ is an arbitrary function. Thus the simplest gauge invariant L_{em} can be represented in the form

$$L_{em} = \frac{1}{8} FMF + \frac{1}{8} F'M'F' + \frac{1}{4} F\Lambda F' \tag{20}$$

where M, M', Λ describe electromagnetic properties of the body (adding higher order terms is not essential for what follows).

The third term L_1 in (17) is purely phenomenological and usually absent in gauge theories of fundamental interactions. There are two simple choices:

$$L_1 = \frac{1}{2} \partial u \, (NF + N'F') \tag{21}$$

and

$$L_1 = (D\psi)^*(NF + N'F') + (D\psi)(N^*F + N'^*F') \tag{22}$$

where N and N' are material parameters which are gauge invariant in (21) and transform via

$$N \to e^{-i\alpha}N \quad , \quad N' \to e^{-i\alpha}N' \tag{23}$$

in (22) to preserve gauge invariance of L_1.

The total Lagrangian depends on four fields u, θ, A, A' or ψ, ψ^*, A, A', hence we should have four Euler equations which can be written in different equivalent forms.

THE COUPLED EQUATIONS

First we obtain equations in complex form which is convenient for the analysis of effects related to the phase variable $\theta(x)$. Let us choose the total Lagrangian (17) as the sum of (13), (20) and (22) and the reanimate the coupling constant η setting $D = \partial + i\eta A$. Neglecting the terms which are nonlinear in small displacements, we obtain

$$D(CD\psi) + D(\eta NF + N'F') = 0, \text{ plus c.c. eqn} \tag{24}$$

$$\eta^2 \partial \, (MF + \eta\Lambda F') = -j \tag{25}$$

$$\partial(M'F' + \eta\Lambda F) = -j' \tag{26}$$

where

$$j = 4\eta \mathrm{Re}\{2\partial[(D\psi)^*M] + i\psi^*(\eta NF + N'F')\} \tag{27}$$

$$j' = 8\mathrm{Re}\, \partial[(D\psi)^*N'] \tag{28}$$

are the corresponding currents for the Maxwell equations (25) and (26).

When considering the Bohm-Aharonov effect, the field A can be viewed as given. Then the equation describing this effect reduces to

$$(\partial + i\eta A)[C(\partial + i\eta A)\psi] = 0 \quad . \tag{29}$$

The real form of the equations is of interest when one considers Bohm-Aharonov type effects related to the displacement u rather than the phase θ. To obtain these equations, it is convenient to take the total Lagrangian (17) as the sum of (13),(20), (21) and then use gauge invariance of L to exclude formally the phase θ. Indeed, this is possible since the transformation

$$\psi \to e^{-i\theta}\psi = u, \quad A \to \tilde{A} = A + \partial\theta \tag{30}$$

is an admissable gauge which makes ψ real. If u, $\tilde{A}$, A' are taken as new variables, then explicit dependence of L on θ is eliminated. The price of doing this is the loss of gauge invariance since the gauge for $\psi = u$ and $\tilde{A}$ becomes fixed. We write down only the equation analogous to (29)

$$\partial(C\partial u) + \eta^2\tilde{A}C\tilde{A}u = 0 \quad . \tag{31}$$

Note that the gauge equations (24)-(29) or the equivalent real equations reduce to classical equations in the limit as $\eta \to 0$ or in the linear case as we would expect.

DISCUSSION

The gauge equations above were derived for the interaction of electromagnetic and elastic fields. In reality, any macroscopic field could be considered instead of displacement. The physical significance of these equations depends on numerical values of the coupling constant η. For extremely small or large η the gauge effects of Bohm-Aharonov type are negligible. This does not mean that gauge effects could be of that type only. More generally, the following problem is crucial from the gauge theory point of view.

Suppose that strong enough magnetic or elastic field exists in a restricted domain of a body. Is there any possibility to find out the presence of these fields by any measurements in the surrounding medium? If the answer is positive, then macroscopic electrodynamics can be treated as a gauge theory. If the answer is negative...we might only regret this.

REFERENCES

[1] Continuum Physics, ed. A. C. Eringen (Academic Press, New York, 1976, Vol. 3).

[2] Maugin, G. A., Eringen, A. C., J. Mécanique 16 (1977) 101-147.

[3] Drechsler, W., Mayer, M. E., Fibre bundle technique in gauge theories (Lecture Notes in Physics No. 67, Springer, Berlin, (1977)).

[4] Aitchison, I. J. R., An informal introduction to gauge field theories (Cambridge University Press, Cambridge, 1982).

[5] Gauge field theories of defects in solids, ed. Kröner, E. (Max-Plank-Institut, Stuttgart, 1982).

Part 7

MAGNETOELASTICITY OF PARAMAGNETS

The Mechanical Behavior of Electromagnetic Solid Continua
G.A. Maugin (editor)
Elsevier Science Publishers B.V. (North-Holland)

HIGH FREQUENCY ELASTIC WAVES ON RARE EARTH SOLIDS

Robert E. Camley
Department of Physics
University of Colorado
Colorado Springs, CO 80907
U.S.A.

Peter Fulde
Max Planck Institute
fur Festkorperforschung
Heisenbergstrasse 1
7000 Stuttgart 80. B.R.D.

We discuss the properties of high frequency elastic waves on rare earth compounds. In these materials there is a magnetoelastic coupling between the electronic states of the rare earth ion and the vibrational motion of the surrounding ions. Here we deal with effects which occur at high frequencies. We find interesting results in the presence of a magnetic field. The most interesting is nonreciprocal Rayleigh wave propagation, i.e., $\omega(k) \neq \omega(-k)$ where ω and k are the frequency and wavevector of the Rayleigh wave.

There has been considerable interest in the properties of elastic waves in rare-earth (RE) compounds (for recent reviews see References 1-3). Both bulk (4) and surface (5-6) waves have been studied. In RE compounds, the deformation caused by an elastic wave interacts with the crystalline-electric field (CEF) split energy levels of the incomplete 4f shell of the RE ions. This magnetoelastic interaction is temperature dependent due to the different thermal population of the CEF levels at different temperatures and leads to a number of interesting effects.

One can distinguish between two different types of effects. One type exists even in the zero frequency limit of the elastic wave. To this group belongs, for example, the temperature dependence of the elastic constants (4), the rotational interaction effects in the presence of an applied magnetic field (7-9), and the acoustic analogue of the Voigt or Cotton-Mouton effect (10-11). The other type of effect exists only at finite frequencies. The Faraday rotation of the transverse acoustic modes (10-11) belongs in this category. We will refer to the first group of effects as "static" and the second group as "finite frequency" effects. In this paper we concentrate on the finite frequency effects.

The Hamiltonian for the system in the presence of an applied external magnetic field is written as

$$H = H_{el} + H_{CEF} + H_{me} + H_{Ze} \tag{1}$$

The elastic energy contribution to the total Hamiltonian is given by

$$\begin{aligned} H_{el} = \; & C_{11}(e_{xx}^2 + e_{yy}^2 + e_{zz}^2) \\ & + C_{12}(e_{yy}e_{zz} + e_{zz}e_{xx} + e_{xx}e_{yy}) \\ & + C_{44}(e_{xy}^2 + e_{yz}^2 \quad e_{zx}^2) \end{aligned} \tag{2}$$

Here C_{11}, C_{12}, and C_{44} are the usual adiabatic elastic constants for a cubic elastic solid in the absence of magnetoelastic interactions. H_{CEF} is the CEF Hamiltonian which for a cubic system has a well-known simple form (1-3). The Zeeman energy H_{Ze} is given by

$$H_{Ze} = \sum_{m=1}^{N} g_L \mu_B \vec{J}^m \cdot \vec{H}_o \tag{3}$$

where g_L is the Lande factor, $\vec{H}_o$ is the applied magnetic field and $\vec{J}^m$ is the total angular momentum of the incomplete 4f shell of the RE ion at site m, N is the number of RE sites per unit volume. Finally for the magnetoelastic portion of the Hamiltonian we take

$$\begin{aligned} H_{me} = \; & g_1 (J_x^2 e_{xx} + J_y^2 e_{yy} + J_z^2 e_{zz}) \\ & + g_2 ([J_x J_y + J_y J_x] e_{xy} \\ & + [J_y J_z + J_z J_y] e_{yz} \\ & + [J_z J_x + J_x J_z] e_{zx}) \end{aligned} \tag{4}$$

We have neglected rotational contributions (7) for simplicity.

The elastic equations of motion are given in the usual way by

$$\rho \frac{\partial^2 u_\alpha}{\partial t^2} = \sum_\beta \frac{\partial}{\partial x_\beta} \left(\frac{\partial H}{\partial e_{\alpha\beta}} \right) \tag{5}$$

where u_α is the displacement in the αth direction and ρ is the density. For static effects we can solve these equations in the following way. We replace every combination of J operators by its thermal average. Thus for example

$$\langle J_x^2 \rangle = \frac{\Sigma_I \langle I | J_x^2 | I \rangle e^{-E_I/kT}}{\Sigma_I \, e^{-E_I/kT}} \tag{6}$$

where $|I\rangle$ are the CEF levels <u>in the presence</u> of the strain and applied magnetic field.

We find the following form for the thermal averages enter the problem

$$\begin{bmatrix} \langle J_x^2 \rangle \\ \langle J_y^2 \rangle \\ \langle J_z^2 \rangle \\ \langle J_x J_y + J_y J_x \rangle \\ \langle J_y J_z + J_z J_y \rangle \\ \langle J_z J_x + J_x J_z \rangle \end{bmatrix} = \begin{bmatrix} A_{11} & A_{12} & A_{12} & & & \\ A_{12} & A_{11} & A_{12} & & & \\ A_{12} & A_{12} & A_{11} & & & \\ & & & A_{44} & & \\ & & & & A_{44} & \\ & & & & & A_{44} \end{bmatrix} \begin{bmatrix} e_{xx} \\ e_{yy} \\ e_{zz} \\ e_{xy} \\ e_{yz} \\ e_{zx} \end{bmatrix} \tag{7}$$

where the explicit forms for the matrix elements will be given elsewhere. The cubic nature of the result is obvious when we compare the form of the A matrix to the matrix of the elastic stiffness constants. If a magnetic field is applied the matrix A has the form

$$A_o = \begin{bmatrix} A_{11} & A_{12} & A_{13} & & & \\ A_{12} & A_{22} & A_{13} & & & \\ A_{13} & A_{13} & A_{22} & & & \\ & & & A_{44} & & \\ & & & & A_{66} & \\ & & & & & A_{66} \end{bmatrix} \qquad (8)$$

From the form of the matrix it is clear the symmetry of the system has been changed from cubic to tetragonal.

As the frequency of vibrations increases, we must calculate the response to a time dependent perturbation. Using linear response theory, we find a frequency dependent contribution to the matrix A of the form

$$A_\omega = i\omega \begin{bmatrix} 0 & 0 & 0 & A_{14} & 0 & 0 \\ 0 & 0 & 0 & -A_{14} & 0 & 0 \\ 0 & 0 & 0 & 0 & 0 & 0 \\ -A_{14} & A_{14} & 0 & 0 & 0 & 0 \\ 0 & 0 & 0 & 0 & 0 & A_{56} \\ 0 & 0 & 0 & 0 & A_{56} & 0 \end{bmatrix} \qquad (9)$$

We note that the matrix elements A_{ij} are also functions of ω. The complete matrix A is then given by

$$A = A_o + A_\omega \qquad (10)$$

Using Eqs. 5-10 we obtain the elastic equations of motion in terms of the displacements and various thermal averages of combinations of J operators. These equations are supplemented by the boundary conditions which may be obtained similarly. These equations may be solved in the usual manner. For bulk waves we assume a plane wave type solution of the form $\vec{u} = \vec{u}_o \exp(i\vec{k}\cdot\overline{x} - i\omega t)$. For surface wave solutions we look for a solution which is a plane wave parallel to the surface but is exponentially damped as one moves away from the surface into the material.

We have solved these equations for various geometries and find that several interesting effects occur. Of these the most interesting is that for a particular geometry the surface acoustic wave (Rayleigh wave) is nonreciprocal, i.e., the dispersion relation $\omega(k) \neq \omega(-k)$ where k and ω are the wavevector and frequency of the Rayleigh wave.

The geometry considered is illustrated in Fig. 1 below.

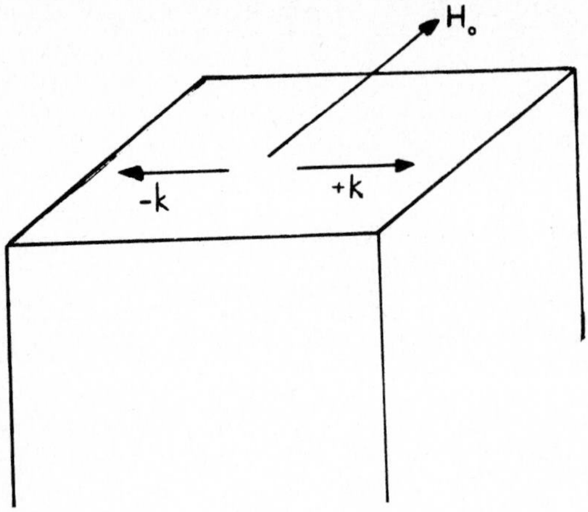

Fig. 1. The geometry for nonreciprocal Rayleigh wave propagation

We have calculated the velocity of Rayleigh waves propagating left to right (V+) across the magnetic field and for those propagating right to left (V-). In Fig. 2 we present the difference of these two velocities divided by the Rayleigh wave velocity in the absence of interactions as a function of applied magnetic field. We see that near .7 Tesla a large difference between the two velocities is found. At this field the energy of the elastic vibration is sufficient to cause a transition from one CEF level to another and the effect is enhanced.

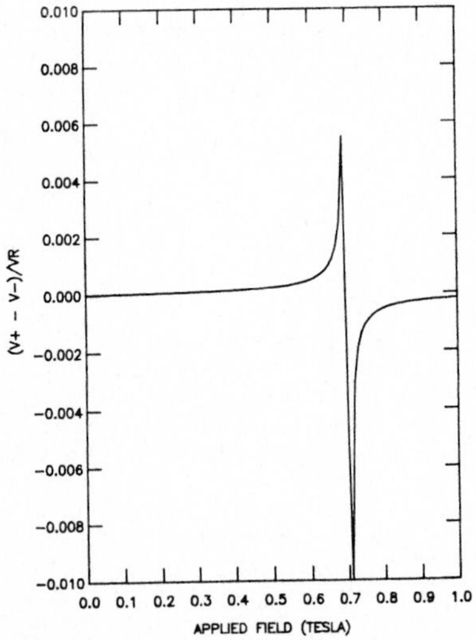

Fig. 2. Difference of Rayleigh wave velocities as a function of applied field. The frequency of the elastic wave is about 20 GHz.

We note that the velocity difference is as large as 1% near the resonance. Such a difference is easily measured by Brillouin scattering which is appropriate for this frequency range. The calculation is done for $CeAl_2$ with a temperature of 70 Kelvin. The elastic and magnetoelastic parameters for $CeAl_2$ can be found in Reference (6).

We mention some other effects which are found only in the finite frequency regime. There is a coupling of bulk longitudinal and bulk transverse waves which leads to a modification of the dispersion relations for these waves. Also we have found that in the presence of an applied magnetic field, a shear polarized surface wave should exist. Such a result has been found previously for elastic waves on a ferromagnetic solid (12). These effects are however small.

REFERENCES

1) P. Fulde in Handbook on the Physics and Chemistry of Rare Earths ed. by K. A. Geschneider and L. Eyring (North Holland, Amsterdam, 1978)

2) B. Lüthi in Dynamical Properties of Solids, ed. by G. K. Horton and A. A. Maradudin (North Holland, Amsterdam 1980) Vol. III

3) P. Fulde in Crystalline electric fields and Structual Effects in f-electron systems, ed. by J. E. Crow (Plenum, New York, 1980)

4) M. E. Mullen, B. Lüthi, S. P. Wang, E. Bucher, L. D. Longinotti, J. P. Maita and H. R. Ott, Phys. Rev. B 10, 186 (1974)

5) C. Lingner and B. Lüthi, Phys. Rev. B 23, 256 (1981)

6) R. Camley and P. Fulde, Phys. Rev. B 23, 2614 (1981)

7) V. Dohm and P. Fulde, Z. Phys. B 21, 369 (1975)

8) P. S. Wang and B. Lüthi, Phys. Rev. B 15,2718 (1977)

9) L. Bonsall and R. L. Melcher, Phys. Rev. B 14, 1128 (1976)

10) B. Lüthi and C. Lingner, Z. Phys. B 34,157 (1979)

11) P. Thalmeier and P. Fulde, Z. Phys. B 29, 299 (1978)

12) R. Q. Scott and D. L. Mills, Phys. Rev. B 15, 3545 (1977)

The Mechanical Behavior of Electromagnetic Solid Continua
G.A. Maugin (editor)
Elsevier Science Publishers B.V. (North-Holland)

WAVE PROPAGATION IN PARAMAGNETIC BODIES EXHIBITING INDUCED LINEAR MAGNETOELASTIC COUPLINGS

A.HAKMI and G.A.MAUGIN

Université Pierre-et-Marie Curie,Laboratoire de Mécanique Théorique,
Tour 66, 4 Place Jussieu, 75230 Paris Cedex 05 , France .

The propagation of bulk and surface magnetoelastic modes is studied in paramagnetic bodies on the basis of a phenomenological,fully nonlinear, rotationally invariant theory of the magnetoelasticity of insulators such as rare-earth compounds. The initial nonlinear feature of the theory associated with rotational invariance allows one to place in evidence the effects of a bias magnetic field:symmetry breaking favoring the existence of magnetoelastic couplings(biased piezomagnetism) which would not exist in the absence of bias field, alteration in the bulk-wave speeds(direct magnetoacoustic effect), an acoustical linear birefringence and the existence, in addition to a Classical Rayleigh mode, of an SH magnetoacoustic mode akin to the Bleustein-Gulyaev mode of acousto-electricity.The influence of viscosity and magnetic relaxation is also examined.

GENERAL NONLINEAR EQUATIONS

The nonlinear equations which govern at time t in the present configuration K_t a nonpolarizable, but magnetizable, material in the quasi-magnetostatic approximation can be deduced from general equations established in a previous work [1].In the material body $\mathcal{B}$ they consist of the following equations:

. equation of continuity:

$$\rho J = \rho_R \quad \text{or} \quad \dot{\rho} + \rho \underset{\sim}{\nabla}.\underset{\sim}{v} = 0 \qquad ,(1)$$

. the Euler-Cauchy equations of motion:

$$\rho \dot{v}_i = t_{ij,j} + f_i^{em} \quad , \quad \dot{\overline{(.)}} = (\frac{\partial}{\partial t} + \underset{\sim}{v}.\underset{\sim}{\nabla})(.) \qquad ,(2)$$

. Maxwell's magnetostatic equations(Lorentz-Heaviside units;no currents):

$$\underset{\sim}{\nabla} \times \underset{\sim}{H} = \underset{\sim}{0} \quad , \quad \underset{\sim}{\nabla}.\underset{\sim}{B} = 0 \quad , \quad \underset{\sim}{H} = \underset{\sim}{B} - \underset{\sim}{M} \qquad (3)$$

while the corresponding boundary conditions at $\partial\mathcal{B}$ (of unit normal $\underset{\sim}{n}$) in general read

$$t_{ij}n_j = T_i + T_i^{em} \quad , \quad \underset{\sim}{n} \times [\![\underset{\sim}{H}]\!] = \underset{\sim}{0} \quad , \quad \underset{\sim}{n}.[\![\underset{\sim}{B}]\!] = 0 \qquad (4)$$

and the local thermodynamics is governed by the Clausius-Duhem inequality

$$-\rho(\dot{\psi} + \eta\dot{\theta}) + \sigma_{ij} D_{ij} - \rho\, {}^L B_i \, \dot{m}_i \geq 0 \quad ; \quad \underset{\sim}{m} = \underset{\sim}{M}/\rho \qquad .(5)$$

In these equations $\rho, \underset{\sim}{B}, \underset{\sim}{H}, \underset{\sim}{M}, \psi$, η and θ are the matter density,the magnetic induction,the magnetic field,the magnetization per unit volume, the free energy and entropy per unit mass and the thermodynamical temperature. If $x_i = \mathscr{X}_i(X_K,t)$ is the motion mapping between a stress-free, field-free configuration K_R and K_t, then we have

$$v_i = \left.\frac{\partial \mathscr{X}_i}{\partial t}\right|_{X_K}, \quad F_{iK} = \left.\frac{\partial \mathscr{X}_i}{\partial X_K}\right|_t, J = \det|F_{iK}|, D_{ij} = v_{(i,j)}, \quad \hat{A}_i = \dot{A}_i - v_{[i,j]}A_j \quad .(6)$$

The magnetostatic ponderomotive force $\underset{\sim}{f}^{em}$ and its surface counterpart $\underset{\sim}{T}^{em}$ are given by

$$f_i^{em} = t_{ij,j}^{em} = M_j B_{i,j} = [(\underset{\sim}{M}.\underset{\sim}{\nabla})\underset{\sim}{B}]_i \quad , \quad t_{ij}^{em} = H_i B_j - \frac{1}{2}(\underset{\sim}{B}^2 - \underset{\sim}{M}.\underset{\sim}{B})\,\delta_{ij} \quad ,(7a)$$

$$T_i^{em} = [\![t_{ij}^{em}]\!] n_j = (\langle \underset{\sim}{M} \rangle . [\![\underset{\sim}{B}]\!])n_i \quad ,(7b)$$

where the symbolisms $[\![.]\!]$ and $\langle . \rangle$ denote the jump and mean value of their anclosure. Finally, the (nonsymmetric) Cauchy stress tensor t_{ij} and $\underset{\sim}{B}$ are related to the fields of intrinsic stress σ_{ij} and local induction ${}^L\underset{\sim}{B}$ by

$$t_{ij} = \sigma_{ij} + {}^L B_{[i}M_{j]} \quad , \quad \underset{\sim}{B} = - {}^L\underset{\sim}{B} \; , \text{ so that } t_{[ij]} = {}^L B_{[i}M_{j]} = M_{[i}B_{j]} = - t_{[ij]}^{em} \quad .(8)$$

Alternately, we can rewrite eqns.(5) and $(8)_1$ as

$$-\rho(\dot{\psi} + \eta\dot{\theta}) + t_{ij}^E D_{ij} - {}^L B_i \overset{*}{M}_i \geq 0 \quad (9)$$

$$t_{ij} = t_{ij}^E + {}^L B_i M_j \quad , \quad t_{ij}^E = \sigma_{ij} + {}^L B_{(i}M_{j)} = t_{ji}^E \quad ,(10)$$

where we have introduced the convected time derivative

$$\overset{*}{M}_i = \dot{M}_i - M_j v_{i,j} + M_i v_{j,j} \quad .(11)$$

Ultimately, on the basis of (5) or (9), one has to give constitutive equations for either σ_{ij} and ${}^L\underset{\sim}{B}$ or t_{ij}^E and ${}^L\underset{\sim}{B}$. Both fields may present recoverable (i.e., derivable from ψ) parts and irreversible parts, the latter yielding viscosity and magnetic relaxation effects. The elaboration of such nonlinear constitutive equations for a prescribed material symmetry in K_R now is a routine work in continuum mechanics and it is not given here for lack of room.

MAGNETOACOUSTIC EQUATIONS ABOUT A BIASED STATE

All above-given equations can be linearized about a rigid-body solution corresponding to a state $K_o \neq K_R$ of spatially uniform magnetic field $\underset{\sim}{H}_o$. The linearized equations are obtained by using a so-called Lagrangian variation (variation at fixed X_K's) effected on all above-given equations once they have been cast in the Lagrangian or material framework by convection back to K_R with the help of the operator F_{iK} or its inverse (see [2] for the general method and [3] for the variation of boundary conditions). All quantities at K_o are labelled with a right subscript "zero" and lower case letters denote perturbations. After a lengthy calculation we obtain:

+ equation of motion : $\rho_o \ddot{u}_i = (\tau_{ij} + \tau_{ij}^{em})_{,j}$,(12)

+ Maxwell's equations :

$$\underset{\sim}{\nabla} \times [\underset{\sim}{h} - (\nabla\underset{\sim}{u}).\underset{\sim}{H}_o] = \underset{\sim}{0} \; ,$$
$$\underset{\sim}{\nabla} . [\underset{\sim}{b} - (\nabla\underset{\sim}{u}).\underset{\sim}{B}_o] = \underset{\sim}{0} \; , \quad (13)$$

with

$$\tau_{ij} = \bar{C}_{ijkl}u_{(k,l)} + D_{ijkl}u_{[k,l]} + \lambda_{ijkl}\dot{u}_{(k,l)} + E_{kij}\, m_k - b_i\, M_{oj} \quad ,(14)$$

$$b_i = \alpha_{ij}m_j + r_{ij}\dot{m}_j + \epsilon_{ikl}\, u_{(k,l)} \quad ; \quad \tau_{ij}^{em} = (b_i - B_{ok}u_{k,i})M_{oj} \quad ,(15)$$

$$h_i = b_i - m_i - u_{(p,q)}[\delta_{pq} M_{oi} - 2\delta_{i(p} M_{oq)}] \qquad ,(16)$$

and

$$\bar{C}_{ijkl} = C^o_{ijkl} + t^E_{oj(l}\delta_{k)i} + B_{o(k}\delta_{l)i} M_{oj} \qquad ,(17)$$

$$D_{ijkl} = t^E_{oj[l}\delta_{k]i} + B_{o[k}\delta_{l]i} M_{oj} \qquad ,(18)$$

$$E_{kij} = \epsilon_{kij} - B_{oi}\delta_{jk} \qquad .(19)$$

Here t^E_{oij}, $\mathbf{B}_o$ and $\mathbf{M}_o$ are initial fields corresponding to K_o, λ_{ijkl} and r_{ij} correspond to viscous and magnetic relaxation, respectively; and $C^o_{ijkl}=C^o_{(ij)(kl)}=C^o_{klij}$, $\epsilon_{kij}=\epsilon_{kji}$ and $\alpha_{ij}=\alpha_{ji}$ are directly defined by second-order derivatives of the energy density with respect to the set (E_{KL}, M_K) of variables which are invariant measures of strain and magnetization in K_R; $\mathbf{u}=\delta\mathfrak{X}$ is the infinitesimal displacement between K_o and K_t. Note that the rotation $u_{[k,l]}$ is involved in the linearized constitutive equations (14). Let $\mathbf{M}_{ot}$ and $P_{pq}=\delta_{pq}-n_{op}n_{oq}$ denote the tangential component of M_o on $\partial\mathcal{B}_o$ and the projection operator onto the local tangent plane to $\partial\mathcal{B}_o$. Then the boundary conditions which accompany eqns.(12)-(13) can be shown to read

$$(\tau_{ij} - t_{oij}P_{pq}u_{p,q})n_{oj} + \frac{1}{2} n_{oi}\left\{\mathbf{M}_{ot}\cdot\mathbf{m} + M_{otj}M_{ok}u_{j,k} - M_{oj}[\![b_j - u_{p,j}B_{op}]\!]\right\}$$

$$- \frac{1}{2}\mathbf{M}^2_{ot}(P_{pq}u_{p,q}n_{oi} + u_{p,i}n_{op}) = \bar{T}_i = \delta T_i \qquad ,(20)$$

$$\mathbf{n}_o \times [\![\mathbf{h} - (\nabla\mathbf{u}).\mathbf{H}_o]\!] - [\![((\nabla\mathbf{u}).\mathbf{n}_o) \times \mathbf{H}_o]\!] = \mathbf{0} \qquad ,(21)$$

$$\mathbf{n}_o \cdot [\![\mathbf{b} - (\nabla\mathbf{u}).\mathbf{B}_o - (\mathbf{B}_o\cdot\nabla)\mathbf{u}]\!] = 0 \qquad .(22)$$

Magnetoacoustics in a quasi-magnetostatic framework usually refers to a problem involving the displacement $\mathbf{u}$ and the "dynamical" magnetostatic potential. On account of eqn.$(13)_1$, we note here that such a potential φ is related to $\mathbf{h}$ through the equation

$$\mathbf{h} = -\nabla\varphi + (\nabla\mathbf{u}).\mathbf{H}_o \qquad .(23)$$

Then we can show that eqns.(12)-(13) and (20)-(21) can be replaced by the following equations (in the absence of dissipation):

. <u>in the bulk</u> :

$$\rho_o\ddot{u}_i = C^H_{ijkl}u_{k,lj} + \epsilon^H_{mij}\varphi_{,mj} \qquad ,(24a)$$

$$\mu_{ik}\varphi_{,ki} - (\epsilon^H_{ipq} + \hat{\epsilon}_{ipq})u_{p,qi} = 0 \qquad ;(24b)$$

. <u>at the boundary</u> :

$$\Big\{C^H_{ijpq} - (\epsilon^H_{ipq} + \hat{\epsilon}_{ipq})M_{oj} - (t_{oij}P_{pq} + \frac{1}{2}\mathbf{M}^2_{ot}\delta_{iq}\delta_{jp}) + \frac{1}{2}\delta_{ij}[M_{otp}M_{oq} - \mathbf{M}^2_{ot}P_{pq}$$

$$+ M_{om}(\epsilon^H_{mpq} + \hat{\epsilon}_{mpq}) + M_{otk}(\epsilon^H_{kpq} + \check{\epsilon}_{kpq})]\Big\} u_{p,q}n_{oj} + \Big\{\frac{1}{2}(M_{oq}[\![\mu_{pq}\varphi_{,p}]\!]$$

$$- M_{otk}\chi_{kp}\varphi_{,p})\delta_{ij} + (\epsilon^H_{pij} + \mu_{ip}M_{oj})\varphi_{,p}\Big\}n_{oj} = \bar{T}_i \qquad ,(25a)$$

$$[\![\underset{\sim}{\nabla}_n \varphi]\!] \times \underset{\sim}{n}_o + [\![\underset{\sim}{H}_o \times ((\nabla_H \underset{\sim}{u}) . \underset{\sim}{n}_o)]\!] = \underset{\sim}{0} \quad ,(25b)$$

$$n_{oi} [\![\mu_{ik} \varphi_{,k}]\!] + (\epsilon^H_{ipq} + \hat{\epsilon}_{ipq} - \delta_{ip} B_{oq}) u_{p,q}\, n_{oi} = 0 \quad ,(25c)$$

where we have set

$$C^H_{ijkl} = C^o_{ijkl} + t^E_{ojl}\,\delta_{ik} - E_{pij}\,\chi_{pm}\,\bar{E}_{mkl} \quad , \quad \epsilon^H_{mij} = -\chi_{km}\,E_{kij} \quad ,(26a)$$

$$\bar{E}_{ipq} = \epsilon_{ipq} + M_{oi}\,\delta_{pq} - \delta_{ip}\,M_{oq} - \delta_{iq}\,B_{op} \quad ,(26b)$$

$$\hat{\epsilon}_{ipq} = -2\,\mu_{im}\,\delta_{p[q}\,M_{om]} \quad , \quad \check{\epsilon}_{kpq} = -2\,\chi_{kl}\,\delta_{p[q}\,M_{ol]} \quad (26c)$$

if (symbolically) $\chi_{ij} = (\alpha_{ij} - \delta_{ij})^{-1}$ and $\mu_{ij} = \delta_{ij} + \chi_{ij}$. ∇_n and ∇_H denote gradient operators in planes orthogonal to $\underset{\sim}{n}_o$ and $\underset{\sim}{H}_o$, respectively. Notice that equation (25b) is much more complexe than the usual continuity condition $[\![\varphi]\!] = 0$.

BULK WAVE MODES

Consider a motion such that $\underset{\sim}{\nabla} = \underset{\sim}{\xi}\mathcal{D}$ where $\underset{\sim}{\xi}$ is a unit vector and assume that $\mu_\xi = \xi_i\,\mu_{ik}\,\xi_k \neq 0$. Then from eqns(24) we have

$$[\,\delta_{ik} \frac{\partial^2}{\partial t^2} - \Gamma_{ik}(\underset{\sim}{\xi})]u_k = 0 \quad ,(27)$$

where the magnetoacoustic Christoffell tensor Γ_{ik} has been defined by

$$\Gamma_{ik}(\underset{\sim}{\xi}) = \rho_o^{-1}[\,C^H_{ijkl} + \frac{1}{\mu_\xi}\,\xi_m\,\epsilon^H_{mij}\,\xi_a(\epsilon^H_{akl} + \hat{\epsilon}_{akl})]\,\xi_l\,\xi_j \quad .(28)$$

This describes the <u>direct</u> magnetoacoustic effect. The analogy with piezoelectricity is obvious(compare [4]).However, <u>pure</u> piezomagnetism is a seldom encountered effect which, if allowed by the material (magnetic) symmetry, yields a very weak magnetomechanical coupling in any event. This is not the case of eqn.(28) because <u>biased piezomagnetism</u> [5](p.103),obtained with the application of an intense magnetic field, yields a much stronger magnetomechanical coupling and it exists for a wider class of materials since it is the symmetry created by this initial field which generates the effect in the last instance, eventhough pure piezomagnetism might not be allowed for the material in its natural (free) state. For instance, had we linearized the nonlinear equations about a free state (such as K_R) and considered isotropy for the material symmetry of the monlinear material, we would have automatically found $\epsilon^H_{mij} = 0$, while if we still consider such an isotropy but linearize about K_o in which $H_o \neq 0$, we do obtain a linear magnetoelastic coupling for the perturbations. For instance, for bulk waves propagating in such an isotropic body in a direction orthogonal to the direction $\underset{\sim}{d}$ of the field $\underset{\sim}{H}_o$ (so-called <u>orthogonal</u> setting), accounting for dissipative processes we obtain the following modes of propagation $[\omega(k) = \Omega(k) + i\,\Gamma(k)]$:

- longitudinal elastic displacement:

$$\Omega^2 \simeq \omega_L^2(1 - \epsilon_{ML}) \;,\; \Gamma \simeq \tfrac{1}{2}\omega_L^2(\delta_L + \epsilon_{ML}\mu_L\tau_L) \;,\; (\omega_L^2 = c_L^2\,k^2) \quad ;(29)$$

- transverse displacement polarized in the direction of $\underset{\sim}{H}_o$:

$$\Omega^2 \simeq \omega_{T2}^2(1 - \epsilon_{MT}) ,\; \Gamma \simeq \tfrac{1}{2}\omega^2_{T2}(\delta_{T2} + \epsilon_{MT}\tau_1) ,\; (\omega_{T2}^2 = c_{T2}^2 k^2) \quad ;(30)$$

transverse displacement polarized in the $(\underset{\sim}{\xi}\times\underset{\sim}{d})$-direction:

$$\Omega^2 = \omega_{T3}^2 , \quad \Gamma \simeq \frac{1}{2}\omega_{T3}^2 \delta_{T3} \quad , (\omega_{T3}^2 = c_{T3}^2 k^2 , c_{T3} \neq c_{T2}) \qquad ,(31)$$

where δ_L, δ_{T2} and δ_{T3} are viscosity coefficients, τ_1 and τ_L are magnetic relaxation constants, μ_L is the longitudinal value of the magnetic permeability at K_o and ϵ_{ML} and ϵ_{MT} are small magnetomechanical coupling parameters. For instance,

$$\epsilon_{MT} = \frac{H_o^2 \chi_o^3}{\rho_o \mu_o c_{T2}^2} \left(f - \frac{\mu_o}{\chi_o}\right)^2 , \quad f = \partial\Sigma/\partial(M_K E_{KL} M_L) \qquad ,(32)$$

where χ_o and μ_o are the isotropic magnetic susceptibility and permeability at the initial state and Σ is the energy per unit volume of K_R. From the above results we can deduce the following <u>acoustical linear birefringence effect</u> :

$$\phi(\Omega) = k(\Omega)\Big|_{[\underset{\sim}{u}//\underset{\sim}{d}]} - k(\Omega)\Big|_{[\underset{\sim}{u}//(\underset{\sim}{\xi}\times\underset{\sim}{d})]} = \frac{\Omega}{2\rho_o c_T^2}\left(\frac{\chi_o M_o \hat{f}^2}{\mu_o} - \alpha_2\right) \qquad ,(33)$$

where α_2 is an anisotropic elastic coefficient. Equation (33) is the elastic analog of the Voigt-Cotton-Mouton effect of electromagnetic optics. It can be shown that $\phi(\Omega;M_o)$ goes to zero with M_o .

SURFACE WAVE MODES

For the sake of illustration we consider the propagation of surface waves on the elastic half-space $(y>0)$. For a material isotropic with respect to K_R and for an orthogonal setting of the bias magnetic field ($\underset{\sim}{H}_o$ orthogonal to the sagittal plane), it is shown <u>exactly</u> that the generalized Rayleigh problem splits into a classical Rayleigh problem governing the elastic displacement component parallel to the sagittal plane and a problem of the Bleustein-Gulyaev type (cf. [6]) governing the elastic component parallel to $\underset{\sim}{H}_o$, hence orthogonal to the sagittal plane, and the potential φ . The following results emerge from the analysis where one has to use the boundary conditions (25) with T_i=0. The elastic displacement u_z , the real frequency Ω , the reciprocal relaxation time τ_S^{-1} and the penetration depth of the wave in the substrate are given by

$$u_z = u^o \exp(-\beta k y)\exp[-t/\tau_S(k)]\exp\{i[\Omega(k)t - kx]\}, \quad y>0 \qquad ,(34)$$

$$\Omega^2 \simeq \omega_{T2}^2 (1-\epsilon_{MT})(1-\beta^2) \qquad ,(35)$$

$$\tau_S^{-1}(k) \simeq \frac{1}{2}\omega_{T2}^2 (1-\epsilon_{MT})(\delta_{T2}+\epsilon_{MT}\tau_1)(1-\beta^2) \qquad (36)$$

and (λ = wavelength)

$$p(\lambda) = \frac{(1+\mu_o)(1-\epsilon_{MT})}{2\pi\chi_o \epsilon_{MT}}\lambda \quad , \beta = \frac{\epsilon_{MT}\chi_o}{(1+\mu_o)(1-\epsilon_{MT})} \qquad .(37)$$

As H_o goes to zero this surface mode degenerates into a face shear wave. Therefore, we have here a magnetic example of a symmetry breaking in the simplest magnetic structure which allows for the existence of a mode which would not otherwise exist. For $\chi_o \simeq 3$ and $\epsilon_{MT} \simeq 0.1$, we obtain $p \simeq 2\text{-}3\lambda$ and $v_{surface} = 0.9970\, v_{bulk}$ .These results are very similar to those pertaining to Bleustein-Gulyaev modes at the surface of a linear piezoelectric half-space of adequate material symmetry (i.e., the z-axis being an axis of order six) when the surface is not shorted [6]. In particular, we note that $\epsilon_{MT} \simeq K^2$ and $\mu_o \simeq \varepsilon_{11}$, where K is the usual electromechanical coupling factor and ε_{11} is a component of the dielectric tensor. However, in the present case the whole phenomenon disappears when the dc-field H_o is switched off.

The present continuum approach, with the effects placed in evidence, seems to be directly applicable to paramagnetic insulators such as rare-earth compounds (e.g., laves phase compounds) which have received considerable attention from physicists in recent years [7].(the detailed version of the present work shall be published elsewhere).

REFERENCES:

[1] Maugin G.A.,The method of virtual power in continuum mechanics:Application to coupled fields, Acta Mechanica,35(1980) 1-70.

[2] Maugin G.A.,Wave motion in magnetizable deformable solids:Recent advances, Int.J.Engng.Sci., 19(1981) 321-388.

[3] Pouget J. and Maugin G.A.,Bleustein-Gulyaev Surface modes in elastic ferroelectrics, J.Acoust.Soc.Amer., 69 (1981) 1304-1318.

[4] Nelson D.F.,Electric,optic and acoustic interactions in dielectrics,(J.Wiley-Interscience, New York,1979).

[5] Auld B.A.,Acoustic fields and waves in Solids (J.Wiley-Interscience, New York, 1973).

[6] Dieulesaint E. and Royer D.,Ondes élastiques dans les solides:Application au traitement du signal(Masson, Paris, 1974);Chapter 5.

[7] Bonsall L. and Melcher R.L., Phys.Rev.,B 14(1976) 1128; Lüthi B. and Lingner C., Zeit.Phys.,B21(1975) 157-163; Dohm V. and Fulde P.,Zeit.Phys.,B21(1975) 369-379; Camley R. and Fulde P.,Phys.Rev., B23(1981) 2614-2619; Lingner C. and Lüthi B., Phys.Rev.,B23(1981) 256-262; Camley R. and Fulde P.,in these Proceedings.

Part 8

MAGNETOELASTICITY OF MAGNETICALLY ORDERED BODIES

The Mechanical Behavior of Electromagnetic Solid Continua
G.A. Maugin (editor)
Elsevier Science Publishers B.V. (North-Holland)

MAGNETOACOUSTICS OF FERRO- AND ANTIFERROMAGNETS

E.A.Turov

Institute of Metal Physics Ac. Sci. USSR,
GSP 170 Sverdlovsk 620219
U.S.S.R

This lecture begins with a historical review of principle ideas and results in magnetoacoustics of magneto-odered materials since there have been discovered experimentally coupled magnetoelastic waves 25 years ago. The second part is devoted to some results of the last years concerning magneto-acoustic effects of the spontaneous symmetry breaking near magnetic phase transition points (a magnetoelastic magnon energy gap and soft magnetoacoustic modes). The latest technical applications are told as well.

INTRODUCTION

This year 1983 has just turned out to be the 25th anniversary of the experimental discovery [1] of coupled magnetoelastic waves in ferromagnets. These waves are connected with an interaction between lattice oscillations (elastic waves, or phonons) and magnetization oscillations (spin waves, or magnons). The reason of that is the modulation of magnetic (dipole, spin-orbit and axchange) interactions by elastic oscillations of the lattice.

The phenomenon had been predicted in 1956 [2-4] and then described theoretically in more detail in 1958 [5,6] as a magnetoelastic (ME) or magnon-phonon resonance. The main results can be formulated as follows.

The dimensionless parameter determining the power of the dynamic ME (magnon-phonon) coupling can be taken as

$$\zeta = \omega_{me} / \omega_o \tag{1}$$

Here $\omega_o = \gamma H_o$ is a gap for the spin wave frequency

$$\omega_k = \omega_o + \omega_e (ak)^2 \tag{2}$$

That is, ω_o corresponds to the frequency of the magnetization uniform precession (with a wave vector $k=0$) in some effective field H_o which includes the applied field H, anisotropy and other internal magnetic fields. Usually $H_o=10^3-10^4$Oe. ω_e is an exchange field, and a is an interatomic parameter. Then

$$\omega_{me} = \gamma H_{me} \qquad (H_{me} = B^2/CM_o) \tag{3}$$

is the ME frequency associated with some effective ME field H_{me} being created by spontaneous ME strains. Inasmuch as the magnetization $M_o \gtrsim 10^2$ Gs, the elastic constants $C \gtrsim 10^{11}$ erg/cm^3, and the ME constants $B \lesssim 10^7$ erg/cm^3, one usually has $\zeta \ll 1$. Just this fact ($\zeta \ll 1$) allowed Akhiezer in his pioneer paper [7] on magnon-phonon interaction to consider the latter as a small perturbation causing a relaxation between magnetization and lattice. However, there is a special (resonance) case of the strong phonon-magnon coupling at the condition when frequencies ω and wave vectors k of both elastic and spin waves of the same symmetry are coinciding (or fairly close) (Figure 1a). Near this point 0 the ME waves are just eigenmodes of the coupled oscillations of the magnetization $\vec{M}$ and the elastic displacement gradients $u_{ij} = \partial u_i/\partial x_j$.

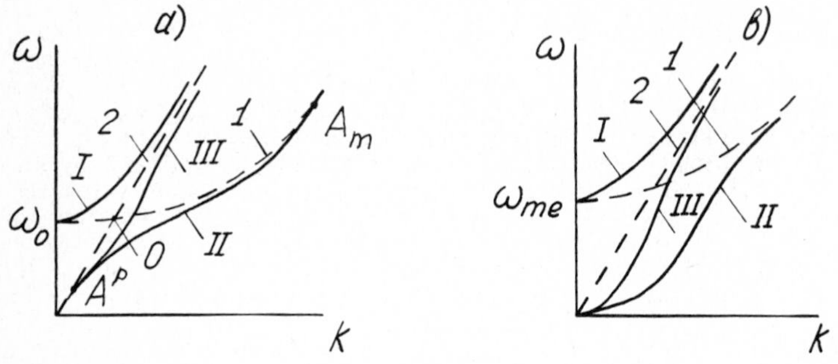

Figure 1

Coupled ME waves with the wave vector $\vec{k} \parallel \vec{M}_o \parallel [001]$ in cubic ferromagnets: a) Far from the point of phase transition (where $\zeta \ll 1$), and b) Just in the phase transition point (where $\zeta = 1$). Dash lines correspond to magnons (1) and transversal phonons (2) without dynamic ME coupling. Unbroken lines - quasimagnons (I) and quasiphonons (II and III) taking into account ME coupling.

Subsequent studies and referencies can be found in the review papers [8,9].

In 1957 Vlasov [8] has shown (by the method of irreversible thermodynamics) that a medium magnetized (spontaneously, or by applied field) in the direction Z must have some additional (as compared with the case $M = H_o = 0$) components of the dynamic elasticity tensor being antisymmetrial in the sense that $C_{yzxz} = -C_{xzyz}$. These components are associated with a magnetic gyrotropy of magnetized medium, and are responsible for acoustic gyromagnetic effects - magnetic rotary polarization, and ellipticity of originally plane - polarized acoustic waves. The first theory and experiment have been done correspondingly in [6] and [11]. Then, see [12].

The next important step was the development of a rigorous theory of ME phenomena both in physical and mathematical aspects on the basis of the first principles of Physics (like conservation laws). The theory led to the discovery of a new mechanism of the ME interaction which is connected with magnetic anisotropy of crystals. The simple physical idea was suggested by Vlasov and Ishmuhametov [13] that for the strained crystals the anisotropic part of the energy density has to be written in local axes of reference which are rotated with a volume element due to the rotation (antisymmetric) components

$$\omega_{ij} = \frac{1}{2}\left(\frac{\partial u_i}{\partial x_j} - \frac{\partial u_j}{\partial x_i} \right)$$

of the tensor u_{ij}. For example, in the case of uniaxial crystals with the anisotropy energy $KM_{z'}^2$ in the local system of axes (x', y', z') after transition to the laboratory system (x,y,z) one obtains

$$KM_{z'}^2 = K\,(M_z^2 + 2\omega_{zx}M_zM_x + 2\omega_{zy}M_zM_y + \ldots). \qquad (4)$$

Here the terms with ω_{ij} just give the additional part of the ME coupling connected with elementary rotations. Such mutual rotations of the crystal axes and the local magnetization causes an additional torque which gives an asymmetric contribution to the stress tensor.

One year later (and I sure indipendently) Tierston [14] developed a strict mathematical theory of ME interactions with the account of rotational invariance for the stored energy (and density) of a strained elastic medium. This invariance ensures the conservation law for the total angular momentum of the system. The first confirmation of this theory seems to be obtained by Eastman [15] for cubic ferromagnet YIG.

A concretized and profound development the ME dynamics of magneto-odered materials obtained in the well-known book [16].

Brilliant studies, both theoretical and experimental, of ME effects associated with rotational invariance have been fulfieled by Melcher [17]. They have a very trasparent physical meaning. He has investigated in detail one of these effects existing in tetragonal antiferromagnet MnF_2. The effect is just due to the ω_{ij} dependent terms coming in particular from magnetic anisotropy as in (4). The old theory of elastic waves taking into account the symmetry part only of tensor u_{ij}, i.e. the deformation tensor

$$e_{ij} = \frac{1}{2}\left(\frac{\partial u_i}{\partial x_j} + \frac{\partial u_j}{\partial x_i} \right), \qquad (5)$$

had given equal sound velocities for two types of transverse waves - for a wave propagating in the direction [001] polarized along the axis [100], and for a wave propagating in the direction [100] polarized along the axis [001]. Such a reciprocity is broken by ME coupling on account of the terms with ω_{ij}, and these velocities turn out to be different in agreement with the experiment.

The line on construction of a rigorous fundamental theory of ME medium was extended in the series of systematic and elegant papers by Maugin with co-workers [18,19]. (Being short of place I give here two referencies only, where [18] is a review paper, and [19] are the newest ones.) The most important results of those seem to concern of a consistent (from the first principles) taking into account dissipative forces in the coupled motion equations for magnetization and elastic displacements. I am under the impression that after these studies the general (phenomenological) problems of the dynamic theory of ME media can be in principle considered as closed. Main efforts must be now directed for the study of specific effects in concrete materials. The ivestigations of nonlinear ME effects are especially promissing in both science and industrial aspects. Surely for those the general theory above spoken is of a great significance.

When proceeding to concrete ME effects, one should note a particular importance of Schlömann's study [20] on propagation of ME waves in applied field H being inhomogeneous along the specimen. Look again at Figure 1a, and take, for example, the ME mode II. Fairly far from the intersection point O this wave is to be a pure spin one in the upper part (point A_m), and a pure elastic one in the lower part (point A_p). At some given frequency ω= const a corresponding point on the curve is of the type A_m or the type A_p in dependence on the magnitude of H (as $\omega_o = \gamma H$). It means that if such a wave propagates in a specimen with inhomogeneous field H, then it will be a spin wave in the specimen region with $H < \omega/\gamma$, and an elastic one at $H > \omega/\gamma$. Such transformations of spin waves into elastic ones and back have found an application in microwave electronic devices such as an acoustic delay line, etc. [21].

Below there will be considered some other special problems of magnetoacoustics of our main interest which concern of spontaneous symmetry breaking (SSB), magnetic phase transitions (PT) and soft ME modes (see also [22] and referencies there).

Before this, in conclusion of the hystorical review it should be mentioned that the first calculation of ME modes in antiferromagnets was undertaken by Peletminskii in 1959 [23].

MAGNETOACOUSTIC EFFECTS OF SYMMETRY BREAKING. CUBIC FERROMAGNET.

Cubic ferromagnet is the best object to demonstrate the ideas and problems under consideration. For that, with an exactness sufficiant for our purposes, the stored energy density can be written as follows:

$$F = F_m + F_e + F_{me} \,, \tag{6}$$

where

$$F_m = (1/2)\, A\, (\partial\vec{m}/\partial x_i)(\partial\vec{m}/\partial x_i) + K(m_x^2 m_y^2 + m_y^2 m_z^2 + m_z^2 m_x^2) - \vec{M}\,\vec{H} - (1/2)\vec{M}\,\vec{H}_{dip} \tag{7}$$

is a magnetic contribution;

$$F_e = (1/2)C_{11}(e_{xx}^2 + e_{yy}^2 + e_{zz}^2) + C_{12}(e_{xx}e_{yy} + e_{yy}e_{zz} + e_{zz}e_{xx}) + 2C_{44}(e_{xy}^2 + e_{yz}^2 + e_{zx}^2) \tag{8}$$

is an elastic contribution; and

$$F_{me} = B_1(m_x^2 e_{xx} + m_y^2 e_{yy} + m_z^2 e_{zz}) + 2B_2(m_x m_y e_{xy} + m_y m_z e_{yz} + m_z m_x e_{zx}) \qquad (9)$$

is a ME one. Here $\vec{M} = \vec{m}\, M_o$ - magnetization ($\vec{m}^2 = 1$); the terms in (7) with the constants A and K represent respectively nonuniform exchange and anisotropy contributions; further, H_{dip} is a dipole field being determined by the magnetostatic equations:

$$\text{rot}\, \vec{H}_{dip} = 0, \quad \text{div}(\vec{H}_{dip} + 4\pi \vec{M}) = 0 . \qquad (10)$$

Lastly, C in (8) and B in (9) are elastic and ME constants.

For thermodynamically equilibrium processes by minimization of F (6) one finds

$$m_i^{(o)} \equiv \alpha_i = \text{const}$$

$$e_{ii}^{(o)} = -\frac{B_1}{C_{11}-C_{12}}\left(\alpha_i^2 - \frac{C_{12}}{C_{11}+2C_{12}}\right), \quad e_{ij} = -\frac{B_2}{2C_{44}}\alpha_i\alpha_j \quad (i \neq j). \qquad (11)$$

The putting of (11) in (6)-(9) gives that at the condition of equilibrium between $\vec{M}$ and e_{ij} the total energy density (6) reduces to

$$F_o = \text{const} + K^*(\alpha_x^2\alpha_y^2 + \alpha_y^2\alpha_z^2 + \alpha_z^2\alpha_x^2) - \vec{M}_o \cdot \vec{H}, \qquad (12)$$

were

$$K^* = K + B_1^2/(C_{11} - C_{12}) - B_2^2/2C_{44} . \qquad (13)$$

Thus, the taking into account magnetostriction in equilibium processes leads only to renormalization of the anisotropy constant:

$$K \rightarrow K^*. \qquad (14)$$

It should be stressed that K* is just the constant, which is measured in static experiments (from magnetization or torque curves). It determines also the orientation magnetic phase transition (PT). If K* changes its sign with temperature in some point T*, then at the transition through this point (at H=0) there happens the replacement of easy axes: the vector $\vec{\alpha}$ is parallel to an edge or a space diagonal of the cube at $K^* > 0$ and $K^* < 0$, respectevely.

Note: equilibrium magnetic properties are described accoringly to (12) by a cubic symmetry even though one takes into account the magnetostriction strains (11). In this sense one has here unbroken symmetry - the symmetry of equilibrium (macroscopic) properties.

At the same time the symmetry of any given ground (equilibrium) state is broken. For instance, the symmetry of the state with [001] is tetragonal. This state is 6-fold degenerated (with the states $\vec{\alpha} \parallel [010], [00\bar{1}]$, etc.), and all degenerated states are separated one from others by energy barriers. Such a rather important property of crystals follows from a discrete symmetry of those.

Consider now small (linear) oscillations of $\vec{M}$ and e_{ij} near the ground state $\vec{\alpha} \parallel \vec{H} \parallel [001] \parallel \vec{z}$ (that is, the state (11) with $\alpha_x = \alpha_y = 0$, $\alpha_z = 1$). Then the coupled motion equations for $\vec{m}$ and elastic displacements $\vec{u}$ in the case of wave vector $\vec{k} \parallel \vec{\alpha} \parallel \vec{z}$ reduce to the system

$$(\omega_k \pm \omega)m_{\pm}(k) + ik\gamma(B_2/M)u_{\pm}(k) = 0, \quad (15)$$

$$(ikB_2/\rho)m_{\pm}(k) + (\omega^2 - \omega_t^2(k))u_{\pm}(k) = 0,$$
$$(\omega_1(k) - \omega)u_1(k) = 0. \quad (16)$$

Here ω_k is the spin wave frequency (2), which does take into account the spontaneous strains in the ground state but does not the dynamic ME coupling. At this, in (2)

$$\omega_o = (2\gamma/M_o)(K^* + B_2^2/2C_{44}) + \gamma H. \quad (17)$$

Further,

$$\omega_t(k) = s_t k \qquad (s_t^2 = C_{44}/\rho)$$
$$\omega_1(k) = s_1 k \qquad (s_1^2 = C_{11}/\rho) \quad (18)$$

are the frequencies of transversal and longitudinal elastic waves respectively (see Figure 1a); and

$$m_{\pm}(k) = m_x(k) \pm i\, m_y(k) \quad \text{and} \quad u_{\pm}(k) = u_x(k) \pm i\, u_y(k)$$

are the Fourie components of $\vec{m}$ and $\vec{u}$ oscillations. (16) gives the longitudinal elastic wave with the frequency $\omega_1 = s_1 k$ to be not coupled with spin waves. Two other independent systems in (15) (for + and -) determine two types of coupled ME waves:

- right-hand polarized waves ($u_x - iu_y = m_x - im_y = 0$) with the dispersion equation

$$(\omega_k + \omega)(\omega_t^2 - \omega^2) - \zeta\omega_o\omega_t^2 = 0, \quad (19')$$

and

- left-hand polarized waves ($u_x + iu_y = m_x + im_y = 0$) with the dispersion equation

$$(\omega_k - \omega)(\omega_t^2 - \omega^2) - \zeta\omega_o\omega_t^2 = 0. \quad (19'')$$

In (19) the parameter of ME coupling (ζ) has been introduced in the accordance with (1), where ω_o is (17), and

$$\omega_{me} = \gamma B_2^2/C_{44}M_o \quad (20)$$

The frequency $\omega = \omega_o$ is the only positive root of equations (19) at $k \to 0$, representing a minimum frequency ("gap") for a quasi-magnon ME mode (Figure 1a).

At H=0 the state with $\vec{\alpha} \parallel [001]$ under consideration is stable only when $K^* > 0$. At a sign change of K^* (13) at the point, where

$$K^* = 0, \text{ or } K + B_1^2/(C_{11} - C_{12}) = B_2^2/2C_{44}, \tag{21}$$

there takes place an orientation PT: from the state $\vec{\alpha} \parallel [001]$ to the state $\vec{\alpha} \parallel [111]$. Directly in the transition point one has a smallest spin wave gap $\omega_o = \omega_{me}$ (Figure 1b) that does not equal zero even though the cubic (thermodynamical) anisotropy (K^*) does. This is just the ME effect of SSB. It reflects the tetragonal distortion of the ground state, and the fact that dynamic strains u_- for this mode I don't follow the oscillations of magnetization: one has from (15) at $k \to 0$ and $\omega \to \omega_{me}$ that $u_- = 0$ ("frozen lattice effect").

The field $H \neq 0$ shifts the PT point to the position

$$2K^* / M_o + H = 0 , \tag{22}$$

but the gap in this point keeps the same ME magnitude (20) ("ME gap").

Consider now solutions of the equations (19) at $k \neq 0$. The Figure 1a represents schematically their nonnegative roots at the condition $\zeta \ll 1$ that is fairly far from the PT point. In this case there is a strong magnon-phonon interaction (for waves of the left-hand polarization (-)) only near the intersection point 0, where the frequency splitting (proportional to $\sqrt{\zeta}$) appears.

However, the ME coupling rises greatly when the parameter ζ increases with approaching to the PT point where $\zeta = 1$. Directly at this point for extremely long waves when $\omega_t \ll \omega_{me}$ one has a single quasi-magnon mode:

$$\begin{aligned} \omega_{\underline{I}}^{(-)} &= \omega_{me} + (\omega_D^2/\omega_{me})(ak)^2, \\ iku_- &= (\omega_t^2/\omega_{me}^2)(B_2/C_{44})\, m_- , \end{aligned} \tag{23}$$

and two quasi-phonon transversal modes:

$$\omega_{\underline{II}}^{(-)} = \omega_E (ak)^2, \qquad \omega_{\underline{III}}^{(+)} = (\omega_D^2/\omega_{me})(ak)^2 , \tag{24}$$

$$m_{\pm} = - (C_{44} / B_{44})\, iku_{\pm} , \tag{25}$$

where $\omega_E = \gamma A/M_o a^2$ is an exchange frequency, and $\omega_D = s_t/a$ is a Debye frequency (we took $\omega_E \omega_{me} \ll \omega_D^2$). The corresponding dispersion curves are shown in Figure 1b.

Thus, the second ME effect of SSB is a softening of acoustic modes: a dispersion law for those changes from linear to quadratic one in wave vector k. This means their velocities $v = \omega/k$ to go to zero at $k \to 0$ when nearing to the PT point ($\zeta \to 1$).

Conclusion: instead of the usual condition $\omega_o \to 0$ for a magnetic PT taking place without ME coupling, one has another condition - the fall to zero the sound velocity for modes coupled with spin waves (that is, the relevant renormalized dynamic elastic constant).

When considering quasi-acoustic modes in the low frequency limit, one can use a thermodynamical approach: to find m through e_{ij} using a minimum condition for the energy sum F_m+F_{me} from (7) and (9), then to insert m obtained in the total energy (6), and lastly to calculate a relevant renormalized (by ME coupling) dynamic elastic constant. In the case under consideration it must be

$$C_{44} = (1/4)\,\partial^2 F/\partial e_{xz}^{\,2} = (1/4)\,\partial^2 F/\partial e_{yz}^{\,2}\,, \qquad (26)$$

which actually comes to zero at $\zeta \to 1$. Certainly an anharmonicity becomes especially important in this case.

If one compares the amplitude relation (25) for quasi-phonons, one will find that to be coinciding with the corresponding equilibrium relation from (11) (at $iku_i \to \partial u_i/\partial z$). Therefore, the quasiacoustic ME modes correspond to joint oscillations of M and e_{ij} so that those have to be excited both by sound and magnetic fields.

A great deal that a sound velocity of these modes through the parameter ζ (1) depends significantly upon magnetic field. It will be shown below when considering antiferromagnets this phenomenon can find important applications in electronics.

The ME effects of SSB spoken above are greater and have been discovered in easy plane (EP) rare-earth ferromagnets such as Dy, Tb and their alloys [24-27]. There are two reasons for this. Firstly, those have very large ME constants B_{ij}. Secondly, an uniaxial anisotropy, which forces magnetization to be in easy plane, is very strong as well, and this enhances effectively the ME frequency ω_{me} in (1). However, the first objects, where the effects were discovered (Borovik-Romanov and Rudashevskii [28], Iida and Tasaki [29]; more references see again in [22]), then were studied in detail and found industrial applications, actually were the EP antiferromagnets α-Fe_2O_3 (hematite) and $FeBO_3$ (iron borate).

ANTIFERROMAGNETS. EXCHANGE ENHANCEMENT

Simplest antiferromagnets are characterized by two vectors:

$$\vec{m} = (\vec{M}_1+\vec{M}_2)/2M_o \quad \text{and} \quad \vec{l} = (\vec{M}_1-\vec{M}_2)/2M_o\,, \qquad (27)$$

where $\vec{M}_1$ and $\vec{M}_2$ - sublattice magnetizations with

$$\vec{M}_1^{\,2} = \vec{M}_2^{\,2} = M_o^{\,2} = \text{const}\ (\vec{l} \perp \vec{m})\,. \qquad (28)$$

Usually (apart from a very large magnetic field H)

$$|\vec{m}| \ll |\vec{l}|, \quad \text{and} \quad |\vec{l}| \approx 2M_o\,. \qquad (29)$$

A very important peculiarity of dynamic ME properties of antiferromagnets as compared ferromagnets is an so-called exchange enhancement of those. To show what is it, consider some isotropic antiferromagnet with the applied field parallel to the axis x and with a shear strain e_{xy}. The energy density is

$$F = (1/2)A_o m^2 + (1/2)A(\partial l/\partial x_i)(\partial l/\partial x_i) - 2M_o m_x H + + 2B_{66} e_{xy} l_x l_y + 2C_{66} e_{xy}^2 . \quad (30)$$

Here the terms with A_o and A represent an exchange energy ensuring antiferromagnetism in the ground state (with m=0 and l=1 at H=0).

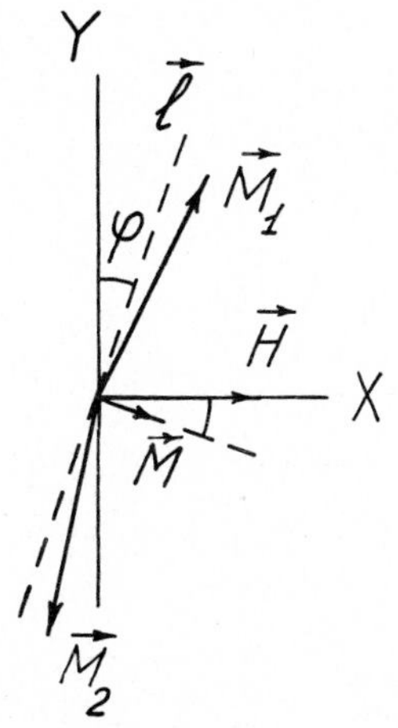

Figure 2
The elastic strain e_{xy} tries to turn l for some angle φ.

The field H induces magnetization $M=2M_o m$ and keeps it along the axis x (therefore the vector $\vec{l}$ being $\perp \vec{m}$ is kept along the axis y, for example, see Figure 2). However, the strain e_{xy} tries to turn $\vec{l}$ (together with $\vec{m}$) for some angle φ in the plane xy.

Taking m and φ as independent variables one obtains from minimum of F (30):

$$M = 2M_o(H/H_E), \qquad \varphi = -(2B_{66}/MH)\, e_{xy} , \quad (31)$$

where $H_E = A_o/2M_o$ is some effective exchange field. After putting these values in (30) and separating the terms with e_{xy}^2 one finds a renormalized (dynamic) elastic constant:

$$C_{66} = C_{66}(1 - \omega_{me}\omega_E/\omega_o^2) , \quad (32)$$

where

$$\omega_{me} = \gamma B_{66}^2/2M_o C_{66}$$

is a ME frequency similiar to (20), $\omega_E = \gamma H_E$ is an exchange frequency, and $\omega_o = \gamma H$.

In ferromagnet for the relevant case the second term in (32) equals $\zeta = \omega_{me}/\omega_o$. Thus, this term exactly plays a role of the effective ME parameter:

$$\zeta_{AFM} = \omega_{me}\omega_E/\omega_o^2 = (\omega_{me}/\omega_o)(\omega_E/\omega_o) . \quad (33)$$

The second factor in (33) just displays the exchange enhancement of dynamic ME effects in antiferromagnet as compared ferromagnet.

The origin of this enhancement is very clear from the expression for φ in (31). The denominator has a very small magnetization $M \ll 2M_o$ (at $H \ll H_E$) by means of which the field H keeps all spin construction. At the same time the elastic field e_{xy} acts directly through the antiferromagnetism vector with $|\vec{M}_1 - \vec{M}_2| \gg M$. In ferromagnet the formula (31) for φ would include its total magnetization $M=M_o$ which was kept by the field H in that case.

A strict calculation must take into account spontaneous ME strains in the ground state, and this gives the only change in ω_o

$$\omega_o^2 = \gamma^2 H^2 + \omega_{me}\omega_E \qquad (34)$$

and maintains (32) and (33). More speaking, these expressions (32)-(34) are right also for EP antiferromagnets if the easy plane is just xy and one treats a quasiacoustic mode with $\vec{k} \parallel \vec{H} \parallel \vec{x}$, polarized along y. An additional term appears in (34) if there is anisotropy in EP, crystallographic or induced by pressure. At any rate, all such a kind terms as a whole together with the first one in (34) turn into zero at a PT point (on temperature, field, pressure, etc.). Thereafter, (34) results in $\omega_o^2 = \omega_{me}\omega_E$ and $\zeta = 1$.

EP antiferromagnets (including α-Fe_2O_3 and $FeBO_3$) are just interested because of a small anisotropy in EP. This provides a great influence of H on ω_o, and thereby on ζ and a sound velocity in such a material. It should be the same for other antiferromagnets near various orientation PT points. More details see in [22].

A great interest has a gigantic elastic anharmonicity in antiferromagnets discovered by Ozhogin and Preobrazhenskii [30]. Effective nonlinear (anharmonic) elastic constants caused by ME coupling can exceed usual ones (without ME coupling) by two and more orders of value. The reason is again an exchange enhancement. For example, one could obtain an anharmonicity if kept in (30) terms of the next (after second) orders in φ. However, a more rigorous theory has to be developed, thereby taking into account the effects of rotational invariance as well.

Among various concrete nonlinear effects of such an origin I will mention only the self-focussing of the sound wave recently predicted in [30b].

APPLICATIONS

Below there is given a short list of some newest perspective applications of the coupled ME modes.

Harmonic magneto-elastic modes

1) A strong field dependence of the velocity of quasi-acoustic waves allows to construct a magneto-acoustic resonator with a smooth change of its resonant frequencies by the applied magnetic field. The EP antiferromagnets (with a weak ferromagnetism) α-Fe_2O_3 and $FeBO_3$ are just perspective materials for such applications. There is the communication [31] that on this principle the resonator has been built with a frequency change up to 60% (that is, within one octave), and a quality Q up to 10^5.

Such kind resonators and sound conductors can find a wide use in Electronics:

- for narrow-band filters and autogenerators [32] in the frequency region 100 kHz to 100 MHz;
- as a magneto-acoustic converter of radio-signals with controllable phase and frequency shifts, etc.[33];
- as portable autogenerator sensors for the measuring of magnetic fields and temperatures [34].

2) Inhomogeneous oscillations of magnetic axes for quasi-acoustic modes can be a reason of the acoustic modulation of the optical magnetic birefringence. This effect has to be rather perspective for magnetooptic applications [35].

Non-linear interactions of sound and spin waves

The materials with a gigantic magneto-elastic anharmonicity (similar to α-Fe_2O_3) can find various and unexpected applications. As an example, in [36] some devices are suggested for the correlation processing of radio-signals (magneto-acoustic convolvers). Many interesting possibilities are connected with a confluence of two acoustic signals which have equal (or close) wave vectors in magnitude but opposite in direction.

CONCLUSION COMMENTS

When considering ME modes, the damping of those has not been paid any attention. Certainly, the damping has been taking into account in original papers by including the simplest relaxation terms in the motion equations (see,[22]). And it occours that in many cases (even near PT points) the damping actually does not cancel ME waves even though a detail picture of the phenomena changes noticeably. Nevertheless, the problem can not be declared as totally solved. The development is necessary in the two directions: an application of the phenomenological first principle's theory by Maugin, and a consideration of concrete mechanisms of relaxation.

In the year 1964 when there has been discovered [28, 29] the ME gap in hematite and proved convincingly its reason to be a spontaneous striction, the paper by Higgs [37] appears concerning a new mechanism of the mass formation in the theory of elementary particles. Being from quite remote fields of physics, these two phenomena turn out to be very similar ones with some general theoretical point of view: the both are effects of SSB in the system of two coupled fields [38]. The first effect of such a kind discovered in physics (1933) seems to be Meissner's effect in superconductors. However, there can be no doubt that many other dynamic phenomena in nonsingle-component systems with phase transitions can be considered as analogies of ME ones spoken above. An especially close analogy exists with a mixing of the soft optic ferroelectric mode and the elastic acoustic one (the pseudospin-phonon coupling) at the phase transition "order ↔ disorder" (see, for example,[39]). Even the existance in itself of acoustic and optic modes in solids can be described as a result of symmetry breaking at the atomic ordering.

And lastly, the conceptions of symmetry breaking seem to be rather useful in dynamics of nonuniform systems such as ME oscillations of domain walls in ferro- and antiferromagnets [40].

REFERENCIES

[1] Spencer, E.G. and Le Craw, R.C., Phys. Rev. Letters 1 (1958), 241.

[2] Turov, E.A. and Irkhin, Yu.P., Fizica Metallov i Metallovedenie 3 (1956) 15-17 (in Russian).

[3] Akhiezer, A.I., Conf. on Physics of Magnetic Phenomena, Summaries of the reports, Moscow, May 1956 (Ac.Sci.USSR, Moscow 1956) p.27.
[4] Turov, E.A., Ibid, p.54-56.
[5] Akhiezer, A.I., Bar'yakhtar, V.G. and Peletminskii, S.V., Zh. Exsp. Teor. Fiz. 35 (1958) 228-239.
[6] Kittel, C., Phys. Rev. 110 (1958) 836.
[7] Akhiezer, A.I., Journ. Phys. USSR 10 (1946) 217-230.
[8] Strauss, W., in Physical Acoustics, Ed. W.P.Mason (Acad. Press, New York - London, 1968), vol.IYB, ch.5.
[9] Lemanov, V.V., in Fizika Magnitnih Dielektrikov, Ed.G.A.Smolenskii (Ac.Sci.USSR, Leningrad 1974), ch.4.
[10] Vlasov, K.B., Fiz.met. i Metalloved 4 (1957) 542-544; Abstracts of Conference on Magnetic material Physics (in Russian) Leningrad 1957 (Ac.Sci.USSR, Leningrad-Moscow 1957), p.3.
[11] Matthews, H. and Le Craw, R.C., Phys.Rev.Letters 8 (1962) 397
[12] Le Craw, R.C. and Comstock, R.L., in Physical Acoustics, Ed. W.P.Mason (Acad.Press, New York - London 1965), p.127.
[13] Vlasov, K.B. and Ishmuhametov, B.H., Zh. Exp. Teor. Fiz. 46 (1964) 201-212.
[14] Tirsten, H.F., J. Math. Phys. 5 (1964) 1298.
[15] Eastman, D.E., Phys. Rev. 148 (1966) 530-542.
[16] Akhiezer, A.I., Bar'yakhtar, V.G. and Peletminskii, S.V., Spinovie volni (Nauka, Moskva 1967); English translation: Spin Waves (Amsterdam 1968) ch. IY.
[17] Melcher, R.L., Phys. Rev. Lett. 25 (1970) 1201; in the book "Lectures in Fermi's school", Varena, Curse LII, Ed.E. Burstein (Acad. Press, New York - London 1972) 258.
[18] Maugin, G.A., Int. J. Eng. Sci. 17 (1979) 1073.
[19] Sioke-Rainaldy, J. and Maugin, G.A., J.Appl.Phys. 54 (1983) 1490-1506; Maugin, G.A. and Sioke-Rainaldy, J., Ibid, 1507-1518.
[20] Schlömann, E., Advances in Quantum Electronics (New York 1961) 444.
[21] Strauss, W., Physical Acoustics (Ed. By W.P.Mason 1968) vol. IYB, ch.5.
[22] Turov, E.A. and Shavrov, V.G., Usp. Fiz. Nauk, SSSR 140 (1983)
[23] Peletminskii, S.V., Zh. Exp. Teor. Fiz. 37 (1959) 452 (Sov. Phys. JETP 10 (1960) 321.
[24] Turov, E.A. and Shavrov, V.G., Fiz.Tverd.Tela 7 (1965) 217.
[25] Nielsen, M., Bjerrum Møller, H., Lindgard, P.A. and Mackintosh, A.R., Phys.Rev. Lett. 25 (1970) 1451.
[26] Chow, H and Keffer, F., Phys. Rev. B7 (1973) 2028.
[27] Jensen, J and Palmer, S.B., J.Phys. C12 (1979) 4573.
[28] Borovik-Romanov, A.S. and Rudashevskii, E.G., Zh. Exp. Teor. Fiz. 47 (1964) 2095.
[29] Iida, B. and Tasaki, A., Proc. ICM, Nottingam (1964) 538.
[30] a) Ozhogin, V.I. and Preobrazhenskii, V.P., Zh. Exp. Teor. Fiz. 73 (1977) 998; b)Ozhogin, V.I., Manin, D.Yu., Petriashvili, V.I. and Lebedev, A.Yu., Degest of Intermag'83 (April 1983, Philadelphia) paper EE5.
[31] Evtikheev, N.N., Pogozhev, S.A., Preobrazhenskii, V.L. and Ekonomov, N.A., Voprosi radioelektroniki, ser. OF 5 (1981) 87.
[32] Evtikheev, N.N., Pogozhev, S.A, Preobrazhenskii, V.L., Savchenco, M.A. and Ekonomov, N.A., Ibid, p.96.
[33] Berezhnov, V.V., Evtikheev, N.N., Preobrazhenskii, V.L. and Ekonomov, N.A., Radiotekhnika i elektronika 26 (1983) 2.
[34] Evtikheev, N.N., Pogozhev, S.A., Preobrazhenskii, V.L., Shumilov, V.N. and Ekonomov, N.A., Novie elementi i metodi rascheta informatzionnikh sistem (Ed. MIREA, Moskva 1980) 35.

[35] Evtikheev, N.N., Moshkin, V.V., Preobrazhenskii, V.L. and Ekonomov, N.A., Pis'ma v ZHETF 35 (1982) No 1, 31.
[36] Berezhnov, V.V., Voprosi radioelektroniki, ser. OF (1982) 121.
[37] Higgs, P., Phys. Rev. Lett. 13 (1964) 508.
[38] Turov, E.A., Taluts, G.G. and Lugovoi, A.A., JMMM 15-18 (1980) 582-584.
[39] Blinc, R. and Žekš, B., Soft modes in ferroelectrics and antiferroelectrics (North-Holland, 1974) ch.5, § 5.
[40] Turov, E.A. and Lugovoi, A.A., Fiz. Met. Metalloved. 50 (1980) 717, 903; Pis'ma v ZHETF 31 (1980) 308.

The Mechanical Behavior of Electromagnetic Solid Continua
G.A. Maugin (editor)
Elsevier Science Publishers B.V. (North-Holland)

GYROMAGNETIC PHENOMENA IN MAGNETO-ELASTICITY

J.B. Alblas

Eindhoven University of Technology
Department of Mathematics and Computing Science
Eindhoven
The Netherlands

The basic equations for the interaction of magnetic and elastic waves are given. Some boundary value problems, e.g. for the plate, are treated. The results may be of interest for the interpretation of magnetic resonance phenomena.

1. INTRODUCTION

In this paper we are concerned with the interaction of spin and elastic waves in an infinite body, a half space and an infinite plate of finite thickness. The body is magnetized to saturation and the magnetization vector $\underline{M}$ is perpendicular to the faces of the plate. We take the 3($\equiv$z)-axis of the Cartesian coordinate system along the direction of magnetization. The crystalline material of the body has cubic symmetry and the edges of the cube are along the $\underline{x}$-axes.

We distinguish three states of the body: the rest state, $\underline{X}$, the intermediate state, $\underline{\xi}$, of constant magnetization and the final state, $\underline{x}$, which is considered as an infinitesimally small dynamic perturbation on the intermediate state. To this state we apply the general theory, as has been given in [1], Chapter 10. For the field and balance equations, we refer to [1], especially (10.58) (with $\underline{J} = \underline{0}$) and (8.31). The boundary and jump conditions are (9.9), (9.17) and (4.8) of [1]. With a few exceptions that will be stated explicitely, we use the notation of [1], see also [2], Section 6.

The perturbations $\underline{m}$ of the magnetization and $\underline{h}$ of the field are given in this paper by

$$m_k(\underline{x}) = M_k(\underline{x}) - M_k^{(0)}(\underline{\xi}) = M_k(\underline{x}) - M_k^{(0)}(\underline{x}) \ , \tag{1.1}$$

$$h_k(\underline{x}) = H_k(\underline{x}) - H_k^{(0)}(\underline{\xi}) = H_k(\underline{x}) - H_k^{(0)}(\underline{x}) \ , \tag{1.2}$$

in contrast to (10.61) and (10.63) of [1], respectively.

We shall introduce a damping of the form (cf. [1], (8.12))

$$G_k = R_k - \eta(\dot{M}_k - e_{klm}\Omega_l M_m) \ , \tag{1.3}$$

as has been proposed by Gilbert [3]. In (1.3) $\underline{\Omega}$ is the material angular velocity.

We derive the constitutive equations from ψ, given by

$$\psi = \tfrac{1}{2}\alpha A_{\alpha\beta}A_{\alpha\beta} + \tfrac{1}{2}k\left[\bar{M}_1^2\bar{M}_2^2 + \bar{M}_2^2\bar{M}_3^2 + \bar{M}_3^2\bar{M}_1^2\right]$$

$$+ \tfrac{1}{2}b_{ijkl}\bar{M}_i\bar{M}_jE_{kl} + \tfrac{1}{2}c_{ijkl}E_{ij}E_{kl} \, , \tag{1.4}$$

with

$$\bar{M}_\alpha = x_{k/\alpha}M_k \, . \tag{1.5}$$

Cubic symmetry is expressed by

$$b_{1111} = b_{2222} = b_{3333} = b_1 \, ,$$

$$b_{1122} = b_{2233} = \;\dots\; = b_2 \, ,$$

$$b_{1212} = b_{2121} = \;\dots\; = b_3 \, , \tag{1.6}$$

the other coefficients being zero, and corresponding coefficients c_1, c_2, c_3.

We have calculated the deformations in the intermediate state. They affect the coefficients in the final state. For simplicity we shall neglect them.

2. THE DIFFERENTIAL EQUATIONS

We assume the quantities to be proportional to $e^{-i\omega t}$ and find then

$$(\tfrac{1}{2}b_2M_0^2+c_1)u_{x,xx} + (\tfrac{1}{2}b_2M_0^2+c_3)u_{x,yy} + (\tfrac{1}{2}b_1M_0^2+c_3-\tfrac{1}{4}i\omega\eta M_0^2)u_{x,zz} +$$

$$+ (c_2+c_3)u_{y,yx} + (c_2+c_3+b_2M_0^2+b_3M_0^2+\tfrac{1}{4}i\omega\eta M_0^2)u_{z,xz} +$$

$$+ (b_3M_0+\tfrac{1}{2}i\omega\eta M_0)m_{x,z} + \mu_0M_0\varphi_{,xz} + \omega^2u_x = 0 \, , \tag{2.1}$$

and a corresponding equation for u_y,

$$(kM_0^4+\tfrac{1}{2}b_2M_0^2+2b_3M_0^2+c_3-\tfrac{1}{4}i\omega\eta M_0^2)(u_{z,xx}+u_{z,yy}) + (\tfrac{5}{2}b_1M_0^2+c_1)u_{z,zz} +$$

$$+ (c_2+c_3+b_2M_0^2+b_3M_0^2+\tfrac{1}{4}i\omega\eta M_0^2)(u_{x,xz}+u_{y,yz}) +$$

$$+ (kM_0^3+b_3M_0-\tfrac{1}{4}i\omega\eta M_0)(m_{x,x}+m_{y,y}) + \mu_0M_0\varphi_{,zz} + \omega^2u_z = 0 \, , \tag{2.2}$$

$$\varphi_{,kk} + \rho_0(m_{x,x}+m_{y,y}) - \rho_0 M_0 u_{p,pz} = 0 \ , \quad \text{in } V^- \ , \tag{2.3}$$

$$\varphi_{,kk} = 0 \ , \quad \text{in } V^+ \ , \tag{2.4}$$

$$\alpha m_{x,kk} + \mu_0\varphi_{,k} - (kM_0^2 + \frac{\mu_0 H_0}{M_0} - i\omega\eta)m_x +$$

$$- (kM_0^3+b_3M_0-\tfrac{1}{2}i\omega\eta M_0)u_{z,x} - (b_3M_0+\tfrac{1}{2}i\omega\eta M_0)u_{x,z} + \frac{i\omega}{\Gamma M_0}m_y = 0 \ , \tag{2.5}$$

with a corresponding equation for m_y.

The boundary and jump conditions are

$$\varphi^+ - \varphi^- = -\rho_0 M_0 u_z \quad ; \quad \left(\frac{\partial\varphi}{\partial z}\right)^+ = \left(\frac{\partial\varphi}{\partial z}\right)^- \ , \tag{2.6}$$

$$m_{x,z} = m_{y,z} = 0 \ , \tag{2.7}$$

$$\delta T_{\alpha 3} + T^*_{\alpha 3} = -\tfrac{1}{2}\mu_0\rho_0^2 M_0^2 u_{z,x} \quad , \quad \alpha = 1,2 \ ,$$

$$\delta T_{33} + T^*_{33} = -\tfrac{1}{2}\mu_0\rho_0^2 M_0^2(u_{p,p}+u_{z,z}) \ . \tag{2.8}$$

3. SPECIAL CASES

a) We first take $\frac{\partial}{\partial x} = \frac{\partial}{\partial y} = 0 \ , \ \frac{\partial}{\partial z} \neq 0 \ .$ (3.1)

We restrict ourselves to the boundary value problem of the plate

$$-h \leq z \leq h \ , \tag{3.2}$$

and neglect damping. With (3.1) the system (2.1) to (2.5) splits up into three systems for (u_z,φ), (m^+,u^+) and (m^-,u^-), where

$$m^{\pm} = m_x \pm im_y \quad ; \quad u^{\pm} = u_x \pm iu_y \ . \tag{3.3}$$

— The eigenvalues for the longitudinal vibrations (u_z,φ) are

$$\omega_k = (2k+1)\frac{\pi}{2h}\sqrt{\frac{5}{2}b_1 M_0^2 + c_1 + \mu_0\rho_0 M_0^2} \ , \quad (k = 1,2,\ldots) \ . \tag{3.4}$$

The dispersion equations for (m^+,u^+) and (m^-,u^-) are respectively

$$(\omega^2 - \omega_1^2)(\omega - \omega_0) - \Gamma M_0^3 b_3^2 \lambda^2 = 0 \ ,$$

$$(\omega^2 - \omega_1^2)(\omega + \omega_0) + \Gamma M_0^3 b_3^2 \lambda^2 = 0 \ , \tag{3.5}$$

where

$$\omega_1^2 = \lambda^2(\tfrac{1}{2}b_1 M_0^2 + c_3) \ ,$$

$$\omega_0 = \Gamma M_0\left(\alpha\lambda^2 + kM_0^2 + \frac{\mu_0 H_0}{M_0}\right) \ , \tag{3.6}$$

and λ is the wave number. We only consider the boundary value problem for (m^+,u^+).

The equation (3.5') is a biquadratic equation for λ. If

$$\omega > \Gamma M_0\left(kM_0^2 + \frac{\mu_0 H_0}{M_0}\right) \ , \tag{3.7}$$

there are two positive roots λ_1^2 and λ_2^2, which may be written as

$$\lambda_1^2 = \frac{\omega^2}{\tfrac{1}{2}b_1 M_0^2 + c_3} + f_1(\omega) \ ,$$

$$\lambda_2^2 = \frac{\omega}{\alpha\Gamma M_0} - \left(\frac{kM_0^2}{\alpha} + \frac{\mu_0 H_0}{\alpha M_0}\right) + f_2(\omega) \ , \tag{3.8}$$

with $|f_k(\omega)|$ very small. We write the solution in the form

$$m(= m^+) = A_1 \cos \lambda_1 z + A_2 \cos \lambda_2 z \ ,$$

$$u(= u^+) = B_1 \sin \lambda_1 z + B_2 \sin \lambda_2 z \ . \tag{3.9}$$

To satisfy the boundary conditions for $|z| = h$, λ_1 and λ_2 have to meet the equation

$$\frac{\lambda_2}{\lambda_1}\,\frac{\tan \lambda_2 h}{\tan \lambda_1 h} = \frac{\omega^2 - \lambda_1^2(\tfrac{1}{2}b_1 M_0^2 + c_3)}{\omega^2 - \lambda_2^2(\tfrac{1}{2}b_1 M_0^2 + c_3)} \ , \tag{3.10}$$

a transcendental equation for ω, which may be expected to have an infinite number of roots. If (3.7) is not satisfied, λ_2 is imaginary and the relation (3.10) becomes

$$\frac{\lambda_2}{\lambda_1}\frac{\tanh \lambda_2 h}{\tanh \lambda_1 h} = \frac{\lambda_1^2(\frac{1}{2}b_1M_0^2 + c_3) - \omega^2}{\lambda_2^2(\frac{1}{2}b_1M_0^2 + c_3) + \omega^2} . \qquad (3.11)$$

b) For

$$\frac{\partial}{\partial x} \neq 0 , \quad \frac{\partial}{\partial y} = \frac{\partial}{\partial z} = 0 , \qquad (3.12)$$

the system splits up into systems for (u_x, u_y) and (u_z, φ, m_x, m_y). The dispersion equation for the latter system is

$$(\omega^2-\omega_2^2)(\omega^2-\omega_0^2-\omega_0\Gamma M_0\rho_0\mu_0) - \lambda^2\Gamma M_0\omega_0(kM_0^3 + b_3M_0)^2 = 0 , \qquad (3.13)$$

where

$$\omega_2^2 = \lambda^2(kM_0^4 + \tfrac{1}{2}b_2M_0^2 + 2b_3M_0^2 + c_3) . \qquad (3.14)$$

c) Boundary value problems for the plate and the half space of a more general type e.g., with

$$\frac{\partial}{\partial x} \neq 0 , \quad \frac{\partial}{\partial z} \neq 0 , \quad \frac{\partial}{\partial y} = 0 , \qquad (3.15)$$

become very involved. They may be treated with the method of 3a. However, approximate solutions may easily be obtained. Results show the existence of magneto-elastic surface waves.

REFERENCES

[1] Alblas, J.B., General Theory of Electro- and Magneto-Elasticity, in Electromagnetic Interactions in Elastic Solids, ed. H. Parkus, C.I.S.M. Courses and Lect. 257, Springer-Verlag, Wien etc. 1979.

[2] Alblas, J.B., Electro-Magneto-Elasticity, Topics in Appl. Cont. Mech., eds. J.L. Zeman and F. Ziegler, Springer-Verlag, Wien etc. 1974.

[3] Gilbert, T.L., Phys. Rev. 100, 1243, (1953); See also W.F. Brown, Micromagnetics, Intersc. Publ., New York etc. 1963.

The Mechanical Behavior of Electromagnetic Solid Continua
G.A. Maugin (editor)
Elsevier Science Publishers B.V. (North-Holland)

INFLUENCE OF STRUCTURAL DEFECTS ON MAGNETOMECHANICAL DAMPING OF HIGH PURITY IRON

J. Degauque, B. Astié

Laboratoire de Physique des Solides,
associé au C.N.R.S.
I.N.S.A., Avenue de Rangueil, 31077 Toulouse-Cedex
France

It is shown that the magnetomechanical damping due to irreversible motions of 90° magnetic domain walls induced by a vibrational stress is strongly attenuated by defects such as interstitial carbon, dislocation tangles and boundaries associated with small grains.

INTRODUCTION

Characteristically most ferromagnetic materials present a strong damping of their mechanical vibrations. This behaviour is due to the non reversible motions of 90° magnetic domain walls (DW's) under the alternating vibrational stress, which induces magnetomechanical hysteresis energy losses ΔU_{mag} (1, 2). This magnetomechanical damping Q^{-1}_{mag} is the difference between the damping in the non magnetized state and the damping under a magnetic field H which is high enough to cancel out any magnetic contribution ; then :

$$Q^{-1}_{mag} = \frac{1}{\pi}(\delta_{H=0} - \delta_{H=1000\ Oe}) = \frac{1}{2\pi}\frac{\Delta U_{mag}}{U}$$

where δ is the logarithmic decrement of the mechanical vibration, U is the vibrational elastic energy.

When it is measured as a function of the alternative shear stress amplitude τ, Q^{-1}_{mag} exhibits a well marked maximum. The position τ(M) and the height Q^{-1}_{mag}(M) of this maximum can be directly related to the average level of the internal stress field σ_i which the 90° DW's have overcome to make large irreversible jumps (3). Actually, σ_i should be considered as an effective internal stress rather than an internal stress in the usual sense because the defects interact with DW's not only by their stress fields but also by their stray magnetic fields (4).

Magnetomechanical damping has a technological application in the use of high damping capacity materials. Such materials are employed in the machine part subjected to high vibration to reduce vibration and noise pollution (5). The aim of this paper is to present the influence of defects of different kind (interstitial and precipitated carbon, dislocations, grain boundaries) on the behaviour of the magnetomechanical damping of high purity iron. As it will be shown here, σ_i and $1/Q^{-1}_{mag}$(M) (which show almost the ability of 90° DW's to perform irreversible jumps) are strongly dependent on the nature, density and arrangement of the structural defects.

EXPERIMENTAL TECHNIQUES

The material studied is polycristalline iron which has undergone zone melting and contains less than 20 weight ppm of impurity atoms. After room temperature (R.T.) rolling, specimens of rectangular shapes (80 x 10 x 0.5 mm^3) are cut with the largest dimension parallel to the rolling direction. Damping experiments are performed at R.T. on a inverted torsion pendulum for almost continuous values of the shear stress ($4 \times 10^{-2} < \tau < 1.4$ daN/mm^2), applied at the surface of specimen. The Q^{-1}_{mag} values are corrected in order to take into account the non uniformity of τ through the sample (6).

EXPERIMENTAL RESULTS

1. Effect of interstitial carbon and precipitated carbides :

Carbon is introduced in iron samples by heating for 12 hours in a mixture of CH_4-H_2 at 750°C. After a solution treatment for 17 hours at 720°C in vacuum, the interstitital carbon content is determined by the Snoek peak method. The average grain size of samples was about 0.8 mm.

Figure 1 shows the behaviour of $Q^{-1}_{mag} = f(\tau)$ for different contents of interstitial carbon : 15 wt ppm (curve a), 130 wt ppm (curve b) and 180 wt ppm (curve c) : the maximum intensity Q^{-1}_{mag} (M) decreases and shifts to higher τ.

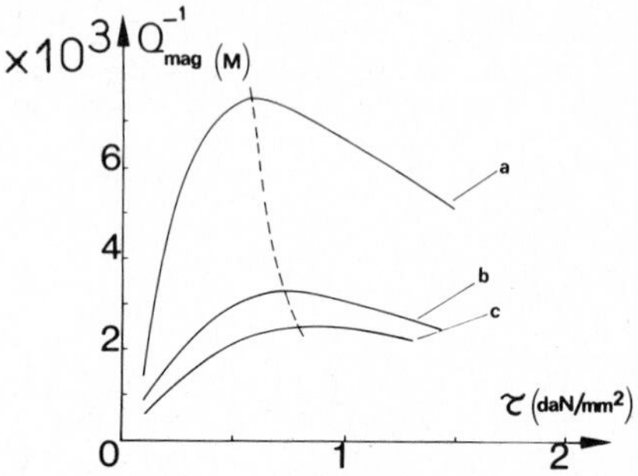

Figure 1
Magnetomechanical damping versus shear stress amplitude for three Fe-C alloys : a : 15 ; b : 130 ; c : 180 wt.ppm interstitial carbon.

For a specimen of 180 wt ppm carbon content and for two distinct temperatures θ_a of isothermal ageing, the plots of $1/Q^{-1}_{mag}$, and σ_i as a function of time are given in figures 2 and 3. At θ_a = 40°C, $1/Q^{-1}_{mag}$(M) shows a marked decrease at $t > 100$ hours, while σ_i shows a gradual increase. At θ_a = 280°C, $1/Q^{-1}_{mag}$(M), exhibits at first a small decrease, followed by a very high maximum at $t \simeq 2$ min ; the variation of σ_i shows a gradual increase followed, for $t > 6$ min by a slight decrease.

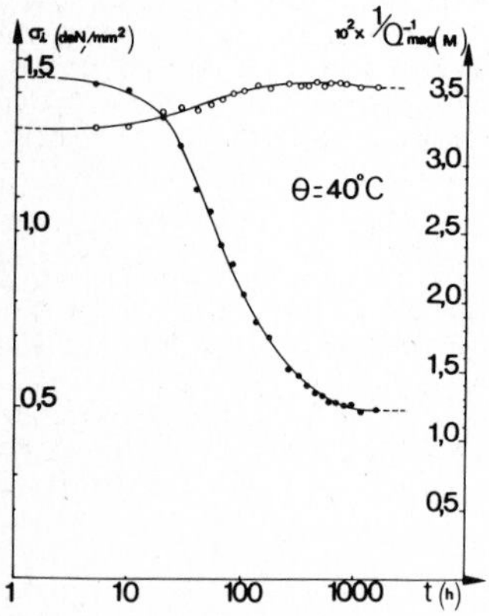

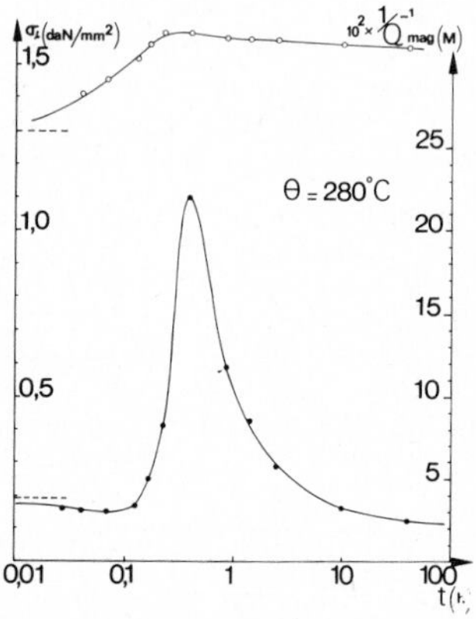

Figure 2 Figure 3
Effective internal stress σ_i (o) and reciprocal magnetomechanical damping maximum $1/Q^{-1}_{mag}$(M) (●) as a function of time for the Fe-180 wt ppm C alloy annealed at 40°C (Fig. 2) and 280°C (Fig. 3).

2. Influence of dislocations

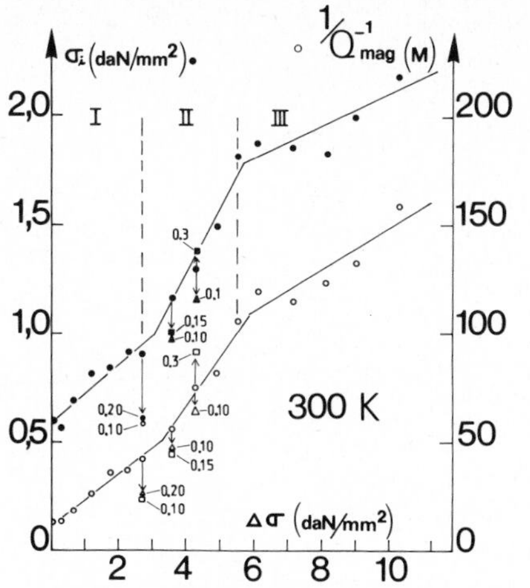

Figure 4
σ_i (●) and $1/Q^{-1}_{mag}$(M) (○) versus stress hardening $\Delta\sigma$ for straining at 300 K and additional straining at 77 K.
The numerical values give the increments (in %) of the additional strain at 77 K.

After an annealing of 12 hours at 750°C in an atmosphere of purified hydrogen, tensile samples (average grain size about 0.8 mm) are strained at a strain rate of 3 x 10^{-5} s^{-1}, following two procedures (Fig. 4) :

a) Straining at 300 K :
The plots of σ_i and $1/Q^{-1}_{mag}$(M) as a function of the stress hardening $\Delta\sigma = \sigma - \sigma_o$ (where σ is the applied stress value, σ_o is the yield stress at the elastic limit) show three linear stages corresponding to three ranges : I ($0 < \Delta\sigma < 2.5$ daN/mm^2) ; II ($2.5 < \Delta\sigma < 5.5$ daN/mm^2) and III ($5.5 < \Delta\sigma < 10.3$ daN/mm^2).

b) Prestraining at 300 K + straining at 77 K :
After prestraining at R.T. in the range II, at $\Delta\sigma$ = 2.6, 3.6 and 4.3 daN/mm^2 respectively, three samples are further strained at 77 K under small applied stresses, lower than the low stress at this temperature. These strain increments at 77 K induce a decrease of σ_i and $1/Q_{mag}$(M) except for one point at $\Delta\sigma$ = 4.3 daN/mm^2 where the applied stress at 77 K exceeds the flow stress.

3. Influence of the grain size

Before the final rolling, the iron ingot undergoes two strong plastic compressions along three perpendicular directions and two annealings, to create a quite homogeneous intermediate grain size. After final rolling, samples are subjected to one of the two following recrystallisation thermal processings :

A : $500 < T < 800$°C during $t \lesssim 10$ hours
B : $440 < T < 550$°C during $t \simeq 60$ hours.

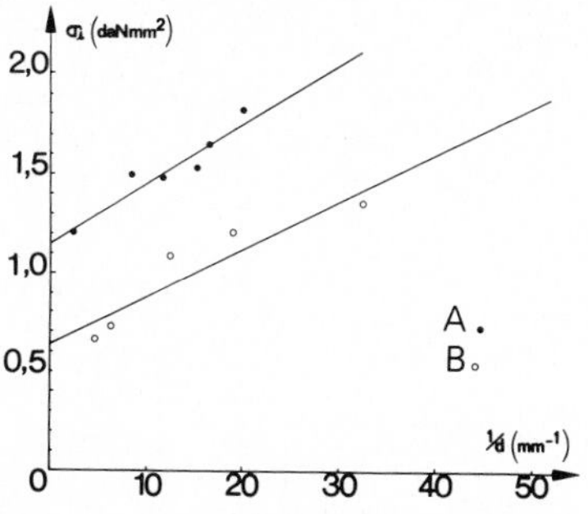

Figure 5

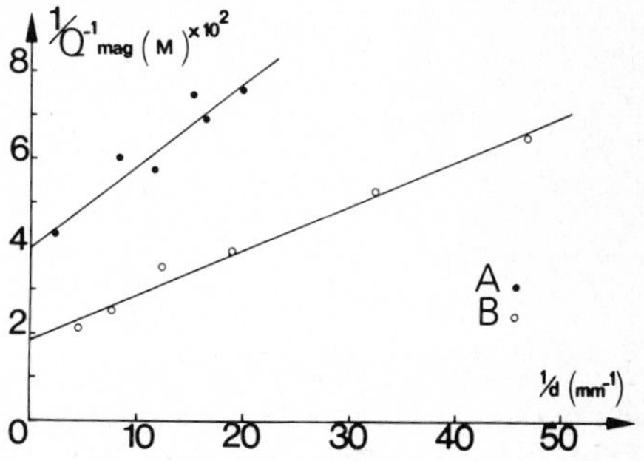

Figure 6

σ_i (Fig. 5) and $1/Q^{-1}_{mag}$(M) (Fig. 6) as a function of reciprocal grain size 1/d for the two sets of samples A and B.

These processings give samples with homogeneous grains size d, ranging between 0.21 and 0.425 mm. Figures 5 and 6 show that σ_i and $1/Q^{-1}_{mag}(M)$ are linear functions of the density of grain boundaries 1/d, but the two sets of samples A and B give clearly different curves.

DISCUSSION

1. Effect of carbon

Interstitial atoms induce a restoring force, proportional to their concentration, which strongly reduces the mobility of 90 DW's for long range irreversible displacements (7, 8) The beginning of precipitation is connected with "release effect" due to decrease of the restoring force and to the DW's stabilization. This effect is not counterbalanced by the anchoring effect of ε carbide precipitation which appears at θ_a = 40°C and during the first stage at θ_a = 280°C. At the second stage of precipitation at 280°C, which corresponds to the cementite, the anchoring effect is efficient because the size of the precipitates becomes comparable with the DW's width and the damping decreases (i.e. $1/Q_{mag}(M)$ increases). The damping increase which appears at the end of ageing may be linked with the growing of large precipitates of cementite, which increase the total area of the 90° DW's because the formation of new closure domains.

2. Effect of dislocations

a) Straining at 300 K : The strain hardening $\Delta\sigma$ is related to the dislocation density N by (9) : $\Delta\sigma = \alpha\,\mu\, b(N^{1/2} - N_0^{1/2})$ where μ is the shear modulus, b the Bürgers vector of dislocations, N_0 the dislocation density at the flow stress σ_0 and $\alpha \simeq 1$. According to our high-voltage TEM observations, stage I is characterized by the creation of mixed and isolated dislocations distributed in a rather homogeneous fashion. Dislocation tangles which appear in stage II, evolve in stage III into a cellular structure made of high density dislocation walls. Dislocation tangles and cellular structure, by reducing the 90° DW's mobility, induce a strong increase of σ_i and $1/Q_{mag}(M)$.

b) Straining at 77 K : Under low stresses applied at 77 K, edge or mixed dislocations segments leave tangles, trailing behind them long sessile screw dipoles (10) $\langle 111 \rangle$. The decrease of dislocation density inside tangles allows an intensification of the irreversible motion of 90° DW's, which induces an increase of the magnetic damping.

3. Effect of grain size

The damping is reduced by decreasing the grain size. The difference in crystallographic orientation between different grains leads to a discontinuity of the normal component of the magnetization accross the grain boundaries and free magnetic poles will appear in the boundaries. These poles give rise to demagnetizing fields which immobilize DW's thereby increasing σ_i and $1/Q_{mag}(M)$. Closure domains (making a 90° angle with each other) would emanate from grain boundaries to relieve the magnetic pole density, which leads to a (1/d) law (11), fitted by our experimental values. Grain boundaries can also interfere through associated dislocations. TEM observations, made on samples A and B, show that the two different types of preparation lead to substantially higher density of isolated dislocations inside the grains of samples A. This leads to two experimental levels of σ_i and $1/Q_{mag}(M)$ observed, as a result of 90° DW's strong sensibility to the long range effect of dislocations. Measurements of magnetic quantities (Rayleigh parameters, coercive force) for which mainly 180° DW's are active, show only a level of variation for the same experimental samples (12). For this case, the pinning action arises mainly from grain boundaries and rarely from the dislocations, which shows a different behaviour for 180° and 90° DW's.

CONCLUSION

Structural defects can strongly attenuate the magnetomechanical damping. Interstitial carbon, dislocation tangles and boundaries associated with small grains are the most efficient defects on the lowering of 90° DW's mobility. This results are confirmed by direct High Voltage Electron Microscopy of the interaction between defects and DW's made by displaced aperture method (13) and by Lorentz method (14). These observations show a very good agreement with the macroscopic behaviour of 90° DW's (15).

REFERENCES

|1| Nowick, A.S. and Berry, B.S., Anelastic relaxation in crystalline solids (Academic Press, New-York, 1972).

|2| Degauque, J. and Zarembowitch, A., J. Physique, Europhys. Jrnl. C5, 42 (1981) 607.

|3| Smith, G.W. and Birchak, J.R., J. Appl. Phys., 41 (1970) 3315.

|4| Néel, L., J. Phys. Radium, 9 (1948) 185.

|5| Sugimoto, K., Mem. Inst. Sci. Ind. Res., Osaka Univ., 35 (1978) 31.

|6| Degauque, J., Thesis, P. Sabatier Univ., Toulouse (February 1977).

|7| Néel, L., J. Phys. Radium, 13 (1952) 249.

|8| Astié, B. and Degauque, J. in : Smith, C.C. (Ed.), Internal Friction and Ultrasonic attenuation in solids (Pergamon Press 1980).

|9| Keh, A.S. and Weissmann, S., Electron microscopy and the strength of crystal (Interscience, New-York, 1963).

|10| Degauque, J., Astié, B. and Kubin, L.P., J. Appl. Phys., 50 (1979) 2140.

|11| Goodenough, J.B., Phys. Rev., 95 (1954) 917.

|12| Degauque, J., Astié, B., Porteseil, J.L. and Vergne, R., J.M.M.M., 26 (1982) 261.

|13| Degauque, J. and Astié, B., Phys. Stat. Sol. (a), 74 (1982) 201.

|14| Taylor, R.A., Jakubovics, J.P., Astié, B. and Degauque, J., J.M.M.M., 31-34 (1983) 970.

|15| Taylor, R.A., Jakubovics, J.P., Astié, B. and Degauque, J., to be published.

The Mechanical Behavior of Electromagnetic Solid Continua
G.A. Maugin (editor)
Elsevier Science Publishers B.V. (North-Holland)

DOMAIN PATTERN VISUALIZATION AND MAGNETOSTRICTION

J. Miltat

Laboratoire de Physique des Solides
Bât. 510 - Université Paris-Sud
91405 Orsay (France)

Experimental techniques commonly used for magnetic domain patterns visualization are reviewed. Specificities of the various methods are briefly discussed and applications examples presented.

INTRODUCTION

Although obviously linked, two aspects in the study of domain patterns may be separated, namely domain pattern recognition as well as domain wall characterization and magnetostrictive deformations analysis.

In materials with non vanishing anisotropy, in order to decrease the magnetostatic energy arising from surface charges, the magnetic body is divided into domains within which the magnetization is more or less uniform. The magnetization distribution within the transition regions between domains is primarily governed by the competition between anisotropy and exchange energies, yielding for instance the classical Bloch or Néel wall structures. Walls, however, often bear spin line defects called Néel or Bloch lines ; lines themselves may bear point defects called Bloch or Néel points. Therefore, three levels in the domain pattern recognition should in principle be considered : domain pattern geometry, wall characteristics, lines and points configurations.

In non zero magnetostriction materials, specific domain patterns may lead to incompatible deformations of the elastic continuum. Additional stresses and strains must relax those incompatibilities. It is obvious that magnetostrictive strains will usually be small because of the low values of the magnetostriction constants. Therefore, only those imaging techniques with a high strain sensitivity will be appropriate.

DOMAIN PATTERN VISUALIZATION

The main techniques used for domain pattern visualization may be listed as follows

- the colloid or Bitter technique,
- Optical methods,
- X-Ray Imaging,
- Electron Microscopy,
- Neutron Imaging.

The Bitter technique

The principle of this technique historically introduced by Bitter [1] is the following : a drop of a colloidal suspension of magnetic particles is deposited between the surface of the sample and a cover slide. When the sample's surface is observed by means of an optical microscope, accumulations of particles may be observed, coinciding with the maxima of the stray fields. If a one to one correspondance between the extrema of the stray fields and domain walls exists, then the observed pattern is the domain pattern.

For the technique to be operative, two conditions have to be fullfilled :

i) the stray fields must be strong enough to overcome thermal agitation within the suspension,

ii) the particle size must be small enough to avoid natural clustering of the magnetic particles. Practical grain sizes are in the 50 to 100 Å diameter range.

A few milestones in the use of this technique are :

a) the observation of domain structures in [100] Fe Si samples by Williams, Bozorth and Shockley [2]. On top of the now classical analysis of the domain pattern as a function of surface disorientation, these authors have shown that the determination of Bloch Wall chirality was possible with the aid of a small bias field.

b) the observation of domain structures in highly perfect Fe and Ni whiskers, as well as in permalloy films by De Blois [3,4,5]. Amongst numerous exciting observations, De Blois was the first to show that 180° walls in iron platelets were segmented (e.g. fig. 1). Using the bias field technique described above, De Blois was able to demonstrate that adjacent segments had opposite chiralities, being therefore separated by a Néel line. Segmentation was later explained by Strikman and Treves according to energetic considerations [6].

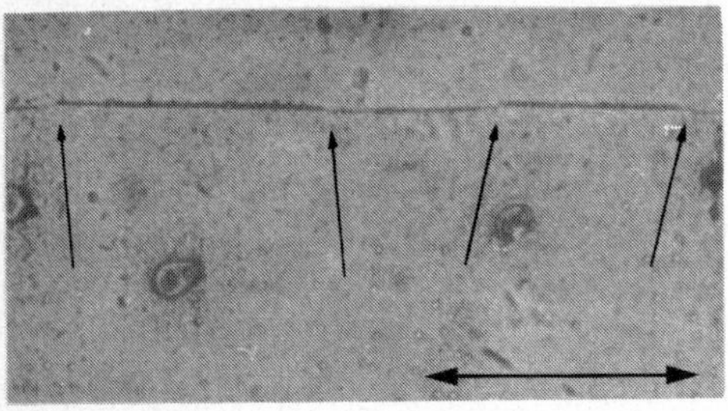

Fig.1 : Néel lines in FeSi. Bitter technique. Scale mark : 50μm.

c) the observation of so-called charged walls in implanted garnets [7,8,9,10]. Here also, the ability of the technique to detect chiralities has been used and a model for the charged walls proposed [11].

Methods utilizing magneto-optical effects.

Magneto-optical effects are associated with the spin-dependent electric dipole moment induced by the incident electromagnetic field. For quasi-transparent media, the permittivity tensor may be decomposed [12] into a real part which is symmetrical and even in M :

$$\varepsilon'_{ij} = \varepsilon'_{oij}\,\delta_{ij} + \sum_{\ell,m} f^{\ell m}_{2ij} M_\ell M_m + \ldots$$

where ε'_{oij} is the spin independant part of the permittivity and an imaging part which is antisymmetrical and odd in M :

$$\varepsilon''_{ij} = \sum_k f^{k}_{1ij} M_k + \sum_{k,\ell,m} f^{k\ell m}_{3ij} M_k M_\ell M_m + \ldots$$

Looking at first order effects, it appears that the polarization caused by an optical incident field E reradiates a field with polarization perpendicular to $\vec{E}$ and $\vec{M}$. The total electric field is therefore rotated by an angle α which is a function of the magnitude of $\vec{M}$. Obviously, the total rotation will be proportional to the thickness of the sample. This effect is the Faraday effect. Second order effects give rise to linear magnetic birefringence or the Cotton-Mouton effect. Observation rules have been summarized by Dillon [13].

Fig. 2 shows an example of domain visualization by means of a polarizing microscope in a ferrimagnetic garnet with appreciable Faraday notation. A nearly perfectly ordered bubble array is observed in this epitaxial layer with perpendicular anisotropy.

If the polarizer and analyser of the microscope are exactly crossed, no domain contrast is observed (fig. 3). Walls however appear black lines. A precise understanding of this type of contrast is not yet available.

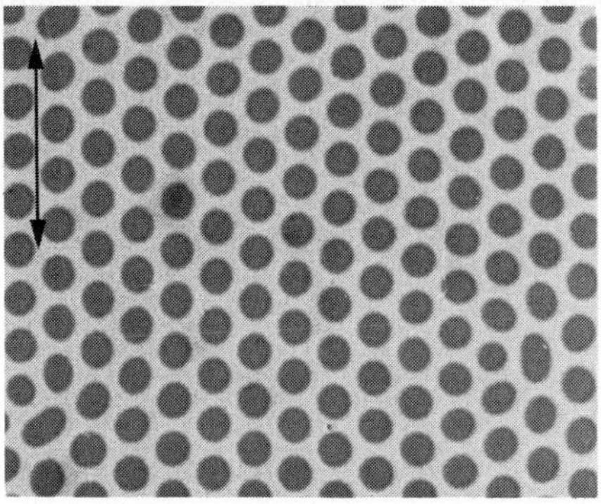

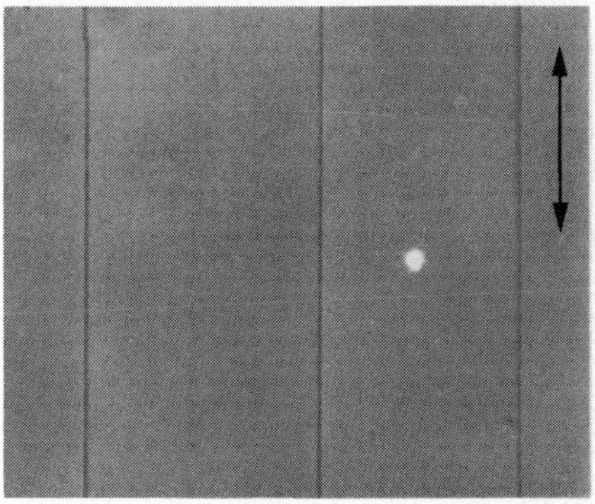

Fig.2 (left) : Bubble array. Faraday rotation. Polarizer and Analyser slightly uncrossed. Scale Mark : 50 μm.
Fig.3 (right) : Stripes. Faraday rotation. Polarizer and Analyser crossed. Scale mark : 100 μm. Courtesy of Ph. Trouilloud.

Fig. 4 shows a domain pattern in a YIG single crystal with in plane anisotropy. Here, no Faraday rotation is expected within the domain since $\vec{M}$ is perpendicular to the light beam. However, a local Faraday effect is observed within the walls, enabling a direct visualization of the wall structure : the zig zag lines are Néel lines [14].

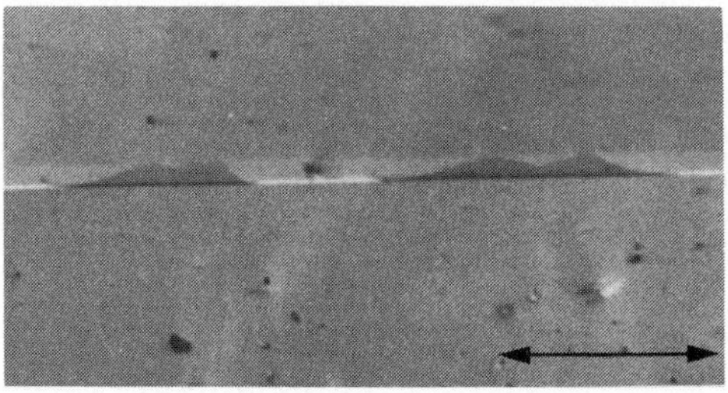

Fig.4 : Néel lines within a 180° Bloch wall in YIG. Faraday rotation. Scale mark 200 μm.

Since defects may be visualized by their stress induced birefringence, optical methods have also been used to study the interaction of moving walls in a quasi static regime and crystal dislocations. Vlasko-Vlasov et al. [15] have studied the interaction of a 180° wall and an edge dislocation in YIG. They observe the nucleation of closure domains when the wall approaches the dislocation. This observed behaviour might however be a surface effect. Dedukh et al.[16] have studied the interaction of a 180° wall with an edge dislocation in a configuration of easy direction perpendicular to the film plane. Here, the 180° wall may bend freely and rather astonishing effects have been observed.

Let us also mention that Faraday rotation has been used in wall dynamics studies, either by means of high speed imaging [e.g. 17] or in a diffraction setting [18].

Finally, in the reflection geometry, domain structures may be observed by means of the various Kerr effects (polar, longitudinal and transverse) [13]. Kerr effects have for instance been utilized successfully in the study of Néel lines in FeSi [19] and domain patterns in amorphous materials [20].

X-Ray diffraction imaging.

In the X-Ray frequency range, the interaction of photons with the magnetization is too weak to allow for a direct visualization of domain structures. However, under diffraction conditions, X-Ray imaging techniques (see for instance [21]) are highly sensitive to minute lattice distortions (sensitivities to strains as low as 10^{-8} may be achieved). Domain patterns may therefore be visualized through their associated magnetostrictive distortions.

Fig. 5 shows a classical domain pattern in a FeSi single crystal recorded by means of the Lang technique. Crystal lattice defects may be seen, such as dislocations D, a subgrain boundary B together with domain walls, wall junctions and Fir tree patterns. Note worthy is the absence of contrast due to 180° walls.

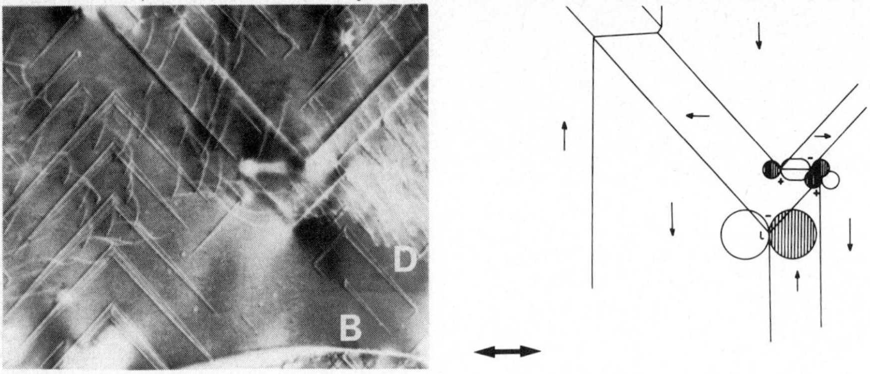

Fig.5 : a) X-Ray Lang topograph of a FeSi single crystal. Scale mark : 200 μm.
b) magnetization distribution.

Actually, a 180° wall embedded in an infinite crystal suffers no magnetostrictive strains as the adjacent domains impose their spontaneous deformation to the wall (22). Stresses however exist. In a finite crystal, those stresses relax at the surfaces, inducing elastic displacements (23).

It is straightforward to show that 180° wall visibility is only expected for $\lambda_{100} > \sim 3.10^{-5}$ which corresponds to the iron value, where λ_{100} is the usual magnetostriction constant.

On the contrary, 90° walls are characterized by a lattice rotation :

$$|\Delta\omega| = \frac{3}{2}\lambda_{100}$$

and are expected to be visible for λ_{100} as small as 10^{-6} as confirmed by the experiments [24,25,26].

Besides, it has been shown that wall junctions were elastically equivalent to disclinations [25,27]. X-Ray imaging experiments have confirmed the general validity of the deformation calculations [28,29,30].

Let us finally mention that the availability of intense X-Ray sources, namely synchrotron sources,has recently allowed for stroboscopic observations of domain wall motion at intermediate inductions in FeSi single crystals [31].

Electron microscopy.

It is common knowledge that an electron propagating in a magnetic field is subjected to a force (Lorentz force) and is therefore deviated. This is the basic contrast forming mechanism in the electron microscopy of magnetic materials [32] .

Fig. 6a indicates, in a simple geometry, that walls will either be converging (excess electron density) or diverging (depleted electron density). Walls will therefore appear black or white as shown in fig. 6b. It is worth pointing out that the observation conditions in the electron microscope impose defocalization of the beam and passivation of the objective lens in order for the sample not to be subjected to a saturating field.Therefore, the resolution is in non way comparable to that of an electron microscope used under usual diffraction conditions. If the incident beam is coherent enough, converging walls act as a Fresnel biprism and interference fringes are observed. It is worth insisting on the fact that the final interference picture depends on the distribution of $\vec{B}$ and therefore on the distribution of $\vec{M}$ within the wall. Such interferences are shown in fig. 7. Their interpretation requires a proper wave optical treatment (for a short review, see [33]).

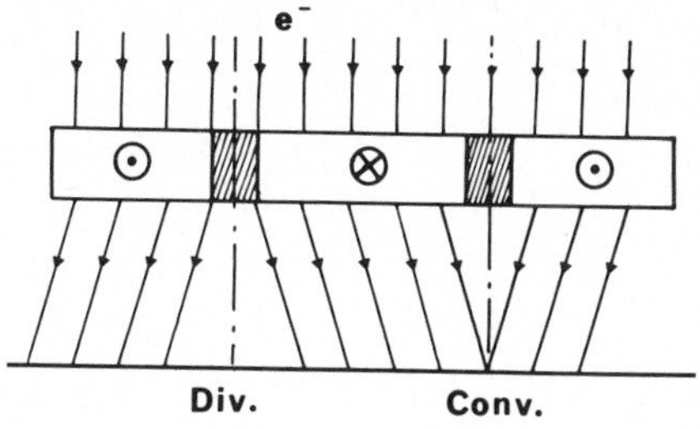

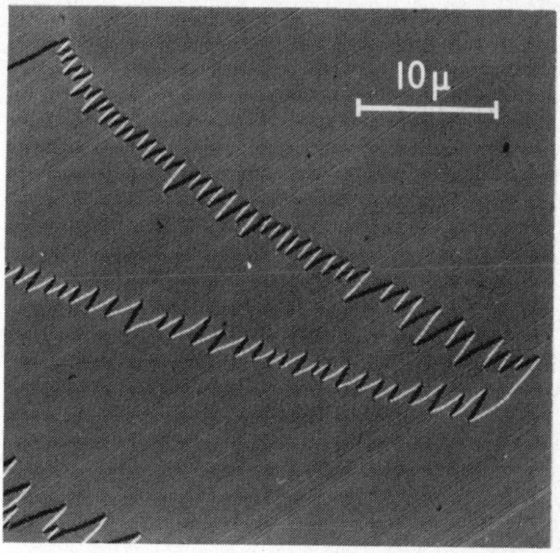

Fig.6 : a) schematic wall contrast formation in an Electron Microscope.
b) wall configuration in a permalloy amorphous film. Courtesy of I.B. Puchalska.

Finally, diffraction contrast may be superimposed onto Lorentz force imaging, allowing for the observation of, for instance, interactions between moving walls and crystal defects [34].

Let us finally mention a very beautiful study of cross-tie walls in Permalloy films under stroboscopic illumination at high frequency [35]. Masses of Bloch lines have been determined directly in this experiment.

Neutron imaging.

Two basic techniques may be utilized here :

i) diffraction contrast from unpolarized neutrons. Similarly to the X-Ray case, magnetic domain patterns may then be imaged through their associated magnetostrictive distortions.

ii) diffraction contrast from polarized neutrons where a direct interaction between neutrons and magnetization now exists.

Those techniques have found a number of applications [36,37], but are still suffering from a very poor spatial resolution.

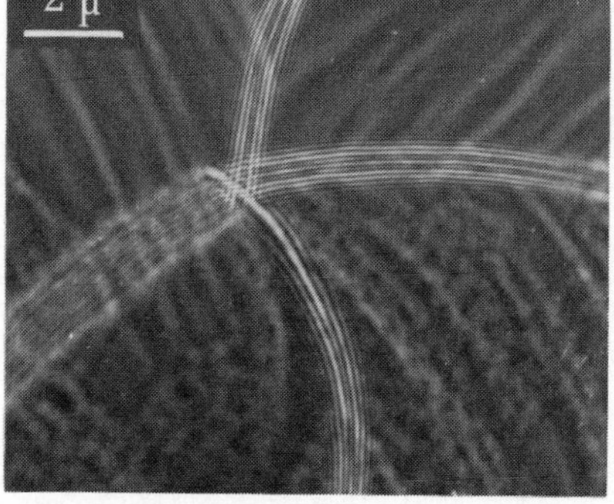

Fig.7 : Interferences along a converging wall under coherent illumination. Courtesy of C. Mory.

DISCUSSION

It should be first noticed that all imaging techniques described above suffer from a fairly limited spatial resolution. The resolution limit is set by the diffraction limit in optical methods ; for electron imaging, it is dictated by the image forming mechanism ; for X-rays and neutrons, a number of factors should be considered, amongst which photon or neutron statistics and detector performances are highly significant. The spatial resolution is in the μm range and therefore is one order of magnitude larger than a typical domain width. Informations gathered from walls are therefore integrated.

Intrinsic limitations of surface observation techniques (Bitter, Kerr and the not yet quoted Scanning Electron microscopy) are obvious and deserve little further comments.

It appears that techniques which probe the bulk of magnetic materials may de facto be separated into methods that sense directly the magnetization distribution and are little sensitive to magnetostrictive strains and methods that are characterized by the opposite potentialities. This is due to the fact that

i) stress birefringence is too weak for the optical visualization of magnetostrictive strains,

ii) although, in electron microscopy, diffraction imaging superimposed onto Lorentz force imaging, allows for strain detection, the strain sensitivity is usually too low. Further, since samples need to be very thin, surface stress relaxation becomes highly significative.

Therefore, X-ray imaging appears to be the more appropriate technique for magnetostrictive strains determination. X-ray imaging techniques which are more quantitative than the conventional Lang technique do actually exist [38]. Plane wave X-ray imaging allows for a local measure of both the strain and lattice rotation components [38,39].

Finally, no direct observation of Bloch points has been presented. The reason is simply that none yet exists.

[1] Bitter F., Phys. Rev. 38 (1931) 1903.
[2] Williams H.J., Bozorth R.M., Shockley W., Phys. Rev. 75 (1949) 155.
[3] De Blois R.W., Graham C.D., J. Appl. Phys. 29 (1958) 931.
[4] De Blois R.W., J. Appl. Phys. 36 (1965) 1647.
[5] De Blois R.W., J. Appl. Phys. 39 (1968) 442.
[6] Strikman S. and Treves D., J. Appl. Phys. 31 (1960) 147S.
[7] Almasi G.S., Giess E.A., Hendel R.J., Keefe G.E. Lin Y.S., Slusarczuk M. AIP Conf. Proc. 24 (1975) 630.
[8] Puchalska I.B., Jouve H., Wade R.H., J. Appl. Phys. 48 (1977) 2069.
[9] Argyle B.E., Kryder M.H., Mundie R.E., Slonczewski J.C., IEEE Trans. Magn. MAG14 (1978) 593.
[10] Lin Y.S., Dove D.B., Schwarzl S., Shir C.C. IEEE Trans. Magn. MAG14 (1978)494.
[11] Kléman M., Puchalska I.B., J. Magn. Mag. Mat. 15 (1980) 1473.
[12] Le Gall H. and Jamet J.P., Phys. Stat. Sol.(b) 46 (1971) 467.
[13] Dillon J.F. Jr., in Smit. J. (Ed), Magnetic Properties of Materials (Mc Graw Hill, 1971).
[14] Labrune M., Miltat J., Kléman M., J. Appl. Phys. 49 (1978) 2013.
[15] Vlasko-Vlasov. V.K., Dedukh. L.M., Nikitenko V.I. Phys. St. Sol(a)29(1975)367.
[16] Dedukh L.M., Indenbom M.V., Nikitenko V.I., Sov. Phys. JETP 53 (1981) 194.
[17] Humphrey F.B., J. Magn. Mag. Mat. 15-18 (1980) 1464 and Loc. Cit.
[18] Slonczewski J.C., Argyle B.E., Spreen J.H. IEEE Trans. Magn. MAG17 (1981)2760.
[19] Schön L. and Buchenau U., Intern. J. Magnetism 3 (1972) 145.
[20] Kronmüller H., J. de Physique C8-41 (1980) 618.
[21] Tanner B.K., X-ray Diffraction Topography (Pergamon, Oxford, 1976).
[22] Néel L., Cahiers de Physique 25 (1944) 1 and J. Phys. et Rad.5(1944) 241,265.
[23] Kroupa F. and Vagera I., Czech. J. Phys. B19 (1969) 1204.
[24] Polcarova M. and Gemperlova J., Phys. Stat. Sol. 32 (1969) 769.
[25] Kléman M. and Schlenker M., J. Appl. Phys. 43 (1972) 3184.
[26] Labrune M., J. de Physique 37 (1976) 1033.
[27] Kléman M., J. Appl. Phys. 45 (1974) 1377.
[28] Miltat J. and Kléman M., Phil. Mag. 28 (1973) 1015.
[29] Miltat J., Phil. Mag. 33 (1976) 225.
[30] Miltat J., IEEE Trans. Magn. MAG17 (1981) 3090.
[31] Miltat J., Kléman M., J. Appl. Phys. 50 (1979) 7695.
[32] Hirsch P.B., Howie A., Nicholson R.B., Pashley D.W., Whelan M.J., Electron Microscopy of thin Crystals (Butterworth, 1965).
[33] Wade R.H., J. de Physique C2-29 (1968) 95.
[34] Degauque J. and Astié B., this Volume.
[35] Bostanjoglo O. and Rosin Th., Phys. Stat. Sol. (a) 57 (1980) 561.
[36] Baruchel J., Schlenker M., Kurosawa K., Saito S., Phil. Mag. B 43 (1981) 853 and 861.
[37] SchlenkerM. and Baruchel J., IEEE Trans. Magn. MAG17 (1981) 3085.
[38] Sauvage M. in Tanner B.K. and Bowen D.K.(Eds) Characterization of Crystal Growth Defects by X-Ray Methods (Plenum Press, 1980).
[39] Bonse U. and Hartman I.,Zeitschrift für Krist. 156 (1981) 265.

The Mechanical Behavior of Electromagnetic Solid Continua
G.A. Maugin (editor)
Elsevier Science Publishers B.V. (North-Holland)

ON THE ELECTRODYNAMICS OF A DEFORMABLE FERROMAGNET UNDERGOING MAGNETIC RESONANCE

Frederic R. Morgenthaler

Department of Electrical Engineering and Computer Science
Research Laboratory of Electronics
MIT Cambridge, MA 02139

The energy-momentum tensor of a deformable ferromagnet is developed using the Chu formulation of electrodynamics which we here review. It is shown that when the magnetization is undergoing ferromagnetic resonance, the new tensor differs from that of conventional theory, principally because large-signal linear momentum due to spin precession can appear in the rest-frame of the solid. Under transient conditions, the time rate of change of that momentum leads to predictions of an altered net force even when the ferromagnet is acting as a rigid body.

INTRODUCTION

Some twenty-five years ago, Lan Jen Chu [1] set forth a theory of the electrodynamics of deformable media that was based upon a magnetic charge model of the continuum magnetization. By banning Amperian currents from what has since become known as the Chu formulation, he also prevented the "hidden-momentum" due to the moving mass of circulating electric charges ordinarilly held responsible for magnetic effects. Because the continuum dielectric polarization is naturally based upon an electric charge model, the Chu formulation enjoys both a symmetry of form and a relative simplicity that follows because the electromagnetic momentum in the rest-frame of the material is purely $\bar{E} \times \bar{H}/c^2$.

However, in light of the fact that quantum theory places electron orbital- and/or spin-angular-momentum at the heart of all accepted microscopic models for magnetization, one naturally wonders if the Chu formulation should be taken seriously. Almost immediately after its introduction, comparisons were made with the rival Minkowski and Abrams formulations and apparent disagreements were found in the predicted force densities. This led to generally unfavorable reviews which inhibited acceptance of the theory. However, several of us at MIT who were privileged to be students and/or colleagues of Lan Chu began to reexamine the overall subject of electrodynamics and, in particular, Penfield and Haus (P-H) showed in their important monograph [2] that the Chu treatment does in fact yield force densities in agreement with the Minkowski formulation provided material subsystems defined by the constitutive laws are properly taken into account. Unless this is done, all electrodynamic models will generally be in error. It is true that the choice of formulation leads to different definitions of the continuum electric and magnetic fields within polarizable matter and thus what one defines as the electromagnetic portion of the force density is somewhat arbitrary. Nevertheless, in the many examples that they treated, P-H found agreement of the total force density. Free to choose the simplest model, they and we opt for the Chu formulation.

Because ferrites with low magnetic-loss can support spin waves and other resonances that have no dielectric counterparts, they provide an opportunity to apply electrodynamic theories to a nonlinear medium with a well understood constitutive law. P-H considered a single domain ferrite within the context of magneto-quasistatics; their results agree with earlier alternate treatments carried out by Tiersten, Brown, and Eastman; these are referenced in [2].

However, this theory did not satisfy the present author because it creates force paradoxes through neglect of linear-momentum associated with excess kinetic-energy of magnetically-resonant electrons. Evidently, the fault does not lie with the Chu formulation, because based upon it and the linearized equations of motion, Morgenthaler was able to find the small-signal stress and momentum [3,4] and subsequently a large-signal theory [5] that removed the paradoxes, is subject to relatively simple interpretation, and sheds light on the age-old controversy concerning the proper form of the electromagnetic momentum in magnetic media. In addition, a particle representation and therefore connection with the microscopic physics is possible--even under large signal conditions. The theory is relativistic and applies to crystals with more than one magnetic lattice when the magnetizations are defined in the spirit of continuum micromagnetics.

July 1983 marks the tenth anniversary of Lan Chu's untimely death; two weeks later would have been his seventieth birthday. It therefore seems particularly fitting to honor the man by reviewing his formulation of electrodynamics and its role in the development of the new energy-momentum tensor for ferrimagnetic insulators (ferrites). This conference affords a most appropriate forum for doing so.

CHU FORMULATION OF ELECTRODYNAMICS

The continuum model that Chu developed was based upon matter considered as small grains each surrounded by free space. Therefore, he simply used the non-controversial free-space Maxwell Equations written in terms of the $\bar{E}$ and $\bar{H}$ fields present outside of the grains and then added field sources inside of them. Since the dominant response of matter to fields is the creation or alteration of electric and magnetic dipoles, any formulation that produces the required moments is tenable; this follows from the field equivalence principle. The natural choice based upon microscopic physics has seemed to be bound electric-charge pairs and Amperian current-loops. Instead, while embracing the former, Chu chose a magnetic-charge model for magnetization. In four-vector notation (defined in the Appendix—along with the other symbols), the polarization $\mathbf{P}$ and the magnetization $\mathbf{M}$ are expressed as

$$\mathbf{P} = \left(\frac{1}{\gamma} \bar{P} + \gamma \frac{\bar{P}\cdot\bar{v}}{c^2} \bar{v} \,,\; i\gamma \frac{\bar{P}\cdot\bar{v}}{c} \right) \tag{1a}$$

$$\mathbf{M} = \left(\frac{1}{\gamma} \bar{M} + \gamma \frac{\bar{M}\cdot\bar{v}}{c^2} \bar{v} \,,\; i\gamma \frac{\bar{M}\cdot\bar{v}}{c} \right) \tag{1b}$$

and give rise to or augment the corresponding electric and magnetic currents $\mathbf{J}_e$ and $\mathbf{J}_m$ through

$$\mathbf{J}_e = \mathbf{J}_f + \Box \cdot (\mathbf{P} \times \mathbf{V}) \tag{2a}$$

$$\mathbf{J}_m = 0 + \Box \cdot (\mu_o \mathbf{M} \times \mathbf{V}) \tag{2b}$$

where $\mathbf{J}_f$ is the free current density (zero for a ferrite insulator) and $\mathbf{V}$ is the usual continuum four-velocity defined in the Appendix.

Maxwell's Eqs in free space with the current sources become

$$\Box \cdot \mathbb{G} = \mathbb{J}_e \qquad (3a)$$

and

$$\Box \cdot \mathbb{K} = \mathbb{J}_m \qquad (3b)$$

where $\mathbb{G}$ and $\mathbb{K}$ are the usual free-space antisymmetric field tensors; the present author finds it helpful to write them as

$$\mathbb{G} = -\frac{i}{c}(\mathbb{H} \times \mathbb{V})^+ - \epsilon_o \mathbb{E} \times \mathbb{V} \qquad (4a)$$

and

$$\mathbb{K} = \frac{i}{c}(\mathbb{E} \times \mathbb{V})^+ - \mu_o \mathbb{H} \times \mathbb{V} \qquad (4b)$$

where

$$\mathbb{E} = \gamma(\bar{E} + \mu_o \bar{v} \times \bar{H}\ ,\ \ i\,\bar{E} \cdot \bar{v}/c) \qquad (5a)$$

and

$$\mathbb{H} = \gamma(\bar{H} - \epsilon_o \bar{v} \times \bar{E}\ ,\ \ i\,\bar{H} \cdot \bar{v}/c) \qquad (5b)$$

The total four-vector force acting upon the electric charges q_i^e within a material grain is

$$\mathbb{F}_e = \sum_i q_i^e \mathbb{E}_i = \mu_o \mathbb{G} \cdot \sum_i q_i^e \mathbb{V}_i \qquad (6a)$$

Since magnetic charges q_i^m have been introduced, an analogous force must also be included, namely

$$\mathbb{F}_m = \sum_i q_i^m \mathbb{H}_i = \epsilon_o \mathbb{K} \cdot \sum_i q_i^m \mathbb{V}_i \qquad (6b)$$

On a per unit volume basis, these reduce to electric and magnetic force densities

$$\mathbb{f}_e = \mu_o \mathbb{G} \cdot \mathbb{J}_e \qquad (7a)$$

and

$$\mathbb{f}_m = \epsilon_o \mathbb{K} \cdot \mathbb{J}_m \qquad (7b)$$

Chu defines the electromagnetic force density as their sum

$$\mathbb{f}_{em} = \mu_o \mathbb{G} \cdot (\Box \cdot \mathbb{G}) + \epsilon_o \mathbb{K} \cdot (\Box \cdot \mathbb{K}) = \Box \cdot \mathbb{T}_{em} \qquad (8)$$

where the electromagnetic energy-momentum tensor $\mathbb{T}_{em}$ is

$$\begin{aligned} \mathbb{T}_{em} &= \frac{1}{2}(\mu_o \mathbb{G} \cdot \mathbb{G} + \epsilon_o \mathbb{K} \cdot \mathbb{K}) \\ &= \frac{i}{c}\,\frac{1}{2}(\mathbb{K}^+ \cdot \mathbb{G} - \mathbb{G}^+ \cdot \mathbb{K}) \\ &= \epsilon_o \mathbb{E}\,\mathbb{E} + \mu_o \mathbb{H}\,\mathbb{H} \\ &\quad - \frac{1}{2}(\epsilon_o \mathbb{E} \cdot \mathbb{E} + \mu_o \mathbb{H} \cdot \mathbb{H})(\mathbb{I} + 2\frac{\mathbb{V}\,\mathbb{V}}{c^2}) \\ &\quad + \frac{i}{c^3}[\mathbb{V}\,(\mathbb{E} \times \mathbb{H})^+ \cdot \mathbb{V} + (\mathbb{E} \times \mathbb{H})^+ \cdot \mathbb{V}\,\mathbb{V}] \end{aligned} \qquad (9)$$

and $\mathbb{I}$ is defined as the four-dimensional Kronecker delta. Notice that if one makes use of the auxilliary four-vectors defined by

$$\mathbb{D} = \epsilon_o \mathbb{E} + \mathbb{P} \quad \text{and} \quad \mathbb{B} = \mu_o (\mathbb{H} + \mathbb{M})$$

and rearranges Eqs.(3a,b), the result is

$$\Box \cdot [-\frac{i}{c}(\mathbb{H} \times \mathbb{V})^+ - \mathbb{D} \times \mathbb{V}] = \mathbb{J}_f \qquad (10a)$$

$$\Box \cdot [-\frac{i}{c}(\mathbb{E} \times \mathbb{V})^+ + \mathbb{B} \times \mathbb{V}] = 0 \qquad (10b)$$

When $\Box \mathbf{V} = 0$, Eqs.(10a,b) can be expressed as

$$-\frac{i}{c}(\Box \times \mathbf{H})^{+} \cdot \mathbf{V} = \frac{\partial \mathbf{D}}{\partial \tau} + \mathbf{J}_f + \frac{\mathbf{J}_f \cdot \mathbf{V}}{c^2}\mathbf{V} \tag{11a}$$

and

$$-\frac{i}{c}(\Box \times \mathbf{E})^{+} \cdot \mathbf{V} = -\frac{\partial \mathbf{B}}{\partial \tau} \tag{11b}$$

which is still the Chu formulation! Notice also that we can express the four-vectors $\mathbf{E}$ and $\mathbf{B}$ in terms of

$$\bar{E}^{o} = \bar{E} + \mu_o \bar{M} \times \bar{v} \quad \text{and} \quad \bar{B} = \mu_o (\bar{H} + \bar{M})$$

Since the four-vectors are unchanged, such transformations are physically irrelevent but the Minkowski energy-momentum tensor

$$\begin{aligned}\mathbf{T}_M &= \frac{i}{c}\,\frac{1}{2}(\mathbf{K}_M^{+} \cdot \mathbf{G}_M - \mathbf{G}_M^{+} \cdot \mathbf{K}_M) \\ &= \mathbf{E}\,\mathbf{D} + \mathbf{H}\,\mathbf{B} \\ &\quad - \frac{1}{2}(\mathbf{E} \cdot \mathbf{D} + \mathbf{H} \cdot \mathbf{B})\left(\mathbf{I} + 2\frac{\mathbf{V}\,\mathbf{V}}{c^2}\right) \\ &\quad + \frac{i}{c^3}\mathbf{V}\,(\mathbf{E} \times \mathbf{H})^{+} \cdot \mathbf{V} + \frac{i}{c}(\mathbf{D} \times \mathbf{B})^{+} \cdot \mathbf{V}\,\mathbf{V}\end{aligned} \tag{12}$$

written in terms of the alternate field tensors

$$\mathbf{G}_M = -\frac{i}{c}(\mathbf{H} \times \mathbf{V})^{+} - \mathbf{D} \times \mathbf{V} \tag{13a}$$

and

$$\mathbf{K}_M = \frac{i}{c}(\mathbf{E} \times \mathbf{V})^{+} - \mathbf{B} \times \mathbf{V} \tag{13b}$$

is different from $\mathbf{T}_{em}$. As they stand, neither tensor is suitable.

A FERROMAGNET SUPPORTING MAGNETIC RESONANCE

We assume that the micromagnetic magnetization can be expressed as $\mathbf{M} = n'\,\mathbf{a}$ where n' is the rest-frame density of the magnetic moment $\mathbf{a}$. Because $\mathbf{M}$ is assumed locally saturated, $\mathbf{a}$ is of constant magnitude but free to change orientation under the influence of whatever net torque acts upon it; therefore, without loss of generality, the constitutive law (torque equation) can be taken as

$$\frac{\partial \mathbf{a}}{\partial \tau} = -\frac{i}{c}\gamma_g'\mu_o(\mathbf{a} \times \mathbf{H}^{E})^{+} \cdot \mathbf{V} + \mathbf{a} \cdot \frac{\partial \mathbf{V}}{\partial \tau}\,\mathbf{V}/c^2 \tag{14}$$

where γ_g' is the rest-frame value of the gyromagnetic ratio (negative for electrons), τ the proper time and

$$\mathbf{H}^{E} = \mathbf{H} + \mathbf{H}^{anis} + \mathbf{H}^{ex} + \mathbf{H}^{me} + \mathbf{H}^{loss} \tag{15}$$

is the effective field that in addition to $\mathbf{H}$ includes anisotropy, exchange, magnetoelastic, and magnetic dissipation components.

The anisotropy field can be expressed in terms of W'_{anis}, the rest-frame anisotropy energy density, as

$$\mathbf{H}_i^{anis} = -\frac{1}{\mu_o}\frac{\partial}{\partial \mathbf{a}_i}(W'_{anis}/n') \tag{16}$$

If the rest-frame exchange energy density is isotropic with exchange constant Λ, the associated field can be shown to be

$$\mathbf{H}^{ex} = \frac{1}{n'}\frac{\partial}{\partial x_j}\left[n'\Lambda\left(\mathbf{I}_{jk} + \frac{\mathbf{V}_j\mathbf{V}_k}{c^2}\right)\frac{\partial \mathbf{a}}{\partial x_k}\right] \tag{17}$$

Magnetoelastic coupling is taken to be quadratic in $\mathbf{a}$ and linear in the mechanical strains calculated from $\Box\mathbf{R}$ where $\mathbf{R}$ is the continuum displacement vector also used to describe the elastic properties of the solid. Higher-order analyses employ finite strains.

When there is no microwave excitation, the moments are in a static equilibrium configuration characterized by a unit vector $\mathbf{U}$ that is assumed known and need not be spatially uniform. Since $\mathbf{U}.\mathbf{V} = 0$, an important identity, based upon Eqs.(14) and (A-9) is

$$-\frac{1}{2}\mu_o \mathbf{a} . \mathbf{H}^E = \frac{-1}{2\gamma'_g \mathbf{a}.\mathbf{U}} \mathbf{U}.(\mathbf{a} \times \frac{\partial \mathbf{a}}{\partial \tau})^+ . \mathbf{V} - \frac{1}{2}\mu_o \frac{\mathbf{a}.\mathbf{a}}{\mathbf{a}.\mathbf{U}} \mathbf{U}.\mathbf{H}^E \qquad (18)$$

The first term on the RHS of Eq.(18) has previously been identified as the excess kinetic energy associated with the precession of $\mathbf{a}$ when the latter is non-parallel to $\mathbf{U}$. That energy is related to linear-momentum within the material; namely,

$$(\mathbf{p}_M)_i = \frac{-i}{c} \frac{1}{2\gamma'_g \mathbf{a}.\mathbf{U}} \mathbf{U}.(\mathbf{a} \times \frac{\partial \mathbf{a}}{\partial x_i})^+ . \mathbf{V} \qquad (19)$$

We therefore expect $-n' \mathbf{p}_M \mathbf{V}$ to appear in the complete energy-momentum tensor of the ferromagnet. The rotational kinetic energy of the magnetic electrons, KE'_{rot} is interrelated with the Zeeman energy. Detailed study of the latter has suggested that $-\mu_o \mathbf{M}^o.\mathbf{H}$ is an appropriate "ground-state" energy density where

$$\mathbf{M}^o = \frac{1}{2}\left(\frac{\mathbf{M}.\mathbf{M}}{\mathbf{M}.\mathbf{U}} + \mathbf{M}.\mathbf{U}\right)\mathbf{U} \qquad (20)$$

The exchange torques arising from Eq(17) also produce stresses; these contribute the term $\mathbf{T}^{ex}$ to the material energy-momentum where

$$\mathbf{T}^{ex}_{ij} = -\mu_o n' \frac{\partial \mathbf{a}}{\partial x_i} . \Lambda(\mathbf{I}_{jk} + \frac{V_j V_k}{c^2}) \frac{\partial \mathbf{a}}{\partial x_k} \qquad (21)$$

The total energy-momentum is given by the sum of $\mathbf{T}_M$ and $\mathbf{T}^m$ where

$$\mathbf{T}^m = \mathbf{T}^{ex} - n' \mathbf{p}_M \mathbf{V} + [\, W'_{anis} + \frac{1}{2}\mu_o(2\mathbf{M}^o.\mathbf{H} + \mathbf{M}.\mathbf{H}^{anis} - \frac{\mathbf{M}.\mathbf{M}}{\mathbf{M}.\mathbf{U}} \mathbf{U}.\mathbf{H}^E)\,]\,\mathbf{I} + (\mu_o \mathbf{M}^o.\mathbf{H} - n'KE'_{rot}) \frac{\mathbf{V}\mathbf{V}}{c^2} + - \frac{i}{c}(\mathbf{D} \times \mu_o \mathbf{M}^o)^+ . \mathbf{V}\,\mathbf{V} + \mathbf{T}^{me} \qquad (22)$$

and $\mathbf{T}^{me}$ is the magnetoelastic contribution.

Loss torques are required in our model if we are to avoid force paradoxes [5]. Here a phenomenological loss torque is provided by

$$\mathbf{H}^{loss} = -\frac{i}{c} \frac{\pm \Delta H'}{\mathbf{M}.\mathbf{U}} (\mathbf{M} \times \mathbf{U})^+ . \mathbf{V} \qquad (23)$$

where $\Delta H'$ is the rest-frame line width. In terms of the lossless force density, $\mathbf{f}$ can be written as

$$\mathbf{f} = \Box . \mathbf{T} = \mp 2\gamma'_g \mu_o \Delta H' n' \mathbf{p}_M + \mathbf{f}_{lossless} \qquad (24)$$

The sign is chosen so that $\mathbf{f} . \mathbf{V}$ remains positive definite.

SUMMARY

When $\partial \mathbf{p}_M/\partial\tau$ is not zero, the total force based upon Eq.(24) can differ from the conventional prediction. In the magnetostatic limit and neglecting exchange and anisotropy, the leading term of $\mathbf{f}_{lossless}$ is given by $\mu_o \mathbf{M}^o.\Box\mathbf{H}$. If $\Box\mathbf{U}$ and $\mathbf{P}$ are zero, and magnetic resonance is not excited, the new and conventional energy-momentum tensors—and thus the force density predictions—agree.

APPENDIX - DEFINITIONS AND IDENTITIES

Double width letters such as $\mathbb{A}$ and $\mathbb{B}$ represent four-vectors

$$\mathbb{A} = (\bar{A}\ ,\ A_4) \qquad \text{and} \qquad \mathbb{B} = (\bar{B}\ ,\ B_4)$$

where $\bar{A},\bar{B}$ are the three-space and A_4,B_4 the temporal components

$$\mathbb{A}\ .\ \mathbb{B} = \bar{A}\ .\ \bar{B} + A_4\ B_4 \qquad \text{(A-1)}$$

$$\mathbb{A} \times \mathbb{B} = \mathbb{A}\ \mathbb{B} - \mathbb{B}\ \mathbb{A}$$

$$= (\bar{A} \times \bar{B});(B_4\bar{A} - A_4\bar{B}) \qquad \text{six-vector} \qquad \text{(A-2)}$$

$$(\mathbb{A} \times \mathbb{B})^+ = (B_4\bar{A} - A_4\bar{B});\ (\bar{A} \times \bar{B}) \qquad \text{dual six-vector} \qquad \text{(A-3)}$$

In terms of the coordinates $x_1 = x$, $x_2 = y$, $x_3 = z$, and $x_4 = ict$, $\Box_j = \partial/\partial x_j$ or in four-vector notation $\Box = (\nabla\ ,\ -i\frac{\partial}{c\partial t})$

$$\Box\ .\ \mathbb{A} = \partial A_j/\partial x_j \qquad \text{(summation implied)} \qquad \text{(A-4)}$$

$$(\Box \times \mathbb{A})_{ij} = \partial A_j/\partial x_i - \partial A_i/\partial x_j \qquad \text{(A-5)}$$

$$\Box\ .\ (\mathbb{A} \times \mathbb{B}) = \mathbb{A}(\Box\ .\ \mathbb{B}) + (\mathbb{B}\ .\ \Box)\ \mathbb{A} - \mathbb{B}(\Box\ .\ \mathbb{A}) - (\mathbb{A}\ .\ \Box)\ \mathbb{B} \qquad \text{(A-6)}$$

The four-velocity $\mathbb{V} = (\gamma\bar{v}, i\gamma c)$ satisfies $\mathbb{V}.\mathbb{V} = -c^2$.
In terms of the proper time differential, $\frac{\partial}{\partial t} = \gamma\frac{d}{dt} = \mathbb{V}.\Box$, and the particle density $n = \gamma n'$ (n' is the rest-frame value),

$$\Box\ .\ (Q\mathbb{V}) = n'\frac{\partial}{\partial\tau}(Q/n') \qquad \text{(A-7)}$$

holds for any quantity Q that can be a scalar, four-vector, etc.
For the special case $Q = n'$ we have particle conservation.

$$(\mathbb{A} \times \mathbb{B})^+\ .\ \mathbb{V} = (\mathbb{B} \times \mathbb{V})^+\ .\ \mathbb{A} = \mathbb{B}\ .(\mathbb{A} \times \mathbb{V})^+ \qquad \text{(A-8)}$$

$$= \gamma(\bar{A} + iA_4\bar{v}/c)\times(\bar{B} + iB_4\ \bar{v}/c),\ \frac{i}{c}\ \gamma(\bar{A} \times \bar{B}).\bar{v}$$

If $\mathbb{U}\ .\ \mathbb{V} = 0$,

$$-\frac{i}{c^2}\ \mathbb{U}\ .\ \{\ \mathbb{A} \times [(\ \mathbb{B} \times \mathbb{C})^+\ .\ \mathbb{V}]\ \}^+\ .\ \mathbb{V} =$$

$$[\mathbb{A}\ .\ \mathbb{C} + \frac{\mathbb{A}\ .\ \mathbb{V}\quad \mathbb{C}\ .\ \mathbb{V}}{c^2}]\ \mathbb{B}\ .\ \mathbb{U} \qquad \text{(A-9)}$$

$$-\ [\mathbb{A}\ .\ \mathbb{B} + \frac{\mathbb{A}\ .\ \mathbb{V}\quad \mathbb{B}\ .\ \mathbb{V}}{c^2}]\ \mathbb{C}\ .\ \mathbb{U}$$

REFERENCES

1. R.M. Fano, L.J. Chu, and R.B. Adler Electromagnetic Fields Energy and Forces-Appendix One p.453 Wiley and Sons (1960)
2. P.P. Penfield and H.A. Haus Electrodynamics of Moving Media M.I.T. Research Monograph #40 M.I.T. Press (1967); contains references to the earlier literature and historical controversies
3. F.R. Morgenthaler, "Exchange Energy, Stress and Momentum in a Rigid Ferromagnet" J. Appl. Physics vol 38, 1069, (1967)
4. F.R. Morgenthaler, "Dynamic Magnetoelastic Coupling in Ferromagnets and Antiferromagnets", IEEE Trans. MAG-8 130, March 1972
5. F.R. Morgenthaler, "Predicted New Components of Magnetic Force on a Ferromagnet Undergoing Resonance" AIP Conf. Proc. No.29 (1975)

This research was sponsored in part by RLE administered JSEP funds under Contract DAAG 29-83-K-0003 and by NSF under Grant 8008628-DAR

The Mechanical Behavior of Electromagnetic Solid Continua
G.A. Maugin (editor)
Elsevier Science Publishers B.V. (North-Holland)

ON SKEW - SYMMETRIC STRESSES IN CUBIC SINGLE - DOMAIN CRYSTALLITES

Jürgen Lenz

Institut für Theoretische Mechanik
Universität Karlsruhe
Federal Republic of Germany

For cubic single-domain particles the order of magnitude of those skew-symmetric stresses is estimated which originate in the magnetic anisotropy energy. It is shown that the maximum skew-symmetric stresses are negligibly small compared to the stresses usually encountered in technical applications. On the micromagnetic scale, however, they are of the same order of magnitude as the symmetric stresses. Furthermore, for certain directions of the applied field, in addition to the conventional magnetization curves further ("metastable") branches are predicted.

In general, a ferromagnetic specimen is composed of many (so-called Weissian) domains. Under certain circumstances, however, the specimen may consist of a single domain, for instance when it is magnetized to saturation by the application of a sufficiently large magnetic field. Another example is present when the specimen is of very small size. Then the quantum mechanical exchange forces may dominate such that the particle is uniformly magnetized even in the absence of any applied field. Such single-domain crystallites have become increasingly important in technical applications during the past three decades. As examples, the production of hard ferromagnetic materials i.e. magnets with large coercive forces (single-domain particles as powder or dispersed in a suitable unmagnetic matrix) and magnetic recording media (tapes, disks and drums for digital, audio and video purposes) shall be called to notice.

In [1] Tiersten derived a system of nonlinear differential equations and boundary conditions governing the macroscopic behavior of non-conducting, arbitrarily anisotropic, magnetically saturated bodies undergoing large deformations, with the saturation condition given by:

$$\bar{\mu}\cdot\bar{\mu} = \mu_1^2 + \mu_2^2 + \mu_3^2 = \mu_s^2 = \text{const.} \tag{1}$$

For the understanding of the subject in question it is necessary to recapitulate those main features of this theory, to which it reduces in the case of quasistationary processes and homogeneous distributions of magnetic moments $\bar{\mu}$ per unit mass. Fig. 1 illustrates the forces and couples, then exerted on each volume element dV of the body $\mathcal{M}$. Herein the following designations have been used:

ρ	: mass density,
$\underline{\sigma} = \underline{\sigma}^S + \underline{\sigma}^A$	: (non-symmetric) mechanical Cauchy stress tensor ($\underline{\sigma}^S$: symmetric portion, $\underline{\sigma}^A$: antisymmetric portion of the stress tensor),
$\bar{\sigma}^A$	: pseudovector (axial vector) associated with the antisymmetric portion $\underline{\sigma}^A$ of the stress tensor according to $\sigma_i^A = e_{ikl}\sigma_{kl}^A/2$ (e_{ikl} : permutation tensor),
$\bar{B}$	: magnetic induction field.

It should be noted that the forces and couples presented in fig. 1 can be derived - under the stated simplified conditions - from Tiersten's integral formulations of the balance equations for linear - and angular momentum when applying the divergence theorem to the surface integrals which describe the contact force and contact couple on the surface $\partial\mathcal{B}$ of a portion $\mathcal{B}$ of the body $\mathcal{M}$. In place of $\overline{B}$, however, Tiersten employs the magnetic field vector $\overline{H}$ (for the procedure preferred here, see e.g. [2]). The field equations then have the form:

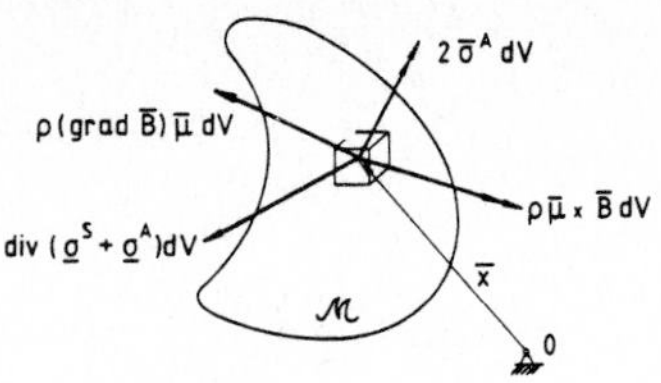

Figure 1
Forces and couples

balance of forces : $\mathrm{div}\,\underline{\sigma} + \rho\,(\mathrm{grad}\,\overline{B})\,\overline{\mu} = \overline{o}$, (2)

balance of couples: $\rho\,\overline{\mu}\times\overline{B} + 2\,\underline{\sigma}^A = \overline{o}$, (3)

Maxwell's equation: $\mathrm{div}\,\overline{B} = 0$. (4)

The accompanying constitutive equation is given by (upper index "T" marks the transposition of the tensor)

$$\underline{\sigma} = \rho\left[\underline{F}\,\frac{\partial U}{\partial \underline{E}}\,\underline{F}^T + \underline{F}\,(\frac{\partial U}{\partial \overline{N}}\otimes\overline{\mu})\right] , \tag{5}$$

where

U : internal energy per unit mass,

$\underline{F} = \mathrm{grad}\,\overline{x}$: deformation gradient,

$\underline{E} = \frac{1}{2}\,(\underline{F}^T\underline{F} - \underline{1})$: Green's strain tensor ($\underline{1}$: unit tensor),

$\overline{N} = \underline{F}^T\overline{\mu}$.

The internal energy is assumed as given function $U = U(\underline{E}, \overline{N})$ of the variables $\underline{E}$ and $\overline{N}$.

From (5) we obtain for the antisymmetric portion of the stress tensor

$$\underline{\sigma}^A = \frac{1}{2}\,(\underline{\sigma} - \underline{\sigma}^T) = \frac{1}{2}\,\rho\left[\underline{F}\,(\frac{\partial U}{\partial \overline{N}}\otimes\overline{\mu}) - (\overline{\mu}\otimes\frac{\partial U}{\partial \overline{N}})\,\underline{F}^T\right]. \tag{6}$$

For the internal energy Tiersten assumes a polynomial approximation of the form

$$U = \frac{1}{2\rho_o}\,c_{ijkl}\,E_{ij}E_{kl} + \frac{1}{2}\,\rho_o\chi_{ij}\,N_iN_j + \varepsilon_{ijk}\,E_{ij}N_k + \rho_o b_{ijkl}\,E_{ij}N_kN_l + \dots \tag{7}$$

where the material coefficients c_{ijkl}, χ_{ij}, ε_{ijk} and b_{ijkl} are referred to as the elastic, anisotropy, piezomagnetic and magnetostrictive constants, respectively. In (7) the exchange, magnetoexchange and exchangestrictive terms have been omitted, since they vanish in the case of a homogeneous distribution of magnetic moments studied here. Furthermore, in ferromagnetic media the piezomagnetic effect is negligibly small if at all existent [3]. According to (6) only derivatives of the internal energy U with respect to the variables N_i contribute to the skew-symmetric stress tensor. Therefore the antisymmetric stresses depend essentially on the specific form of the anisotropy- and magnetostriction energy.

In this paper we solely study the contribution of the magnetic anisotropy energy to the skew-symmetric stresses in single-domain particles. As distinguished from (7), we consider the following more general polynomial form for the anisotropy energy:

$$U^{aniso}(N_i) = U_0^{aniso} + \rho_o\mu_s\tau_iN_i + \frac{1}{2}\,\rho_o\chi_{ij}N_iN_j + \frac{\rho_o}{\mu_s}\,\eta_{ijk}N_iN_jN_k + \frac{1}{2}\,\frac{\rho_o}{\mu_s^2}\,\omega_{ijkl}N_iN_jN_kN_l + \frac{\rho_o}{\mu_s^3}\,\zeta_{ijklm}N_iN_jN_kN_lN_m + \frac{1}{3}\,\frac{\rho_o}{\mu_s^4}\,\psi_{ijklmn}N_iN_jN_kN_lN_mN_n \tag{8}$$

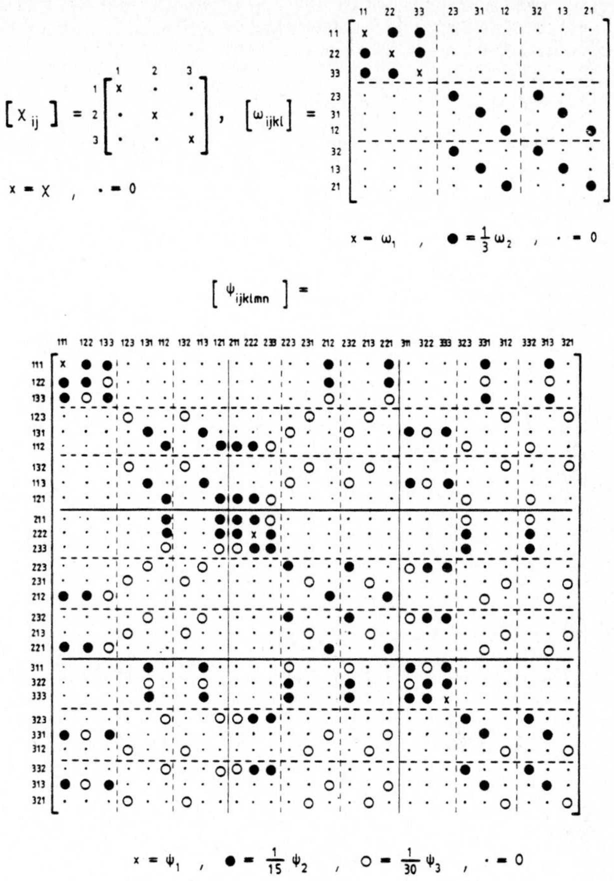

Figure 2
Non-vanishing material tensors

and turn our interest upon materials with cubic symmetry, of which e.g. iron and nickel are representatives. By consequently carrying out the symmetry operations which leave invariant the anisotropy energy of a cubic crystal, we can show that the material tensors of uneven rank vanish, i.e. $\tau_i = 0$, $\eta_{ijk} = 0$, $\zeta_{ijklm} = 0$, whereas the material tensors of even rank are given by the presentations shown in fig. 2.

If we collect all terms in (8) we arrive at

$$
\begin{aligned}
U^{aniso}(N_i) = U_0^{aniso} &+ \frac{1}{2}\rho_o \chi (N_1^2 + N_2^2 + N_3^2) + \\
&+ \frac{1}{2}\frac{\rho_o}{\mu_s^2}\Big[\omega_1 (N_1^4 + N_2^4 + N_3^4) + 2\omega_2 (N_1^2N_2^2 + N_2^2N_3^2 + N_3^2N_1^2)\Big] + \\
&+ \frac{1}{3}\frac{\rho_o}{\mu_s^4}\Big\{\psi_1 (N_1^6 + N_2^6 + N_3^6) + \psi_2\Big[N_1^4(N_2^2 + N_3^2) + N_2^4(N_3^2 + N_1^2) + \\
&\qquad + N_3^4(N_1^2 + N_2^2)\Big] + 3\psi_3 N_1^2 N_2^2 N_3^2\Big\} ,
\end{aligned}
\tag{9}
$$

containing the six material constants χ; ω_1, ω_2; ψ_1, ψ_2, ψ_3. Unfortunately, measurements of these coefficients are not known. In the case of small deformations, however, when $\bar{N} \approx \bar{\mu}$, because of the saturation condition (1) we additionally have

$$\mu_1^4 + \mu_2^4 + \mu_3^4 = \mu_s^4 - 2(\mu_1^2\mu_2^2 + \mu_2^2\mu_3^2 + \mu_3^2\mu_1^2),$$

$$\mu_1^4(\mu_2^2 + \mu_3^2) + \mu_2^4(\mu_3^2 + \mu_1^2) + \mu_3^4(\mu_1^2 + \mu_2^2) = \mu_s^2(\mu_1^2\mu_2^2 + \mu_2^2\mu_3^2 + \mu_3^2\mu_1^2) - 3\mu_1^2\mu_2^2\mu_3^2,$$

$$\mu_1^6 + \mu_2^6 + \mu_3^6 = \mu_s^6 - 3\mu_s^2(\mu_1^2\mu_2^2 + \mu_2^2\mu_3^2 + \mu_3^2\mu_1^2) + 3\mu_1^2\mu_2^2\mu_3^2,$$

and if we set

$$U_0^{aniso} = -\rho_o\mu_s^2\left(\frac{1}{2}\chi + \frac{1}{2}\omega_1 + \frac{1}{3}\psi_1\right),$$

we obtain the (linearized) anisotropy energy per unit mass for cubic crystals in the form

$$U_{cubic}^{aniso}(\mu_i) = \frac{\rho_o}{\mu_s^2}\left[(\omega_2 - \omega_1) + \frac{1}{3}(\psi_2 - 3\psi_1)\right](\mu_1^2\mu_2^2 + \mu_2^2\mu_3^2 + \mu_3^2\mu_1^2) + \frac{\rho_o}{\mu_s^4}(\psi_1 - \psi_2 + \psi_3)\mu_1^2\mu_2^2\mu_3^2 \tag{10}$$

(which also holds in the nonlinear case for vanishing deformations, i.e. $\underline{F} = \underline{1}$ or $\underline{E} = \underline{0}$). If we compare (10) with the corresponding result well-known in the physical-technical literature [4, 5]:

$$U_{cubic}^{aniso}(\alpha_i) = \frac{1}{\rho_o}\left[K_1(\alpha_1^2\alpha_2^2 + \alpha_2^2\alpha_3^2 + \alpha_3^2\alpha_1^2) + K_2\alpha_1^2\alpha_2^2\alpha_3^2\right], \tag{11}$$

where the $\alpha_i = \mu_i/\mu_s$ are the direction cosines of the magnetization and K_1, K_2 the so-called anisotropy constants, after all we find

$$K_1 = \rho_o^2\mu_s^2\left[(\omega_2 - \omega_1) + \frac{1}{3}(\psi_2 - 3\psi_1)\right], \quad K_2 = \rho_o^2\mu_s^2(\psi_1 - \psi_2 + \psi_3) \tag{12}$$

and thus:

$$U_{cubic}^{aniso}(\mu_i) = \frac{K_1}{\rho_o\mu_s^4}(\mu_1^2\mu_2^2 + \mu_2^2\mu_3^2 + \mu_3^2\mu_1^2) + \frac{K_2}{\rho_o\mu_s^6}\mu_1^2\mu_2^2\mu_3^2. \tag{13}$$

At room temperature the following approximate values for these coefficients have been measured [4, 5]: $K_1 = 4{,}4\cdot10^4$ N/m^2, $K_2 = 2\cdot10^4$ N/m^2 for iron; $K_1 = -4{,}5\cdot10^3$ N/m^2, $K_2 = 2\cdot10^3$ N/m^2 for nickel. From (11) or (13) we conclude, that the anisotropy energy takes the value 0 for magnetization parallel to the [100]-direction, the value $K_1/4\rho_o$ for magnetization parallel to the [110]-direction and the value $(K_1/3 + K_2/27)/\rho_o$ for magnetization parallel to the [111]-direction (cf. fig. 3). In the case $K_1 > 0$, $(K_1/3 + K_2/27) > K_1/4$ (satisfied for iron) the anisotropy energy takes an absolute minimum upon magnetization in [100]-direction ("easy axis") and an absolute maximum upon magnetization in [111]-direction ("hard axis") (for nickel the easy axis is hence given by the [111]-direction).

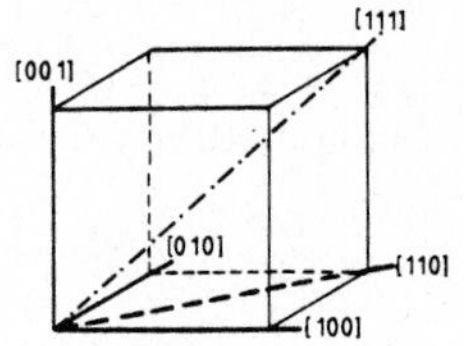

Figure 3
Crystal directions

In the case of small deformations, the expression (5) for the stress tensor reduces to

$$\underline{\sigma} = \rho_o\left[\frac{\partial U}{\partial \underline{\varepsilon}} + \frac{\partial U}{\partial \bar{\mu}} \otimes \bar{\mu}\right] \quad \text{or} \quad \sigma_{ik} = \rho_o\left[\frac{\partial U}{\partial \varepsilon_{ik}} + \frac{\partial U}{\partial \mu_i}\mu_k\right] \tag{14), (15}$$

in cartesian components, where $\underline{\varepsilon} = \varepsilon_{ik}\,\bar{e}_i \otimes \bar{e}_k$ is the infinitesimal strain tensor. Likewise, the skew-symmetric stress tensor (6) is then given by

$$\underline{\sigma}^A = \frac{1}{2}\,\rho_0 \left[\frac{\partial U}{\partial \bar{\mu}} \otimes \bar{\mu} - \bar{\mu} \otimes \frac{\partial U}{\partial \bar{\mu}} \right] \quad \text{or} \quad \sigma^A_{ik} = \frac{1}{2}\,\rho_0 \left[\frac{\partial U}{\partial \mu_i}\,\mu_k - \frac{\partial U}{\partial \mu_k}\,\mu_i \right] \qquad (16),\ (17)$$

in cartesian components.

We now consider two (quasistatic) processes in cubic single-domain crystallites (with the easy [100]-direction):

1) Rotation of magnetization by the angle φ about the $\bar{e}_3$-axis (cf. fig. 4)

With

$$\bar{\mu} = \mu_1\bar{e}_1 + \mu_2\bar{e}_2 + \mu_3\bar{e}_3 = \qquad (18)$$

$$= \mu_s(\cos\varphi\,\bar{e}_1 + \sin\varphi\,\bar{e}_2)\,,$$

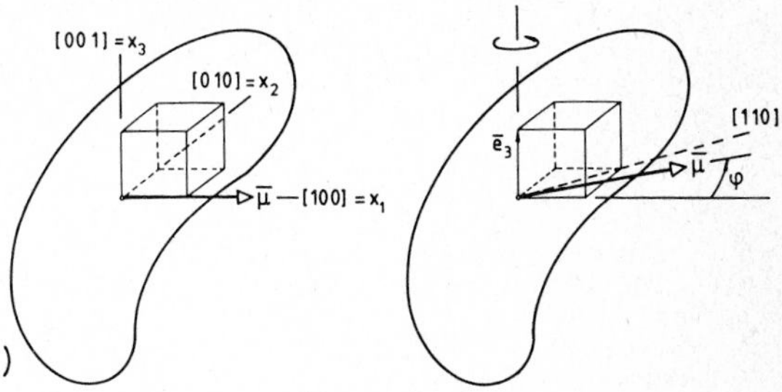

Figure 4
Magnetization rotation about $\bar{e}_3$-axis

we at once obtain from (13)

$$U^{aniso}_{cubic}(\mu_i) = \frac{K_1}{\rho_0\,\mu_s^4}\,\mu_1^2\mu_2^2 \qquad (19)$$

or

$$U^{aniso}_{cubic}(\varphi) = \frac{K_1}{\rho_0}\,\sin^2\varphi\,\cos^2\varphi = \frac{1}{4}\,\frac{K_1}{\rho_0}\,\sin^2 2\varphi\ . \qquad (20)$$

According to (17) we find the anti-symmetric stresses

$$\sigma^A_{12} = -\,\sigma^A_{21} = \frac{1}{2}\,\rho_0 \left[\frac{\partial U}{\partial \mu_1}\,\mu_2 - \frac{\partial U}{\partial \mu_2}\,\mu_1 \right] = \frac{K_1}{\mu_s^4}\,\mu_1\mu_2(\mu_2^2 - \mu_1^2)\ ,$$

$$\sigma^A_{23} = -\,\sigma^A_{32} = 0\ ,\quad \sigma^A_{31} = -\,\sigma^A_{13} = 0\ , \qquad (21)$$

or

$$\sigma^A_{12}(\varphi) = -\,\sigma^A_{21}(\varphi) = -\,\frac{1}{4}\,K_1\,\sin 4\varphi\ .$$

Fig. 5 illustrates these skew-symmetric stresses as function of the rotation angle φ: They vanish for magnetization parallel to an easy ([100]- and equivalent) axis as well as for magnetization parallel to a mean ([110]- and equivalent) axis and take their extreme value for the intermediate directions. Since

$$\left|\sigma^A_{12,max}\right| = \left|\sigma^A_{21,max}\right| = \frac{1}{4}\,K_1\ ,$$

the order of magnitude of these skew-symmetric stresses for iron is given by

$$\left|\sigma^A_{ik,max}\right| \approx 1{,}1\cdot 10^{-2}\ \text{N/mm}^2\ .$$

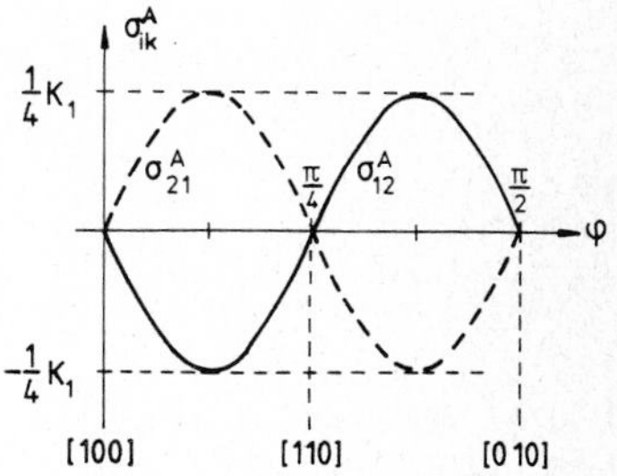

Figure 5
Skew-symmetric stresses for the rotation [100]→[110]

If we compare this value with the 0,2% - proof stress of iron (as a rough measure for elastically-supportable stresses), $\sigma_{0,2} \approx 10^2$ N/mm², we have

$$\left|\sigma^A_{12,max}\right| \approx 10^{-4}\ \sigma_{0,2}\ . \qquad (22)$$

In view of the order of magnitude of the stresses we have to deal with in conventional technical applications, the skew-symmetric stresses, quoted above, are therefore negligibly small. As we shall see, however, on a "micromagnetic scale" they are of the same order of magnitude as the symmetric stresses.

2) Rotation of magnetization by the angle ψ about the $(-\bar{e}_2+\bar{e}_3)/\sqrt{2}$ - axis (cf. fig. 6)

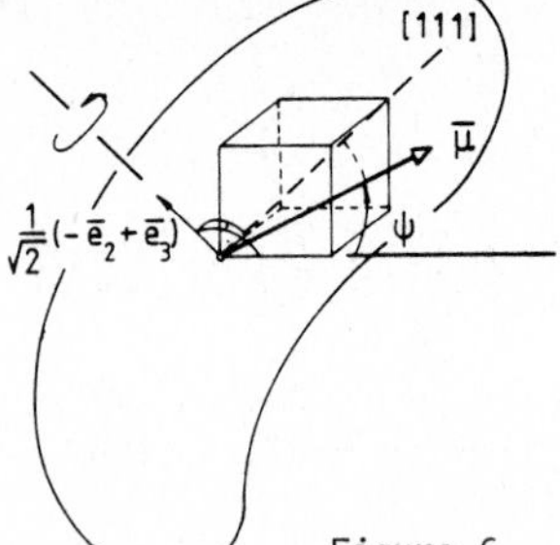

Figure 6
Magnetization rotation about $(-\bar{e}_2+\bar{e}_3)/\sqrt{2}$ - axis

With

$$\bar{\mu} = \mu_1\bar{e}_1 + \mu_2\bar{e}_2 + \mu_3\bar{e}_3 = \tag{23}$$

$$= \mu_s\left[\cos\psi\,\bar{e}_1 + \frac{1}{\sqrt{2}}\sin\psi\,\bar{e}_2 + \frac{1}{\sqrt{2}}\sin\psi\,\bar{e}_3\right],$$

we obtain from (13) the anisotropy energy in the form

$$U^{aniso}_{cubic}(\psi) = \frac{K_1}{\rho_0}\sin^2\psi(\cos^2\psi + \frac{1}{4}\sin^2\psi) + \frac{1}{4}\frac{K_2}{\rho_0}\sin^4\psi\cos^2\psi \quad . \tag{24}$$

This leads to the following skew-symmetric stresses:

$$\sigma^A_{12} = -\sigma^A_{21} = -\sigma^A_{31} = \sigma^A_{13} = \frac{\sqrt{2}}{4}\sin 2\psi\,(\frac{3}{2}\sin^2\psi - 1)(K_1 + \frac{1}{2}K_2\sin^2\psi) \quad ,$$
$$\sigma^A_{23} = -\sigma^A_{32} = 0 \quad . \tag{25}$$

As fig. 7 visualizes for iron (with the above mentioned values for K_1 and K_2), the anti-symmetric stresses vanish for magnetization in the easy as well as in the hard direction ($\psi \approx 54{,}74^0$). Their extreme value occurs at $\psi \approx 26{,}15^0$ (i.e. slightly closer to the easy direction) and amounts to

$$\left|\sigma_{ik,max}\right| \approx 0{,}9\cdot 10^{-2}\ \text{N/mm}^2 \ . \tag{26}$$

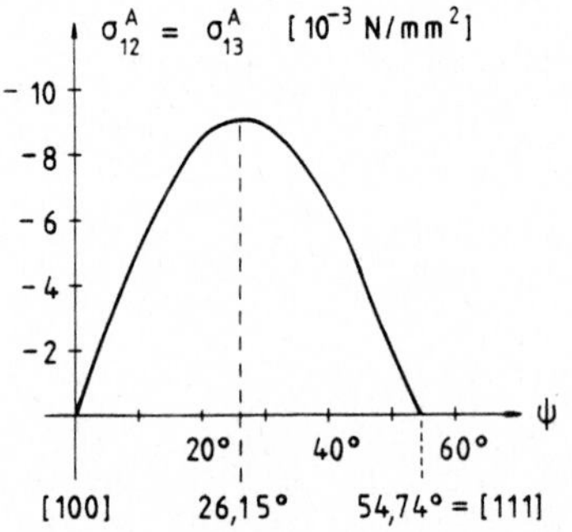

Figure 7
Skew-symmetric stresses in iron for the rotation [100]→[111]

We thus arrive at the same order of magnitude for the skew-symmetric stresses as in the rotation process 1).

Next, we compute the entire stress tensor, brought about in a cubic crystallite by the magnetic anisotropy in the rotation process 1): [100]→[110]. According to (14) or (15), we obtain the plane stress state

$$[\underline{\sigma}] = [\underline{\sigma}^S] + [\underline{\sigma}^A] = \frac{1}{2}K_1\left(\begin{bmatrix}\sin^2 2\varphi & \sin 2\varphi & 0\\ \sin 2\varphi & \sin^2 2\varphi & 0\\ 0 & 0 & 0\end{bmatrix} + \begin{bmatrix}0 & -\frac{1}{2}\sin 4\varphi & 0\\ \frac{1}{2}\sin 4\varphi & 0 & 0\\ 0 & 0 & 0\end{bmatrix}\right). \tag{27}$$

In fig. 8 the stress components $\sigma_{11} = \sigma_{22}$, $\sigma^S_{12} = \sigma^S_{21}$ and $\sigma^A_{12} = -\sigma^A_{21}$ are plotted as function of the rotation angle φ. It becomes evident that the absolute extremal value of the skew-symmetric stresses is half of the absolute extremal value of the symmetric stresses. Thus, as mentioned before, symmetric and antisymmetric stresses, due to magnetic anisotropy, are of the same order of magnitude.

For magnetization parallel to the [110]-direction ($\varphi = \pi/4$) we especially have the plane symmetrical stress tensor

$$\left[\underline{\sigma}\right]^{[110]} = \frac{1}{2} K_1 \begin{bmatrix} 1 & 1 \\ 1 & 1 \end{bmatrix} = \tag{28}$$

$$= \frac{1}{2} K_1 \begin{bmatrix} 1 & 0 \\ 0 & 1 \end{bmatrix} + \frac{1}{2} K_1 \begin{bmatrix} 0 & 1 \\ 1 & 0 \end{bmatrix},$$

the superposition of an axial-symmetric stress state and a simple shear stress state.

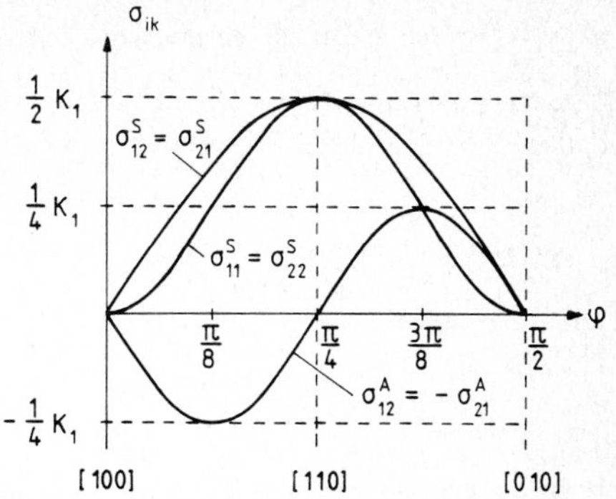

Figure 8
Stresses for the rotation [100]→[110]

For the rotation process 2): [100]→[111] we can likewise calculate the stress state, arriving at

$$\begin{aligned}
\sigma^S_{11} &= \frac{1}{2}\sin^2 2\psi \left[K_1 + \frac{1}{4}K_2 \sin^2\psi\right], \\
\sigma^S_{22} &= \sigma^S_{33} = \sigma^S_{23} = \sin^2\psi\left[K_1(1 - \frac{1}{2}\sin^2\psi) + \frac{1}{8}K_2 \sin^2 2\psi\right], \\
\sigma^S_{12} &= \sigma^S_{13} = \frac{\sqrt{2}}{4}\sin 2\psi\left[K_1(1 + \frac{1}{2}\sin^2\psi) + \frac{1}{2}K_2 \sin^2\psi(1 - \frac{1}{2}\sin^2\psi)\right], \\
\sigma^A_{12} &= -\sigma^A_{21} = \sigma^A_{13} = -\sigma^A_{31} = \frac{\sqrt{2}}{4}\sin 2\psi(\frac{3}{2}\sin^2\psi - 1)\left[K_1 + \frac{1}{2}K_2 \sin^2\psi\right], \\
\sigma^A_{23} &= -\sigma^A_{32} = 0 .
\end{aligned} \tag{29}$$

For magnetization parallel to the [111]-direction ($\psi = \arcsin(\sqrt{2/3}) \approx 54{,}74°$), we particularly obtain the symmetric stress state

$$\left[\underline{\sigma}\right]^{[111]} = \frac{4}{9}(K_1 + \frac{1}{6}K_2)\begin{bmatrix} 1 & 0 & 0 \\ 0 & 1 & 0 \\ 0 & 0 & 1 \end{bmatrix} + \frac{4}{9}(K_1 + \frac{1}{6}K_2)\begin{bmatrix} 0 & 1 & 1 \\ 1 & 0 & 1 \\ 1 & 1 & 0 \end{bmatrix}, \tag{30}$$

the superposition of a hydrostatic stress state and a general shear stress state.

Finally, we determine the induction field $\overline{B}$ requisite for the magnetization rotation of the process 1): [100]→[110]. We set (cf. fig. 9)

$$\overline{B} = B(\cos\vartheta\,\overline{e}_1 + \sin\vartheta\,\overline{e}_2), \tag{31}$$

which delivers for the balance of couples (3), together with (18) and $\sigma^A_3 = (\sigma^A_{12} - \sigma^A_{21})/2 = -(K_1 \sin 4\varphi)/4$ (cf. (21)) the equation

$$\rho_o \mu_s B \sin(\vartheta - \varphi) - \frac{1}{2}K_1 \sin 4\varphi = 0 \tag{32}$$

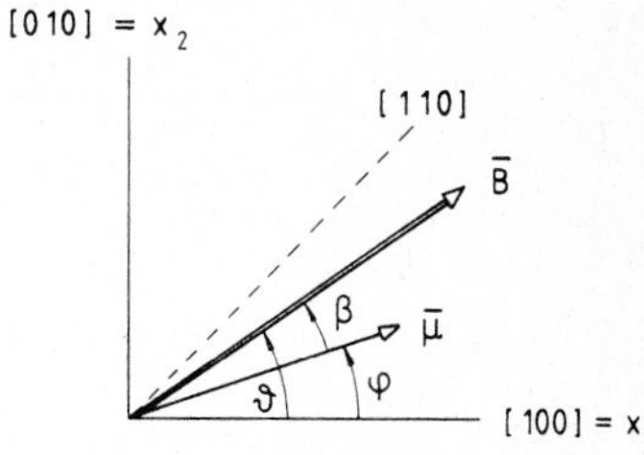

Figure 9
Notation for rotation [100]→[110]

for $B = B(\varphi;\vartheta)$. This equation can also be produced by starting from the "potential energy per unit mass"

$$U^* = U^{aniso}_{cubic} + U^{field} = \frac{K_1}{\rho_o \mu_s^4}\mu_1^2\mu_2^2 - \overline{\mu}\cdot\overline{B} = \frac{1}{4}\frac{K_1}{\rho_o}\sin^2 2\varphi - \mu_s B\cos(\vartheta - \varphi), \tag{33}$$

where (19) and (20), respectively, have been used. The extreme values of U^* are given by

$$\frac{dU^*}{d\varphi} = \frac{1}{2}\frac{K_1}{\rho_o} \sin 4\varphi - \mu_s B \sin(\vartheta - \varphi) = 0 \quad , \tag{34}$$

clearly equation (32); the minima follow from

$$\frac{d^2U^*}{d\varphi^2} = 2\frac{K_1}{\rho_o} \cos 4\varphi + \mu_s B \cos(\vartheta - \varphi) > 0 \quad . \tag{35}$$

Fig. 10 shows, for field directions $\vartheta = \pi/8$ and $\vartheta = 0$ or $\pi/2$, the magnetization curves computed from (34) and (35), when introducing the variables $\beta = \vartheta - \varphi$ (angle between field and magnetization) and $b = B/(2K_1/\rho_0\mu_s)$: The piercing curves belong to the absolute minimum, the interrupted curves to the remaining relative minima of the potential energy and may therefore be regarded as "metastable equilibrium configurations". For completeness, fig. 11 illustrates the dependence of the potential energy U^* upon the field variable b and the angle β.

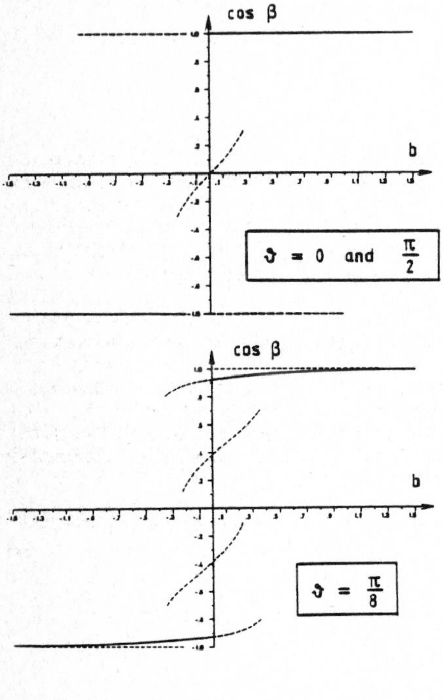

Figure 10 Magnetization curves

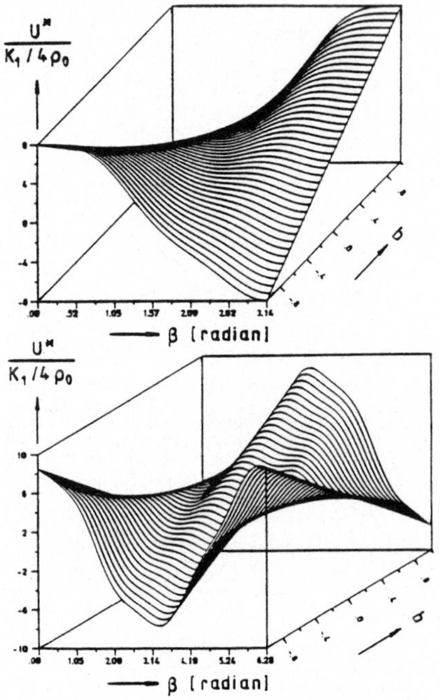

Figure 11 Potential energy

The author wants to thank E. Waigel and M. Grafmüller for the numerical calculations and the production of the figures.

[1] Tiersten, H.F., Coupled magnetomechanical equations for magnetically saturated insulators, J. Math. Phys. 9 (1964) 1298 - 1318.

[2] Stratton, J.A., Electromagnetic theory (McGraw - Hill, New York, 1941).

[3] Bhagavantam, S., Crystal symmetry and physical properties (Academic Press, London, New York, 1966).

[4] Chikazumi, S., Physics of magnetism (John Wiley & Sons, New York, 1966).

[5] Morrish, A.H., The physical principles of magnetism (John Wiley & Sons, New York, 1965).

The Mechanical Behavior of Electromagnetic Solid Continua
G.A. Maugin (editor)
Elsevier Science Publishers B.V. (North-Holland)

ΔE EFFECT IN Ni-Mn-Co FERRITES

Zbigniew Kaczkowski

Polish Academy of Sciences Institute of Physics
02-668 Warszawa, Al. Lotników 32/46, Poland

The dependence of the modulus of elasticity at constant magnetic field H and at constant induction B, i.e. E_H and E_B, on static magnetic field H and temperature T in $Ni_{0.953}Mn_{0.02}Co_{0.027}Fe_2O_4$ ferrites are presented. The minimum of E_H vs H ("negative ΔE effect) and minima E_H and E_B vs T near the room temperature, in the compensation range of magnetocrystalline anisotropy constants, are observed.

INTRODUCTION

In consideration of basic magnetomechanical dependences occuring in piezomagnetic (magnetostrictive) materials two relations are taken as a starting point: one magnetic ($B = \mu H = \mu_o H + J = \mu_o (H + M)$) and another one mechanical, i.e. Hooke's law ($T = \sigma = \varepsilon E^o = SE$) supplemented by the terms connected with magnetomechanical interactions.

It can be verified experimentally, that on the one hand majority of magnetics shows in the presence of magnetic field (lower than that of saturation) deviation from the conventional Hooke's law, and on the other hand these materials exhibit, in most cases, variations of their magnetic properties under the influence of applied mechanical stresses. These interactions manifest themselves by variation of elasticity moduli E, G or K or variation of the such magnetic properties as magnetization M (induction B or polarization J or I), permeability μ, (susceptibility $\varkappa$), remanence B_r or coercive force H_c. For these reasons a supplementary term ΔB accounting for the influence of mechanical stresses on magnetic properties and the term ΔS accounting for the influence of magnetic field on mechanical properties should be added, i.e.

$$B = \mu_T H + \Delta B = \mu_o \mu'_T H + d\,T \quad , \tag{1}$$

$$S = \varepsilon = \frac{T}{E_H} + \Delta S = \frac{T}{E_H} + d\,H \quad , \tag{2}$$

where B, H, μ_T, μ'_T are magnetic induction field, permeability of the free vibrating sample (at constant stress T) and its relative value ($\mu'_T = \mu_T/\mu_o$, where $\mu_o = 4\pi \cdot 10^{-7}$ H/m respecively, T or σ S or ε and E_H are mechanical stress, strain and Young's modulus at constant magnetic field H respectively, d is piezomagnetic (stress) sensitivity coefficient and increments ΔB and ΔS, representing the influence of mechanical and magnetic interactions, can assume positive or negative values, according to the sign of magnetostriction.

For very small variations, magnetomechanical processes become reversible and increments can be replaced by differentials [1].

When the tension is applied to the unmagnetized magnetics, its length increases as a result of a purely elastic elongation acording to the Hooke's law and an additional elongation resulting from the orientation of magnetic domain under stress. The application of the magnetic field to the magnetics produces the process of ordering of domain magnetization towards the direction of the applied field. The magnetic moment of any one domain is specified by the magnitude and direction of its magnetization and by its volume. The moments of the domains are changed by usualy reversible rotation or by change in volume of the domain realised by the reversible and irreversible Bloch walls (boundaries) displacements (shifting, moving).

Changes of the domain volumes and rotations of magnetization vectors, caused by the applied magnetic field, change the form and volume of materials (linear and volume magnetostriction).

Δ E EFFECT

With magnetostriction, i.e. changes of dimensions and form with magnetization, is connected ΔE effect, i.e. change of elasticity modulus with this magnetization. Measure of the Δ E effect is its relative difference of the elasticity moduli at magnetic saturation E_s and at demagnetization state E_o, i.e. $\Delta E/E_s = (E_s - E_o)/E_s = 1 - E_o/E_s$ or $\Delta E/E_o = (E_s - E_o)/E_o = (E_s/E_o) - 1$. In the other pair of piezomagnetic equations are another magnetomechanical (piezomagnetic) coefficients [1] and

$$S = \frac{T}{E_B} + g\,B \quad , \tag{3}$$

$$T = E_B S - hB \quad , \tag{4}$$

$$T = E_H S - eH \quad , \tag{5}$$

where E_B is modulus of elasticity at constant induction (or E_M - at constant magnetization M or E_I - at constant polarization), g, h, e are piezomagnetic coefficients [1] analogical to the piezoelectric coefficients.

The magnetomechanical terms added to the conventional Hooke's law [(2) - (5)] may be equal zero in three cases, when magnetomechanical coupling vanishes, i.e. at demagnetization state H=0, B=0, at magnetic saturation M_s or B_s and at (or near) coercitive force (H_c, B=0). In these cases moduli of elasticity at constant field E_H and induction E_B are equal to each other, i.e.: $E_{Ho} = E_{Bo} = E_o$, $E_{Hs} = E_{Bs} = E_s$, $E_{Hc} = E_{Bc} = E_c$.
The effects of tensile forces (stress T or σ) are in addition to normal elongation ε_n, according to the Hooke's law, the additional

elongations produced by the striction (mechanostriction ε_m or magnetostriction λ) and so the resultant elasticity modulus E is smaller

$$E = \frac{T}{S} = \frac{\sigma}{\varepsilon} = \frac{\sigma}{\varepsilon_m + \varepsilon_n} = \frac{\sigma}{\lambda + \sigma/E_n} . \tag{6}$$

After transformations

$$\varepsilon_m = \lambda = \left(\frac{1}{E} - \frac{1}{E_n}\right)\sigma, \tag{7}$$

whence

$$\frac{\Delta E}{E} = \frac{E_n - E}{E} = \frac{\varepsilon_m}{\varepsilon_n} . \tag{8}$$

For magnetic saturation $E_s = E_n$, $\varepsilon_m = \lambda_s$ (saturation magnetostriction) and Δ E effect will be

$$\frac{\Delta E}{E} = \frac{E_s - E_o}{E_o} = \frac{E_s}{E_o} - 1 = \frac{\lambda_s}{\varepsilon_n} . \tag{9}$$

EXPERIMENTAL

Piezomagnetic $Ni_{.953}Mn_{.02}Co_{.027}Fe_2O_4$ ferrites (E1) were investigated. Their density was equal 5.11 Mg/m^3, saturation induction B_s = = 0.284 T saturation magnetostricion $\lambda_s = -29.10^{-6}$ relative static initial permeability $\mu_i' = 63$ and maximum permeability 300, remanence $B_r = 0.163$ T and coercitivity $H_c = 300$ A/m (3.7 Oe) [2] and relative maximum reversible permeability at 0.2 A/m $\mu_{Ti}' = 64$ [3]. Testing was performed on toroidal samples with mechanical resonance frequencies changing from 70 to 100kHz, outer diameter 20.7 mm, inner diameter 13.7 mm and thickness 5.3 to 6.0 mm [4-6].

Dynamic elasticity moduli are determined at constant magnetic field H and at constant induction B

$$E_H = \pi^2 d_s^2 f_H^2 \rho = c_H^2 \rho \approx \pi^2 d_s^2 f_r^2 \rho , \tag{10}$$

$$E_B = \pi^2 d_s^2 f_B^2 \rho = c_B^2 \rho \approx \pi^2 d_s^2 f_a^2 \rho . \tag{11}$$

Here f_H, f_B, c_H and c_B are the respective frequencies and ultrasonic velocities at a constant H or B respectively, f_r and f_a - frequencies of resonance and antiresonance, d_s is mean diameter of toroid $(d_{out} + d_{in})/2$ and ρ is density.

In practice and, in particular, in case of the ferrites, it may be assumed approximately that the maximum of impedance moduli Z occurs at the mechanical resonance what under sufficiently large quality factor Q, corresponds to the constant magnetic field (constant current source), i.e. $f_H \approx f_r$. In the case of antiresonance, minimum corresponds to the constant voltage source and hence the magnetic induction is then constant, $f_B \approx f_a$ [7].

Impedance of transducer with respect to the nonlinear phenomena and losses occuring in core and winding of transducer depends both on the constant and alternating magnetic field (Fig. 1) [5,8].

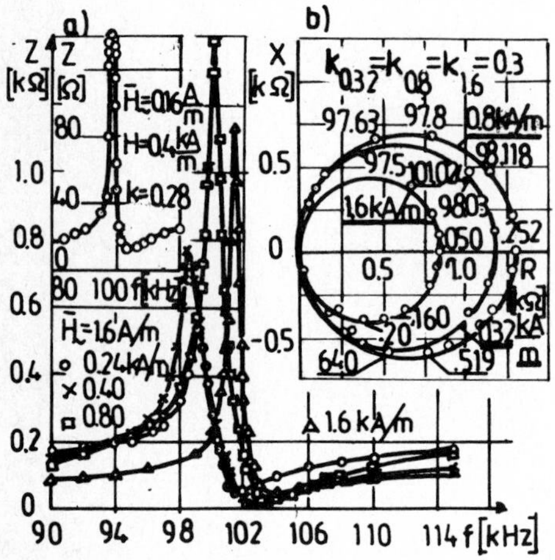

Figure 1
Impedance Z vs frequency f (a) [5] and impedance motional circle (b) [8] of toroidal sample of the investigated ferrite E1 at various polarization H and amplitude $\bar{H}$

RESULTS

Fig. 2 presents the family of initial curves of elasticity moduli of investigated ferrite. The minima of E_H were observed for all temperatures higher than 200 K in E1 ferrite and also in other Ni-Co ferrites, e.g. [4-6,8-15]. The minimum deepness in E1 ferrite is near room temperature what is connected with temperature compensation range of magnetocrystalline anisotropy constants.

In 1846 Guillemin gave out the hypothesis upon the magnetization influence on material elasticity [16]. First experimental results on ΔE effect was presented in 1902 [17]. In 1955 it was observed by Ochsenfeld that in the case of nickel and its alloys the modulus of elasticity initially decreases with increases polarization and after reaching a minimum it begins to increase so as to reach a stable maximum value in the range of magnetic saturation [18].

The range in which the modulus E_H is smaller than the value in the state of demagnetization E_o is called the negative increment range of E, and the effect itself - the negative ΔE effect. In this case as a relation point is chosen the value of E_o insted at saturation E_s (Fig. 3). The temperature dependencies of moduli at various magnetic polarization are presented in Fig. 4 [5,6,20,21].

DISCUSSION AND CONCLUSIONS

Under ideal reversible conditions the work done during elastic deformation is completly transformed into the potential energy of internal elasticity forces. The specific elastic energy in magnetics subjected to the magnetic field is changed due to the changes of the elasticity modulus. The values of the moduli E_H and E_B are equal to each other in the state demagnetization (Fig. 5), but differ in

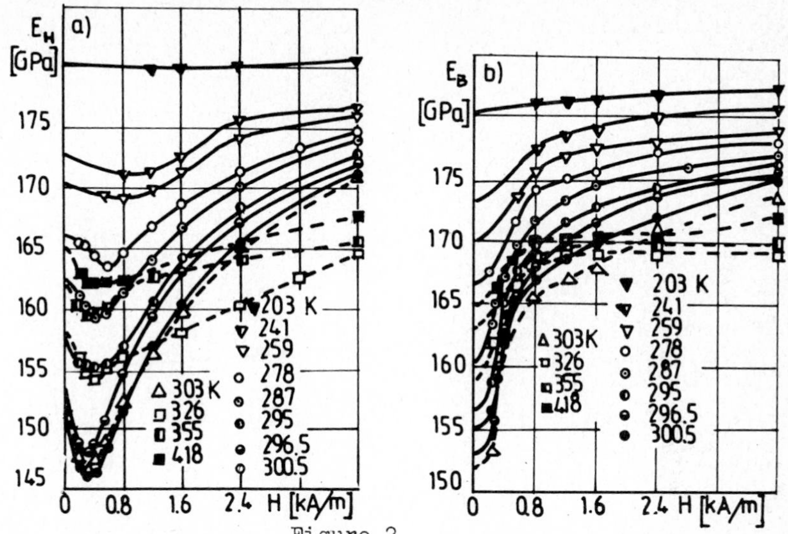

Figure 2
Dependence of moduli E_H (a) and E_B (b) on magnetic polarization H at various temperatures

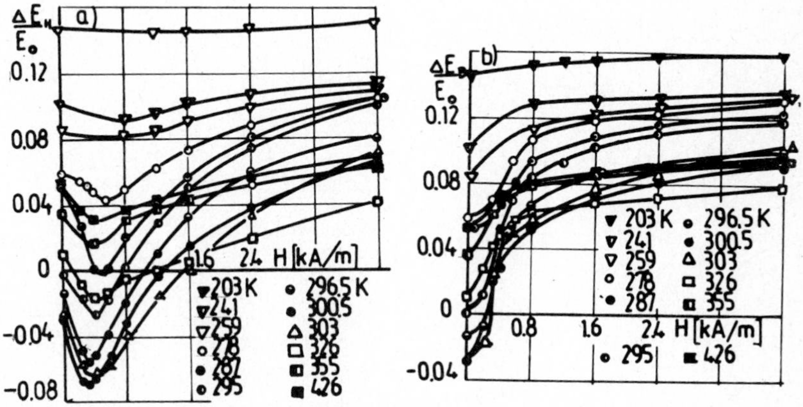

Figure 3
ΔE_H and ΔE_B effects at various temperatures

the case of polarization. The ratio of mechanical energy transformed into magnetic energy W_μ to the total energy W_H stored in piezomagnetic material is called square of magnetomechanical coupling coefficient

$$k^2 = \frac{W_\mu}{W_H} = \frac{W_H - W_B}{W_H} = 1 - \frac{W_B}{W_H} = 1 - \frac{E_H}{E_B} . \tag{12}$$

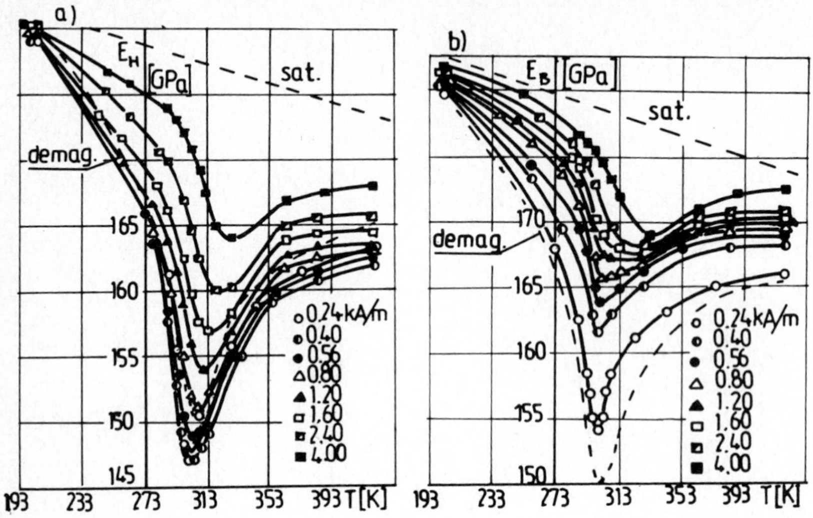

Figure 4
Temperature dependencies of E_H (a) and E_B (b) moduli at various polarizations

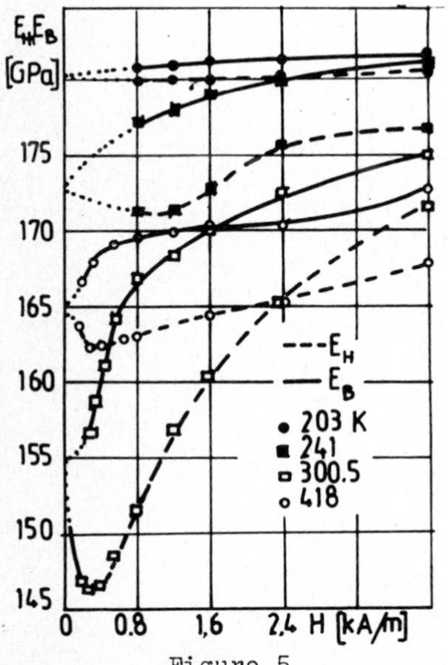

Figure 5
Primary curves of the E_H and E_B moduli below, at and above compensation temperature of K

Values of k coefficient are given in Fig. 1. When magnetomechanical coupling does not occur the elastic moduli E_H and E_B are equal to each other. According to the theory of Akulov and Kondorsky [22] it is associated with the domain structure of magnetics. The additional change of the magnetics form which does not result from Hooke's law was called mechanostriction [22] and its theory was developed by Brown [23]. The interpretations was given by author in previous papers, e.g. [6,10]. The ΔE effect is an even phenomenon and does not occur in the case of 180° Bloch walls motion. It is connected with rotation of the magnetization vectors and 180° displacements of domain walls. The negative ΔE effect occurs when $k^2 > \Delta E_B/E_B$. If the magnetocrystalline anisotropy constant are compensated, e.g. at 300 K in E1 ferrite [24], the moduli of elasticity reache minimum values [25]. Slight changes of cobalt contents produces considerable variations of the moduli [6,10,12,25,26].

REFERENCES

[1] Kaczkowski, Z., The coefficients of magnetostrictive equations and their relationships, Proc. Vibr. Probl. 2 (1961) 457-468
[2] Kaczkowski, Z., Static magnetic properties of piezomagnetic Ni-Co-Cu-Mn ferrites, Electron Technology 6,3 (1973) 103-116
[3] Kaczkowski, Z., Dynamic magnetic properties of piezomagnetic Ni-Co-Cu-Mn ferrites, Electron Technology 7,1 (1974) 93-107
[4] Kaczkowski, Z., Dependence of mechanical resonance frequency on magnetic polarization and temperature in Ni-ferrite with Mn and Co admixtures, Bull. Acad. Polon. Sci., Ser. sci. techn. 11 (1963) 443-448
[5] Kaczkowski, Z., Ekstrema przenikalności i modułów sprężystości ferrytów Ni-Co-Mn, Arch. Elektrot. 15 (1966) 3-28
[6] Kaczkowski, Z., Współczynniki piezomagnetyczne ferrytów magnetostrykcyjnych i ich histereza magnetyczna, Pr. Inst. Fizyki PAN, 39 (1972) 1-378
[7] Kaczkowski, Z., Uproszczony układ pomiarowy do wyznaczania częstotliwości rezonansu mechanicznego i współczynnika sprzężenia magnetomechanicznego k materiałów magnetostrykcyjnych, Arch. Elektrot. 11 (1962) 635-639
[8] Kaczkowski, Z., Impedance of some piezomagnetic ferrites in Proc. 7th Int. Congress on Acoustics, 4, 341-344 (Akademiai Kiado, Budapest, 1971)
[9] Kaczkowski, Z., Hysteresis of the magnetomechanical parameters of type E1 magnetostrictive ferrite, Sov. Phys. Acoust. 9 (1963) 29-36
[10] Kaczkowski, Z., Dependence of dynamic elasticity moduli of ferrites on magnetic polarization, Arch. Mech., 33 (1981) 401-414
[11] Kuznetsov, W.N., ΔE effekt and internal friction in ferrites at ultrasonic frequencies, Fiz. Metals Metallography, 19 (1965) 123-128
[12] Kataev, G.I., Magnetostriction and negative ΔE effect in ferrites, in Physical and physicochemical properties of ferrites 161-168 (in Russian)
[13] Kaczkowski, Z., and Walecki, T., Influence of magnetic history on some mechanical properties in $Ni_{0.968}Co_{0.012}Mn_{0.02}Fe_2O_4$ magnetostrictive ferrite, J. Magn. Magn. Mat. 1984 (6th Conf. on Soft Magnetic Materials in Eger)
[14] Ponkrateva, R.I., and Presnova, Magnetic annealing of magnetostrictive ferrites, Sov. Phys. Acoust. 17 (1971) 220-223
[15] Golyamina I.P. and Chulkova, Dependence of Young's modulus and the mechanical quality factor of magnetostrictive ferrites on the magnetization, Sov. Phys. Acoust., 12 (1966) 374-379
[16] Kikuchi, Y., Magnetostrictive materials and applications, IEEE Trans. Magn., MAG-4 (1968) 107-117
[17] Guillemin, A.V., Observation relatives an changement qui se produit dans l'élasticité d'un barreau de fer doux sous l'influence de l'électricité, C. r. hebd. Seanc. Acad. Sci. Paris, 22 (1846) 264-265, 433-443
[18] Honda, K., Shimizu, S., and Kusakabe, S., Change of the modulus of ferromagnetic substances by magnetization, Phil. Mag. 4 (1902) 459-468
[19] Ochsenfeld, R., Über die magnetoelastischen Eigenschaften einiger ferromagnetischer Eisen-Nickel Legirungen, Z.Phys. 143 (1955) 375-391
[20] Kaczkowski, Z., Piezomagnetic properties of the Ni-Mn-Co-Cu ferrites in the range of anisotropy compensation, J. Physique, 38 (1977) C1-203-206

[21] Kaczkowski, Z., Temperature dependence of piezomagnetic coefficients of ferrites, Sov. Phys. Acoust., 14 (1968) 183-190
[22] Akulov, N.S., and Kondorsky, E.I., Über einen magnetomechanischen Effect, Z. Phys. 78 (1932) 801-807
[23] Brown, W.F., Theory of magnetoelastic effect in ferromagnetism, J. Appl. Phys., 36 (1965) 994-1000
[24] Kaczkowski, Z., The effect of temperature upon the dynamic magnetic properties of the piezomagnetic Ni-Co-Cu-Mn ferrites, Electron Technology, 7,3(1974) 19-38
[25] van der Burgt, C.M., Performance of ceramic ferrite resonators as transducers and filter elements, J. Acoust. Soc. Am. 28 (1956) 1020-1032
[26] Kaczkowski, Z., and Walecki, T., Magnetic hysteresis of the elastic moduli and internal friction in piezomagnetic Ni-Mn-Co-Cu ferrites, in Ferrites, Proc. Int. Conf. Japan, (1980) 229-232

The Mechanical Behavior of Electromagnetic Solid Continua
G.A. Maugin (editor)
Elsevier Science Publishers B.V. (North-Holland)

OSCILLATIONS OF A BLOCH WALL IN ELASTIC FERROMAGNETS

S. MOTOGI+ and G.A. MAUGIN

Université Pierre et Marie Curie, Laboratoire de Mécanique Théorique associé au C.N.R.S., Tour 66-4, place Jussieu 75230 Paris, FRANCE

+Present address; Osaka Municipal Technical Research Institute 1-6-50, Morinomiya Joto-ku, Osaka, JAPAN

Dynamic magnetoelastic couplings in a Bloch wall are investigated theoretically. The initial magnetostriction gives a wall stiffness and the dynamic couplings increase the wall mass, hence, the resonance frequency of wall oscillations is decreased. The localized strains excited by wall oscillations are shown to be greater than traveling strains.

INTRODUCTION

Dynamic magnetoelastic couplings in magnetic domain walls contain two different interactions; one is the coupling between static strains and wall oscillations or spin waves in the wall, the other is the interaction between dynamic strains and spin oscillations in the wall. Winter (1961) showed that a Bloch wall is unstable for small perturbations of spins, and then he introduced an additional energy to represent a restoring force (wall stiffness) which yields the resonance of wall oscillations. Kittel & Galt (1976) have suggested that the wall stiffness arises from spin-defects interactions. Hence, we may regard that the wall oscillation dynamics is essentially magnetoelastic or magneto-mechanical. We have shown in another paper that the static initial magnetostriction plays the role of restoring force for small displacements of Bloch and Néel walls and, consequently, it yields a wall stiffness (Motogi & Maugin (1983)). On the other hand, the fully dynamical interactions between strains and walls were intensively studied in the last several years in relation with the hypersound excitations by wall oscillations since the resonance of wall oscillations is rather easier to achieve than that of uniform precessions of spins in domains. (See, e.g., Mitin & Tarasov (1977) and Lugovoi & Turov (1981) in the case of anti-ferromagnets). These theories are mainly concerned with spin wave-elastic wave interactions in the wall. The excited strains are obtained in the form of traveling waves. However, these traveling waves are quite small in their intensities in the long wavelength limit which may be applicable to the frequency range, even in the case of the resonance of wall oscillations.

Here we study the fully dynamical interactions of strains and wall oscillations taking account of the spin damping. First, we shall survey briefly the field equations and the results concerning the effect of initial magnetostriction, and we seek the localized strains excited by Bloch wall oscillations in uniaxial cubic ferromagnets. The whole system of equations is satisfied by these localised strains within the limit of the long wavelength approximation. Moreover, we shall show that the localized strains are much greater than the traveling strains in their intensities.

GENERAL EQUATIONS

The phenomenological theory of magnetoelastic interactions in ferromagnets gives the fundamental field equations as follows in the case of small deformations (Maugin (1976));

Mass Conservation: $$\rho = \rho_R(1 - u_{i,i}), \tag{1}$$

Momentum Conservation: $$\rho_R \frac{\partial^2 u_i}{\partial t^2} = t_{ji,j} + {}_M f_i - (2\gamma)^{-1} \epsilon_{ijk} (\rho \mathcal{R}_k)_{,j}, \tag{2}$$

Magnetization Precession: $$\frac{\partial \mu_i}{\partial t} = \gamma \epsilon_{ijk} \mu_j (H_k + {}_L H_k + \rho_R^{-1} \mathcal{B}_{mk,m}) + \mathcal{R}_i = \gamma \epsilon_{ijk} \mu_j H_k^{eff} + \mathcal{R}_i , \tag{3}$$

where the meaning of symbols is as follows; ρ is the mass density in a deformed configuration, ρ_R is the mass density in a reference configuration, u_i is the displacement vector, t_{ji} is the stress tensor, ${}_M f_i$ is the magnetic body force (${}_M f_i = M_j H_{i,j}$ in quasi magnetostatics in insulators), μ_i is the magnetization per unit mass, γ is the gyromagnetic ratio, ϵ_{ijk} is the permutation symbol, H_k is the Maxwellian field, ${}_L H_k$ is the local (anisotropy) field, $\mathcal{B}_{mk}$ is the spin-interaction tensor, $\mathcal{R}_i$ is the spin damping vector and H_k^{eff} is the effective field ($H_k^{eff} = H_k + {}_L H_k + \rho_R^{-1} \mathcal{B}_{mk,m}$). The magnetization per unit volume is defined as $M_j = \rho \mu_j$. The constitutive quantities are derived from the free energy Σ per unit volume;

$$t_{ji} = \frac{\partial \Sigma}{\partial e_{ji}}, \quad {}_L H_i = -\frac{1}{M_s} \frac{\partial \Sigma}{\partial \alpha_i}, \quad \rho_R^{-1} \mathcal{B}_{mk} = \frac{1}{M_s} \frac{\partial \Sigma}{\partial \alpha_{k,m}}, \tag{4}$$

where e_{ij} is the strain tensor, and $M_s = \rho_R \mu_s$ (μ_s: saturation magnetization per unit mass), and α_i is the director cosine of μ_i.

BLOCH WALL IN UNIAXIAL CUBIC CRYSTALS

We shall consider an isolated 180° Bloch wall with the easy axis in the x-direction and with the angle θ between magnetizations and the easy axis in the static ground state ($\theta = \theta(z)$). The energy expressions for uniaxial cubic ferromagnets are given by

$$\begin{aligned}
\Sigma^{ex} &= \tfrac{1}{2} \lambda \alpha_{i,j} \alpha_{i,j} , \\
\Sigma^{anis} &= \tfrac{1}{2} K(\alpha_y^2 + \alpha_z^2) \\
\Sigma^{me} &= B_1 (\alpha_x^2 e_{xx} + \alpha_y^2 e_{yy} + \alpha_z^2 e_{zz}) + 2B_2 (\alpha_x \alpha_y e_{xy} + \alpha_y \alpha_z e_{yz} + \alpha_z \alpha_x e_{zx}), \\
\Sigma^{el} &= \tfrac{1}{2} c_{11} (e_{xx}^2 + e_{yy}^2 + e_{zz}^2) + c_{44} (e_{xy}^2 + e_{yz}^2 + e_{zx}^2) \\
&\quad + c_{12} (e_{xx} e_{yy} + e_{yy} e_{zz} + e_{zz} e_{xx}) ,
\end{aligned} \tag{5}$$

where Σ^{ex} is the exchange energy, Σ^{anis} is the magnetic anisotropy energy, Σ^{me} is the magnetoelastic energy, Σ^{el} is the elastic energy. The free energy Σ is given by the sum of these energies.

We shall now distinguish two states of the material; i.e., a time-independent ground state (with suffix 'o') and a dynamic perturbed state (with upper bars). The variables can be decomposed into two parts:

$$e_{ij} = e_{oij} + \bar{e}_{ij}, \quad \alpha_i = \alpha_{oi} + \beta_i, \text{ etc.}$$

$\underset{\sim}{\beta}$ is the perturbation component of the director cosine of equilibrium magnetization. The spin damping vector $\mathcal{R}_i$ is written as

$$\mathcal{R}_i = - \frac{1}{\rho_R \gamma \tau} \epsilon_{ijk} \alpha_{oj} \left[\beta_{k,t} - \bar{u}_{[k,m]t} \alpha_{om} \right] , \tag{6}$$

where τ is the relaxation time. This is a rotationally invariant generalization of the spin damping proposed by Gilbert (1956) to the case of deformable ferromagnets (see Maugin (1975)). We note that in the linear theory the first term in the r.h.s. of eqn(6) need not to be replaced by the effective field.

A simple perturbation procedure yields the following equations in the one dimensional case;

in the ground state:

$$\theta'' - K\sin\theta\cos\theta + 2\left\{B_1 \sin\theta\cos\theta(e_{oxx} - e_{oyy}) + B_2(\sin^2\theta - \cos^2\theta)e_{oxy}\right\} = 0.$$

in the dynamic state: (7)

$$\rho_R \bar{u}_{x,tt} = c_{44}\bar{u}_{x,zz} + 2B_2(\cos\theta\, \beta_z)_{,z} + (2\gamma^2\tau)^{-1}\left[\cos\theta \left\{\beta_z + \tfrac{1}{2}(\bar{u}_{x,z}\cos\theta + \bar{u}_{y,z}\sin\theta)\right\}\right]_{,zt} , \tag{8}$$

$$\rho_R \bar{u}_{y,tt} = c_{44}\bar{u}_{y,zz} + 2B_2(\sin\theta\, \beta_z)_{,z} + (2\gamma^2\tau)^{-1}\left[\sin\theta \left\{\beta_z + \tfrac{1}{2}(\bar{u}_{x,z}\cos\theta + \bar{u}_{y,z}\sin\theta)\right\}\right]_{,zt} , \tag{9}$$

$$\rho_R \bar{u}_{z,tt} = c_{11}\bar{u}_{z,zz} , \tag{10}$$

$$\left(\frac{M_s}{\gamma}\right) \beta_{p,t} = - \lambda \beta_{z,zz} + K(\cos^2\theta - \sin^2\theta)\, \beta_z + 4\pi M_s \beta_z + B_2(\bar{u}_{x,z}\cos\theta + \bar{u}_{y,z}\sin\theta) - 2\Sigma_o^{me} \beta_z + (\tau\gamma^2)^{-1}\left[\beta_z + \tfrac{1}{2}(\bar{u}_{x,z}\cos\theta + \bar{u}_{y,z}\sin\theta)\right]_{,t} , \tag{11}$$

$$\left(\frac{M_s}{\gamma}\right) \beta_{z,t} = \lambda \beta_{p,zz} - K(\cos^2\theta - \sin^2\theta)\, \beta_p + M_s(\bar{H}_y\cos\theta - \bar{H}_y\sin\theta) + 2\Sigma_o^{me} \beta_p - (\tau\gamma^2)^{-1} \beta_{p,t} , \tag{12}$$

where we have assumed that the external rf field $\bar{H}$ lies in the plane of the wall, and we have defined the new variable β_p (Figure 1);

$$\beta_p = \beta_y\cos\theta - \beta_x\sin\theta.$$

Σ_o^{me} is the magnetoelastic energy in equilibrium. The terms in Σ_o^{me} are the direct influences of initial magnetostriction on the magnetization equations.

The initial magnetostriction is uniform in the case of a Bloch wall in an infinite medium. Using the expression of initial magnetostriction in Motogi & Maugin (1983), eqn(7) is integrated to give a Landau-Lifshitz type distribution of initial magnetization;

$$\sin\theta = \mathrm{sech}(z/\delta'), \quad \cos\theta = -\tanh(z/\delta'), \quad (13)$$

where the thickness of a Bloch wall δ' is given by

$$\delta' = \left\{\frac{\lambda}{K + \dfrac{2B_1^2}{c_{11}-c_{12}}}\right\}^{\frac{1}{2}}, \quad (14)$$

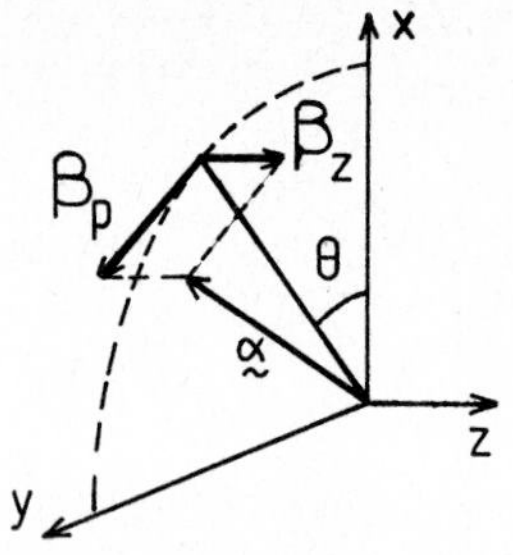

Figure 1

which shows that the wall thickness is diminished by magnetostriction as compared to the case when there are no magnetoelastic interactions since the magnetoelastic factor in the denominator of eqn(14) is always positive.

We shall now consider the effect of initial magnetostriction on wall oscillations. The magnetoelastic energy in equilibrium is given by

$$2\Sigma_0^{me} = -\bar{K} + \frac{2B_1^2}{c_{11} - c_{12}} \sin^2\theta,$$

where

$$\bar{K} = \frac{2(c_{11}+c_{12})B_1^2}{(c_{11}-c_{12})(c_{11}+2c_{12})} .$$

Using (13) and (14), and discarding dynamic coupling and damping terms in eqns (11) and (12), we obtain the resonance frequency of wall oscillations as far as the perturbations are localized in the wall (i.e., wall type excitations);

$$\omega_0 = \left(\frac{\gamma}{M_s}\right)\left\{\bar{K}(4\pi M_s^2 + \bar{K})\right\}^{\frac{1}{2}}. \quad (15)$$

We may say that the initial magnetostriction gives the resonance of wall oscillations, and the resonance frequency is entirely determined by several known material constants. Hence, we may conclude that $2\Sigma_0^{me}$ is approximately replaced by $-\bar{K}$ in eqns (11) and (12).

WALL OSCILLATIONS AND LOCALIZED STRAINS

We examine dynamic couplings without spin damping. Assuming a time dependence $\exp(-i\omega t)$ for all variables and normalizing the spatial coordinate by the wall thickness δ, we obtain spatially non-dimensionalized equations (we omit them for the lack of space). With a few manipulations of the equations of motion and the definition of parameters

$$c_e^2 = c_{44}/\rho_R \quad (c_e\text{: velocity of transverse sound wave})$$

$$\epsilon_e = \frac{B_2}{c_{44}}, \quad \tilde{k} = \left(\frac{\omega}{c_e}\right)\delta = k\delta \quad (k\text{: wave number})$$

we obtain the following two equations (note that we now neglect the damping terms);

$$\cos\theta\left[u_x^{o\prime} + \tilde{k}^2\int^{\xi} u_x^o(\xi')d\xi'\right] + \sin\theta\left[u_y^{o\prime} + \tilde{k}^2\int^{\xi} u_y^o(\xi')d\xi'\right] + 2\epsilon_e\beta_\xi^o$$
$$= C_1\cos\theta + C_2\sin\theta, \qquad (16)$$

$$\sin\theta\left[u_x^{o\prime} + \tilde{k}^2\int^{\xi} u_x^o(\xi')d\xi' - C_1\right] = \cos\theta\left[u_y^{o\prime} + \tilde{k}^2\int^{\xi} u_y^o(\xi')d\xi' - C_2\right], \quad (17)$$

where

$$u_x^o(\xi) = \bar{u}_x(z)/\delta \quad , \quad u_y^o(\xi) = \bar{u}_y(z)/\delta \quad , \beta_\xi^o = \beta_z^o \ ,$$

and '$\prime$' stands for the differentiation with respect to $\xi(=z/\delta)$, and C_1 and C_2 are constants of integration. The upper superscript 'o' means the amplitude function. From (17), we may assume the form of solutions as follows;

$$u_x^{o\prime} + \tilde{k}^2\int^{\xi} u_x^o(\xi')d\xi' - C_1 = U_o\cos\theta\Phi(\xi),$$
$$u_y^{o\prime} + \tilde{k}^2\int^{\xi} u_y^o(\xi')d\xi' - C_2 = U_o\sin\theta\Phi(\xi), \qquad (18)$$

where U_o is a constant and $\Phi(\xi)$ is an arbitrary function of ξ. Introducing (18) into (16), we find

$$U_o\Phi(\xi) = -2\,\epsilon_e\beta_\xi^o \ . \qquad (19)$$

On the other hand, as is seen from eqns (11) and (12), the external rf field in the x-direction can excite the localized solution

$$\beta_\xi^o \ , \ \beta_p^o \propto \sin\theta$$

Then, in this case, we may take

$$\Phi(\xi) = \sin\theta = \text{sech}(\xi).$$

This choise is consistent with our formulations for the effect of initial magnetostriction. The equations (18) can be solved by the Green's function method (Morse & Ingard (1968)) to yield the strains in the limit of the long wavelength approximation;

$$\bar{u}_x(\xi,t)_{,\xi} = U_o(\sin\theta)_{,\xi}\exp(-i\omega t) + \frac{\pi}{2}U_o\tilde{k}^2\text{sign}(\xi)\exp\{i(k|\xi| - \omega t)\},$$
$$\bar{u}_y(\xi,t)_{,\xi} = -U_o(\cos\theta)_{,\xi}\exp(-i\omega t), \qquad (20)$$

The first terms of these solutions are localized strains and the second term in (20-1) is a traveling wave of the order of $\tilde{k}^2$, which shows that it is much smaller than the localized strains. The traveling strains for $\bar{u}_{y,\xi}$ is at most of the third order of $\tilde{k}$.

Discarding small terms in (20), we obtain the strain component in the direction of the local magnetization as follows;

$$\tfrac{1}{2}(u_{x,\xi}^o\cos\theta + u_{y,\xi}^o\sin\theta) = \tfrac{1}{2}U_o\sin\theta.$$

On the other hand, the strain component perpendicular to the local magnetization is given by

$$\tfrac{1}{2}(u_{x,\xi}^o\sin\theta - u_{y,\xi}^o\cos\theta) = 0.$$

Thus, we may say that the localized strain is polarized in the direction of the local magnetization. Moreover, we see, from eqns (11), (12) and (21), that the magnetization equations are strictly satisfied by these localized solutions. This is none other than the dynamic interaction between wall oscillations and localized strains.

The excited strain amplitude U_o and the magnetization amplitude are determined from (11), (12) and (19). Setting the uniform amplitude of the external rf field h_x^o in the x-direction, we have

$$B_\xi^o = -\frac{i\gamma\omega}{\omega^2 - \omega_w^2} h_x^o , \quad U_o = \frac{2i\gamma\omega}{\omega^2 - \omega_w^2} \frac{B_2}{c_{44}} h_x^o , \tag{21}$$

where

$$\omega_w = \left(\frac{\gamma}{M_s}\right)\left\{\bar{K}(4\pi M_s^2 + \bar{K} - \frac{2B_2^2}{c_{44}})\right\}^{\frac{1}{2}} , \tag{22}$$

and we have set $\beta_\xi^o = B_\xi^o \sin\theta \exp(-i\omega t)$. ω_w is the modified resonance frequency of wall oscillations. Comparing the expressions (15) and (22), the static and dynamic magnetoelastic effects are set in evidence; except for the factor (γ/M_s), the constant $\bar{K}$ is referred to the wall stiffness and $(4\pi M_s^2 + \bar{K})^{-1}$ is the wall mass in the absence of dynamic couplings. Hence, defining the modified wall mass as

$$m_w^{-1} = (4\pi M_s^2 + \bar{K} - \frac{2B_2^2}{c_{44}}) ,$$

we may conclude that the wall stiffness arises from the interactions between static magnetostriction and wall motions, and the fully dynamic magnetoelastic coupling increases the wall mass to decrease the wall resonance frequency. This phenomenon is akin to the magnetoelastic gap in the spin wave spectrum in the wall.

When the spin damping is accounted for, the procedure is almost the same and we have the same conclusions which are obtained in the case without spin damping if the denominators in (21) are replaced by

$$(\omega^2 - \omega_w^2) + 4\pi i(\tau)^{-1}\omega .$$

REFERENCES

[1] Winter, J.M., Bloch wall excitation.Application to nuclear resonance in a Bloch wall, Phys.Rev. 124 (1961) 452-459.
[2] Kittel, C. & Galt, J.K., in Solid state physics, vol.3, Seitz, F. & Turnbull, D.(eds.), (Academic Press, New York, 1976)
[3] Motogi, S. & Maugin, G.A., Effects of magnetostriction on vibrations of Bloch and Néel walls, to be published in 1983.
[4] Mitin, A.V. & Tarasov, V.A., Sound generation in a multidomain ferromagnet, Sov.Phys.JETP 45 (1977) 414-419.
[5] Lugovoi, A.A. & Turov, E.A., Magnetoelastic oscillations of domain walls in antiferromagnets, Sov.Phys. Solid State 23 (1981) 1556-1561.
[6] Maugin, G.A., Micromagnetism, in: Eringen, A.C.(ed.), Continuum Physics, vol.3, (Academic Press, New York, 1976)
[7] Gilbert, T.L., A phenomenological theory of ferromagnetism, Ph.D. Thesis, Ill. Inst. Techn. (1956)
[8] Maugin, G.A., On the spin relaxation in deformable ferromagnets, Physica, 81A (1975) 454-468.
[9] Morse, P.M. & Ingard, K.U., Theoretical acoustics, (McGraw-Hill, New York, 1968)

The Mechanical Behavior of Electromagnetic Solid Continua
G.A. Maugin (editor)
Elsevier Science Publishers B.V. (North-Holland)

MULTIPOLAR LATTICE DEFECTS IN ELECTROMAGNETIC ELASTIC CONTINUA

R.K.T. Hsieh

Department of Mechanics
Royal Institute of Technology
S-100 44 Stockholm
Sweden

It is shown that lattice defects in electromagneto-elasticity can be modelled by point force arrays leading to the notion of elastic multipole forces with different strength. It is also shown that line defects such as dislocations can be represented by surface and volume distribution of elastic monopole forces with strength P_{ij} .

INTRODUCTION

Lattice defects are always present in real crystalline materials and their presence influences numerous properties of the material. But the determination of the atomic displacements due to a single defect and thus of its quantitative properties with account of the discrete arrangement of the atoms is still an open problem. These quantitative properties are tracked in terms of the continuum model, see e.g. [4]. However each defect needs a different treatment. Kröner [1] was first to introduce multipolar forces to describe lattice defects and thus gave the possibility to unify and rationalize the different treatments of the lattice defects. Siems [2] applied these concepts to point defects in elastic materials. Hsieh et. al [3] extended the treatment to elastic micropolar solids. The basic idea is to replace the effect of the defects by fictituous multipole body forces. Similar ideas can be found by regarding the defects as a source of elastic internal stresses which is a standard way to treat a distribution of dislocations in elastic solids. Kléman [5] and Maugin [6] extended this idea of the representation of dislocations to magnetic materials by also treating this latter contribution as quasi-plastic processes. There are also other ways to treat the problem e.g. through polar theories, see Hsieh [7]-[8].

The purpose of this paper is to give an unified theory for the treatment of the quantitative properties of lattice defects in electromagneto-elasticity by modelling them as point force arrays. Thus, from the mechanical view, here too, both the electromagnetic contribution and the part due to defects are considered either as sources of elastic internal stresses or as fictituous multipole forces. We shall start by giving some results of electromagneto-elasticity without and with defects. To render the equations amenable we restrict ourselves to the static theory, see Paria [9]. For more complicated models we refer to e.g. Hsieh [8], Brown [10]. The basic equations of the mechanics of point force arrays are then derived and at last, the results are applied to dislocations.

ELECTROMAGNETO-ELASTICITY

In the linear theory of electromagneto-elasticity, the condition of equilibrium is given by

$$T_{ji,j} + f_i^o = 0 \qquad (1)$$

where T_{ij} is the sum of the mechanical stresses due to the elastic deformations $\bar{t}_{ij}$ and the electromagnetic stress $\bar{\bar{t}}_{ij}$

$$T_{ij} = \bar{t}_{ij} + \bar{\bar{t}}_{ij} \tag{2}$$

with $\bar{t}_{ij}$ related to the displacement u_i by

$$\bar{t}_{ij} = C_{ijmn}\, u_{m,n} \tag{3}$$

where C_{ijmn} is the tensor of the elastic constants. The stresses $\bar{\bar{t}}_{ij}$ are related to the electric field E_i and to the magnetic induction B_i by [9]

$$\bar{\bar{t}}_{ij} = \varepsilon\, [E_i E_j - \tfrac{1}{2} E_k E_k \delta_{ij}] + \frac{1}{\mu_e}\, [B_i B_j - \tfrac{1}{2} B_k B_k \delta_{ij}] \tag{4}$$

with coefficients ε and μ_e being the electric permittivity and the magnetic susceptibility and δ_{ij} the Kronecker symbol. The force f_i^o is composed of the mechanical forces $\bar{f}_i$ and the electromagnetic force $\bar{\bar{f}}_i$

$$f_i^o = \bar{f}_i + \bar{\bar{f}}_i \tag{5}$$

with $$\bar{\bar{f}}_i = \varepsilon_{imn}\, j_m\, B_n + \rho_e\, E_i \tag{6}$$

where j_m is the electric current density and ρ_e the electric charge density.

The electromagnetic fields have also to verify the Maxwell equations. In the case of statics they write

$$\varepsilon_{imn}\, H_{n,m} = j_i \qquad B_{i,i} = 0 \tag{7}$$

$$\varepsilon_{imn}\, E_{n,m} = 0 \qquad D_{i,i} = \rho_e \tag{8}$$

(in MKS units) together with the relations

$$D_i = \varepsilon\, E_i\ , \qquad B_i = \mu_e\, H_i \tag{9}$$

Here D_i is the electric displacement, H_i is the total magnetic field vector. The current j_i is determined by Ohm's law within the same assumptions

$$j_i = \sigma\, E_i \tag{10}$$

where σ is the electrical conductivity.

Substituting expression (3) into eq. (1) we get the differential equations for the displacements

$$C_{jinm}\, u_{n,mj} + f_i = 0 \tag{11}$$

with f_i defined as $f_i = f_i^o + \bar{\bar{t}}_{ji,j}$

These equations can be satisfied by the solution

$$u_l(\underline{x}) = \int f_i(\underline{x}') \, G_{li} \, (\underline{x},\underline{x}') \, dV' \tag{12}$$

Here $G_{li}\,(\underline{x},\underline{x}')$ is the Green's function tensor defined by the equation

$$C_{ijlm} \, G_{ln,mj} \, (\underline{x},\underline{x}') + \delta_{in} \, \delta \, (\underline{x}-\underline{x}') = 0 \tag{13}$$

where $\delta\,(\underline{x}-\underline{x}')$ is the Dirac δ function.

Equations (1) - (13) form the basis of electromagneto-elasticity and are to be solved with prescribed initial and boundary conditions.

CONTINUUM MODEL OF A DISLOCATION

It is known that real materials always contain defects. Let us now analyze the case of a line defect. The plastic deformation of a crystalline material takes place by slips along crystallographic planes generally with a slip vector of the atomic distance. The slip along a certain slip-plane is generally only partial and therefore divides the slip plane into two parts, the slipped and unslipped region. The boundary between these regions is called a dislocation and the slip vector belonging to it, the Burgers vector, see e.g. Kovács & Zsoldos [4].

The problem of determining the atomic displacements due to a dislocation taking into account the discrete arrangement of the atoms is still unsolved. However excluding the region closely around the dislocation line the problem can be solved to a good approximation by linear elasticity.

Applied to a continuum, the Volterra model can be described in the following way. First let us make a cut along the surface S bounded by the curve C and slip the two sides of the cut relative to each other by a constant vector b_i. Then one obtains a dislocation line lying along the curve C.

To describe the displacement field $u_i(\underline{x})$ in the matrix due to the so created dislocation (for $f_i^o \equiv 0$), let us multiply the Green's equation (13) by $u_i(\underline{x}')$ and integrate over the volume of the body, excluding the cut surfaces, by a surface S'. Then for an infinite medium when $G_{ln}(\underline{x},\underline{x}') = G_{ln}(\underline{x}-\underline{x}')$ one obtains

$$u_n(\underline{x}) = -\int_{V'} C_{ijlm} \, G_{ln,mj} \, (\underline{x}-\underline{x}') \, u_i(\underline{x}') \, dV' \tag{14}$$

Applying the Gauss theorem we get

$$u_n(\underline{x}) = \int_{S'} [\bar{\bar{t}}_{kl} \, G_{ln} + C_{iklm} \, u_i \, G_{ln,m}] \, dS'_k + \int_{V'} \bar{\bar{t}}_{ml,m} \, G_{ln} \, dV' \tag{15}$$

Now let S' tend to the cut surface S_c. In equilibrium the mechanical part of the stresses must be continuous, the displacement has a discontinuity $u_i|_{S^+} - u_i|_{S^-} = b_i$ and the electromagnetic stresses have the discontinuity

$\bar{\bar{t}}_{lk}|_{S^+} - \bar{\bar{t}}_{lk}|_{S^-} = [\bar{\bar{t}}_{lk}]$. The resulting displacement then writes

$$u_n(\underline{x}) = C_{iklm} b_i \int_{S_c} G_{ln,m}(\underline{x}-\underline{x}') \, dS'_k + \int_{V'} \bar{\bar{t}}_{ml,m} \, G_{ln} \, dV' + \int_{S_c} [\bar{\bar{t}}_{kl}] \, G_{ln} dS'_k \tag{16}$$

Therefrom one can calculate the strain and the stress field due to a dislocation.

POINT FORCE ARRAYS

Let us consider N point forces acting in a small volume centered at $\underline{x}'$ of an arbitrary electromagneto-elastic continuum. The resulting displacement fields can then be written as

$$u_1(\underline{x}) = \sum_{q=1}^{N} f_k^q \, G_{1k} \, (\underline{x}, \underline{x}' + \underline{d}^q) \tag{17}$$

where $\underline{x}' + \underline{d}^q$ is the position vector of the force f_k^q. Expanding the Green's function in a Taylor series about $(\underline{x}, \underline{x}')$ we obtain

$$u_i(\underline{x}) = \sum_{p=1}^{\infty} \frac{(-1)^p}{p!} G_{ij,sm_1 \ldots m_{p-1}} (\underline{x}-\underline{x}') \, P_{jsm_1 \ldots m_{p-1}} \tag{18}$$

where we introduced multipole forces of order p (p=1,...,∞)

$$P_{jsm_1 \ldots m_{p-1}} = \sum_q f_j^q d_s^q \ldots d_{m_{p-1}}^q \tag{19}$$

and we used the relations

$$P_j = 0 \text{ . We also have } \varepsilon_{jmn} P_{mn} = 0 \tag{20}$$

i.e. the point force arrays are in equilibrium. The Green's function in (18) has been derived, see e.g. [8], and is the continuum expression of the discrete Green's functions for long-wavelength. If we consider a single multipole of order p, eq. (18) can be interpreted as the corresponding elastic displacement $u_i^p(\underline{x})$ due to the strength of the p-th order multipole force at $\underline{x}'$ with the p-th order Green's function as the displacement at $\underline{x}$ due to a unit p-th order multipole force.

If now a defect can be represented by such a point force array, from eqs. (18) and (3) we can calculate the therefrom originated elastic stresses and also define a corresponding fictituous body force:

$$f_i = \sum f_i^p = \sum_{p=1}^{\infty} \frac{(-1)^p}{p!} P_{ikk_1 \ldots k_{p-1}} \, \delta_{,kk_1 \ldots k_{p-1}} (\underline{x}-\underline{x}') \tag{21}$$

We also can calculate the force which acts on such an array due to its interaction with an additional strain field ε_{ik}^A, this writes

$$f_n^{int} = - U_{,n} \tag{22}$$

where the interaction energy U is given by

$$U = \int f_i \, u_i^A \, dV$$

$$= \frac{(-1)^p}{p!} P_{ikk_1 \ldots k_{p-1}} \, \varepsilon^A_{ik,k_1 \ldots k_{p-1}} \tag{23}$$

and f_i is given by eq. (21).

As an application of this method, let us consider the case of a line defect. We first note that the displacement, due to the electromagnetic fields, in the defects-free elastic material writes, see eq. (12)

$$u_k(\underline{x}) = \int_V \bar{\bar{t}}_{lp,l}\, G_{kp}\, dV' \tag{24}$$

We now introduce a definition of defect which is based on Volterra's model. Let a surface S be chosen in an arbitrary way in the medium and let it suffer a transformation such that the element $dS'_m(\underline{x}')$ has a displacement $a_l(\underline{x}')$. The surface thus becomes a surface of elastic internal stresses which will be defined by

$$dP_{ik} = C_{iklm}\, a_l\, dS'_m + \bar{\bar{t}}_{ik} dV' \tag{25}$$

The displacement due to this elastic monopole (because it can be seen that no torque is exerted on it) is given by eq. (18) and

$$u_l(\underline{x}) = \int_{S'} \left[C_{iklm}\, a_l\, G_{li,k'} + \bar{\bar{t}}_{mk}\, G_{lk} \right] dS'_m \tag{26}$$

For dislocation $a_l = b_l$ = constant, S' is replaced by the cut surface S_c and $\bar{\bar{t}}_{mk}|^+ - \bar{\bar{t}}_{mk}|^- = [\bar{\bar{t}}_{mk}]$ is the discontinuity of the stresses $\bar{\bar{t}}_{mk}$ across S_c. The total elastic displacement is then

$$u_l(\underline{x}) = \int_V \bar{\bar{t}}_{np,n}\, G_{lp}\, dV + \int_{S_c} \left(C_{iklm}\, b_l\, G_{li,k'} + [\bar{\bar{t}}_{mk}]\, G_{lk} \right) dS'_m \tag{27}$$

which is equal to expression (16) of the displacement derived by the classical Green's function method. Introducing eq. (25) into eq. (22), we can find the force exerted on a dislocation due to its interaction with another field.

DISCUSSION

This paper illustrates the modelling of line defects (cut surface) in electro-magneto-elastic continuum as point force arrays, in particular as a distribution of surface and volume monopole forces with strength P_{ij}. We could have included magneto-strictive effects,... and have more general constitutive relations (3) and (9). Though the calculations would have been more tedious, the principle would have remained the same. The method can also be generalized to volume defects (close surface) by applying the Gauss theorem to eq. (25). Such theory has the advantage to have unified approach and expressions for the quantitative physical properties of the defects, e.g. the interaction energy, the Peach-Kohler force,...

REFERENCES

[1] Kröner, E., *Allgemeine Kontinuumstheorie der Versetzungen und Eigenspannungen*, Arch. Rational. Mech. Anal. 4 (1959) 275-334

[2] Siems, R., *Mechanical Interactions of Point Defects*, Phys. Stat. Sol. 30 (1968) 645-658

[3] Hsieh, R.K.T., Vörös, G. and Kovács, I., *Stationary Lattice Defects as Sources of Elastic Singularities in Micropolar Media*, Physica 101 B+C, 2 (1980) 201-208

[4] Kovács, I. and Zsoldos, L., *Dislocations and Plastic Deformations* (Pergamon Press, London, 1974)

[5] Kléman, M., *Dislocations, Disclinations and Magnetism* in: Nabarro, F.R.N. (ed.), Dislocations in Solids 5 (North Holland Pub. Co, Amsterdam, 1980) 349-402

[6] Maugin. G.A., *Classical Magnetoelasticity in Ferromagnets with Defects* in: Parkus, H. (ed.), Electromagnetic Interactions in Elastic Solids (Int'l Centre Mech. Sci. Courses, Udine; Springer Verlag, 1979) 243-324

[7] Hsieh, R.K.T., *Dislocation in a Ferromagnetic Continuum*, Arch. Mech, 33, No 6 (1981) 957-960

[8] Hsieh, R.K.T., *Micropolarized and Magnetized Media* in: Brulin, O. and Hsieh, R.K.T. (eds.), Mechanics of Micropolar Media (Int'l Centre Mech. Sci. Courses, Udine; World Pub. Co, 1981) 187-279

[9] Paria, G., *Magneto-Elasticity and Magneto-Thermo-Elasticity*, Advances in Applied Mechanics 30, 1 (1967) 73-112

[10] Brown, W.F.jr., *Magnetoelastic Interactions* (Springer Verlag, Berlin, 1966)

Part 9

MAGNETOELASTICITY OF CONDUCTORS

The Mechanical Behavior of Electromagnetic Solid Continua
G.A. Maugin (editor)
Elsevier Science Publishers B.V. (North-Holland)

A SURVEY OF LINEAR AND NONLINEAR WAVE MOTION IN A PERFECT MAGNETOELASTIC MEDIUM

J. Bazer

Courant Institute of Mathematical Sciences
New York University
New York, New York
U.S.A.

INTRODUCTION. BASIC SYSTEM OF CONSERVATION LAWS

We present an account - a necessarily impressionistic one - of linear and nonlinear wave motion in a perfect magnetoelastic medium. Our survey is based on the papers [1; (1971)], [2; (1971)] and [3; (1974)] by the author and his colleagues, W. B. Ericson and F. Karal. The reader is referred to these works for additional details and references and a fuller and more general treatment. Additional and/or more recent references are given in [4]-[20].[1]

We shall be concerned with one-dimensional wave motion unless explicitly stated otherwise. A perfect magnetoelastic medium is by definition an infinitely conducting, electrically neutral, nonmagnetizable, elastically perfect solid, subject to magnetic body forces on induced currents. Heat conduction and heat generating mechanisms, elastic or magnetic in origin, are ignored so that thermodynamic changes of state may be assumed to take place adiabatically - except, possibly, where the strain is discontinuous. A perfect isotropic elastic medium is characterized by a specific potential (internal) energy ε[2] [energy per unit mass] which depends upon the strain and specific entropy alone. By one-dimensional wave motion, we mean wave motion in which the displacement vector $\underset{\sim}{\xi} = \underset{\sim}{\xi}(x,t)$, and all state variables depend upon one space variable x [x representing the first component, say, of the position of a material point in the unstrained state of the medium] and the time variable t; however, no assumptions are made regarding the presence or absence of components of nonscalar quantities. The state of a magnetoelastic medium experiencing one-dimensional wave motion is specified by the ten-component state vector $U = [\underset{\sim}{B},\underset{\sim}{u},\underset{\sim}{w},S]$ consisting of three components each of the magnetic induction $\underset{\sim}{B}$, velocity vector $\underset{\sim}{u}$ and strain vector $\underset{\sim}{w}$ $(\equiv \partial\underset{\sim}{\xi}(x,t)/\partial x)$ and one scalar component of specific entropy $\tilde{S}$.

Temporal and spatial changes of state of a perfect magnetoelastic medium are governed by the following system of conservation laws - we employ the material (Lagrangian) description:

$$(A_0)\quad B_1 = \text{const. (indep. of } x \text{ \& } t,\ B_1 \neq 0 \text{ throughout)},$$

$$(\underset{\sim}{A}_1)\quad 0 = \partial_t[(1+w_1)\underset{\sim}{B}] - B_1\partial_x\underset{\sim}{u}\ ,$$

$$(1.1)\qquad (\underset{\sim}{A}_2)\quad 0 = \bar{\rho}\partial_t\underset{\sim}{u} - \underset{\sim}{\partial}_x\cdot\underset{\sim}{T} - \underset{\sim}{\partial}_x\cdot\underset{\sim}{M},$$

$$(\underset{\sim}{A}_3)\quad 0 = \partial_t\underset{\sim}{w} - \partial_x\underset{\sim}{u}\ ,$$

$$(A_4)\quad 0 = \partial_t S\ ,$$

$$(A_5)\quad 0 = \bar{\rho}\partial_t\left[\varepsilon + \frac{u^2}{2} + \frac{(1+w_1)B^2}{2\mu\bar{\rho}}\right] - \underset{\sim}{\partial}_x\cdot(\underset{\sim}{T}\cdot\underset{\sim}{u}) - \underset{\sim}{\partial}_x\cdot\left[\underset{\sim}{u}\,\frac{B^2}{2\mu} + \frac{\underset{\sim}{B}}{\mu}\times(\underset{\sim}{B}\times\underset{\sim}{u})\right]$$

$$(a)\quad T_{1\alpha} = \bar{\rho}\ \partial\varepsilon(\underset{\sim}{w},S)/\partial w_\alpha\ ,\quad \alpha = 1,2,3\quad \text{(eqns. of state)}$$

(1.2)
$$\text{(b)}\quad \underset{\sim}{M} = \underset{\sim\sim}{BB}/\mu - (\tfrac{1}{2}B^2/\mu)\underset{\sim}{I}\,, \qquad \underset{\sim}{I} = \sum_{i=1}^{3} \underset{\sim}{i}_\alpha \underset{\sim}{i}_\alpha\,,$$

All vectors $\underset{\sim}{Q}$ and tensors $\underset{\sim}{R}$ in (1.1) and (1.2) are referred to a fixed orthonormal basis $\underset{\sim}{i}_1$, $\underset{\sim}{i}_2$, $\underset{\sim}{i}_3$ so that $\underset{\sim}{Q} = \sum_{\alpha=1}^{3} Q_\alpha \underset{\sim}{i}_\alpha$ and $\underset{\sim}{R} = \sum_{\alpha=1}^{3}\sum_{\beta=1}^{3} R_{\alpha\beta}\underset{\sim}{i}_\alpha \underset{\sim}{i}_\beta$ and we call $\underset{\sim}{Q}_{tr} = \underset{\sim}{Q} - Q_1\underset{\sim}{i}_1$ the transverse and $Q_1\underset{\sim}{i}_1$ the longitudinal component of the vector $\underset{\sim}{Q}$. The operator $\underset{\sim}{\partial}_x$ is a shorthand for $\underset{\sim}{i}_1\partial_x$ so that $\underset{\sim}{\partial}_x\cdot\underset{\sim}{Q}$ and $\underset{\sim}{\partial}_x \times \underset{\sim}{Q}$ define the divergence and curl of $\underset{\sim}{Q}$, respectively. $\bar{\rho}$, assumed everywhere constant, denotes the density before deformation, $\underset{\sim}{T}$ and $\underset{\sim}{M}$ denote the elastic and magnetic stress tensors. The medium is nonmagnetic in that we take μ, the magnetic inductive capacity, to have the free space value [Giorgi MKS units are used throughout this exposition]. To shorten and simplify the presentation,[3] we shall assume from the outset that ε is the special Hookean potential

$$\varepsilon = \bar{\tau}S + \frac{1}{2}c_L^2 w_1^2 + \frac{1}{2}c_T^2 w_{tr}^2, \qquad c_L^2 > 2c_T^2\,, \tag{1.3}$$

where c_L, c_T and $\bar{\tau}$ are suitable positive constants; c_L and c_T will be recognized as the longitudinal and transverse disturbance speeds of conventional elasticity.

Equation $\underset{\sim}{A}_1$ is the form taken by Faraday's law of induction in infinitely conducting media. To show this, we need only recall that (i) $\underset{\sim}{E}$, the electric intensity, is $\underset{\sim}{u} \times \underset{\sim}{B}$ in infinite conductors and (ii), starting with Faraday's law of induction in the more familiar spacial (Eulerian) representation, pass over to material coordinates. The result is $\underset{\sim}{A}_1$. Equations $\underset{\sim}{A}_2$ and A_5 are, respectively, the momentum and energy conservation equations of conventional elasticity with added terms to account for effects of magnetic stresses [associated with magnetic body forces on induced currents]. Equation A_4 ensures adiabaticity of the wave motion, while equation A_3, a compatability condition, ensures the existence of a displacement vector $\underset{\sim}{\xi}(x,t)$ such that $\underset{\sim}{u} = \partial_t\underset{\sim}{\xi}$ and $\underset{\sim}{w} = \partial_x\underset{\sim}{\xi}$ -- provided, of course, the motion is sufficiently smooth.

Equations (1.1) $\{A_0,\underset{\sim}{A}_1,\underset{\sim}{A}_2,\underset{\sim}{A}_3,A_4\}$ and (1.1)-$\{A_0,\underset{\sim}{A}_1,\underset{\sim}{A}_2,\underset{\sim}{A}_3,A_5\}$ imply, and hence are equivalent to, each other in regions of x,t space where the wave motion is smooth - continuously differentiable say. Further equation $\underset{\sim}{A}_1$ represents two, not three, independent equations because its longitudinal component is equivalent to the longitudinal component of $\underset{\sim}{A}_3$. Equations $\{\underset{\sim}{A}_1,\underset{\sim}{A}_2,\underset{\sim}{A}_3,\underset{\sim}{A}_4\}$ or equivalently $\{\underset{\sim}{A}_1,\underset{\sim}{A}_2,\underset{\sim}{A}_3,A_5\}$ therefore constitute a determined nonlinear system of nine equations for the nine components of the state vector $U = [B_2,B_3,u_1,u_2,u_3,w_1,w_2,w_3,S] \equiv [\underset{\sim}{B}_{tr},\underset{\sim}{u},\underset{\sim}{w},S]$. Note each equation is in strict conservation form - that is, a sum of derivatives of functions the dependent variables alone. 'Strict' because each term is of the form $\partial_t Q(U)$ or $\partial_x Q(U)$ - no undifferentiated terms are present. This strict conservation property, incidently, carries over naturally to the relativistic magnetoelasticity [G. A. Maugin, [7a,b]]. Note that although in special Hookean media the stress depends linearly on the strain, the system remains highly nonlinear owing to the interaction of the magnetic field with itself and the strain and velocity fields.

The strict conservation property is the keystone of our survey. It implies for elastic energy densities satisfying Hadamard's condition [automatically satisfied in special Hookean media] that the system (1.1) $\{\underset{\sim}{A}_1,\underset{\sim}{A}_2,\underset{\sim}{A}_3,\underset{\sim}{A}_4\}$ may be cast in symmetric-hyperbolic form [K. O. Friedrichs and P. D. Lax, [4]]. From this fact, one may infer in advance that magnetoelasticity possesses a geometrical theory in n dimensions, n = 1,2,3, [[1] & M. F. McCarthy [18a]], and a theory of shocks [3], and simple waves [2] in one dimension analogous to, and in every way as rich as, that of, say, geometrical (magneto-)acoustics and (magneto-)gas dynamics. We shall sketch some results from each of these theories in the ensuing sections and apply them to a magnetoelastic piston and closely related shear-flow problem. The latter, as was pointed out in [1], may be considered a primitive mathematical model of a 'quake' in a solid star - in a pulsar, for example. The quake mechanism was originally adduced by M. Ruderman [5a,b], as an explanation for small changes in the rotational speed of pulsars that have been inferred from certain observations. This and other applications of magnetoelastic wave propagation are discussed further in [1]-[3].

II. LINEARIZED EQUATIONS. CHARACTERISTICS AND SMALL DISTURBANCE WAVES.

If, in equations (1.1)-$\{A_0,\ldots,A_4\}$, we linearize about a constant state U_0 [this assumption suffices for the applications we have in mind here], we obtain a 9×9 system of the form

$$\partial_t \dot{U} + M(U_0)\partial_x \dot{U} = 0 \tag{2.1}$$

where $\dot{U} = [\dot{B}_2,\dot{B}_3,\dot{u}_1,\dot{u}_2,\dot{u}_3,\dot{w}_1,\dot{w}_2,\dot{w}_3,\dot{S}]$, $\dot{Q}$ signifying a small perturbation of a quantity Q. $M(U_0)$ is a 9×9 matrix which, owing to hyperbolicity, of the system has only real eigenvalues.

To lay the ground work for our subsequent treatment of simple waves and shocks and for the applications treated in Sections III and IV, we discuss here the propagation of small jump discontinuities in $\dot{U}$. According to the general theory of linear hyperbolic equations, discontinuities in $\dot{U}$ are carried on special surfaces [in one dimension 'points'] in space, called wave fronts, which, moving with finite speed c, sweep out characterstic curves of the equation in space-time. Further, according to the well known formalism of this theory, the characteristic relations for (2.1) are

$$0 = (cI - M)\delta\dot{U}\ , \tag{2.2}$$

this system being obtained from (2.1) simply by replacing all quantities of the form $\partial_t\dot{Q}$ and $\partial_x\dot{Q}$ by $-c\delta\dot{Q}$ and $\delta\dot{Q}$, respectively. Here, $\delta\dot{Q}$ denotes the jump in Q across the characteristic curve at time t. If, for the sake of brevity, we ignore all nonpropagating surfaces of discontinuity [corresponding to values of c = 0] we obtain nontrivial solutions of the eigensystem (2.2) which may be described as follows: There are 3 pairs of propagating fronts corresponding to the eigenvalues [disturbance speeds] $\pm c_S$, $\pm c_I$ and $\pm c_F$, $c_S < c_I < c_F$ [S, I and F for Slow, Intermediate and Fast]. Specifically,

$$\begin{aligned} &c_I = [c_T^2 + b_1^2]^{1/2}\ , \qquad b_1 = [B_1^2/\mu\bar{\rho}(1+w_1)]^{1/2}\ ,\\ &c_{S,F} = [(c_L^2+c_T^2+b^2 \pm D)/2]^{1/2},\ b = \sqrt{(B_1^2+B_2^2)/\rho\bar{\rho}(1+w_1)}\ ;\ (+ \text{ with F};\ - \text{ with S})\\ &D = b^2[\Delta^2 + 1 - 2\Delta\cos\chi]^{1/2}\ ,\quad \Delta = (c_L^2-c_T^2)/b^2\ . \end{aligned} \tag{2.3}$$

Here and below, we take $B_3 \equiv 0$; χ is the angle between $\underset{\sim}{B}$ and the x-axis. Corresponding to these 3 pairs of disturbance speeds, we have the following 3 pairs of eigenvector solutions of (2.2):

Intermediate: $\delta B_3 = \alpha B,\ \delta B_1 = \delta B_2 = 0 \qquad (B = [B_1^2+B_2^2]^{1/2})$,

$$\begin{aligned} &\delta w_3 = \alpha\frac{B(1+w_1)}{B_1}\ ,\quad \delta w_1 = \delta w_2 = 0\ ,\\ &\delta u_3 = \mp c_I\delta w_3\ ,\quad \delta u_1 = \delta u_2 = 0\ ,\quad \delta S = 0\ , \end{aligned} \tag{2.4}$$

Slow and Fast: $\delta B_1 = \delta B_3 = 0;\ \delta B_2 = -\dfrac{\alpha}{1+w_1}\dfrac{c^2-c_L^2}{b_2^2}B_2$

$$\begin{aligned} &\delta w_1 = \alpha,\ \delta w_2 = -\alpha\frac{B_2}{B_1}\left[\frac{c^2-c_L^2-b_2^2}{b_2^2}\right],\qquad (b_2 = \sqrt{b^2-b_1^2})\\ &\delta w_3 = 0\ ,\quad \delta u_1 = \mp c\delta w_1\ ,\quad \delta u_2 = \mp c\delta w_2\ ,\ \delta u_3 = 0,\ \delta S = 0. \end{aligned} \tag{2.5}$$

In (2.5), c is c_S or c_F according as the solution in question is fast or slow; in all modes, α is an amplitude factor to be determined from initial conditions. The presence of the underlying magnetic field $\underset{\sim}{B} = B_1\underset{\sim}{i}_1 + B_2\underset{\sim}{i}_2$ is reflected in the fact that the propagation speeds are direction-dependent; specifically, these speeds depend upon the angle between the x-axis (the direction of propagation) and the direction of $\underset{\sim}{B}$. The orientation of the field quantities on the discontinuity fronts - here planes perpendicular to the x-axis - are, according to equation (2.4) and (2.5), unambiguously determined by the direction of propagation and the underlying magnetic field provided these are not parallel. Moreover, only in the intermediate wave are the velocity and strain vectors purely transverse, that is, parallel to the wave fronts. In the slow and fast fronts both longitudinal and transverse components are present. These results are in agreement with the corresponding results of magnetogasdynamics and in sharp contrast with those of conventional elasticity.

Each of the above modes as well as the three nonpropagating modes are of the form $\delta U = \alpha R$, where R is a 9-vector whose components depend solely on the unperturbed field quantities, and α is a strength factor. If $\delta U^0 = [\delta B_2^0, \delta B_3^0, \ldots, \delta S^0]$ denotes the jump across a plane at x_0 at t = 0, then we may write $\delta U^0 = \sum \alpha_j R_j$, where the sum is carried over all nine admissible modes. This initial decomposition fixes the strength factors, the α_j's, for all later times [discontinuities carried off on propagating planar fronts have a fixed strength and speed because the unperturbed state is, by assumption, homogeneous] and provides the basis for solving initial-value and mixed initial-boundary-value problems with sufficiently simple initial and boundary data.

Finally, it should be mentioned, that the propagation of small-amplitude time-harmonic magnetoelastic waves and of discontinuities in 3-dimensional special Hookean medium are treated in [1] employing the methods of geometrical optics - eikonals, rays, etc. The corresponding theory of acceleration waves was worked out even earlier in [18a] by H. M. McCarthy for Cauchy elastic media. In these theories, the location of the wave fronts are obtained from a first-order partial differential equation corresponding to the eikonal equation of geometrical optics. This equation is solved by means of six ray systems [bicharacteristic systems of ordinary differential equations], each system being associated with one of the six different disturbance speeds. These rays are then used to determine the variation of the disturbance amplitude along the rays, the amplitudes of each kind being determined at t = 0 on the initial wave front, as in the one-dimensional case. In the special case where the medium is homogeneous, the rays are straight lines and $\alpha_j^2(t)d\sigma_j(t)$ is constant along the j-th ray system. Here, $\alpha_j(t)$ is the amplitude of the j-th disturbance wave and $d\sigma_j(t)$ is the area on the j-th wave front cut out by a tube of j-type rays determined entirely by the local geometry of the initial wave front. When the initial wave front is planar each of the ray systems consists of parallel rays and $d\sigma_j(t)$ is constant along the rays of type j.

III. MAGNETOELASTIC PISTON PROBLEM. SHEAR-FLOW PROBLEM.

Assume, as in Figure 1a, the half space x < 0 is filled with an infinitely conducting, perfectly rigid magnet which is to serve as our 'piston'. Assume the half-space x > 0 is an infinitely conducting special Hookean medium and that the prevailing magnetic field $\underset{\sim}{B}_0 = B_{01}\underset{\sim}{i}_1 + B_{02}\underset{\sim}{i}_2$ ($B_{01} \neq 0$, $B_{02} > 0$) due to the magnet is everywhere constant. Suppose that in the absence of the magnetic field the elastic medium may slide without friction on the face $\underset{\sim}{P}_0$ of the magnet, but may not separate from it. Suppose finally that at time t = 0, the magnet is instantaneously set in motion with velocity $\underset{\sim}{v} = -\hat{v}\underset{\sim}{i}_2$, this velocity being maintained for all t > 0. The problem is to determine the wave motion in the region x > 0. That a wave motion does in fact ensue, is a consequence of the distortion of the magnetic lines of force just to the right at the interface

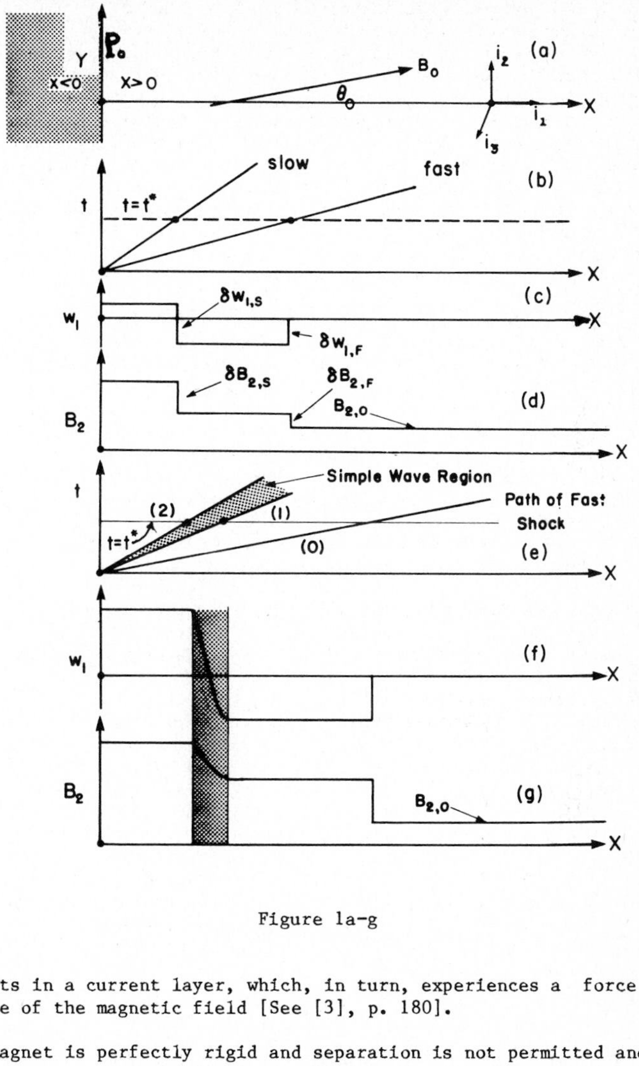

Figure 1a-g

which results in a current layer, which, in turn, experiences a force owing to the presence of the magnetic field [See [3], p. 180].

Since the magnet is perfectly rigid and separation is not permitted and, finally, since the tranverse electric field, parallel to the interface, must be continuous [Recall $\underset{\sim}{E} = \underset{\sim}{u} \times \underset{\sim}{B}$ in an infinitely conducting medium] we find for the three components of velocity just to the right of the interface that

(3.1) $$u_2 = -\vartheta, \qquad u_1 = u_3 = 0 \text{ on } \underline{P}_0, \text{ for all } t > 0.$$

Mathematically, our piston problem is a mixed initial-boundary-value problem. At time $t = 0$, the state of the medium is $U_0 = [\underset{\sim}{B}_{tr,0}, \underset{\sim}{u}_0, \underset{\sim}{w}_0, S_0]$ where $\underset{\sim}{B}_{tr,0}$ is the transverse component of $\underset{\sim}{B}_0$ and $\underset{\sim}{w}_0 = 0$ and $\underset{\sim}{u}_0 = 0$ (the medium is unstrained and motionless). But, this initial state is incompatible with the requirement that the transverse velocity on P_0 be $-\hat{v}i_2$. A wave motion must therefore ensue to resolve this initial incompatibility.

To solve this within the context of small disturbance theory, we view the initial motion of the piston as causing an initial disturbance $\delta\hat{U} = [\delta\hat{B}_{tr},\delta\hat{u},\delta\hat{w},0]$ with $\delta\hat{u} = -\hat{v}i_2$ on the interface P_0. At this stage $\delta\hat{B}_{tr}$ and $\delta\hat{w}_{tr}$ are to be regarded as unknowns to be determined after the boundary conditions have been met. Next, we write $\delta\hat{U}$ as the superposition $\delta U_S + \delta U_I + \delta U_F$ of slow, intermedite and fast propagating waves and adjust these amplitudes to satisfy the three boundary conditions (3.1) on the interface P_0 . It turns out that $\delta U_I = 0$ [if B_0, i_1 and the velocity of the piston were not linearly dependent, we would have $\delta U_I \neq 0$] but slow and fast waves are present. The initial incompatibility is therefore resolved by a fast wave followed by a slow wave. Explicit formulas for the components of δU_S and δU_F are given in [1] and [3]. Here, we need only note that $\delta w_{1,S} = \kappa_S\hat{v}$, $\delta B_S = \beta_S\hat{v}i_2$, $\delta w_{1,F} = -\kappa_F\hat{v}$, $\delta B_F = \beta_F\hat{v}i_2$ where κ_S , κ_F ($\kappa_S > \kappa_F$), β_S , β_F are positive constants depending upon the state U_0 of the unperturbed medium. Because we have supposed $B_{02} > 0$, w_1 decreases across the fast front ($\delta w_{1F} < 0$) and increases across the slow front ($\delta w_{1S} > 0$) [See Fig. 1(c)] as one crosses from right to left. Since $\rho = \bar{\rho}/(1+w_1)$, ρ and $\bar{\rho}$ denoting the densities in the strained and unstrained states, it follows that the medium behind (left side of) the fast front is compressed while that behind the slow is expanded. The general effect of the wave motion is to adjust the motion of the medium to that of the magnet. In the final steady state, which is achieved after the wave has propagated out to infinity, the medium moves downward, in the negative y-direction, with the same velocity as the magnet, the tangential component of the magnetic field is increased [See Fig. 1(d)] and the density near the wall is decreased.

If we take the limit $B_{02}\to 0$ of the solution just described, add $\hat{v}i_2$ to all velocities and finally extend $\delta b_2,\delta w_2,\delta w_1$ evenly and $\delta u_1,\delta u_2$ oddly as functions of x into the region $x < 0$ we get a solution of a shear-flow problem in which at time $t = 0$, $u_2 = \hat{v}$ when $x > 0$ and $u_2 = -\hat{v}$ when $x < 0$ - that is, in which there is a velocity discontinuity across the surface P_0 , initially. Note here, the fact that the initial velocity in $x < 0$ is precisely the negative of that in $x > 0$, is not a genuine loss of generality, since the governing equations are Galilean invariant.

IV. CONSTANT STATES. SIMPLE WAVES, SHOCKS. THE PISTON AND SHEAR FLOW PROBLEM REVISITED.

If each of the nine scalar components of the state vector $U = [B_{tr},u,w,S]$ is constant in a region of x,t-space, then U is a solution of (1.1); constant states are therefore solutions of the governing equations. Next, in order of complication are the simple wave solutions. Let us first observe that the system (1.1) may be written in the form

$$\partial_t U + M(U)\partial_x U = 0 , \tag{4.1}$$

this equation differing from the linearized system (2.1) only in that M is a function of the variable U, not a constant depending on a fixed value of U. A simple wave solution of (4.1) is a smooth [continuously differentiable] solution which is constant on the lines of a one-parameter family covering a region at x,t-space - U varies smoothly from line to line, which turn out to be characteristics of the (hyperbolic) system. A region in x,t space of nonconstant state adjacent to a region of constant state is always a simple-wave region [Jeffrey, A. & Taniuti [6], p. 90]. Together with shocks and constant states, simple waves, therefore, serve as building blocks in the construction of solutions of nonlinear 1-dimensional problems with sufficiently simple initial conditions, boundaries and boundary values.

To shorten the exposition, we shall consider here only centered and simple waves (fan waves) with center at the origin of x,t-space. Such waves have the form $U = U(c)$, $c = x/t$, and lead with the aid of equation (4.1) to the system of ordinary differential equations $0 = [cI - M(U)]dU/dc$. This system, evidently, has nontrivial solutions if and only if $c = c(U)$ is an eigenvalue of $M(U)$, in which case

$$dU/dc = \kappa(U)R(U) \tag{4.2}$$

where $R(U)$ is a right eigenvector of $M(U)$ and $\kappa(U)$ is the factor of proportionality consistent with the choice of c as a parameter [The same first-order differential equation - with a different choice of parameter, in general - governs general simple wave motion] The variable c has the dimensions of speed; we call it a <u>characteristic speed</u> since the lines $x/t = c$ are easily seen to be characteristics of the original hyperbolic system. In special Hookean media it turns out that there are three pairs of nonzero characteristics speeds $\pm c_S$, $\pm c_I$, $\pm c_F$ ($c_K > 0$, $K = S,I,F$) where the notation and terminology is precisely that of the small disturbance theory (Section II). Here, however, the characteristic speeds depend upon U. As is to be expected, in the small amplitude limit, that is, in the limit of narrower and narrower fans, these speeds reduce in a continuous manner to the slow, intermediate and fast disturbance speeds of the medium into which they propagate [Actually, the fan in an intermediate <u>centered</u> wave degenerates to a single characteristic - general intermediate waves have parallel characteristics and propagate without change of shape into regions of constant state]

The existence of simple waves being guaranteed by (4.2), there remains the problem of obtaining these solutions in tractable form. In special Hookean media, contact layers and intermediate waves present little difficulty. Though the problem is much more complicated in the case slow and fast waves, it is nevertheless possible to obtain an almost completely explicit solution. Specifically, if we put

$$q = (c^2 - c_T^2)/(c_L^2 - c_T^2)\ , \quad s = (c_L^2 - c_T^2)/(B_1^2/\mu\bar{\rho}(1+w_1)) \tag{4.3}$$

then it can be shown that $s = f(q,k_*)$ and

$$U = \hat{U}(s;U_*)\ , \quad (x/t)^2/c_T^2 = q((c_L^2/c_T^2) - 1) \tag{4.4}$$

where (i) $f(q,k_*)$ is an explicitly known monotonically decreasing function of q, whose graph is confined to the region $0 < q < 1$, $s < q^{-1}$ in slow waves and to the region $q > 0$, $s > q^{-1}$ in fast waves; (ii) k_* is determined from the knowledge of state U_* at the head or tail of the wave; (iii) the components of $\hat{U}(s;U_*)$ are known explicitly up to the performance of a pair of quadratures. Formulas for these components are derived and listed in [2] and [3]. Here, we shall note only the following qualitative features of general (not necessarily centered) simple wave motion: (a) In slow and fast simple wave motion the linear characteristics may converge or diverge with increasing time. In intermediate simple waves, as we indicated above, the linear characteristics are parallel. (b) In both slow and fast simple waves, the plane determined by $\tilde{B}_{tr}$ and the x-axis remains fixed across the wave and the component of $\tilde{u}_{tr}$ perpendicular to this plane is a constant. (c) Let $m = 1/c$, with $c = c_S$ or c_F, denote the slope of a typical characteristic line in x,t space. Then, as one proceeds across a fast simple wave in the direction of increasing $|m|$, the longitudinal stress T_{11} and strain w_1 increase while the density ρ and transverse magnetic field strength $|\tilde{B}_{tr}|$ and $|c|$ all decrease. (d) On the other hand, as one crosses a slow simple wave in the direction of increasing $|m|$, the variables $|\tilde{B}_{tr}|$, T_{11} and w_1 increase while ρ and $|c|$ decrease. (e) In both slow and fast waves, it can happen that the medium, initially in a state of longitudinal compression ($T_{11} < 0$), passes over to a state of longitudinal tension ($T_{11} > 0$ as $|m|$ increases. (f) $|\tilde{B}_{tr}|$

increases without bound as one crosses a fast (slow) wave in the direction of increasing longitudinal compression (tension). Thus, simple magnetoelastic wave motion in a special Hookean medium affords a mechanism for generating intense magnetic fields. [All of the above results hold without change in nonspecial Hookean media. See footnote 3.]

At the most primitive level of description, shocks are discontinuous solutions of the conservation laws (1.1)-$\{\underset{\sim}{A}_1, \underset{\sim}{A}_2, \underset{\sim}{A}_3, A_5\}$ [note A_4 has been replaced by A_5]. This system has the form

$$(4.5) \qquad 0 = \partial_t F_0(U) + \partial_x F_1(U)$$

where F_0 and F_1 are 9-vectors depending on U. More precisely, the discontinuity is a jump discontinuity in the state vector $U \equiv [\underset{\sim}{B}_{tr}, \underset{\sim}{u}, \underset{\sim}{w}, S]$,[4] the discontinuity occurring across a curve in x,t-space. Not all discontinuous solutions of this kind are, however, physically admissible: only those in which the states U_1 and U_0 on the two sides of the discontinuity front are related by the generalized form,

$$(4.6) \qquad 0 = C(F_0(U_1) - F_0(U_0)) - (F_1(U_1) - F_2(U_0)) ,$$

of the Rankine-Hugoniot relations of gas dynamics. These must be supplemented by 'generalized entropy' or 'evolutionary' conditions to ensure stability of the discontinuity with respect to small incident disturbance waves [See [6]]. C, provided it is not zero, is the shock velocity, more accurately, the shock shift rate or rate which the shock shifts from particle to particle. [Recall we are using the Lagrangian description]. Assuming the state U_0 is given, as it is in our piston problem [Section III], then (4.6) is a nonlinear system of 9 equations for the ten unknowns $U_1 = [\underset{\sim}{B}_{tr,1}, \underset{\sim}{u}, \underset{\sim}{w}, S]$ and C. By a solution of (4.6) we mean a pair of functions $U_1 = \underset{\sim}{u}(\varepsilon, U_0)$, $C = c(\varepsilon, U_0)$, $\varepsilon_0 < \varepsilon < \varepsilon_1$ which expresses the state U_1 and speed C as continuous functions of U_0 and a suitable strength parameter ε, $\varepsilon_0 < \varepsilon < \varepsilon_1$, ε_0 corresponding to a continuous transition. Which strength parameter is most suitable depends upon the kind of discontinuity and the problem. It turns out, in special Hookean media, that the shock wave solutions of (4.6) are closely analogous to those obtained for an ideal polytropic gas in magnetogasdynamics. Thus, there are slow, intermediate, and fast shocks which reduce in the weak-shock limit to the corresponding S, I and F small disturbance waves discussed in Section II. Furthermore, as in magnetogasdynamics, on crossing intermediate shocks the transverse magnetic field $\underset{\sim}{B}_{tr}$ rotates through a fixed angle ε ($> \varepsilon_0 \equiv 0$) while $|\underset{\sim}{B}_{tr}|$, the density ρ, the specific entropy S remain constant. On crossing slow and fast shocks, on the other hand, $|\underset{\sim}{B}_{tr}|$ and ρ vary but the direction of $\underset{\sim}{B}_{tr}$ remains fixed. In addition, slow and fast shocks are compressive - the density behind [the side into which there is a flux of mass] these shocks exceeds that in front. This is an immediate consequence of the magnetoelastic Hugoniot relation [[3], p. 146] and the thermodynamic condition that the entropy behind the shock shall exceed that in front. Moreover, the relations of the slow, intermediate and fast disburbance speeds to each other and to the slow, intermediate and fast shock velocities mirror in detail the corresponding relations in magnetogasdynamics. Finally, it turns out that U_1 and c in slow and fast shocks may be expressed in terms of simple algebraic functions of $\varepsilon = |B_{tr,1} - \underset{\sim}{B}_{tr,0}|/B_0$ [recall that $\underset{\sim}{B}_{tr}$ does not change direction in slow and fast shocks] which, as in magnetogasdynamics, makes possible a detailed analysis of these shocks. In particular, it can be shown that $S_1 - S_0 = O(\varepsilon^2)$ at worst as $\varepsilon \to 0$. Without such a result, we would not be justified in subsuming the theory of weak shocks under the theory of weak discontinuities as we did in Section II.

Finally, let us return to the piston problem of Section III. The linear solution tells us that only slow and fast discontinuities are present. In addition, the fact that $\delta w_{1,F}$ is negative in the fast wave and $\delta w_{1,S}$ is positive in the slow wave suggests that the nonlinear solution should consist of a fast (compressive) shock followed by a slow simple (expansion) wave centered at the origin of x,t-space - only this combination reduces continuously to the small disturbance solution and we wish to build contionuous dependence on initial data into our solution. Without entering into the details, it turns out that this nonlinear solution [fast shock followed by a slow centered simple wave [See Figures 1,e,f, and g]] does indeed provide a solution to the piston problem. In the final steady state which is achieved after the waves have propagated out to infinity, the magnet moves with the same downward speed ϑ as the piston and when ϑ is large, we find $w_{1,2}/\vartheta^{3/2} = O(1)$ and $|\underset{\sim}{B}_{tr,2}|/\vartheta^{1/3} = O(1)$, $w_{1,2}$ and $\underset{\sim}{B}_{tr,2}$ denoting the longitudinal strain and transverse magnetic field near the piston. Clearly, this region is in a state of high tension and is subject to an intense transverse magnetic field. Both $B_{tr,2}$ and $w_{1,2}$ may, evidently, increase without bound as $\vartheta \to \infty$. However, in real media we would expect the effects of plasticity to supervene and place bounds on the magnitudes of w_{12} and $|\underset{\sim}{B}_{tr,2}|$.

V. PRECURSORS AND SOME NEW WORK

Magnetoelasticity has its sources in the fundamental work of H. Alfvén [8] and S. Lundquist [9] in magnetofluid-dynamics. A survey of the literature up to the year 1969 has been given by B. Paria [10]. A good sampling of the papers of S. Kaliski on magnetoelasticity is given in [10]. Also relevant here, are the interesting papers of M. F. McCarthy on acceleration waves [See [16a] and pertinent self-references in [16b].] On the purely mathematical side, the papers [1] - [3], described above, are patterned after the works of K. O. Friedrichs [11], J. Bazer [1], [12] and J. Bazer and W. B. Ericson [13] which deal with magnetogasdynamic shocks, simple waves and small disturbance waves in infinitely conducting gases. On the side of pure elasticity, this work is indebted to the Handbuch articles of C. Truesdell & R. H. Toupin [15], C. Truesdell & W. Noll [19], as well as to the books [15], [16] and [20], by D. Bland, B. E. Green & J. E. Adkins and W. Prager. For relativistic formulations and extensions of many of the above results the reader is referred to the interesting papers by G. A. Maugin [See [7 a,b] and the self-references cited therein].

FOOTNOTES

1. Numbers in square brackets refer to items in the reference section at the end of the paper. Lack of space has, regretfully, precluded our citing more of the relevant literature. Apologies are offered beforehand for omissions.

2. Only compressible media are discussed here. Incompressible media are treated in the report [17] by F. Karal and the author. Some copies of the report are still available.

3. More general potentials (energy densities) considered in [3] include (i) nonspecial Hookean potentials which have the same form as the right member of equation (1.3) plus terms proportional to wS and S^2, (ii) generalized Hookean potentials, so-called because they yield results similar to those governed by nonspecial Hookean potentials but, unlike these potentials, lead to genuinely nonlinear stress-strain and temperature-strain relations and (iii) the most general potentials for isotropic and transversely isotropic materials.

4. The strain $\underset{\sim}{w}$ but <u>not</u> the displacement $\underset{\sim}{\xi}$ may be discontinuous.

REFERENCES

[1] Bazer, J., Geometrical magnetoelasticity, Geophys. J. R. Astro. Soc. **25**, 207-238 (1971).

[2] Bazer, J. & Karal, F., Simple wave motion in magnetoelasticity, Geophys. J. R. Astro. Soc. 25, 127-156 (1971).
[3] Bazer, J. & Ericson, W.B., Nonlinear wave motion in magnetoelasticity, Arch. Rat. Mech. Anal. Vol. 55, No. 3, 124-192 (1974).
[4] Friedrichs, K.O. & Lax, P.D., Systems of conservation equations with a convex extension, Proc. Nat. Acad. Sci. USA, Vol. 68, No. 8, 1686-1688 (1971).
[5] Ruderman, M., (a) Superdense matter in stars, J. de Physique, C3, supplement No. 11-12, Tome 30, C3-152 - C3-160 (1969). (b) Solid stars, Scientific American, pp. 22-31, Feb. 1971.
[6] Jeffrey, A. & Taniuti, T. Nonlinear Wave Propagation, NewYork, Academic Press, New York (1964).
[7] Maugin, G. A., (a) Nonlinear wave propagation in relativistic continuum mechanics, Helvetica Physica Acta, Vol. 52, 150-170 (1979). (b) Ray theory and shock formation in relativistic solids, Phil. Trans. R. Soc. Lond. A. Vol. 302, 189-215 (1981).
[8] Alfvén, H., Cosmical Electrodynamics, Clarendon Press, Oxford (1950).
[9] Lundquist, S., Studies in magneto-hydrodynamics, Arkiv Fysik 5, 297-347, (1952).
[10] Paria, G., Magnetoelasticity and magneto-thermoelasticity, Advances in Applied Mechanics 10, 78-112 (1967).
[11] Friedrichs, K. O., Nonlinear wave motion in magnetohydrodynamics, Los Alamos Sci. Lab. Report LAMS2105 (Physics), written September 1954, distributed March 1957. [See also Notes on Magnetohydrodynamics, VIII, Nonlinear wave motion, AEC Computing and Aplied Mathematics Center, Inst. Math. Sci., NYU, Report No. NYO-6486 (1958) - with H. Kranzer].
[12] Bazer, J., Resolution of an initial shear flux discontinuity, Astrophys. J. 128, 686-721 (1958).
[13] Bazer, J. & Ericson, W. B., Hydromagnetic shocks, Astrophys. J. 129, 758-785 (1959).
[14] Truesdell, C. & Toupin, R. A., Classical field theories, Handbuch der Physik, p. 226-793, Springer, New York (1960).
[15] Bland, D. R., Nonlinear Dynamic Elasticity, Blaisdell Publ. Co., Waltham, Mass. (1969).
[16] Green, A.E. & J.E. Adkins, Large Elastic Deformations, Clarendon Press, Oxford (1970).
[17] Karal, F. & Bazer, J., Nonlinear magnetoelastic wave motion in incompressible infinitely conducting solids -- a report.
[18] McCarthy, M. F., (a) The growth of magneto elastic waves in a Cauchy elastic material of fininte electrical conductivity, Arch. Rat. Mech. Anal. Vol. 23, 191-217 (1966). (b) In Continuum Physics (ed. A. C. Eringen) Vol. 2, pp. 449-521, Academic Press, New York (1975).
[19] Truesdell, C. & W. Noll, The nonlinear field themes of mechanics, Handbuch der Physik, pp. 1-602, Springer, New York (1965).
[20] Prager, W., Introduction to Mechanics of Continua, Ginn & Co.,Boston (1961).

ACKNOWLEDGMENT

This work was supported by the U. S. National Science Foundation under Grant No. MCS-8104243.

The Mechanical Behavior of Electromagnetic Solid Continua
G.A. Maugin (editor)
Elsevier Science Publishers B.V. (North-Holland)

INTRINSIC SYMMETRIES OF MATERIAL TENSORS IN GENERALIZED OHM AND FOURIER CONDUCTION LAWS FOR ANISTROPIC SOLIDS

Yaşar Ersoy +
Middle East Technical University
Ankara, Turkey.

In this paper, intrinsic symmetries and interrelations between some material tensors denoting certain cross-coupling effects in the generalized Ohm and Fourier conduction laws for conducting anisotropic solids are investigated. It is shown that the existence of some relations among the material tensors is linked to a particular form of the powerless term in the decomposition of the thermodynamical fluxes.

INTRODUCTION

One of the prominent problems in contemporary continuum mechanics consists in determining and refining constitutive equations which reveal interactions of electromagnetic fields in solid continua. Therefore, among rather old but relatively important problems, a general constitutive theory of electric and/or thermal conduction has been also a research topic in recent years, e.g. Pipkin and Rivlin /1/, Borghesani and Morro /2/. Employing a rather new physico-mathematical model of materials in our previous papers /3,4/, the Ohm and Fourier conduction laws are generalized for bicoupled thermo-electrical anisotropic solids. A complete characterization of the material tensors (or moduli) such as intrinsic symmetries and interrelations between some material tensors denoting certain cross-coupling effects is not, however, exhibited therein.
In the viewpoint of modern thermodynamics, the very nature of material tensors in the generalized equations of conductivity mentioned above is investigated in the present paper.

THERMODYNAMICAL FORCES AND FLUXES

We consider an electrically and thermally conducting magnetothermorigid continuous solids with a finite volume in the 3-dimensional Euclidean space.

In /4/, the Ohm and Fourier conduction laws are generalized in a rather rational way that they are*

+ Present Address: Planetenlaan 1, 5632 DG Eindhoven, The Netherlands.
* Throughout this paper, the Cartesian tensor notation together with the summation convention is employed. The superposed dot, (.), denotes the derivative with respect to the time, t.

$$\dot{J}_k + \Gamma_{kl}^{(e)} J_l = \pi_{kl}^{(e)} E_l + \chi_{kl}^{(e)} G_l - \Delta_{kl}^{(e)} Q_l$$

(1)

$$\dot{Q}_k + \Gamma_{kl}^{(t)} Q_l = \chi_{kl}^{(t)} E_l - \pi_{kl}^{(t)} G_l - \Delta_{kl}^{(t)} J_l$$

respectively. Although the symbols are fairly standart, we list their meanings for the convenience. $\underset{\sim}{J}$ and $\underset{\sim}{Q}$ are respectively the electric current and heat flux vectors, $\underset{\sim}{E}$ the effective electric field, $\underset{\sim}{G}$ is the temperature gradient. Furthermore $\underset{\sim}{\Gamma}$'s, $\underset{\sim}{\pi}$'s, $\underset{\sim}{\chi}$'s and $\underset{\sim}{\Delta}$'s are the material tensors denoting a lot of thermo-electrical effects and functions of the temperature, θ, and the effective magnetic field, $\underset{\sim}{H}$. It is worthy to remark that these constitutive equations govern bicouplings in thermo-electrical materials, and electrical and thermal relaxation phenomena.

In the derivation of the generalized equations of conductivity above, a thermo-electrical equilibrium (T-EE) state, i.e.,

(2) $$\bar{\Omega} \equiv \Omega|_{eq} = 0$$

is introduced, and a polynomial expression for the modified Helmholtz free energy (MHFE) density in the form

(3) $$\Psi(\Lambda,\Gamma) = \frac{1}{2} \Sigma_{kl}^{(e)} J_k J_l + \frac{1}{2} \Sigma_{kl}^{(t)} Q_k Q_l + \Sigma_{kl}^{(et)} J_k Q_l$$

is employed. In (2) and (3), Ω, Λ and Γ are the sets of variables defined by

(4) $$\Omega = \{E_k,\ G_k,\ J_k,\ Q_k\} \quad ; \quad \Lambda = \{\theta,\ H_k\} \quad ; \quad \Gamma = \{J_k,\ Q_k\}$$

and $\underset{\sim}{\Sigma}$'s are also the material tensors being the function of Λ.

Concerning the complete characterization of Onsager fluxes specified in a series of papers by Edelen, e.g. /5/ and the very nature of the constitutive theory in our papers /3,4/, a particular set of thermodynamical forces, κ (or $\underset{\sim}{K}^{\alpha}$), and of fluxes, L (or $\underset{\sim}{L}^{\alpha}$), can be assigned in the 4x3 = 12 dimensional space. They are defined as

(5) $$\kappa \equiv [K_k^{\alpha}] = \{E_k,\ J_k,\ G_k,\ R_k\} \quad ; \quad L \equiv [L_k^{\alpha}] = \{Y_k,\ \check{J}_k,\ Z_k,\ \check{Q}_k\},$$

where the superscript α (α = 1,2,3,4) stands for each vector-valued quantity in the braces. In (5), $\underset{\sim}{Y}(\Gamma,\Lambda)$ and $\underset{\sim}{Z}(\Gamma,\Lambda)$ are respectively the differentiable functions denoting the partial derivative of the MHFE density with respect to $\underset{\sim}{J}$ and $\underset{\sim}{Q}$, $\check{\underset{\sim}{J}}(\Omega,\Lambda)$ and $\check{\underset{\sim}{Q}}(\Omega,\Lambda)$ the response functions governing the temporal evolution of $\underset{\sim}{J}$ and $\underset{\sim}{Q}$ while $\underset{\sim}{R}$ is introduced for a coincise notation as $\underset{\sim}{R} = -\theta^{-1}\underset{\sim}{Q}$.

The expressions in (5) are naturally different from the conventional thermodynamical forces and fluxes known in the literature. However, they are the generalization of the expressions given by Lebon /6/.

The new thermodynamical forces and fluxes defined in (5) are imposed to fulfill the requirements: (i) κ do vanish in the T-EE state, and (ii) L are continuously differentiable functions of κ and Λ, i.e.,

(6) $$\overline{K}^{\alpha}_{k} \equiv K^{\alpha}_{k}|_{eq} = 0 \quad ; \quad L^{\alpha}_{k} = L^{\alpha}_{k}(\kappa,\Lambda).$$

THE DECOMPOSITION OF THERMODYNAMICAL FLUXES

According to the decomposition theorem presented by Edelen /5/, the thermodynamical fluxes in (5) can be expressed as the sum of two uniquely determined parts, i.e.,

(7) $$L^{\alpha}_{k} = M^{\alpha}_{k} + N^{\alpha}_{k} \quad (\text{or } L = M + N).$$

In (7), $\underset{\sim}{M}^{\alpha}$ are the gradient of a scalar-valued function, $\phi(K,\Lambda)$, while $\underset{\sim}{N}^{\alpha}$ fulfill an orthogonality relation. In other words, ϕ is a potential function for $\underset{\sim}{M}^{\alpha}$, and $\underset{\sim}{N}^{\alpha}$ are the nondissipative fluxes leading to no contribution in the expression of a supplementary concept associated with the physical system here. Thus we write

(8) $$M^{\alpha}_{k} = \frac{\partial\phi(\kappa,\Lambda)}{\partial K^{\alpha}_{k}} \quad ; \quad K^{\alpha}_{k}\, N^{\alpha}_{k} = 0.$$

Furthermore, $\underset{\sim}{L}^{\alpha}$ satisfy a full system of reciprocity relations

(9) $$\frac{\partial L^{\alpha}_{k}(\kappa,\Lambda)}{\partial K^{\beta}_{l}} = \frac{\partial L^{\beta}_{k}(\kappa,\Lambda)}{\partial K^{\alpha}_{l}}.$$

The residual entropy inequality (or entropy production), $D\ (\Lambda,\Omega)$, in /3,4/ can be also expressed in terms of $\underset{\sim}{K}^{\alpha}$ and $\underset{\sim}{L}^{\alpha}$. The result thus obtained is not convenient to specify ϕ. However, concerning the concept of power , $P(\kappa,\Lambda)$, in general, $\phi(\kappa,\Lambda)$ the so called flux-potential function is determined.

On the other hand, the general requirements

(10) $$P(\kappa,\Lambda)\ \varepsilon\ C^{1,0} \quad ; \quad P(0,\Lambda) = 0 \quad ; \quad \frac{\partial P}{\partial K^{\alpha}_{k}}\Big|_{eq} = 0$$

should be considered while expressing P. It will be noted, in contrast to D, that P is not necessarily positive definite ouside the equilibrium state. In view of (10), we then propose the following expression

(11) $$P(\kappa,\Lambda) = \delta^{\beta\sigma}_{\alpha\gamma}\ |C_{\alpha\beta}|\ L^{\gamma}_{k}\ (\kappa,\Lambda)\ K^{\sigma}_{k}$$

for the considered material, where $\delta^{\beta\sigma}_{\alpha\gamma}$ is the generalized Kroncker delta and $|C_{\alpha\beta}|$ are the coefficients employed for the dimensional homogeneity.

In view of (5) and (11), we now express the flux-potential function in (8) as

(12) $$\phi(\kappa,\Lambda) = \int_{0}^{1} \{c_1[Y_k(\lambda\Gamma,\Lambda)E_k + c_2\, \check{\mathfrak{J}}_k(\lambda\kappa,\Lambda)\ J_k] + c_3[Z_k(\lambda\Gamma,\Lambda)\ G_k + c_4\, \check{\mathfrak{Q}}_k(\lambda\kappa,\Lambda)R_k]\}\ d\lambda,$$

where c's are the new coefficients replaced instead of $|C_{\alpha\beta}|$.

Taking derivative of (12) with respect to $\underset{\sim}{K}^{\alpha}$ the expressions, which are necessary to identify the terms at the right hand side of (7), can be obtained after lengthy calculation. These expressions then lead to

$$\text{(13)}\qquad \begin{aligned} Y_k^* &= \frac{\partial\phi}{\partial E_k} + U_k \quad ; \quad Z_k^* = \frac{\partial\phi}{\partial G_k} + V_k \quad , \\ \check{J}_k^* &= \frac{\partial\phi}{\partial J_k} + W_k \quad ; \quad \check{Q}_k^* = -\,\theta\frac{\partial\phi}{\partial Q_k} + T_k \end{aligned}$$

wherein the argument of all functions have been omitted and, for the sake of simplicity in notation, some new quantities defined as

$$\text{(14)}\qquad \begin{aligned} Y_k^* &= c_1\, Y_k \quad ; \quad Z_k^* = c_3\, Z_k \quad , \\ \check{J}_k^* &= (c_1c_2)\, \check{J}_k \quad ; \quad \check{Q}_k^* = (c_3c_4)\, \check{Q}_k \end{aligned}$$

while $\underset{\sim}{N}^{\alpha}$ been replaced by $\underset{\sim}{U}$, $\underset{\sim}{W}$, $\underset{\sim}{V}$ and $\underset{\sim}{T}$.

Making use of (5) and the explicit statement for each vector-valued functions above, the orthogonality relation in (8) can be verified easily and reexpressed in the form

$$\text{(15)}\qquad U_kE_k + W_kJ_k + V_kG_k + T_kR_k = 0.$$

Furthermore, it is also observed that

$$\text{(16)}\qquad \bar{U}_k = \bar{W}_k = \bar{V}_k = \bar{T}_k = 0$$

in the T-EE state. As noticed before the relation in (15) means that the second part of the decomposed fluxes in (7) is of powerless essence.

INTRINSIC SYMMETRIES AND SOME RELATIONS

Either the extended form of the reciprocity relations or well-known Casimir-Onsager relations are, in general, employed to investigate intrinsic symmetries of proper material tensors, which connect the conjugate flux-force pairs, and interrelations between interferencing material tensors, which provide cross-coupling effects between forces of one type and fluxes of another type. As seen in the previous section, if the supplementary concept, i.e., the power and associating flux-potential functions, is taken into account the expressions in (13) are obtained. Instead of referring to (9), we can now employ (13) to search for intrinsic symmetries and interrelations of some material tensors prescribing the generalized Ohm and Fourier conduction laws in (1) and the MHFE density in (3).
Derivative of (13) with respect to $\underset{\sim}{K}^{\alpha}$ and the property of second partial derivatives of ϕ in the resulting expressions thereby lead to the following relations

$$\text{(17)}\qquad \frac{\partial U_k}{\partial E_l} = \frac{\partial U_l}{\partial E_k} \quad ; \quad \frac{\partial V_k}{\partial G_l} = \frac{\partial V_l}{\partial G_k} \quad ; \quad \frac{\partial U_k}{\partial G_l} = \frac{\partial V_l}{\partial E_k} \quad ,$$

and

(18)
$$\frac{\partial \mathfrak{J}^*_k}{\partial E_l} - \frac{\partial Y^*_l}{\partial J_k} = \frac{\partial W_k}{\partial E_l} - \frac{\partial U_l}{\partial J_k} \; ; \quad \frac{\partial \mathfrak{Q}^*_k}{\partial E_l} - \frac{\partial Y^*_l}{\partial R_k} = \frac{\partial T_k}{\partial E_l} - \frac{\partial U_l}{\partial R_k} ,$$
$$\frac{\partial \mathfrak{J}^*_k}{\partial G_l} - \frac{\partial Z^*_l}{\partial J_k} = \frac{\partial W_k}{\partial G_l} - \frac{\partial V_l}{\partial J_k} \; ; \quad \frac{\partial \mathfrak{Q}^*_k}{\partial G_l} - \frac{\partial Z^*_l}{\partial R_k} = \frac{\partial T_k}{\partial G_l} - \frac{\partial V_l}{\partial R_k} ,$$
$$\frac{\partial \mathfrak{J}^*_k}{\partial J_l} - \frac{\partial \mathfrak{J}^*_l}{\partial J_k} = \frac{\partial W_k}{\partial J_l} - \frac{\partial W_l}{\partial J_k} \; ; \quad \frac{\partial \mathfrak{Q}^*_k}{\partial R_l} - \frac{\partial \mathfrak{Q}^*_l}{\partial R_k} = \frac{\partial T_k}{\partial R_l} - \frac{\partial T_l}{\partial R_k} ,$$
$$\frac{\partial J^*_k}{\partial R_l} - \theta^{-1} \frac{\partial \mathfrak{Q}^*_l}{\partial J_k} = - \frac{\partial W_k}{\partial R_l} + \theta^{-1} \frac{\partial T_l}{\partial J_k} .$$

Furthermore, if

(19)
$$\frac{\partial Y^*_k}{\partial J_l} = \frac{\partial Y^*_l}{\partial J_k} \; ; \quad \frac{\partial Z^*_k}{\partial R_l} = \frac{\partial Z^*_l}{\partial R_k} \; ; \quad \frac{\partial Y^*_k}{\partial Q_l} = - \theta (\frac{c_1}{c_3}) \frac{\partial Z^*_l}{\partial J_k}$$

are substituted into some expressions in (18), the relations

(20)
$$\frac{\partial \mathfrak{J}^*_k}{\partial E_l} - \frac{\partial \mathfrak{J}^*_l}{\partial E_k} = \frac{\partial U_k}{\partial J_l} - \frac{\partial U_l}{\partial J_k} + \frac{\partial W_k}{\partial E_l} - \frac{\partial W_l}{\partial E_k} ,$$
$$\frac{\partial \mathfrak{Q}^*_k}{\partial G_l} - \frac{\partial \mathfrak{Q}^*_l}{\partial G_k} = \frac{\partial T_k}{\partial G_l} - \frac{\partial T_l}{\partial G_k} + \frac{\partial V_k}{\partial R_l} - \frac{\partial V_l}{\partial R_k} ,$$
$$\frac{\partial \mathfrak{J}^*_k}{\partial G_l} + \theta^{-1} (\frac{c_3}{c_1}) \frac{\partial \mathfrak{Q}^*_l}{\partial E_k} = \frac{\partial W_k}{\partial G_l} - \frac{\partial V_l}{\partial J_k} + (\frac{c_3}{c_1}) \theta^{-1} (\frac{\partial T_l}{\partial E_k} - \frac{\partial U_k}{\partial R_l})$$

are also deduced.

For our purpose mentioned above, the relations in (17), (18) and (20) or their special forms can be employed. To do this, each vector-valued function in the set N should be, however, known explicitly.

In view of (16) if the vector-valued functions $\underset{\sim}{N}^\alpha$ are expressed in terms of the sixteen new material tensors of rank 2, $\underset{\sim}{\mathfrak{Z}}^{(\cdot)}$ and $\underset{\sim}{K}^\alpha$ at the same linear order of approximation, the orthogonality relation (15) then imposes very significant restriction on those expressions. It is shown that there may need five scalar coefficients and the two axial material tensors of rank 2 to specify $\underset{\sim}{N}^\alpha$.

To illustrate certain implication of analysis in the present paper, we will now consider a special case for the form of $\underset{\sim}{N}^\alpha$, i.e., the nonexistence of terms at the r.h.s. of the relations (18) and (20). This assumption is fulfilled if $\underset{\sim}{N}^\alpha = \underset{\sim}{0}$. Concerning the last assumption and noticing the values of $\underset{\sim}{\mathfrak{J}}$, $\underset{\sim}{\mathfrak{Q}}$, and of $\underset{\sim}{Y}$, $\underset{\sim}{Z}$, if (1) and (3) are substituted into (18-20), the following intrinsic symmetries

(21)
$$\pi^{(e)}_{[kl]} = 0 \; ; \quad \pi^{(t)}_{[kl]} = 0 \; ; \quad \Gamma^{(e)}_{[kl]} = 0 \; ; \quad \Gamma^{(t)}_{[kl]} = 0 ,$$
$$(\Sigma^{(e)}_{[kl]} = 0 \;, \quad \Sigma^{(t)}_{[kl]} = 0)$$

and interrelations

(22) $\chi_{kl}^{(t)} = -\theta\, c_o(\frac{c_1}{c_3})\, \chi_{lk}^{(e)} \quad ; \quad \Delta_{kl}^{(t)} = -\theta\, c_o\, \Delta_{lk}^{(e)}$

and

(23) $\Sigma_{kl}^{(e)} = -c_2\, \pi_{kl}^{(e)} \quad ; \quad \Sigma_{kl}^{(t)} = -\theta^{-1} c_4\, \pi_{kl}^{(t)}$

$$\Sigma_{kl}^{(et)} = -c_o\, c_4\, \chi_{kl}^{(e)} = \theta^{-1} c_o^{-1}\, c_2\, \chi_{lk}^{(t)}$$

are obtained, where c_o is a new coefficient defined as $c_o = c_1 c_2 / c_3 c_4$.

As would be expected, the expressions in (21-23) are nothing but the relations which are equivalent to those implied by the reciprocity relations (9) or by the Casimir-Onsager relations.

If a few or all necessary components of $\underset{\sim}{\mathfrak{z}}^{(.)}$'s are not considered to be zero, then we obtain some relations which are, in principle, similar to those in (22) and (23), but they are interrelated throughout the nonvanishing components of $\underset{\sim}{\mathfrak{z}}^{(.)}$'s.

REFERENCES

/1/ Pipkin, A.S. and Rivlin, R.S., Galvanomagnetic and thermomagnetic effects in isotropic solids, J. Math Phys 1 (1960),542-546.

/2/ Broghesani, R. and Morro, A., Thermodynamics and isotropy in thermal and electrical conduction, Meccanica 9(1974), 63-69.

/3/ Ersoy, Y., A New constitutive theory for magnetizable rigid conductors, in: Tezel, A. and Inan, E.E. (eds.), Proc. National Congress of Mechanics 2 (Tumtmk, Istanbul, 1981), 332-343 (in Turkish).

/4/ Ersoy, Y., A Thermodynamic development of generalized Fourier and Ohm conduction laws for anisotropic materials, Letters in Appl. and Engng Sci. 21 (1983), 835-840.

/5/ Edelen, D.B.G., On the characterization of fluxes in nonlinear irreversible thermodynamics, Int. J. Engng Sci. 12 (1974), 397-411.

/6/ Lebon, G., Thermodynamic analysis of rigid heat conductors, Int. J. Engng Sci. 18 (1980), 727-739.

The Mechanical Behavior of Electromagnetic Solid Continua
G.A. Maugin (editor)
Elsevier Science Publishers B.V. (North-Holland)

THE INFLUENCE OF ELECTROMAGNETIC INTERACTION ON THE SUPERCONDUCTIVITY OF DEFORMABLE SOLIDS

K.P. Sinha

Indian Institute of Science
Bangalore 560 012
INDIA.

The radiation induced interband mixing effects along with the scattering of conduction electrons by appropriate phonon modes when considered for a multiband solid lead to interesting new terms in the effective electron-electron interaction which depend on the density of quanta impressed on the system. These suggest the possibility of radiation controlled pairing of electrons resulting in a non-equilibrium superconducting state in some suitable systems. The stability of such structures in relation to the lifetime of various processes is also discussed.

INTRODUCTION

Among the several theoretical proposals [1-4] to achieve a high temperature superconducting state, the non-equilibrium pairing mechanism [4-8] shows great promise in virtue of its ability to enhance the appropriate parameters through the agency of real boson modes (photons, phonons or excitons). Suitable conditions can be best realised in multiband (valley) systems (e.g. semiconductors, semimetals etc.). Indeed recent observations of large diamagnetic anomalies in some solids (CuCl at temperature $T \sim 200$ K under high pressure [9], CdS at $T \sim 77$K and pressure quenched [10]) seem to be encouraging pointers. Furthermore, in multiband (valley) systems, the optical phonons might play an important role in interband (valley) scattering processes of conduction electrons owing to large electron-phonon coupling constant. If these phonon modes connect states which are made degenerate or near degenerate in the conduction bands, one can take advantage of a large electronic density of states [11]. While pressure can induce relative changes in the minima of conduction band at various symmetry points, it is desirable to have direct dynamical realisation by the injection of appropriate boson modes in the system.

In what follows, we consider a multiband semiconducting solid in the presence of electromagnetic and phonon fields. This will involve the interplay of two boson fields, with one of them having macroscopic occupation in the system.

NON-EQUILIBRIUM PAIRING MECHANISM

In the equilibrium pairing mechanism the attractive electron-electron interaction is established by the virtual exchange of phonons. For some systems, the virtual exchange of excitons has also been suggested [1-3] but the dominant established mechanism involves phonons. Here we shall be concerned with the role of one real boson field in addition to the virutal ones. Thus the Hamiltonian of our model system will consist of electron fields in a multiconduction band and phonon fields in the presence of a coherent radiation field which connects electronic states of the bands. Owing to the macroscopic occupation of the radiation field mode considered here, a classical treatment for this is adequate. Nevertheless, we can express the field intensity in terms of the number of quanta n_o for the mode in question as

$$A_o^2 = (2\pi\hbar c^2/\Omega)\, n_o \quad , \tag{1}$$

where Ω is the radiation frequency, c and $\hbar$ being the usual constants. The interaction of the radiation field with the electron field can be represented as

$$H_{er} = \Sigma\, D\, [C^+_{km\sigma}\, C_{kn\sigma} \exp(-i\Omega t) + C^+_{kn\sigma}\, C_{km\sigma} \exp(i\Omega t)] \tag{2}$$

where ($C^+_{km\sigma}$, $C_{km\sigma}$) are the electron creation, annihilation operators for the Block state $|\underline{k}m, \sigma)$, $\underline{k}$ being the wavevector, m the band index and σ the spin. Here D is the coupling coefficient for the electric dipole interaction. Explicitly,

$$D = (2\pi V)^{-\frac{1}{2}} \left(-\frac{\Omega}{C}\right) A_o\, d_{mn} \tag{3}$$

$$d_{mn} = 2\pi\, (\phi_m |\, e\underline{r}\, | \phi_n) \cdot \underset{\sim}{\varepsilon} \quad , \tag{4}$$

where $\underset{\sim}{\varepsilon}$ is the polarization vector of the radiation field, and $(\phi_m | e\underline{r} | \phi_n)$ is the dipole matrix element connecting the Wannier states $|\phi_m)$ and $|\phi_n)$; V is the volume [4]. In terms of the electron field operators $\Psi(x) = \Sigma\, \underline{k}m) C_{\underline{k}m\sigma}$, the single particle electron Hamiltonian is

$$H_e = \int d^3x\, \Psi^+(\underline{x}) K(\underline{x})\, \Psi(\underline{x}) = \Sigma\, \varepsilon^{(m)}_{\underline{k}}\, C^+_{\underline{k}m\sigma}\, C_{\underline{k}m\sigma} \ , \tag{5}$$

with $\varepsilon^{(m)}_k$ single particle energies. Similarly for the phonons, we have

$$H_{ph} = \Sigma\, \hbar\omega_{\underline{q}}\, b^+_{\underline{q}}\, b_{\underline{q}} \ , \tag{6}$$

where (b^+_q, b_q) are the phonon (creation, annihilation) operators (of mode branch $\underline{q}$ and frequency ω_q). The coupling of the electron and the phonon fields (inclusive of interband (valley)) can be expressed as

$$H_{e\text{-}ph} = \sum_{\underline{qk\sigma}} P_{\underline{q}}\, (C^+_{\underset{\sim}{k}-\underset{\sim}{q},\, m\sigma}\, C_{\underset{\sim}{k}m\sigma}\, b^+_{\underline{q}} + \text{h.c}) \quad , \tag{7}$$

where

$$P_{\underline{q}} = (\hbar/2\rho\, \omega_q V)^{\frac{1}{2}}\, F_{mn} \quad ; \quad \rho \tag{8}$$

being the density of the system and F_{mn} is the deformation potential field connecting states of bands (valleys) m and n. To these interaction we must add the two-body Coulomb interaction H_c between electrons (holes). Its explicit form is not needed at this stage.

In the event the radiation field is such that it produces an exciton field (macroscopic occupation), it will be necessary to consider their interaction with electrons in the conduction bands. This interaction can be written as

$$H_{ex} = \Sigma\, B_{\underline{Q}}\, (C^+_{\underline{k+Q},\, m\sigma}\, C_{\underline{k}n\sigma}\, a^+_{-\underset{\sim}{Q}} + \text{h.c}) , \tag{9}$$

where ($a^+_{\underline{Q}}$, $a_{\underline{Q}}$) are the exciton operators and $B_{\underline{Q}}$ is exciton-electron coupling constant; $\underline{Q}$ is the exciton wave vector.

Let us first consider the situation with the electron field interacting with radiation and phonon fields. The case of excitons and phonons will be taken next. By a suitable unitary transformation the Hamiltonian consisting of several parts is brought to a time-independent form [8] . This leads to a shift in the energy of the electrons in conduction bands and the phonons. For example the phonon energy goes to $\hbar\tilde{\omega}_q = \hbar(\omega_q + \Omega)$, and

$\varepsilon_k^{(m)} = \varepsilon_k^{m} + \frac{1}{2}\hbar\Omega$; $\varepsilon_k^{(n)} = \varepsilon_k^{(n)} - \frac{1}{2}\hbar\Omega$. The role of the time-independent form of $H_{er} = \Sigma D(C_{km\sigma} C_{kn\sigma} + h.c)$ is to produce interband mixing of conduction states. The diagonalization of conduction states taking into account the mixing effect is best achieved by a canonical transformation [8]. Confining our attention to the lowest conduction state thus diagonalized and the modified electron-phonon interaction, we have

$$H = \Sigma E_k^{(c)} C_{k\sigma}^{+} C_{k\sigma} - \Sigma \frac{\tilde{P}_q D}{E_o} (C_{k-q\sigma}^{+} C_{k\sigma} b_q^{+} + h.c) + \tilde{H}_c \qquad (10)$$

where

$$E_o = [(\hbar\Omega - E_{mn})^2 + D^2]^{\frac{1}{2}}, \quad E_k^{(c)} = (\tfrac{1}{2}E_{mn} - E_o) + \frac{(\hbar k)^2}{2m^*} \qquad (11)$$

The energy gap between the two conductions band now is given by $E_{cd} = 2E_o$. It should be noted that the modified electron-phonon coupling constant now carries the additional factor (D/E_o) and depends on the amplitude A_o and hence $(\sqrt{n_o})$ of the radiation field. The above form obtains in a steady state situation in the presence of the influx of quanta of the electromagnetic field. In this non-equilibrium situation one has to invoke the concept of an effective temperature T^* which may be different from the ambient temperature. The fermions and bosons have their usual distribution function but at temperature T^*. Having this steady state situation in mind one can follow either the Bardeen-Cooper Schrieffer (BCS) procedure or the Gorkov-Nambu-Eliashberg formalism as modified by Keldysh for non-equilibrium situation [7,12,13].

For a general spin configuration so as to have both singlet and triplet pairings included in the steady state condition, the reduced Hamiltonian is written as

$$H_r = \Sigma_\sigma E_{\underline{k}}^{(c)} C_{\underline{k}\sigma}^{+} C_{\underline{k}\sigma} - \Sigma\, U_{\underline{kk'}} C_{\underline{k}'\sigma}^{+} C_{-\underline{k}'\sigma'}^{+} C_{-\underline{k}\sigma'} C_{k\sigma}, \qquad (12)$$

where

$$U_{kk'} = U_{rp} - U_c \qquad (13)$$

with

$$U_{rp} = \frac{\tilde{P}_q^2 D^2}{E_o^2} \frac{2(\hbar\omega_q)}{[(\hbar\omega_q)^2 - (E_k^{(c)} - E_{k'}^{(c)})^2]} \qquad (14)$$

the radiation admixed phonon mediated attractive (for $(E_k^{(c)} - E_{k'}^{(c)}) < \hbar\tilde{\omega}_q$) electron-electron interaction and U_c is the screened Coulomb interaction.

NUMERICAL ESTIMATES

A Green's function formalism for the superconducting gap function yields

$$\Delta = W_c \,\mathrm{cosech}\,(8\pi^3/UV\, N(E_F)) \qquad (15)$$

where W_c is BCS type energy cutoff within which $U_{kk'} = U$ constant and zero outside- N(EF) is the density of states at the Fermi level. The ciritical temperature T_c^* is obtained as

$$T_c^* = 1.13 \left(\frac{W_c}{k_B}\right) \exp\left[-\frac{1 + b(n_o/V)}{ab(n_o/V)}\right] \qquad (16)$$

and depends on the density of quanta of the radiation field. The quantities a and b have the explicit forms

$$a = (F_{mn}^2 / \rho\langle\tilde{\omega}^2\rangle) N(E_F) \tag{17}$$

$$b = 4 d_{mn}^2 (\hbar\Omega)/(\hbar\Omega - E_{mn})^2 \tag{18}$$

A convenient form for the density of states in the semiconductor is $N(E_F) = 3/2 (n_e/E_F)$, where n_e is the electron density and E_F is the Fermi energy in the steady state. As remarked earlier we are considering systems in which intraband electron-phonon interactions are rather weak. Accordingly, we have chosen new mechanisms which involve interband (valley) interaction for which the corresponding couplings are strong. In the present case the deformation potential field F_{mn}. We estimate the the transition temperature T_c^* as a function of radiation density for some typical parameters. These are

$F_{mn} \sim 5 \times 10^{-3}$ dynes, $n_e = 10^{21} cm^{-3}$, $E_F = 4.55 \times 10^{-13}$ erg

$(\hbar\Omega - E_{mn}) = 1.6 \times 10^{-15}$ erg. , $\hbar\Omega = W_c = 1.6 \times 10^{-14}$ erg

$d_{mn} \sim 10^{-17} (erg\ cm^3)^{\frac{1}{2}}$, $\rho \sim 3$ g cm^{-3}.

We get $T_c^* = 7.2$ K for $(n_o/V) = 5 \times 10^{14}$ cm^{-3}

and $T_c^* = 30.7$ K for $(n_o/V) = 10^{15}$ cm^{-3}.

Thus one gets considerable enhancement of T_c on increasing the radiation density by a factor of two. Although the values noted above are tentative, they fall within observable range say from 0.1 K to 100 K depending on the quanta of radiation impressed on the system. Beyond $(n_o/V) = 5 \times 10^{15}$ cm^{-3} heating effect may destroy the ordered state and the quasiparticle lifetime may not remain favourable for the maintenance of a coherent state. Below this density, the coherent mixing time for radiation $\tau_r \sim 10^{-11}$ sec, and the corresponding pair relaxation $\tau_{pair} \sim 10^{-11}$ to 10^{-12} sec. The quasiparticle lifetime (electrons and phonons) will remain in the range 10^{-9} to 10^{-11} sec. and hence will not disturb the coherence needed for the ordered and coherent state to be maintained.

EXCITONIC CASE

We now turn our attention to a situation when the radiation field (connecting valence and one of the conduction bands) produces real excitons which are sufficiently long lived to take part in the scattering of conduction electrons along with phonons. The effective pairing interaction arising from exciton and phonon exchange can be expressed as

$$U_{kk'}(ex) = \Sigma \frac{N_Q\ P_q^2\ B_Q^2\ (\hbar\omega_{ex} + \hbar\omega_q)}{(\hbar\omega_{ex} - E_{mn})^2\ [(E_k^m - E_{k'}^m)^2 - (\hbar\omega_{ex} + \hbar\omega_q)^2]}$$

where $\hbar\omega_{ex}$ is the exciton energy, N_Q is the exciton concentration which is maintained at a certain level by means of an appropriate radiation field impressed on the system. The above interaction is attractive for $(E_k^m - E_{k'}^m) < (\hbar\omega_{ex} + \hbar\omega_q)$. The strength of the interaction and hence T_c^* will again depend on the exciton density (N_{ex}/V). Furthermore, the lifetime of excitons should be of the order of 10^{-8} sec or more so that the superconducting order is maintained. The mechanism puts a limit on the carrier concentration in the conduction band. For $n_e > 10^{21}$ cm^{-3}, the excitons will be screened out and the mechanism will become insignificant. In contradistinction to the photon processes wherein the volume of the momentum space spanned by photons is small, the exciton processes will cover a large momentum space. However, in the foregoing we have considered two boson processes in which the real occupancy of one of the boson modes is involved. Thus in the exciton-phonon case also the exciton density actually present in the system will determine the strength of interaction and T_c^*. This in turn is determined by the flux of the radiation field that produces the excitons and the electron-exciton coupling constant. The expression for T_c^* will be similar to that for the photon case

$T_c = 1.13 \frac{W_c(ex)}{k_B} \exp\left[\frac{-1}{U(ex)N(E_F)}\right]$. In the absence of a precise knowledge of electron-exciton coupling, a numerical estimate is not attempted.

REMARKS

In the foregoing discussion the Coulomb repulsion has been taken to be almost screened out. This is further justified in that under the condition stipulated a p-wave (triplet) pairing is strongly indicated and for this Coulomb repulsion is considerably less than the s-wave (singlet) pairing.

In conclusion it is worth noting that the non-equilibrium mechanisms described in the present paper appear promising in the context of attaining a high temperature superconducting state.

REFERENCES

[1] Little, W.A., Phys. Rev. 134A,(1964) 1416.

[2] Ginzburg, V.L., Ann.Rev. Mat.Sci. 2 (1972) 663.

[3] Allender, D., Bray, J. and Bardeen, J., Phys. Rev. B7 (1973) 1020.

[4] Kumar, N. and Sinha, K.P., Phys.Rev. 174 (1968) 482.

[5] Shankar, R.K. and Sinha, K.P., Phys. Rev. B7 (1973) 4291.

[6] Shankar, R.K. and Sinha, K.P., Phys.Stat.Soli(b) 57 (1973) 377

[7] Sinha, K.P., in Interactions of Radiation with Condensed Matter Vol II, (International Atomic Energy Agency, 1977).

[8] Sinha, K.P., J.Low.Temp.Phys. 39 (1980) 1.

[9] Geballe, T.H. and Chu, C.W., Comments Solid State Phys.9 (1979) 115 and references therein.

[10] MacCrone, R.K. and Homan, C.G., Solid State Commun. 35 (1980) 615.

[11] Collins, T.C., Seel, M., Ladik, J.J., Chandrasekhar, M. and Chandrasekhar, H.R., Phys. Rev. B27 (1983) 140.

[12] Schrieffer, J.R., Theory of superconductivity (Benjamin, New York, 1964).

[13] Keldysh, L.V., Zh. ETF 47 (1964) 1515, Sov. Phys. JETP 29 (1965) 1018.

The Mechanical Behavior of Electromagnetic Solid Continua
G.A. Maugin (editor)
Elsevier Science Publishers B.V. (North-Holland)

EDDY-CURRENTS IN A SYSTEM OF MOVING CONDUCTORS

A. Bossavit

Electricité de France, E&R, IMA,
92141 Clamart
France

A variational formulation for the problem of eddy-currents in three dimensions is established. This allows a finite-element treatment where the unknown is the field h itself. The proposed formulation is also valid for a set of separate conductors which move with respect to each other, with elastic coupling. The full coupled system of equations is given for this case. Its structure is such that parametric resonance is possible in specific situations.

A VARIATIONAL FORMULATION OF THE EDDY-CURRENTS PROBLEM

We wish to compute eddy-currents induced in a solid, fixed--for the time being--conductor Ω by a given current distribution $j_g(t)$. For mathematical convenience, Ω will be a bounded open set with smooth boundary Γ, but it has not to be connected, and each of its connected components can be multiply connected (Fig. 1). Let $\omega = \mathbb{R}^3 - \mathrm{cl}(\Omega)$. We assume the support of j_g, $\omega_g(t) = \mathrm{cl}(\{x \in \mathbb{R}^3 |\ j_g(x, t) = 0\})$, is contained in ω. One can think of ω_g (which may depend on time) as a moving coil, for instance a rotor, where a known alternating current is imposed. The resistivity is ρ in Ω. We assume constant permeability μ_0 everywhere (a non essential restriction). Let h_g be some field, of no particular physical significance, satisfying

(1) $$\mathrm{curl}\ h_g = j_g.$$

(The divergence of h_g may be chosen at will.) Our aim: compute the total (physical) magnetic field $h + h_g$.

Let us write the field equations. First, Faraday's law:

(2) $$\mu_0\ \partial/\partial t\ (h_g + h) + \mathrm{curl}\ e = 0$$

(in the whole space $\mathbb{R}^3$). Next, Ampère's law:

(3) $$j + j_g = \mathrm{curl}\ (h + h_g).$$

Remark. By (1) and (3), the support of j is in $\mathrm{cl}(\Omega)$. A natural functional space to look for j in is

(4) $$J = \{j \in \mathbf{L}^2(\Omega) |\ \mathrm{div}\ j = 0 \text{ in } \Omega,\ j.n = 0 \text{ on } \Gamma\}$$

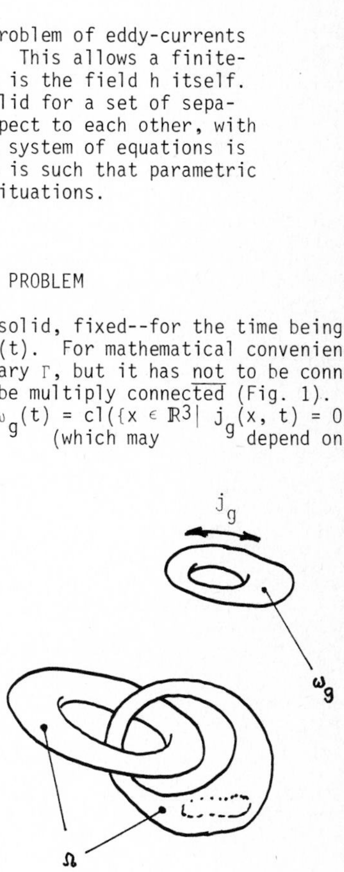

Figure 1

The modelled situation

where n denotes the outwards normal. This implies $\operatorname{curl} h \in \mathbb{L}^2$, and $\operatorname{curl} h = 0$ in ω, which justifies the choice of the functional space H we shall make in a moment.

A third relation ("constitutive law"), namely Ohm's law,

(5) $$e = \rho j \quad \text{in } \Omega,$$

completes the system of equations.

We shall derive a variational formulation for (1)(2)(3)(5) which, besides giving the right set-up for an existence-and-uniqueness proof, also suggests a finite-element approximation. First take the dot-product of (2) by some test vector field h', of the same nature as h, i.e. curl-free in ω. Integrating by parts, and using (5) and (1)(3), one obtains

$$0 = \mu_0 \frac{d}{dt}\left[\int_{\mathbb{R}^3} (h + h_g).h'\right] + \int_\Omega e.\operatorname{curl} h'$$

(6) $$= \mu_0 \frac{d}{dt}\left[\int_{\mathbb{R}^3} (h + h_g).h'\right] + \int_\Omega \rho \operatorname{curl} h.\operatorname{curl} h'$$

So if $h + h_g$ (the physical field) satisfies (1)(2)(3)(5), equation (6) is valid for all h' in the following vector space:

(7) $$H = \{h \in \mathbb{L}^2(\mathbb{R}^3) \mid \operatorname{curl} h \in \mathbb{L}^2(\mathbb{R}^3),\ \operatorname{curl} h = 0 \text{ in } \omega\}$$

(a Hilbert space when equipped with the natural scalar product

$$((h, h')) = \int_{\mathbb{R}^3} h.h' + \int_\Omega \operatorname{curl} h.\operatorname{curl} h' \quad).$$

Theorem 1. If h_g depends smoothly on t on the interval [0, T], there exists a unique function $t \to h(t) \in H$ on [0, T] such that

(8) $$\mu_0 \frac{d}{dt}\left[\int_{\mathbb{R}^3}(h + h_g(t)).h'\right] + \int_\Omega \rho \operatorname{curl} h.\operatorname{curl} h' = 0 \qquad \forall h' \in H$$

(9) $$h(0) + h_g(0) = h_0$$

where h_0 is given in H with the condition

(10) $$\operatorname{div} h_0 = 0.$$

(Cauchy condition (9) may be replaced by the periodicity one, h(0) = h(T), assuming h_g is itself T-periodic.)

The proof is easy, but we now have to proceed backwards: show that this h indeed is a solution to (1)(2)(3)(5), and the process does not go as smoothly as in the case of the heat equation (with which (8) has a strong similarity), because of the constraint curl h = 0 in ω.

Theorem 2. Let $t \to h(t)$ be the solution of (8)(9)(10). Given a smooth function f with compact support in ω, there exists a unique time-varying electric field $t \to e(t)$, with

(11) $$\operatorname{div} e = f \quad \text{in } \omega,$$

and such that (1)(2)(3)(5) hold.

Proof. Step 1. Since (8) is valid in particular for all $h' = \operatorname{grad} \phi'$, $\phi' \in H^1(\mathbb{R}^3)$, $\mu_0\, \partial/\partial t\, (h + h_g)$ is orthogonal to gradients in $\mathbb{L}^2(\mathbb{R}^3)$, and thus, using (10),

(13) $$\operatorname{div} (h + h_g) = 0 \quad \text{in } \mathbb{R}^3, \quad \forall t \in [0, T].$$

Step 2. Now choose h' in the space $\mathcal{D}(\Omega)$ of C^∞-vector fields with compact support in Ω. We obtain, in the sense of distributions in Ω,

$$\mu_0 \frac{\partial}{\partial t}(h + h_g) + \mathrm{curl}(\rho\ \mathrm{curl}\ h) = 0 \quad \text{in } \Omega, \tag{14}$$

which is half of Faraday's law (2).

Step 3. The other half proves much more difficult to establish, in part because we do not take advantage of simplifications that would occur if Ω was a simply-connected domain. We need two lemmas.

Lemma 1. The operator curl, composed with the restriction on Ω, maps H onto J (these spaces as defined in (7) and (4) respectively).

Proof. Take j in J, form $\tilde{j}$ = j on Ω and 0 on ω. Since j.n = 0, div $\tilde{j}$ = 0. Let $a = k * \tilde{j}$, where * is the convolution operator and

$$k(x, y) = \frac{1}{4\pi}\ \frac{1}{|x - y|} \qquad (x, y \in \mathbb{R}^3)$$

is the usual kernel of potential theory. Take h = curl a. One checks that div a = 0 and curl h = $\tilde{j}$, thus h $\in$ H and curl h = j on Ω.

Lemma 2. The orthogonal of J in $\mathbb{L}^2(\Omega)$ is

$$J^{\perp} = \{\mathrm{grad}\ \phi \mid \phi \in H^1(\Omega)\}$$

Proof. By Green's formula:

$$\int_\Omega j.\mathrm{grad}\ \phi = \int_\Gamma j.n\ \phi - \int_\Omega \mathrm{div}\ j\ \phi \qquad \forall\ \phi \in H^1(\Omega),$$

any j orthogonal to the gradients is in J. On the other hand, if grad $\phi \in$ J, $\Delta\ \phi = 0$ and $\partial\phi/\partial n = 0$, so ϕ is constant on each component of Ω, thus grad ϕ = 0.

We now have enough to complete Step 3. Let $a = \mu_0\ k * \mathrm{curl}(h + h_g)$. Equation (8) may be rewritten

$$\frac{d}{dt}\int_{\mathbb{R}^3} \mathrm{curl}\ a.\ h' + \int_\Omega \rho\ \mathrm{curl}\ h.\ \mathrm{curl}\ h' = 0 \qquad \forall\ h' \in H,$$

or else, integrating by parts in the first sum,

$$\int_\Omega (\partial a/\partial t + \rho\ \mathrm{curl}\ h).\mathrm{curl}\ h' = 0 \qquad \forall\ h' \in H.$$

By Lemmas 1 and 2, there exists $\phi \in H^1(\Omega)$ such that

$$\partial a/\partial t + \rho\ \mathrm{curl}\ h = \mathrm{grad}\ \phi.$$

This ϕ has a trace in $H^{1/2}(\Gamma)$, so (by solving the exterior Dirichlet problem $\Delta\phi = f$) we may extend it into a function on the whole space. Now the field

$$e = -\ \partial a/\partial t + \mathrm{grad}\ \phi$$

is equal to curl h in Ω, div-free in ω, and satisfies Faraday's law (2) by construction. Q.E.D.

Remark. We were free to choose any right-hand side f in (11), with only technical restrictions (f should in fact be in the dual of the Beppo-Levi space BL(ω)). This f should be interpreted as a given density of electric charges in the non-conducting region. As we see, the overall solution h does not depend on f. So such charges have no influence on the magnetic field. This is not surprising, since from the beginning, by assuming Ampère's law in the form of Eq. (1), we set on neglecting the displacement currents. Of course this is physically valid only for slow variations of h_g with time.

Remark. It thus appears that the overall electro-magnetic field has two separate sources: the given currents j_g and the outside electric charges f. If we are only interested in the magnetic field, it is enough to know j_g. Now if e is needed, some information on it is available, since e is known in Ω by Ohm's law, but to say more would request the knowledge of f. Notice that nothing is said about the relationship between e and j_g in the coil ω_g. Ohm's law has no reason to be valid there, since the currents are generated in ω_g in some way, and a generator is precisely a place where charges go up the electric field.

Remark. We propose to dub (8) "the electricity equation", in view of the similarity between (8) and the variational form of the heat equation. Just as the heat equation implies infinitely fast propagation of the heat, our model is consistent with instantaneous propagation of electromagnetic effects.

INTERMEZZO

We have no intention to dwell upon the dicretization of (8), but a few words on it are mandatory as motivation for what follows. To discretize a variational equation is a process by which linear and bilinear forms that play a role in it are associated with forms on some finite vector space, the elements of which can be considered as parameters from which some approximation of the generic element in the initial Hilbert space can be reconstructed. In (8), we have an integral, namely the one on $\mathbb{R}^3$, which is not easily dealt with: one should either pave the whole space with finite elements, or find something more clever. A reasonable idea is to take advantage of the a priori information we have on the solution h in ω, by splitting the integral in two and treating the two parts differently. The part on Ω is treated by finite elements (of a special kind), and the part on ω by boundary elements. This requires some transformations of (8) that we now describe, with the simplifying assumption that ω is simply connected.

We assume div $h_g = 0$ in ω, an easily met with condition, since $j_g = \text{curl } h_g$ is the data, not h_g itself. Let us write

$$(15) \qquad \int_{\mathbb{R}^3} (h + h_g).h' = \int_\Omega (h + h_g).h' + \int_\omega (h + h_g).h'$$

and deal with the second part of the right-hand side. Since curl $h' = 0$, there exists ϕ' such that $h' = \text{grad } \phi'$ in ω. Thus

$$\int_\omega (h + h_g).h' = \int_\omega (h + h_g).\text{grad } \phi' = \int_\Gamma \phi' \, (h + h_g).n$$

by Green's formula and (13). Now $h = \text{grad } \Phi$ in ω, since curl $h = 0$, and $\Delta\Phi = 0$ since div $h = 0$. Let $\phi = \Phi|_\Gamma$ and introduce the boundary operator S defined by

$$(16) \qquad S\phi = \partial\Phi/\partial n \qquad (S \in \mathcal{L}(H^{1/2}(\Gamma), H^{-1/2}(\Gamma))).$$

We finally get

$$\int_\omega (h + h_g).h' = \int_\Gamma h_g.n \, \phi' + \int_\Gamma S\phi \, \phi'$$

Let us summarize at this point.

Proposition 1. Assume div $h_g = 0$ and ω simply connected. Equation (8) is equivalent to the problem of finding a pair $\{h, \phi\}$ in

$$(17) \qquad H\Phi = \{\{h, \phi\} \in H \times H^{1/2}(\Gamma) \mid h_\Gamma = \text{grad}_\Gamma \phi\}$$

(the subscript Γ denotes the tangential component) such that

$$(18) \qquad \mu_0 \frac{d}{dt}\left[\int_\Omega (h + h_g).h' + \int_\Gamma (S\phi + h_g.n)\, \phi'\right] + \int_\Omega \rho \text{ curl } h . \text{curl } h' = 0$$

for all pairs $\{h', \phi'\}$ in $H\Phi$.

After this transformation, all integrals are on bounded domains. A tetrahedral paving of Ω is introduced. Degrees of freedom are associated with the edges (not the vertices) of the tetrahedra which lie inside Ω, and with boundary nodes.

Detailed explanations can be found in [1]. The technical difficulty of dealing with multi-connectedness (or, from another point of view, with "multivalued magnetic potentials") is also treated there.

On (18) we see clearly the possibility of a Lagrangian approach to the discretization of the eddy-currents equation when the different connected components of Ω move with respect to each other. Provided the components are rigid, the same discretization network and matrices will be used as time varies, since the integrals on Ω are invariant with respect to displacements in space. The only part of (18) where time occurs explicitly is S. We shall now follow this idea.

MOVING CONDUCTORS

In this section, Ω has p connected components. Let us introduce the manifold U (of dimension 6p) of all imbeddings $u : \Omega \to \mathbb{R}^3$ whose restrictions on all separated components are isometries. A "movement" will be a mapping from [0, T] into U. The physical interpretation is as follows: $u_t(\Omega)$ is the domain occupied by the moving conductors at time t; these keep the same shape (isometry) and don't intersect (imbedding). It is natural to assume $u_0(x) = x \ \forall \ x \in \Omega$ (the position at t=0 is the reference configuration).

Let $\Omega_u = u(\Omega)$, $\Gamma_u = u(\Gamma)$, $\omega_u = \mathbb{R}^3 - cl(\Omega_u)$. To any function ϕ defined on Γ will correspond by "push-forward" the function $\phi_u = u_* \phi$ on Γ_u:

(19) $$\phi_u(x) = \phi(u^{-1}(x))$$

Similarly, from vector fields on Ω, we derive vector fields on $u(\Omega)$ by push-forward:

(20) $$h_u(u(x)) = u_* h(u(x)) = Du_x h(x)$$

where Du is the Fréchet derivative of u, evaluated at the subscripted x.

Given ϕ on Γ, let Φ be the solution to the exterior Dirichlet problem

(21) $$\Delta\Phi = 0 \text{ in } \omega_u, \quad \Phi = \phi_u \text{ on } \Gamma_u.$$

We define $S_u \in (H^{1/2}(\Gamma); H^{-1/2}(\Gamma))$ by the "pull-back" on Γ of $\partial\Phi/\partial n$ on Γ_u, or in a slightly simpler but equivalent way, by

(22) $$\int_\Gamma S_u \phi \, \phi = \int_{\omega_u} |\text{grad}\, \Phi|^2$$

Let finally be given some inductive field $h_g(t)$, div $h_g = 0$, curl $h_g(t) = 0$ on $u_t(\Omega)$, for a specified movement $t \to u(t)$, and consider its pull-back by u_t:

(23) $$\tilde{h}(t, x) = u_t^* h_g(t, x) = [Du_{u_t(x)}]^{-1} \, h_g(t, u_t(x))$$

(There is no special problem to compute $\tilde{h}$, in spite of its ugly definition !) We now state (the proof is easy):

Theorem 3. Given h_g and a Cauchy (or periodicity) condition, there is a unique h, in H such that

(24) $$\mu_0 \frac{d}{dt}\Big[\int_\Omega (h + \tilde{h}).h' + \int_\Gamma (\tilde{h}.n + S_{u_t} \phi)\, \phi'\Big] + \int_\Omega \rho \text{ curl}\, h.\text{curl}\, h' = 0$$

holds for all pairs $\{h', \phi'\}$ in $H\Phi$.

So we have existence of a field for a given movement and a given exciting current, and a Lagrangian numerical procedure to compute it by a mixed finite-elements/boundary-elements method. It remains to treat the case when the movement itself has its origin in the effect of Laplace forces.

Remark. The solution of (24) gives the field in the local frame of reference, as an observer who moves with the material particle would measure it, is the push-forward of the computed h:

$$h_u(u(x), t) = u_* h = (Du)_x h(x, t)$$

for u = u(t).

We now embark in a calculation where the notations should be understood as follows. $T_u U$ is the tangent space to U at u, $T_u^* U$ its dual (both are isomorphic with $\mathbb{R}^{6p}$). Brackets denote the duality between these two spaces. By (21) and (22) we defined a mapping S: $U \to \mathcal{L}(H^{1/2}(\Gamma); H^{-1/2}(\Gamma))$. Thus $\partial S/\partial u$ is in $\mathcal{L}(T_u U; \mathcal{L}(H^{1/2}; H^{-1/2}))$. Now, let $t \to u(t)$ be a movement. We have, for a fixed ϕ,

(25) $$\frac{d}{dt} <S_u \phi, \phi> = 2 <S_u \phi, \frac{d\phi}{dt}> + <\frac{\partial S}{\partial u} \frac{du}{dt} \phi, \phi>$$

and also

(26) $$<\frac{d}{dt} S_u \phi, \phi> = <S_u \frac{d\phi}{dt}, \phi> + <\frac{\partial S}{\partial u} \frac{du}{dt} \phi, \phi>$$

Comparing (25) and (26), and taking h' = h and $\phi' = \phi$ in (24), we get

(27) $$\frac{\mu_0}{2} \frac{d}{dt} [\int_\Omega |h|^2 + <S_u \phi, \phi>] + \frac{1}{2} <\frac{\partial S}{\partial u} \frac{du}{dt} \phi, \phi> + \int_\Omega \rho |\text{curl}\, h|^2 = \ldots$$

outer magnetic energy *inner mag. energy* *mechanical power* *Joule losses*

$$\ldots = -\mu_0 [\int_\Omega (\frac{\partial \tilde{h}}{\partial t} . h + \int_\Gamma \frac{\partial \tilde{h}}{\partial t} . n \, \phi]$$

injected power

In italics, we have indicated the physical meaning of the different parts of this "energy balance" equation. In particular, since du/dt is a (generalized) speed, the quantity

(28) $$\frac{1}{2} <\frac{\partial S}{\partial u} \phi, \phi> \in T_u^* U$$

is a generalized force (the resultant of the Laplace forces).

Now suppose some elastic coupling exists between the conductors, characterized by a potential W(u). Let M be the (6p X 6p) mass-matrix. The mechanical equation is

(29) $$M \frac{\partial^2 u}{\partial t^2} + \frac{\partial W}{\partial u} = \frac{1}{2} < \frac{\partial S}{\partial u} \phi, \phi>$$

Therefore, the dynamics of the system as a whole, given $h_g(t)$, is described by the coupled system (24)(29).

By differentiating (29) with respect to u in the vicinity of an equilibrium position, we get an equation with time-periodic coefficients, so parametric resonance may appear. We refer to [2] for a detailed treatment and an example.

REFERENCES

[1] Bossavit, A. and Vérité, J.C. , The TRIFOU code: solving the 3-D eddy-currents problem by using h as state variable, COMPUMAG conference, Genoa, 1983 (to appear in IEEE Trans. on Magnetics).

[2] Bossavit, A., Dynamique des systèmes de conducteurs élastiquement couplés; application à un câble au SF6, in: Innovative Numerical Analysis for the Applied Engineering Sciences, Shaw et al., eds, U.P. Virginia, 1980, p. 23.

The Mechanical Behavior of Electromagnetic Solid Continua
G.A. Maugin (editor)
Elsevier Science Publishers B.V. (North-Holland)

DIFFRACTION OF MAGNETOELASTIC WAVES BY INHOMOGENEITIES

Igor T. Selezov

Wave Process Hydrodynamics Department
Institute of Hydromechanics of the
Academy of Sci., Ukrainian SSR, Kiev
U.S.S.R.

The linearized equations of motion of a spatially inhomogeneous electroconducting medium in magnetic field are presented. On the basis of improving slight coductivity approximation the two problems of a stationary wave diffraction are considered: of plane elastic waves by a cylindrical magnetoelastic inclusion; of cylindrical magnetoelastic waves by a dilatational magnetoelastic inhomogeneity with a radially variable density.

A behaviour of the magnetoelastic medium is governed here by a coupling system of equations of the electromagnetic field and the theory of elasticity at the follow assumptions: it is not taken into account the electrical charges ρ_e, the electrical forces $\rho_e\vec{E}$, the motion of charges $\rho_e\dot{\vec{u}}$ ($\vec{u}$ is the displacement vector), the electrical stresses in Maxwell tensor; there is undisturbed state characterizing by uniform constant magnetic field H_o. It is supposed also that the corresponding problem of magnetostatics is solved; the disturbances are small; in undisturbanced state the density ρ_c, Lamé parameters λ and G, the electroconductivity γ are depend on the spatial coordinates X_i (i = 1,2,3); the medium is isotropic.

With above assumptions and the conditions of a smothness all of the functions the linearized equations of disturbanced motion of the inhomogeneous magnetoelastic medium are

$$L(\vec{u}) = \rho_c(\vec{x})\,\ddot{\vec{u}} - P_h(\vec{\nabla}\times\vec{h})\times\vec{H}_o \tag{1}$$

$$\begin{aligned} L(\vec{u}) = &\left\{\vec{\nabla}\left[\lambda(\vec{x}) + 2G(\vec{x})\right]\ \vec{\nabla}\cdot\vec{u}\right\} - \vec{\nabla}\times\left[G(\vec{x})\ \vec{\nabla}\times\vec{u}\right] + 2\left\{\left[\vec{\nabla}G(\vec{x})\right]\cdot\vec{\nabla}\right\}\ \vec{u} - \\ &- \left[\vec{\nabla}G(\vec{x})\right](\vec{\nabla}\cdot\vec{u}) + \left[\vec{\nabla}G(\vec{x})\right]\times(\vec{\nabla}\times\vec{u}) \end{aligned} \tag{2}$$

$$\begin{aligned} &\dot{\vec{h}} = -\vec{\nabla}\times\left[R_m^{-1}(\vec{x})\ (\vec{\nabla}\times\vec{h})\right] + \vec{\nabla}\times(\dot{\vec{u}}\times\vec{H}_o) \\ &\vec{\nabla}\times\vec{h} = \vec{j}, \quad \vec{\nabla}\times\vec{e} = -P_H\dot{\vec{h}}, \quad \vec{\nabla}\cdot\vec{h} = 0, \ \vec{\nabla}\cdot\vec{e} = 0, \\ &\vec{j} = R_m(\vec{x})(P_H^{-1}\vec{e} + \mu\dot{\vec{u}}\times\vec{H}_o), \quad \vec{b} = P_H\vec{h}, \quad \vec{d} = \varepsilon\vec{e} \end{aligned} \tag{3}$$

The conditions at the interface of the obstacle and surrounding (outer) medium were given by Selezov (1975). The system (1)-(3) is presented in dimensionless form with character values: 1 is the length, V the velocity, ρ the mass density, H the magnetic field. The dimensionless parameters: $P_H=\mu H^2/(\rho V^2)$ is the magnetic pressure, $R_m(\vec{x})=R_o f(\vec{x})=lV\mu\gamma(\vec{x})$ is the electroconductivity number ($R_o=|R_m(\vec{x})|_{max}$).

The improved slight conductivity approximation is based upon the following assumptions: a strong magnetic field $P_H \gg 1$; $R_o \sim P_H^{-1} \ll 1$; the disturbed electromagnetic field is generated by elastic disturbances. As in the case of a homogeneous medium (Selezov and Selezova (1975) we use the expansions)

$$\vec{u} = \vec{u}^o + \vec{u}^1 R_o + \dots, \qquad \vec{h} = 0 + \vec{h}^1 R_o + \dots$$

After substitution of these expansions into the system (1)-(3) we obtain the first approximation of the form

$$L(\vec{u}) = \rho_c(\vec{x})\ddot{\vec{u}} - P_H R_m(\vec{x})(\vec{u}^o \times \vec{H}_o) \times \vec{H}_o \tag{4}$$

$$\vec{\nabla} \times \vec{h}^1 = f(\vec{x})(\dot{\vec{u}} \times \vec{H}_o), \quad \vec{\nabla} \times \vec{e}^1 = -\dot{\vec{h}}^1, \quad \vec{\nabla} \cdot \vec{h}^1 = 0, \quad \vec{\nabla} \cdot \vec{e}^1 = 0 \tag{5}$$

The system (4), (5) must be integrated by such a way: the general solution is found from the equation (4), then only partial solutions are derived from the equations (5).

The strong electroconductivity approximation $R_o \gg 1$ is derived from expansions $\bar{u}$ and $\bar{h}$ in the powers of R_o^{-1}. In the first approximation we obtain the equations corresponding to perfect conductivity

$$L(\vec{u}^o) = \rho_c(\vec{x})\, \ddot{\vec{u}}^o - P_H[\vec{\nabla} \times \vec{\nabla} \times (\vec{u}^o \times \vec{H}_o)] \times \vec{H}_o \tag{6}$$

$$\vec{h}^o = \vec{\nabla} \times (\vec{u}^o \times \vec{H}_o), \qquad \vec{e}^o = -P_H \dot{\vec{u}}^o \times \vec{H}_o, \tag{7}$$

$$\vec{j}^o = \vec{\nabla} \times \vec{\nabla} \times (\vec{u}^o \times \vec{H}_o) \qquad \vec{\nabla} \cdot \vec{h}^o = 0, \qquad \vec{\nabla} \cdot \vec{e}^o = 0 \tag{8}$$

In this case it is enough to find only the value $\bar{u}^o$ from the equation (6).

Today there is no a general theory to separate the vector equations of inhomogeneous elastic medium. The most advanced contribution in this field was given by Hook (1961) separated the equations for some classes of inhomogeneities. In magnetoelasticity the situation is more complicated due to a magnetoelastic anisotropy.

In the case of a dilatation deformation we must introduce into the equations (1)-(3) or (4)-(8) the value $\vec{v} = \dot{\vec{u}}$ and $-\vec{\nabla}p$ instead of $L(\vec{u})$, add the equations

$$\dot{\rho} + \vec{v} \cdot \vec{\nabla}\rho_o(\vec{x}) + \rho_o(\vec{x})\, \vec{\nabla} \cdot \vec{v} = 0, \quad \dot{p} = c^2[\dot{\rho} + \vec{v} \cdot \vec{\nabla}\rho_o(\vec{x})] \tag{9}$$

The wave diffraction problem is formulated as follow. Consider in an infinite domain $\Omega \in E^3$ the finite domain $\Omega_1 \subset \Omega$. The inner domain Ω_1 and the outer domain $\Omega_2 = \Omega/\Omega_1$ are filled in by magnetoelastic mediums with different physical characteristics. In domain Ω_2 the medium is homogeneous, in domain Ω_1 the medium is homogeneous or inhomogeneous. Each of the mediums is governed by the systems (1)-(3) or (4)-(5), or (6)-(7). In outer domain Ω_2 the incident field is given (known solution for plane, cylindrical or spherical waves).

Consider in nonconducting medium Ω_2 the inclusion - magnetoelastic circular cylinder with which the cylindrical coordinate system $r, \theta, z \to x_1, x_2, x_3$ is connected. There is a undisturbed constant uniform longitudinal magnetic field $(0, 0, H_o)$. The plane incident dilatational wave comes from infinity.

The motion of outer medium Ω_2 is governed by the equations with scalar and vector potentials $\varphi = \varphi(r, \theta, z, t)$ and $\vec{a} = \vec{a}(r, \theta, z, t)$

$$\nabla^2\varphi - c^{-2}\ddot{\varphi} = 0, \qquad \nabla^2\vec{a} - c_e^{-2}\ddot{\vec{a}} = 0 \tag{10}$$

and equation with magnetic field $\vec{h}$

$$\nabla^2\vec{h} - c_s^{-2}\ddot{\vec{h}} = 0 \tag{11}$$

in $\Omega_2 = \{(r, \theta, z):\ r \in (r_o, \infty),\ \theta \in [0, 2\pi],\ z \in (-\infty, \infty), t \in (0, \infty)\}$. The motion of inner medium is governed by the equations

$$G^p \nabla^2\vec{u}^p + (G^p + \lambda^p)\vec{\nabla}(\vec{\nabla}\cdot\vec{u}) = \rho^p\ddot{\vec{u}}^p - P_H^p R_m^p(\vec{u}^p \times \vec{H}_o) \times \vec{H}_o$$
$$\vec{\nabla} \times \vec{h}^p = \dot{\vec{u}}^p \times \vec{H}_o, \quad \vec{\nabla} \times \vec{e}^p = -\dot{\vec{h}}^p, \quad \vec{\nabla}\cdot\vec{h}^p = 0, \quad \vec{\nabla}\cdot\vec{e}^p = 0 \tag{12}$$

in $\Omega_1 = \{(r, \theta, z):\ r \in [0, r_o],\ \theta \in [0, 2\pi], z\ (-\infty, \infty),\ t \in (0, \infty)\}$.

It is supposed that magnetic and electric permeabilities the two of the mediums equal $\mu = \mu^p$, $\varepsilon = \varepsilon^p$. The conditions at the interface $r = r_o$ are

$$\vec{n}\cdot(\vec{u} - \vec{u}^p) = 0, \qquad \vec{n} \times (\vec{u} - \vec{u}^p) = 0,$$

$$n_i[(\sigma_{ik} + t_{ik}) - (\sigma^p_{ik} + t^p_{ik})] = 0,$$
$$i, j, k = 1, 2, 3$$

$$t_{ik} = P_H(H_{oi}\, h_k - h_i\, H_{ok} - \delta_{ik}\, H_{oj}\, h_j), \tag{13}$$

$$\vec{n}\cdot(\frac{\mu}{\mu^p}\vec{h} - \vec{h}^p) = 0, \qquad \vec{n} \times (\vec{h} - \vec{h}^p) = 0$$

$$\vec{n}\cdot(\frac{\varepsilon}{\varepsilon^p}\vec{e} - \vec{e}^p) = 0, \qquad \vec{n} \times (\vec{e} - \vec{e}^p) = 0$$

The conditions for electromagnetic values $\vec{h}, \vec{h}^p, \vec{e}, \vec{e}^p$ are given also. Here δ_{ik} is the Kroneker symbol, σ_{ik} is the component of elastic stress tensor, t_{ik} is Maxwell tensor component, at the repeated indexes the summation is assumed. Besides the conditions of radiation for variables in Ω_2 are implied. The plane incident wave is given by

$$u_y = u_o \exp\ i(sy + \omega t)$$

and we have

$$\exp\ (isy) = \sum_{n=o}^{\infty} \varepsilon_n\ i^n\ J_n(sr)\ \cos n\theta$$

where $J_n(\xi)$ is Bessel function, $\varepsilon_n = 1$ at $n = 0$ and $n = 2$ at $n \geq 1$.

Taking into account the properties of symmetry the nonzero components remain only in (10)-(13): u_r, u_θ, h_z, e_r, e_θ. For example, the solution u_r is

$$u_r = \sum_{n=0}^{\infty} [f_n\, p\, J'_n(pr) + A_n p\, H_n^{(2)'}(pr) + -r^{-1}\, n\, B_n\, H_n^{(2)}(qr)]\cos n\theta$$

Unknown coefficients A_n, A_n^p, B_n, B_n^p are determined as a result of solving the system of equations following from the conditions at the interface (13). Analising the results of calculations it is established that the increasing of the magnetic field P_H and conductivity R_m shifts the resonance frequencies and decreases the scattering wave amplitudes and smoothes out their local extremums.

The problem of the cylindrical wave diffraction by a magnetoelastic cylindrical inhomogeneity* is governed by the equations (4), (5) in outer domain Ω_2 (in further characterized by index g) and the equations (9) in inner domain Ω_1. The co conditions at the interface $r = r_o$ are similar to (13). The functions considered are presented in the form

$$w(r, \theta, z, t) = \tilde{w}(r, \theta, z) \exp(-i\omega t)$$

In further the symbols tilde and exp(-i t) are omitted. The vector $\vec{u}$ is presented in the form

$$\vec{u} = \vec{\nabla}\varphi + \vec{\nabla} \times \vec{a}, \qquad \vec{\nabla}\cdot\vec{a} = 0, \qquad \vec{a} = \vec{e}_z \Psi$$

Then from the equations (4), (5) we derive the two uncoupled equations

$$(2 + æ)\nabla^2\varphi + (\omega^2 + i\omega P_H R_m^g) = 0$$

$$\nabla^2\Psi + (\omega^2 + i\omega P_H R_m^g) = 0$$

In a coordinate system (r_1, θ_1, z_1) connected with the source radiating equivoluminal and dilatational cylindrical waves we find the solutions

$$\varphi = \varphi_o H_o^{(1)}(k_1 r_1), \qquad \Psi = \Psi_o H_o^{(1)}(k_2 r_1)$$

where $k_1 = (\omega^2 + i\omega P_H R_m)/(2 + æ)$, $k_2 = \omega^2 + i\omega P_H R_m$.

By using the additional theorem of cylindrical functions we determine the solutions of a diffraction field at $r < b$

$$\varphi = \sum_{m=o}^{\infty}\left[\varphi_o\, \varepsilon\, J_m(k_1 r)\, H_m^{(1)}(k_1 b) + A_m H_m^{(1)}(k_1 r)\right] \cos m\theta$$

$$\Psi = \sum_{m=o}^{\infty}\left[\Psi_o\, \varepsilon_m J_m(k_2 r)\, H_m^{(2)}(k_2 b) + B_m H_m^{(1)}(k_2 r)\right] \sin m\theta$$

Introducting the potential $(-i\omega\rho_o + P_H R_m)\vec{V} = \vec{\nabla}\Phi$, the three equations of the system (4), (5), (9) are reduced to a single equation

$$\nabla^2\Phi + \frac{i\omega\rho_o}{-i\omega\rho_o + P_H R_m}\frac{\partial\Phi}{\partial r} + \frac{\omega^2\rho_o + i\omega P_H R_m}{c_o^2\rho_o}\Phi = 0$$

The general solution of this equation is

$$\Phi = \sum_{m=0}^{\infty} C_m Q_m(r) \cos m\theta$$

Here $Q_m(r)$ is the solution of equation

---

*) This problem was considered together with V.V. Yakovlev.

$$\frac{d^2Q}{dr^2} + \left(\frac{1}{r} + \frac{i\omega\rho_o^1}{-i\omega\rho_o + P_H R_m}\right)\frac{dQ}{dr} + \left[k^2(r) - m^2 r^{-2}\right] Q = 0 \qquad (14)$$

where $k^2(r) = (\omega^2\rho_o - i\omega P_H R_m)/c^2\rho_o$.

If the $\rho_o(r)$ is polinomial we can obtain the exact solution of the equation (14) by generalized power series method (Selezov and Yakovlev (1978)). Here the case $\rho_o(r) = \rho_1 + (\rho_2 - \rho_1)r^2$ is considered and $Q_m(r)$ is presented by

$$Q_m(r) = \sum_{n=0}^{\infty} \alpha_n \, r^{n+m} \qquad (15)$$

The coefficients α_n are found from the recurrence relationships derived after substitution of the solution (15) into the equation (14). Unknown coefficients A_m, B_m are found from the conditions at the interface.

As an example the calculations were carried out with following parameters: $\rho_2 = 0.102$; $c = 1.36$; $\nu = 0.33$; $b = 5$; $P_H R_m^g = P_H R_m = 0 \div 0.25$; $\rho_1 = \rho_2 \cdot$ $\cdot (1 \pm 0.4)$. The analysis of the calculated polar diagrams of the dilatational and equivoluminal scattering wave fields shows that the field dependence in a space point on a wave number has a form of the oscillating curve having a mean line increasing with a wave number growth. These mean values decrease with the increasing of the conductivity R_m in the outer medium. It is shown also for the dilatation waves in the direction $\theta = 0$ that an influence the conductivity R_m is revealed as the smoothing out of a resonance behaviour of the scattering field. At the great values of wave numbers there are selfsimilarity points in which the amplitudes of scattering field does not depend on the value of applied magnetic field H_o.

REFERENCES:

1 Селезов, И.Т., Некоторые приближенные формы уравнений движения магнитоупругих сред, Известия АН СССР, Механика Твердого Тела. 5 (1975) 86-91.

2 Селезов, И.Т., Селезова, Л.В., Волны в магнитогидроупругих средах (Наукова думка, Киев, 1975).

3 Hook, J.F., Separation of the vector wave equation of elasticity for certain types of inhomogeneous isotropic media. Jrnl. of Acoust. Soc. Amer. 3 (1961) 302-313.

4 Селезов, И.Т., Яковлев, В.В., Дифракция волн на симметричных неоднородностях (Наукова думка, Киев, 1978).

Part 10

MAGNETOELASTICITY OF STRUCTURES AND BUCKLING

The Mechanical Behavior of Electromagnetic Solid Continua
G.A. Maugin (editor)
Elsevier Science Publishers B.V. (North-Holland)

SOME NEW PROBLEMS OF MAGNETOELASTICITY OF THIN SHELLS AND PLATES

Sergey A. Ambartsumian

University of Erevan
Armenian SSR
USSR

Many modern problems of advanced technology and physical experiments are closely connected with the problems of interaction of electromagnetic fields with thin electroconductive elastic bodies such as shells and plates. Some new problems of magnetoelasticity of shells and plates which have not received sufficient attention in the known surveys [1-6] will be elucidated.

THE MAGNETIC FIELD IS A FUNCTION OF THE COORDINATE. In all the earlier studies on magnetoelasticity of electroconductive shells and plates the magnetic field in the shell or plate domain is assumed to be independent of the coordinates of the middle surface x (the plate is considered in a Cartesian system of coordinates so that the middle surface of the plate coincides with the coordinate plane xOy). However, in real problems there are many cases when the inducted magnetic field is a function of the coordinate x.

In [7] an attempt is made to consider the problem of the electroconductive plate vibrations in the case when the inducted magnetic field is a function of the coordinate x. The problem of vibrations in transverse magnetic field is considered assuming that the coefficient of magnetic permeability μ is equal to one. The linear problem is solved neglecting the displacement currents as compared with the conduction ones, but at the same time taking into account transverse shear deformations[9, 10] .

The hypotheses on which the problems in the present paper are based are the following:
a) the hypothesis of the magnetoelasticity of thin bodies [8], which is defined as follows: the tangential components of the intensity vector of the perturbed electric field in the plate (shell), and the normal component of the intensity vector of the perturbed magnetic field through the plate (shell) thickness remain unchanged.
b) the hypothesis of the exacted theory [9], when transverse shear deformations are taken into account.

For the subsequent discussion to be more clear and concrete, let us consider a particular case of plate-strip transverse vibrations in the form of cylindrical surface with the generant parallel to the long edges of the plate. The plate-strip is simply supported on the long edges (x = 0, x = a) and is under the influence of the inducted transverse magnetic field locally acting at the interval $c \leqslant x \leqslant b$.

In this case the problem is reduced to a system of differential equations for the normal displacement $w(x,t)$ and the function of transverse shear $\Phi(x,t)$. Using the Bubnov-Galerkin method, the characteristic equation for defining the relative frequency ($Im\Omega$) and the decrement of oscillations ($\delta = -2\pi Re\Omega/Im\Omega$) may be obtained.

Using a numerical analysis let us consider the nature of the logarithmic decrement of the oscillations δ depending on the magnetic field intensity B and the parameter values c and b.

Table 1 gives the values of δ for a copper plate ($\nu = 0.3$, $E = 1.1 \cdot 10^{12} gmcm^{-1}sec^{-1}$, $E/G' = 2.6$, $\sigma = 5.3 \cdot 10^{17} sec^{-1}$, $\rho = 8.89 gmcm^{-3}$) with the relative thickness $h/a = 0.1$.

Table 1

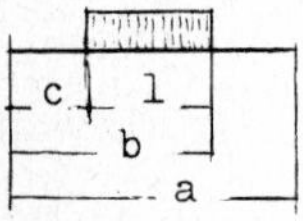	c	b	l	B (in gausses)				
				10^3	10^4	$2\cdot10^4$	$3\cdot10^4$	$4\cdot10^4$
	0	a	a	0.000	0.028	0.110	0.247	0.441
	0	a/2	a/2	0.000	0.014	0.055	0.124	0.220
	0	a/3	a/3	0.000	0.013	0.052	0.116	0.207
	0	a/5	a/5	0.000	0.010	0.039	0.087	0.154
	a/4	3a/4	a/2	0.000	0.005	0.020	0.045	0.080
	a/3	2a/3	a/3	0.000	0.002	0.006	0.014	0.025
	2a/5	3a/5	a/5	0.000	0.000	0.001	0.003	0.006

Considering Table 1 we note that in all the considered cases the logarithmic decrement of oscillations essentially increases with the increase of the magnetic field intensity. Further on, a completely new phenomenon is observed: the absolute values of the logarithmic decrement of vibrations in the presence of the same specific magnetic intensity are much more greater in a peripheral arrangement of the magnetic field. For example, when $l = a/3$, $B=3\cdot10^4$gauss, the logarithmic decrement is 8 times more when the magnetic field is arranged in the periphery than when it is placed in the centre.

With a greater localization of the magnetic field, this phenomenon becomes more intense: for example, when $l = a/5$, $B = 3 \cdot 10^4$ gausses, the logarithmic decrement of vibrations in the peripheral arrangement of the magnetic field is 29 times more than that in the central arrangement.

This "strange" result may be explained in the following way: the point is that the magnetic field, normal to the plate middle plane, enters into magnetoelastic interaction with the tangential displacement $u(x,t)$, which reaches its maximum near the edges of the simply-supported plate ($x = 0$, $x = a$), and equals to zero in the centre of the plate ($x = a/2$).

THE PROBLEM OF PLATE VIBRATIONS IN A TRANSVERSE MAGNETIC FIELD. The problem of the vibration of a plate-strip with the thickness a in a transverse magnetic field of constant intensity B, the intensity vector of which is parallel to the axis Ox, in case of cylindrical-form vibrations is reduced to the system of differential equations for the normal displacement $w(x,t)$ and the components $e = \psi(x,t)$ of the perturbed electromagnetic field [8,10].

A detailed numerical analysis of the characteristic equation in the case of a copper plate ($E = 10^{12}$ gmcm^{-1}sec^{-2}, $\nu = 0.3$, $\rho = 8.9$gmcm^{-3}, $\sigma = 5.3 \cdot 10^{17}$sec^{-1}) is made.

Some characteristic values of $\mathrm{Im}\,\Omega$ and $\mathrm{Re}\,\Omega$ depending on the magnetic field intensity B and the geometric parameters of the plate are presented. Let us first consider comparatively thick plates: $h = 0.05$cm, $a = 10$cm ($\omega_0 \approx 1000$sec^{-1}) (Table 2) and $h = 0.02$cm, $a = 10$cm ($\omega_0 \approx 200$sec^{-1}) (Table 3).

Table 2

$B \cdot 10^{-3}$	0.5	1.0	1.5	2.0	2.5	3.0	3.5	4.0
$\mathrm{Re}\Omega_1$	-0.84	-0.82	-0.78	-0.74	-0.68	-0.62	-0.55	-0.49
$\mathrm{Re}\Omega_2$	-0.003	-0.01	-0.03	-0.05	-0.08	-0.11	-0.14	-0.17
$\mathrm{Im}\Omega_2$	1.00	1.01	1.03	1.06	1.11	1.16	1.22	1.30
$\mathrm{Re}\Omega_3$	-0.003	-0.01	-0.03	-0.05	-0.08	-0.11	-0.14	-0.17
$\mathrm{Im}\Omega_3$	-1.00	-1.01	-1.03	-1.06	-1.11	-1.16	-1.22	-1.30

Table 3

$B\cdot 10^{-3}$	0.5	1.0	1.5	2.0	2.5	3.0	3.5	4.0
$\mathrm{Re}\Omega_1$	-4.17	-4.06	-3.87	-3.56	-3.06	-1.93	-0.68	-0.45
$\mathrm{Re}\Omega_2$	-0.01	-0.07	-0.16	-0.32	-0.57	-1.14	-1.76	-1.87
$\mathrm{Im}\Omega_2$	1.00	1.01	1.02	1.03	1.02	0.94	1.75	2.38
$\mathrm{Re}\Omega_3$	-0.01	-0.07	-0.16	-0.32	-0.57	-1.14	-1.76	-1.87
$\mathrm{Im}\Omega_3$	-1.00	-1.01	-1.02	-1.03	-1.02	-0.94	-1.75	-2.38

Examining Tables 2 and 3 one may notice that the frequency of vibrations of the plates increases with the increase of the magnetic field intensity. For thinner plates ($h = 0.02$cm) an essential increase in frequency is observed, however, in this case, for some values of magnetic field intensity ($B = 2\cdot 10^3 - 3\cdot 10^3$ Oe) one can see a certain decrease of vibration frequency. This phenomenon of frequency localization, which was first discovered in survey [10], causes new qualitative results when the plate thickness is further decreased.

Let us consider copper plates of the following dimensions: $h = 0.01$cm $a = 10$cm ($\omega_0 \approx 200\mathrm{sec}^{-1}$) and $h = 0.01$cm, $a = 20$cm ($\omega_0 \approx 50\mathrm{sec}^{-1}$). The calculation results are given in Table 4, merely the values of $\mathrm{Im}\Omega_i$ depending on B being written.

Table 4

$B\cdot 10^{-3}$	0.5	1.0	1.5	2.0	2.5	3.0	4.0	4.5	5.0
$\mathrm{Im}\Omega_{2,3}$ $a = 10$	±1.00	±0.99	±0.94	±0.75	0.00	0.00	0.00	±5.38	±7.91
$\mathrm{Im}\Omega_{2,3}$ $a = 20$	±0.99	±0.75	0.00	0.00	0.00	±7.36	±21.2	±26.2	±30.1

Considering Table 4 we notice that the above-discovered tendency of decrease (localization) of oscillation frequency is intensified with the reduction of the plate thickness and leads to radically new phenomena. The plate vibration frequency greatly decreases with the increase of the magnetic field intensity and reaches a zero value. The zero value of the vibration frequency is retained in a certain range of magnetic field intensity changes. Within this range it is evident that the perturbations of the plate damp without oscillations. Further on, as is seen from Table 4, the increase of the magnetic field intensity leads to a radical increase in the plate vibration frequency.

The newly-found phenomenon in the nature of the plate vibration frequency changes in longitudinal magnetic field, takes place also in other boundary conditions of the plate fixing.

This phenomenon in a somewhat different manner is observed in the case of plate vibrations in transverse magnetic field B. In this case it takes place only when transverse shear deformations are taken into account [10,11]. These phenomena cannot be observed when using the classical theory [8].

Note also that this phenomenon takes place when considering the vibrations of electroconductive circular cylindrical shell in azimuthal magnetic field generated by a linear current wire placed along the axis of the shell [12].

THE VIBRATIONS OF THE NONLINEAR ELASTIC PLATES IN MAGNETIC FIELD. Electroconductive plates and shells as elements of apparatus and constructions acting in strong magnetic fields are often made of nonlinear elastic materials (copper, aluminium, etc.). If physical nonlinearity of the material is taken into account, the pattern of plate or shell vibrations in the magnetic field may essentially change [15].

These problems are based on the following assumptions and hypotheses [9,13-15]: The hypothesis of undeformable normals; The hypothesis of magnetoelasticity of thin bodies; Incompressibility of the plate material; The coincidence of stress and strain directional tensors; The following nonlinear relation between the intensities of stresses (T_i) and deformations (E_i): $T_i = aE_i - bE_i^m$, where a,b and m are some constants which do not depend on the magnetic field intensity and are determined in a pure compression-tension experiment.

Let us consider the case of cylindrical-form vibrations of a plate-strip with a width of l simply supported on the edges ($x = 0$, $x = l$) in a transverse magnetic field with the intensity B=const when the parameter of nonlinearity $m = 3$. In this case the problem is reduced to the nonlinear equation. The main formulae show that the vibration frequency of the nonlinear system depends on the vibration amplitude which, in its turn, essentially depends on the magnetic field intensity. On the other hand, the vibration frequency depends on the magnetic field intensity. This phenomenon takes place also in the case of the linear problem. However, in the considered case both of the "improved" members lead to the reduction of vibration frequency. One should mention though that, depending on the magnetic field intensity, the specific weight of these members changes, the influence of the second member decreasing and becoming negligible with the increase of the time (t).

Here the improvement in vibration frequency resulting from nonlinear elasticity is positive and leads to an increase of frequency. In this case the improvements in vibration frequency somehow compensate each other. However, the compensation rate depends on time and magnetic field intensity.

Let us consider a numerical example for a copper plate with the following initial parameters: $a = 10^{12} gmcm^{-1}sec^{-2}$, $b = 10^{17} gmcm^{-1}sec^{-2}$, $m = 3$, $\rho = 8.9 gmcm^{-3}$, $\sigma = 5.3 \cdot 10^{17} sec^{-1}$, $c = 3 \cdot 10^{10} cmsec^{-1}$, $h = 0.3cm$, $l = 20cm$, $D_e = 0.024 \cdot 10^{12} gmcm^2 sec^{-2}$, $D_2 = 1.728 \cdot 10^{14} gmcm^3 sec^{-2}$, $\omega = 1.6542 \cdot 10^3 sec^{-1}$.

Table 5 gives the values of the amplitude of damping vibrations A(t) for different values of magnetic field intensity B in different moments in time (in sec.), multiple to the initial period of vibrations $\tau = 2\pi/\omega = 0.0038sec$ when the initial amplitude $A_0 = 0.3cm$.

Table 5

B \ t	0.0	0.004	0.019	0.095	0.475	2.375	11.86	59.38	296.9
$5 \cdot 10^2$	0.300	0.299	0.299	0.299	0.299	0.296	0.279	0.209	0.049
10^3	0.300	0.299	0.299	0.299	0.297	0.283	0.224	0.070	0.000
10^4	0.300	0.297	0.286	0.238	0.094	0.001	0.000		
10^5	0.300	0.118	0.003	0.000					

Referring to Table 5 we note that the magnetic field essentially influences the damping of vibrations. If the values of the magnetic field intensity are not so great ($5 \cdot 10^2 - 10^3$ Oe), visible decrease in the vibration amplitude is observed only one or two minutes after the beginning of the vibrations. While for strong magnetic fields of the intensity equal to 10^5 Oe the vibration damping is so intense that notable decrease of the amplitude is observed in parts of a second since the beginning of the vibrations. This phenomenon affects the nature of changes in vibration frequency. The point is that in the formulae for vibration frequency the "improved" term caused by nonlinearity essentially depends on the vibration amplitude, and for great values of magnetic field intensity or for higher t it extends to zero. Thus, in the presence of great t or B the effect of non-linearity disappears and only that part of "improved" frequency remains, which takes place in the linear case also.

We can note that the stronger the magnetic field intensity, the sooner the vibration frequency becomes steady, containing an improvement in damping only which does not depend on t. At small values of magnetic field intensity the specific weight of the "improvement" of nonlinearity increases, the influence of nonlinearity in the neighbourhood of great t becoming negligible.

Consider the problem of transverse vibrations of a plate-strip simply supported on its long edges in a longitudinal magnetic field with the intensity B = const, directed along the axis Ox ($0 \leqslant x \leqslant l$, $-\infty < y < \infty$). Only cylindrical-form vibrations are considered (the vibrations do not depend on y). Using the Bubnov-Galerkin method, we obtain the equation for defining the general characteristics of the plate vibration in a longitudinal magnetic field with constant intensity B.

Table 6 presents the time values t in seconds during which the oscillation amplitude of the copper plate falls n times:

Table 6

B \ n	2	4	6	8	10	50	100
$5 \cdot 10$	8.35	16.7	21.6	25	27.7	47	55.5
10^2	2.09	4.18	5.4	6.3	6.94	11.8	14
$5 \cdot 10^2$	0.08	0.17	0.22	0.25	0.28	0.47	0.55
10^3	0.02	0.04	0.05	0.06	0.07	0.12	0.14

Studying Tables 6 and 5 we notice that the magnetic field causes damping of vibrations, the damping being stronger in the case of the longitudinal magnetic field than in the case of the transverse field.

It is shown that at the initial moments of vibrations the nonlinearity contribution to the general value of frequency far exceeds the contribution of damping. However, because of strong damping, the influence of nonlinearity is essential during very small parts of a second. This took place also in the case of the transverse magnetic field.

THE PROBLEM OF STABILITY OF CURRENT-CARRYING SHELLS. Of special interest are the problems of stability of current-carrying shells and plates taking into account pondermotor forces which are related to the changes in the form of the middle surface [8,16].

Let a stationary electric current with the density I, uniformly distributed through the thickness, flow through a circular cylindrical shell with the length L, the curvature radius of the middle surface R, and the thickness 2h. The static problem of stability is solved. The Joule heat and inducted electric fields are neglected.

The equation for the shell stability is derived taking into account not only the initial effort, but also the static pondermotor forces conditioned by the variations of the form of the shell middle surface.

Table 7 presents some minimal values of H_* and $\breve{H}_*$ (in oersteds) and the corresponding values of n for a copper plate. In the last column the values of the critical current density I_* (in kA/cm^2) are given.

Table 7

R/L	R/h	H_*	$\breve{H}_*$	n	n	I
2	200	13360	20680	11	12	10.6
2	500	3620	6580	14	16	2.88
1	200	7380	14600	8	9	5.86
1	500	1940	4640	10	10	1.54
0.5	200	4050	10410	6	6	3.22
0.5	500	1030	3290	7	7	0.81
0.2	200	1760	6500	3	4	1.4
0.2	500	420	2040	5	5	0.33

On the basis of the obtained results we note the possibility of loss of the shell stability within the real limits of variation of the shell sizes and the quantity of the current. We note also that the account of perturbed pondermotor forces, conditioned by the shell deformation, is essential in determining the critical parameters of stability.

REFERENCES

[1] Ambartsumian, S.A., On some problems of development of investigations in the field of electro-magneto-elasticity of thin bodies. Izv. AN SSSR, MTT, N2, 1974.

[2] Nowacki, W., Coupled fields in mechanics of solids. Theoretical and applied mechanics, IUTAM Congress, Delft, 1976.

[3] Moon, F.C., Problems in magneto-solid mechanics. Mechanics today, v.4, 1978.

[4] Kudriavtsev, B.A. and Parton, V.Z., Magneto-thermo-elasticity. Science and engineering results, Series: The Mechanics of deformable solid bodies, v.14, 1981.

[5] Ambartsumian, S.A., Magneto-elasticity of thin plates and shells. AMR, v.35, N1, 1982.

[6] Ambartsumian, S.A. and Belubekian, M.V., The interaction of conductive shells and plates with electromagnetic field. Mechanics, Intercollege transactions, v.1, Yerevan, 1982.

[7] Ambartsumian, S.A., Belubekian, M.V. and Sarkissian, S.V., The oscillation of electroconductive plate in transverse magnetic field of variable intensity. Mechanics, Intercollege transactions, v.2, Yerevan, 1982.

[8] Ambartsumian, S.A., Baghdasarian, G.E. and Belubekian, M.V., Magnetoelasticity of thin shells and plates. Moscow, "Nauka", 1977.

[9] Ambartsumian, S.A., The theory of anisotropic plates. Moscow, "Nauka", 1967.

[10] Ambartsumian, S.A., Some peculiarities of plate vibrations in magnetic field. Izv. AN SSSR, MTT, 1983.

[11] Ambartsumian, S.A., On the problem of oscillations of electroconductive plate in transverse magnetic field. Izv. AN SSSR, MTT, N3, 1979.

[12] Vardanian, L.V. and Kazarian, K.B., The problem of magnetoelastic oscillations of cylindrical shell in azimuthal magnetic field. Izv. AN Arm.SSR. Mekhanika, N3, 1983.

[13] Ambartsumian, S.A., On the bending of nonlinear-elastic three-layer plates. Izv. AN SSSR, OTN, Mechanics and Machine-engineering, N6, 1960.

[14] Ilyushin, A.A., Plasticity. Moscow, Gostekhizdat, 1948.

[15] Ambartsumian, S.A., Belubekian, M.V. and Minassian, M.M., On the oscillations of nonlinear-elastic plate in transverse and longitudinal magnetic fields. (In print).

[16] Belubekian, M.V. and Kazarian, K.B., The stability of current-carrying cylindrical shell. Intercollege transactions, v.1, Yerevan, 1982.

[17] Ambartsumian, S.A., The general theory of anisotropic shells. Moscow, "Nauka", 1974.

The Mechanical Behavior of Electromagnetic Solid Continua
G.A. Maugin (editor)
Elsevier Science Publishers B.V. (North-Holland)

EARNSHAW'S THEOREM AND MAGNETOELASTIC BUCKLING OF SUPERCONDUCTING STRUCTURES

F.C. Moon

Theoretical and Applied Mechanics
Cornell University
Ithaca, New York
U.S.A.

Magnetoelastic buckling of current carrying structures is shown to be a consequence of Earnshaw's theorem for magnetostatic fields when the currents are held constant. Examples from magnetic fusion technology are reviewed. Exceptions to Earnshaw's theorem are noted, especially for superconducting structures under constant flux conditions. The magnetoelastic buckling of multiturn coils is discussed. An analysis using a periodic array of current carrying filaments is presented which suggests that internal buckling of coils may be possible for low transverse stiffness magnets. Experiments are cited which seem to verify the prediction of internal buckling in coils.

INTRODUCTION

One of the classes of problems that emerged from the rebirth of magnetoelasticity in the 1960's was the phenomenon of magnetoelastic buckling. This early work focused on the stability of soft ferromagnetic structures in steady uniform magnetic fields. (See [1] for a review of these problems.) While of no direct technological importance at the time, magnetoelastic buckling offered a way to test axiomatic theories against experimental results. However in the 1970's a new set of magnetoelastic stability problems arose, this time out of the important technological area of magnetic fusion energy. It was discovered that the large current carrying magnets which were to stabilize plasmas could themselves become structurally unstable. At the time, it was thought that buckling of superconducting and other high current structures was unique to toroidal arrays of fusion magnets (Figure 1). However it has now been established that many topological arrangements of superconducting coils can exhibit magnetoelastic buckling. Not only has this been demonstrated experimentally ([2],[3]) but it has been proposed that such instabilities are the rule and not the exception. This result is a consequence of Earnshaw's theorem (1842). Until recently these instabilities have been studied for the structural magnet as a whole. However we will demonstrate that magnetoelastic instabilities may occur within the windings of magnets. While overall structural analysis is well understood, the internal deformations of conductors in multiturn magnets are not well understood theoretically. Conductor motion can result in frictional heating which can lead to a thermal "quench" of the magnet from the superconducting to the normal state.

Earnshaw's theorem states that a system of constant current circuits cannot be in stable equilibrium under magnetic and other inverse square law forces. An important concept in the study of magnetoelastic buckling is the idea of magnetic stiffness. When a magnetic structure deforms, the magnetic forces will change to either restore the circuit to its original configuration or act to move it further from equilibrium. Earnshaw's theorm states that while there may be deformation modes for which the magnetic stiffness is positive or restoring, there will always be one mode for which the magnetic stiffness is negative and destabilizing.

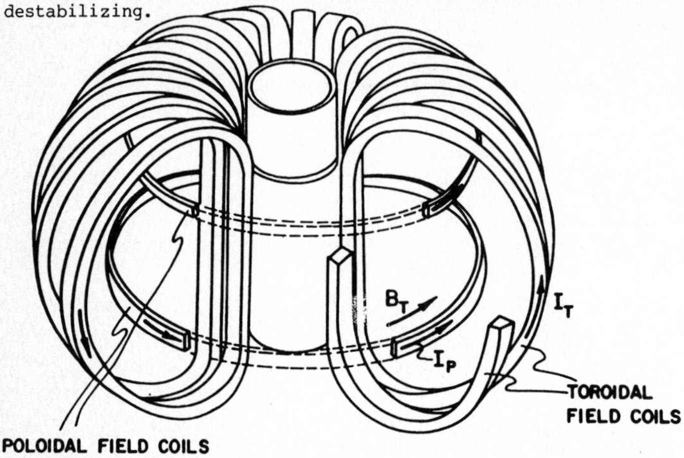

Fig. 1: Sketch of toroidal and poloidal field coils for a tokamak fusion reactor.

Another idea unique to these magneto-mechanical instabilities is the dependence of the critical buckling current on the electrical circuit parameters. In fact it will be shown that when the magnetic flux and not the current is controlled, it is possible to avoid the consequences of Earnshaw's theorem and achieve stable equilibrium under magnetic and gravitational forces alone.

When the magnetic energy of a set of circuits is given by

$$W = \tfrac{1}{2} \Sigma \Sigma L_{ij} I_i I_j \qquad (1)$$

the inductances will depend on the coil deformation and relative motion of the coils. For a given generalized deformation u_α the modal magnetic force is given by

$$F_\alpha = \tfrac{1}{2} \Sigma\Sigma \frac{\partial L_{ij}}{\partial u_\alpha} I_i I_j \ . \qquad (2)$$

This force can be expanded in a Taylor series in the modal displacements $\{u_\alpha\}$

$$F_\alpha = F_\alpha^o - \Sigma \kappa_{\alpha\beta} u_\beta \qquad (3)$$

where $[\kappa_{\alpha\beta}]$ is called the modal magnetic stiffness matrix. (See Chattopadhyay [4].) Unlike the elastic stiffness matrix, $[\kappa_{\alpha\beta}]$ is not positive definite when the currents are fixed and static stability is not possible without elastic stiffness or dynamic control.

REVIEW OF SUPERCONDUCTING STRUCTURAL BUCKLING

A number of experiments and analyses of buckling of superconducting structures have been performed in the United States and Japan. A brief summary of these problems are reviewed below.

Planar Coils in a Partial Torus

One of the most elementary examples of these instabilities is the problem of a planar elastic coil between two rigid coils. When all carry currents in the same direction, the attraction of the center coil to its two neighbors is restrained by the elastic bending of the flexible coil. At a critical value of the current however, the center coils will snap toward one of its neighbors. This experiment has been performed by the Author [5] and by Miya et al. [6]. A similar experiment has been performed on five coils arranged in a partial torus, [7].

Discrete Toroidal Field Magnets

In an attempt to simulate a full torus of magnets for a tokamak fusion reactor superconducting models have been built and tested using eight coils and sixteen planar coils in a full toroidal configuration [8], [9]. These experiments have demonstrated that the mode with alternate pair buckling has the lowest buckling current which agrees with predictions of theory (Figure 2).

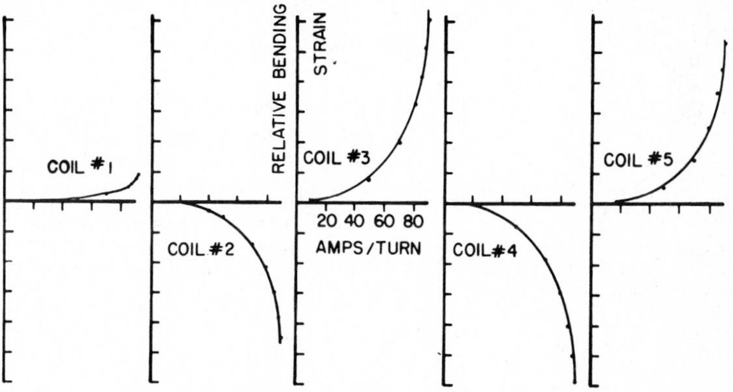

Fig. 2: Bending pattern due to buckling of five coils of an 8 coil torus.

Circular Coil in a Toroidal Magnetic Field

Tokamak fusion reactors require superconducting coils to create vertical or poloidal magnetic fields. Such coils are circular and lie in a plane parallel with the toroidal magnetic field. Although the initial force $\underset{\sim}{I} \times \underset{\sim}{B}$ is zero, out of plane deformation creates destabilizing forces and torques similar to those in plasma physics. The case of buckling of a rigid superconducting circular coil on elastic supports in a toroidal magnetic field has been observed in experiments [2]. The buckling mode involves two degrees of freedom, translation in the plane of the coil, and rotation about a planar axis. The case of an elastic ring shaped coil on rigid supports has been examined analytically. For this case the predicted buckling deformation of the coil is a helical shape around the toroidal field lines (Figure 3).

Yin-Yang Magnets for Tandem Mirror Reactors

A pair of baseball coils can create a mimimum $|\underset{\sim}{B}|$ region. Such a pair is called yin-yang coils after the Chinese and is used in a tandem mirror fusion reactor shown in Figure 4. The small radii of these coils shown in Figure 4 can be thought of as magnetic dipoles. These four dipoles can be shown to be unstable with respect to rotation. This rigid body instability has been observed experimentally [3]. The implication for practical design of these magnets is that special attention be given to the connecting structure between the two coils.

Quadrapole Coil in a Solenoid Field

Another example of a rotational rigid body instability is the case of a quadrapole coil in the solenoidal field of two circular coils. Such a configuration has been proposed for a small compact torus fusion reactor. Experiments done by the author have shown that the coil is unstable with respect to rotation about an axis normal to the long axis of the coil.

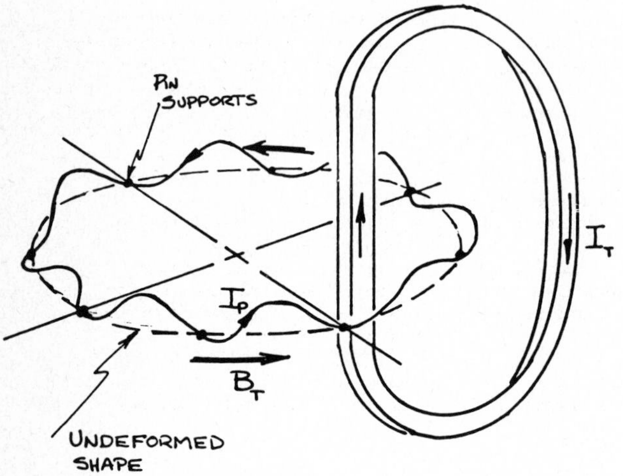

Fig. 3: Sketch of helical buckling pattern of a poloidal field coil in a toroidal magnetic field.

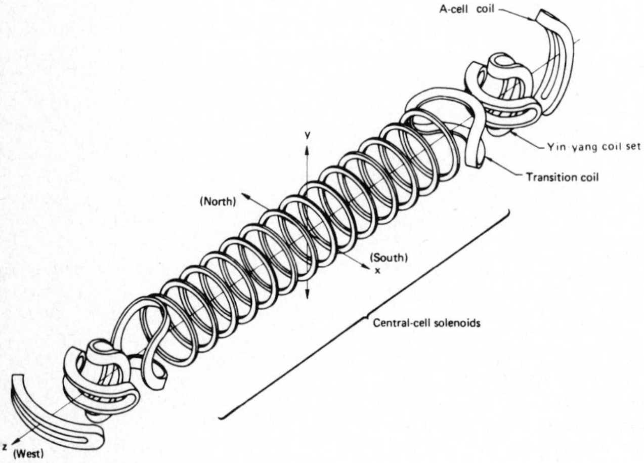

Fig. 4: Magnets for a tandem mirror fusion reactor.

Discrete Solenoid Coil

A linear array of co-axial circular coils arranged to form a discrete coil solenoid has been analyzed with respect to magnetic stability. Such an array is proposed for a tandem-mirror reactor (Figure 4). This calculation shows that these coils are unstable for rotation about any diameter, and axial translation, but are stable with respect to lateral translation.

EARNSHAW'S THEOREM AND MAGNETOELASTIC BUCKLING

Earnshaw's theorem for steady charges states that a given particle cannot be in stable equilibrium under electric forces alone. This result is a consequence of the inverse square law for the electric forces and is related to the fact that solutions of Laplace's equations do not have an absolute maximum or minimum, except on the boundary of the region in question. Statement of the law for electric circuits is often made but a proof is difficult to find (See e.g. [10], [11]).

In this section we want to prove that magnetic circuits cannot be in stable equilibrium under magnetic forces alone. When the magnetic forces are in equilibrium, the equilibrium position has a saddle type behavior. This means that for some degrees of freedom departures from equilibrium will result in restoring forces (positive magnetic stiffness) but there will be at least one degree of freedom for which the perturbed magnetic forces are destabilizing (negative magnetic stiffness). These modes of deformation will require some mechanical constraint. When this constraint is elastic, magnetoelastic buckling is possible since the negative magnetic stiffness can always be increased to cancel out the elastic stiffness.

To demonstrate these results we look at a special case of two rigid circuits (Figure 5) in which the "deformation" is a relative translation of one coil with respect to the other.

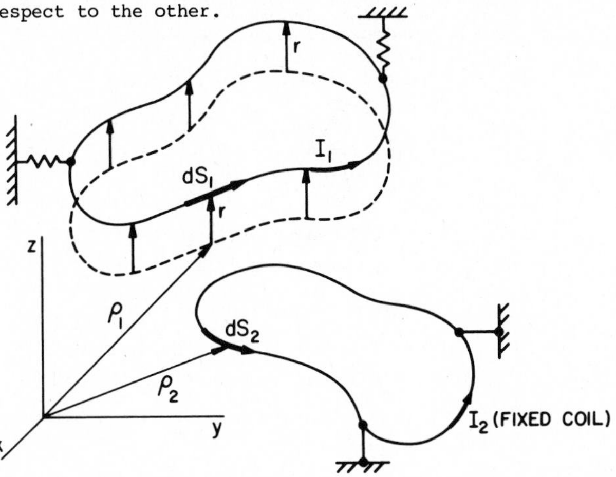

Fig. 5: Two interacting circuits.

The generalized force on circuit I_1 , due to a generalized displacement u can be derived from the mutual inductance between the circuits representing the stored magnetic energy of the circuits (1), (2). If W is the magnetic energy, then -W represents a force potential. If V represents other conservative forces the total force is given by

$$\underline{F} = -\nabla V_1 \; , \quad V_1 = V - W \; . \tag{4}$$

When $V = 0$, W must be maximum about the equilibrium point $\underset{\sim}{u} = 0$. For small departures from equilibrium where $\underset{\sim}{u} = (u_1, u_2, u_3)$ represents components of a translation vector, the magnetic energy function can be represented as a quadratic form

$$W = \tfrac{1}{2} \Sigma \Sigma W_{ij} u_i u_j . \tag{5}$$

For W to have a maximum at $\underset{\sim}{u} = 0$, the matrix W_{ij} must be negative definite. Since W_{ij} is symmetric and real, we can find coordinates for which W_{ij} is diagonal with terms $(\lambda_1, \lambda_2, \lambda_3)$. For negative definiteness $\lambda_1 < 0$, $\lambda_2 < 0$, $\lambda_3 < 0$. This implies that the trace of W_{ij} is negative, i.e.

$$\mathrm{Tr}[W_{ij}] < 0 . \tag{6}$$

However, this necessary condition for stability, cannot be satisfied for constant current circuits. When $\underset{\sim}{u}$ represents a translation, then

$$\mathrm{Tr}[W_{ij}] = \nabla^2 W\big|_{\underset{\sim}{u} = 0} . \tag{7}$$

To evaluate (7) we use the definition of the inductance L_{12}

$$L_{12} = \frac{\mu_o}{4\pi} \int\int \frac{d\underset{\sim}{s}_1 \cdot d\underset{\sim}{s}_2}{|\underset{\sim}{u} + \underset{\sim}{\rho}_1 - \underset{\sim}{\rho}_2|} \tag{8}$$

where ρ_2 is the position vector to the I_2 circuit, and $\rho_1 + \underset{\sim}{u}$ is the position vector to the movable I_1 circuit. Using (1), (8) one can show that

$$\nabla^2 W = 0 \tag{9}$$

Thus a relative maximum of W is impossible since (6) cannot be satisfied and for translation deformation, at least one mode is unstable. Exceptions to Earnshaw's theorem can be found when the assumptions stated in its proof are violated. One example is the case of a superconducting coil with persistent currents. A closed superconducting circuit has the property that the flux threading the circuit is conserved. The flux in the circuit with I_1 is given by

$$\phi_1 = L_{11}I_1 + L_{12}I_2 = \text{constant} . \tag{10}$$

Suppose that the distance between the coils is measured by the variable u , then the mutual inductance L_{12} can be expanded in a Taylor series about the equilibrium value.

$$L_{12} = L_0 + L_1 u + L_2 u^2 . \tag{11}$$

Likewise the current can also be expanded to first order in u ,

$$I_1 = I_{10} - L_1 I_2 u/L_{11} . \tag{12}$$

The magnetic force then becomes

$$F = I_{10}I_2L_1 + [2I_{10}I_2L_2 - L_1^2 I_2^2/L_{11}]u . \tag{13}$$

The first term represents the initial magnetic force which may be required to equilibrate the gravity force for example. The second term in brackets represents the negative magnetic stiffness. The first term can be positive or negative depending on the sign of $I_{10}I_2L_2$. This term is present in the constant current case. The second term always represents a positive stiffness

($L_{11} > 0$) and is a consequence of the flux conservation requirement of the superconductor.

The following example is similar to one given by Tenney [12]. Consider a long current filament below which is a circular superconducting coil under gravitational forces (Figure 6). The coil is neutrally stable along the filament

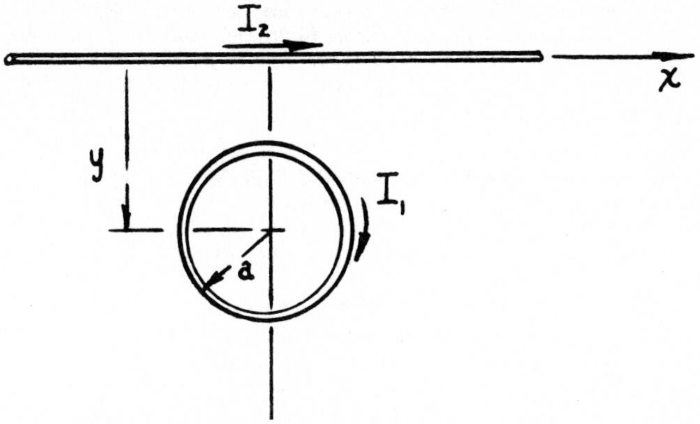

Fig. 6: Stable levitation configuration of a superconducting coil with constant flux operation.

direction, stable with respect to motions tranverse to the filament and is stable with respect to rotation about three axes. To determine the vertical stability one uses the mutual inductance function

$$L_{12} = \mu_o \, [y - (y^2 - a^2)^{\frac{1}{2}}] \, . \qquad (14)$$

Then it is easy to show that $L_1 < 0$, and $L_2 > 0$.

At equilibrium the currents are chosen to satisfy $I_{10} I_2 L_1 = mg$. Clearly if there is constant current control the system is unstable since the magnetic stiffness is negative. However the constant flux condition due to the superconducting state can provide enough positive magnetic stiffness for sufficiently high I_2 to stabilize the system;

$$I_2 > 2I_{10}L_{11}L_2/L_1^2 \, . \qquad (15)$$

BUCKLING OF FILAMENTS IN A MULTI-TURN COIL

From the above discussion we can conclude that for constant current coil systems magnetoelasic buckling of at least one mode of deformation is a possiblity. All the examples thus far have concerned interaction of separate coils. Of particular interest however is the problem of interaction of the different turns or wires of a given coil with one another. This problem is important because a few large superconducting magnet designs have had problems with internal wire motion. This motion can cause frictional heating which at liquid helium temperatures can result in thermal instability of the coil (called a quench). One major example was the superconducting magnets for ISABELLE, a multimillion dollar physics research failure in which the magnets could not consistently achieve their design currents without quenching. (See e.g. [13].)

To simplify the problem consider a cross section of a magnet and neglect the curvature of the wires along the current direction. We replace each wire cross section by a current filament whose self field is assumed to vary inversely with the distance from the filament. We also consider the interaction of one wire with its nearest eight neighbors. We first show that the interaction of one moveable filament is unstable with respect to all rigid neighbors and then show how one solves the case of all moveable filaments or turns.

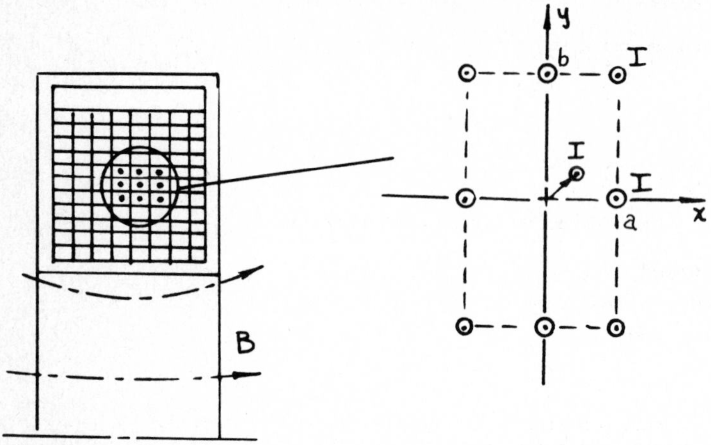

Fig. 7: Left; Cross-section of a multi-turn solenoid coil, Right; Periodic array of current filaments.

Consider the interaction of the center wire with the four corner wires. If the center wire displacement is (u,v) then the perturbation force per unit length is given by

$$F = \frac{2\mu_o I^2 [a^2 - b^2]}{\pi (a^2 + b^2)^2} (u, -v) . \qquad (16)$$

Thus if $a > b$, the x motion is unstable and the y motion is stable.

The magnetic force on a displaced center filament and its nearest neighbors along the x,y axes can be shown to be given by

$$F = \frac{-\mu_o I^2 [a^2 - b^2]}{\pi a^2 b^2} (u, -v) . \qquad (17)$$

In this case the x motion is stable when $a > b$ and the y motion is unstable. The combined effect of all eight neighbors shows that the x motion produces positive stiffness and the y motion produces negative magnetic stiffness for $a > b$. It is clear that adding effects from the next eight neighbors does not change the stability question. Thus elastic constraints are required to maintain the turns in their relative position. For a high enough current, however, the negative magnetic stiffness will prevail and magnetoelastic buckling will result.

To examine the case of all elastically restrained filaments we use the theory of periodic structures. The original position of each wire will be given a position vector

$$\underset{\sim}{p} = na\underset{\sim}{i} + mb\underset{\sim}{j} . \qquad (18)$$

Now we assume a magnetic and elastic interaction between the wire in the $\underset{\sim}{p}$ cell and neighbor $\underset{\sim}{p}'$. The sum of these forces will be assumed to be linear in the displacement of the $\underset{\sim}{p}$ cell $\underset{\sim}{u}_{\underset{\sim}{p}}$.

Further the interaction force is assumed to depend on the difference $\underset{\sim}{h} = \underset{\sim}{p}' - \underset{\sim}{p}$. Thus the equation of motion of the wire in the $\underset{\sim}{p}$ cell is given by

$$m \ddot{\underset{\sim}{u}}_{\underset{\sim}{p}} = - \sum_{\underset{\sim}{h}} G(\underset{\sim}{h}) \cdot u_{\underset{\sim}{p}+\underset{\sim}{h}} \; . \tag{19}$$

where $G(\underset{\sim}{h})$ is a matrix.

To obtain a solution we use a theorem from difference equations which says that we can write the solution in terms of a constant phase difference between each cell

$$\underset{\sim}{u}_{\underset{\sim}{p}} = e^{i\underset{\sim}{q}\cdot\underset{\sim}{p}} \, \underset{\sim}{U}_{\underset{\sim}{q}}$$

or

$$m_1 \ddot{\underset{\sim}{U}}_{\underset{\sim}{q}} = \hat{G}(\underset{\sim}{q}) \cdot \underset{\sim}{U}_{\underset{\sim}{q}} \tag{20}$$

where $\hat{G}(\underset{\sim}{q})$ is the discrete Fourier transform of G, i.e.

$$\hat{G}(\underset{\sim}{g}) = \sum_{\underset{\sim}{h}} G(\underset{\sim}{h}) e^{i\underset{\sim}{g}\cdot\underset{\sim}{h}} \; . \tag{21}$$

For buckling $\ddot{U}_{\underset{\sim}{q}} = 0$, and

$$\det \hat{G}(\underset{\sim}{q}) = 0 \; . \tag{22}$$

For pure elastic interactions with no initial stress effects, the matrix $\hat{G}(\underset{\sim}{g})$ can be shown to be positive definite and (22) cannot be satisfied. Thus buckling is impossible. However, if magnetic stiffness terms are added to $\hat{G}(\underset{\sim}{g})$ as in the previous example, then the condition (22) can be satisfied and internal magneto-elastic buckling may be possible.

Fig. 8: Photograph of circular singular layer coil showing buckling of outer turns.

Experiments on single pancake or radially wound copper coils have shown a buckled shape under high transient current pulses (Figure 8). An early example has been discussed by Daniels [14] who attempted to account for the magnetic stiffness effect. Part of the explanation of buckling in radially wound coils however may be attributed to compression in the outer turns. To analyze this effect one must use a ring or curved beam model which includes axial compression and curvature effects. One cannot predict buckling from classical elasticity. Experiments at Cornell using both pulsed currents (Figure 8) in copper coils and steady currents in superconducting pancake coils suggest that both magnetic stiffness and compressive circumferential stress may be important in the analysis of internal coil buckling.

REFERENCES

1 Moon, F.C., Magneto-Solid Mechanics, (J. Wiley & Sons, New York, 1984).

2 Moon, F.C., Buckling of a Superconducting Ring in a Toroidal Magnetic Field, J. Appl. Mech. 46 (1979a) 151-155.

3 Moon, F.C. and Hara K., Elastic Buckling of Superconducting Yin-Yang Magnets for Fusion, IEEE Trans. on Magnetics MAG-17 No. 1 (1981).

4 Chattapadhyay, S., Magnetoelastic Buckling and Vibration of a Rod Carrying Electric Current, Int. J. Solids and Structures 15 (1979).

5 Moon, F.C. and Swanson C., Experiments on Buckling and Vibration of Superconducting Coils, J. Appl. Mech. 44 No. 4 (1977) 701-713.

6 Miya, K., Takayi, T., and Uesaka, M., Finite Element Analysis of Magnetoelastic Buckling and Experiments on a Three-Coil Superconducting Partial Torus, Mechanics of Superconducting Structures, American Soc. Mech. Engrs. Publ. No. AMD-Vol. 41 (1980) 91-107.

7 Swanson, C. and Moon, F.C., Buckling and Vibrations in a Five Coil Superconducting Partial Torus, J. Appl. Phys. (1977) 3110-3115.

8 Miya, K. and Uesaka, M., An Application of Finite Element Method to Magnetomechanics of Superconducting Magnets for Magnetic Fusion Reactors, Nuclear Engineering and Design (1982).

9 Moon, F.C., Experiments on Magnetoelastic Buckling in a Superconducting Torus, J. Appl. Mech. 46, No. 1 (1979b) 145-150.

10 Thornton, R.D., Design Principles for Magnetic Levitation, Proc. IEEE 61, No. 5 (1973) 586-598.

11 Moon, F.C., Magnetoelastic Instabilities in Superconducting Structures and Earnshaw's Theorem, Mech. of Superconducting Structures, ASME Appl. Mech. Monograph, Vol. 41 (1980) 77-90.

12 Tenney, F.H., On the Stability of Rigid Current Loops in an Axisymmetric Field, Princeton Plasma Physic Laboratory Report MATT-693 (1969).

13 MIT Magnet Technology Group, ISABELLE Superconducting Magnets, F. Bitter National Magnet Lab. Report, Massachusetts Institute of Technology (1980).

14 Daniels, J.M., High Power Solenoids-Stresses and Stability, British J. Appl. Physics 4, No. 2 (1953) 50-54.

The Mechanical Behavior of Electromagnetic Solid Continua
G.A. Maugin (editor)
Elsevier Science Publishers B.V. (North-Holland)

SOME RECENT RESULTS IN THE NONLINEAR THEORY OF SHELLS WITH ELECTROMAGNETIC EFFECTS

A.E.Green[+] and P.M.Naghdi[++]

+Mathematical Institute, University of Oxford,24-29 St.Giles, Oxford, OX1 3LB, United Kingdom

++Department of Mechanical Engineering, University of California, Berkeley , California 94720, U.S.A.

The objective of this lecture is to present an account of a recent formulation of the nonlinear theory of deformable shell-like bodies in the presence of electromagnetic effects [1]. After some preliminaries and a description of shell-like bodies, the main features of the construction of a general bending theory of shells are indicated in the presence of electromagnetic effects. More specifically, employing a direct approach based on a two-dimensional continuum model known as Cosserat (or directed surfaces with K directors, a general theory of shells in the context of thermomechanics and in the presence of electromagnetic effects is constructed. The electromagnetic aspect of the subject is dealt with as an appropriate nonrelativistic theory in which the conservation laws represent two-dimensional analogs of the corresponding conservation laws in the three-dimensional theory for moving media.

For many purposes, instead of a general theory with K directors, it will suffice to utilize a theory with a single director corresponding to the usual bending theory of shells. The latter includes as a special case a constrained or a restricted theory which, in the absence of electromagnetic effects, corresponds to the classical Kirchhoff-Love theory of shells [1,Sec.8].

In the absence of the directors the results reduce to a nonlinear membrane theory of shells with electromagnetic effects. In the latter part of the lecture, we illustrate some of the electromagnetic effects with the use of the membrane theory. In particular, we consider circular tubes regarded as an elastic membrane and study the response of a finitely deforming thin, magnetic, polarized elastic membrane. The case of propagation of electromagnetic waves in an isotropic circular tube is discussed and a comparison is made with the parallel development which can be obtained from the three-dimensional equations [2].

REFERENCES

[1] Green A.E. and Naghdi P.M., On electromagnetic effects in the theory of shells and plates, Phil.Trans.Roy.Soc.Lond.(in press,1983).

[2] Green A.E. and Naghdi P.M., Electromagnetic effects in an elastic circular cylindrical waveguide ,to appear.

The reader is referred to References 1 and 2 for a full development of the contents of the lecture

The Mechanical Behavior of Electromagnetic Solid Continua
G.A. Maugin (editor)
Elsevier Science Publishers B.V. (North-Holland)

THE VIBRATION AND STABILITY OF THE CURRENT-CARRYING PLATE IN THE EXTERNAL MAGNETIC FIELD

M.V.Belubekian

Vice-director, Mechanics Institute
Armenian Academy of Sciences
Yerevan, Armenia
USSR

The investigations on the stability and vibration problem of the rod with the electrical current and in the presence of the external magnetic field are well-known [1,2]. The problems of the current-carrying plate stability have been also considered [2-4]. In the report the equations and boundary conditions are brought which describe the behaviour of the current-carrying plate in the presence of ex - ternal magnetic field parallel to plates electric current direction. It's shown that external magnetic field has the stabilising effect.

In rectangular orthogonal system (x_1, x_2, x_3) the electroconductive thin plate take the region $\{0\leq x_1\leq a,\ 0\leq x_2\leq b,\ -h\leq x_3\leq h\}$ On the edges $x_1=0$, $x_2=0$ the plate is included into an electrical circuit of constant current. Due to it the uniformly distributed current with density J_0 is flowing in the plate. In addition it's supposed that the plate is in an uniform stationary magnetic field of external devices. External magnetic field is directed along axis.

On the basis of assumptions presented in [5] equations of magneto-elastic vibrations are simplified for the plate under consideration From assumptions the principal ones are the hypothesis of magneto-elasticity of thin bodies and the supposition concerns the beha - viour of outward induced electromagnetic field near the plate surfaces $x_3=\pm h$.

Simplified equations of the problem in the case of one-dimension vibration of current-carrying plate (plate-beam) have the form

$$D\frac{\partial^4 w}{\partial x_1^4}+2\rho h\frac{\partial^2 w}{\partial t^2}=\frac{8\pi h}{c^2}J_0^2 w-\frac{h}{2\pi}\left(\frac{h_1^+-h_1^-}{2h}-\frac{\partial f}{\partial x_1}\right),$$

(1)

$$\frac{\partial^2 f}{\partial x_1^2}-\frac{4\pi\sigma}{c^2}\frac{\partial f}{\partial t}=\frac{1}{2h}\frac{\partial}{\partial x_1}(h_1^+-h_1^-)-\frac{4\pi\sigma}{c^2}H_0\frac{\partial^2 w}{\partial t\partial x_1},$$

$$\left(\frac{\partial^2}{\partial x_1^2} - \frac{1}{c^2}\frac{\partial^2}{\partial t^2}\right)(h_1^+ - h_1^-) = \frac{2}{\lambda}\left[\frac{\partial f}{\partial x_1} + \frac{1}{4\pi\sigma}\frac{\partial}{\partial t}\left(\frac{h_1^+ - h_1^-}{2h} - \right.\right.$$

$$\left.\left. - \frac{\partial f}{\partial x_1}\right) - \frac{H_0}{c^2}\frac{\partial^2 w}{\partial t^2}\right]$$

Here w -deflection, ρ -plate mass per unit volume, D -flexural rigidy, H_0 -intensity of external magnetic field, f -normal component of induced magnetic field, h_1^+ , h_1^- -tangential components of induced magnetic field on the surfaces $x_3 = \pm h$, c -electrodynamical constant (speed of light in vacuum), σ - electrical conductivity, λ -specified size, which henceforth is taken equal to wave length. Equations (1) are presented in Gauss's system.

For simply supported plate on edges $x_1 = 0$ and $x_1 = a$ boundary conditions are 6

$$w\Big|_{x_1=0,a} = 0, \qquad (h_1^+ - h_1^-)\Big|_{x_1=0,a} = 0$$

$$\left.\frac{\partial^2 w}{\partial x_1^2}\right|_{x_1=0,a} = 0, \quad f\Big|_{x_1=0,a} = H_0 \left.\frac{\partial w}{\partial x_1}\right|_{x_1=0,a} \tag{2}$$

Solutions of equations (1) satisfying boudary conditions (2) are expressed in the form

$$w = \sum_{n=1}^{\infty} w_n e^{i\omega_n t} \sin\frac{n\pi x}{a}, \quad f = \sum_{n=1}^{\infty} f_n e^{i\omega_n t} \cos\frac{n\pi x}{a},$$

$$h_1^+ - h_1^- = \sum_{n=1}^{\infty} d_n e^{i\omega_n t} \sin\frac{n\pi x}{a} \tag{3}$$

Substituting (3) into (1) we obtained the following dispersion equation respect to vibration dimensionless frequency.

$$\theta_n^3 + a_{0n}\theta_n^2 + (1 + a_{1n} - b_n)\theta_n + (1 - b_n)a_{0n} = 0 \tag{4}$$

$$\theta_n = \frac{i\omega_n}{\omega_{0n}}, \quad \omega_{0n}^2 = \frac{\mathcal{D}}{2\rho h}\left(\frac{n\pi}{a}\right)^4, \quad \beta_n = \frac{8\pi h}{c^2\mathcal{D}}\left(\frac{a}{n\pi}\right)^4 J_0^2,$$

$$a_{0n} = \frac{\frac{n\pi}{a}\left(1+\frac{n\pi h}{a}\right)c^2}{4\pi\sigma h\omega_{0n}}, \quad a_{1n} = \frac{\frac{n\pi}{a}\left(1+\frac{n\pi h}{a}\right)}{h\omega_{0n}^2}\,\frac{H_0^2}{4\pi\rho}$$

In deducing equation (4) it have been taken into account

$$(\omega_n a)^2 (n\pi c)^{-2} \ll 1$$

Using of Routh-Gurvitz's criterion to equation (4) results in that for stability of plate it's necessary and sufficient the fulfilment of the following inequalities

$$\beta_n < 1 + a_{1n}$$

Equalities $\beta_n = 1 + a_{1n}$ determine current critical density, beyond which plate is unstable. Substituting in these equalities values of a_{1n} and β_n shows that minimal critical density J_* achieves at $n=1$

$$J^* = J_*\left[1 + \frac{3(1-\nu^2)(1+\pi h/a)}{4\pi E(\pi h/a)^3} H_0^2\right]^{1/2} \qquad (5)$$

Here E -elasticity module, ν -Poisson's coefficient, J_* corresponding critical density of electrical current in the case of exterminal magnetic field absence

$$J_*^2 = \frac{c^2\mathcal{D}}{8\pi h}\left(\frac{\pi}{a}\right)^4 \qquad (6)$$

From (5) it follows that the presence of external magnetic field reduces current-carrying plate stabilization. The increasing of the magnetic field intensity bring to increasing of the electrical current critical density.

The effect of magnetic field is essentialy depended on the relative thikness of the plate. The smaller is the ratio h/a the stronger is the influence of the magnetic field.

Let consider example of the plate with $a = 40\,cm$, $2h = 0{,}1\,cm$. In the table the ratious J^*/J_* are given for different values of external magnetic field in cases of two different materials of plate.

From this table one can see that for the plate under consideration when magnetic field's intensity is order $10^3 oe$ the critical cur-

$H_0 \cdot 10^3\,oe$		0	0,6	0,8	I,0	I,2
$\mathcal{T}^*$	Al	I	I,683	2,063	2,467	2,886
$\mathcal{T}_*$	Cu	I	I,438	I,703	I,992	2,297

rent increases by a factor two.

The investigation of the dispersion equation (4) shows that the electrical current brings to decreasing of the plate vibration frequency. The external magnetic field brinds to frequency increasing.

REFERENCES:

1 Prudnikov V.V., Elastic Oscillations of Current-carrying Rod in a Longitudinal Magnetic Field. Zh.Prik.Mekh. i Techni.Fiziki, 1(1968) 168-172.

2 Moon F.C., Problems in Magnetosolid Mechanics. Mechanics Today. 4(1978) 307-390.

3 Belubekian M.V., Kazarian K.B., About Unstability of Current-carrying Plate-beam. Zh.Mathem.Metodi i Fiziko-mechan. Polia. 16(1982) 40-44.

4 Belubekian M.V., About Statical Stability of Current-carrying plate. Doklady AN Arm.SSR. 5(1982) 208-212.

5 Ambartsumian S.A., Baghdasarian G.E., Belubekian M.V. The Magneto-elasticity of Thin Shells and Plates. (Science, Moscow, 1977).

6 Ambartsumian S.A., Belubekian M.V., Sarkissian S.V., The Magneto-elasticity Oscillation of Plate in the Longitudinal magnetic field with the Account of Information Displacement, in: Sarkissian V.S. (ed.), The Transactions of the 12th All-Union Conference of Theory of Shells and Plates (Yerevan, 1981).

The Mechanical Behavior of Electromagnetic Solid Continua
G.A. Maugin (editor)
Elsevier Science Publishers B.V. (North-Holland)

INTERACTION OF TWO NEARBY FERROMAGNETIC PANELS ON THE MAGNETOELASTIC BUCKLING

Junji Tani and Kikuo Otomo

Institute of High Speed Mechanics
Tohoku University
Sendai
Japan

This study is concerned with an interactive effect of two ferromagnetic panels close to each other on the magnetoelastic buckling. These panels with wide surfaces normal to a uniform magnetic field are supported parallel to each other and composed of magnetically soft material. The analysis is based on the small bending theory of thin plates and a dipole model for the force exerted by the magnetic fields. The Galerkin method and Fourier transforms are used. It is found that the decrease in the distance between two plates considerably reduces a buckling magnetic field and a buckling mode is symmetry. The results as predicted from this theoretical analysis is confirmed experimentally also.

INTRODUCTION

Since Moon and Pao [1] first investigated the magnetoelastic buckling of a ferromagnetic plate, there has been a considerable amount of works [2-11] done on the subject. Recently a close agreement between theoretical and experimental values was obtained by Miya, Takagi and Ando [12]. All these studies are confined to the problem of only one ferromagnetic plate set in a uniform magnetic field. The magnetoelastic buckling of two nearby ferromagnetic plates seems to be a subject of considerable interest, but the author is not aware of any published study.

The purpose of this paper is to examine the interactive effect of two ferromagnetic plates close to each other on the magnetoelastic buckling. First a theoretical study is made on a following model. Two plates of finite length and infinite width are set parallel to each other with their wide surfaces normal to a uniform magnetic field and simply supported at both edges. The thin plates are composed of soft linear magnetic material. On the basis of the small bending theory of thin plates and a dipole model for the magnetization, the problem is solved by means of the Galerkin method and Fourier transforms. Numerical results reveal that the buckling magnetic field is lower for the two nearby ferromagnetic plates than for the single one and that the former buckles in a symmetric mode. Next an experiment is conducted for cantilever beam-plates of magnetically soft material with a great care to minimize initial imperfection and demagnetize fully the speciment. It is shown that the experimental result supports the theoretical one.

MATHEMATICAL FORMULATION

Consider that two thin elastic ferromagnetic plates with the same configuration and property are set parallel to each other in a static, uniform transverse magnetic field of the magnetic induction B_0. The two plates of thickness h, finite length L and infinite width are simply supported at both edges by nonmagnetic material and their midsurfaces are separated by the distance $2d + h$. The coordinate system is taken as shown in Figure 1. The deflections and the magnetic body couples of two plates are denoted by W_j and P_j $(j = 1,2)$, respectively. The basic equations and boundary conditions of two plates are given from the small deflection theory by

$$DW_{j,xxxx} + P_j = 0, \quad (j = 1, 2) \tag{1}$$

$$W_j = W_{j,xx} = 0 \quad \text{at} \quad x = 0, L \tag{2}$$

where $D = Eh^3/12(1 - \nu^2)$ is the flexural rigidity, E and ν are Young's modulus and Poisson's ratio of the plates, respectively, while subscripts following a comma stand for differentiation.

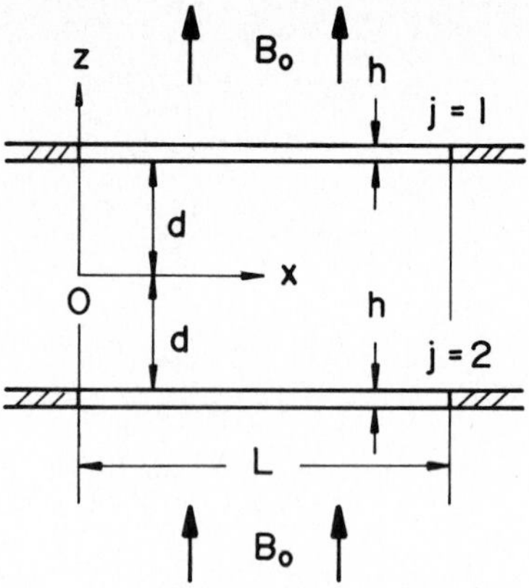

Figure 1
Geometry of Two Plates

The static magnetic field in absence of electric field, charges and conduction currents is governed by Maxwell's equations and a magnetic potential can be defined [3]. Hence the disturbed magnetic field due to the deformation of the ferromagnetic plate can be described by the disturbed magnetic potential Φ. Since the plates bend in a two-dimensional mode, this potential satisfies

$$\Phi_{,xx} + \Phi_{,zz} = 0 \tag{3}$$

According to the dipole model for the magnetization, the magnetic body couple per unit area on the soft linear magnetic plate is given by

$$P_1 = \chi B_0 \int_d^{d+h} \Phi^-_{,xx} dz \ , \quad P_2 = \chi B_0 \int_{-d-h}^{-d} \Phi^-_{,xx} dz \tag{4}$$

where χ is the susceptibility of the plate, while super-signs - and + denote the field quantities inside and outside the plates, respectively. Considering that W and Φ are very small quantities, the boundary conditions for the disturbed field quantities which must vanish at $|z| \to \infty$ are expressed by

$$\Phi^+_{,x} - \Phi^-_{,x} = QW_{1,x} \ , \quad \Phi^+_{,z} - (1 + \chi)\Phi^-_{,z} = 0 \quad \text{at} \quad z = d, \ d+h \tag{5}$$

$$\Phi^+_{,x} - \Phi^-_{,x} = QW_{2,x} \ , \quad \Phi^+_{,z} - (1 + \chi)\Phi^-_{,z} = 0 \quad \text{at} \quad z = -d-h, \ -d \tag{6}$$

where μ_0 is a universal constant and $Q = \chi B_0/\mu_0(1 + \chi)$.

Here considering two buckling modes, i.e., (a) a symmetric and (b) an antisymmetric ones with respect to x axis, we obtain the following conditions.

$$\text{(a)} \ W_2 = -W_1 \ , \quad \text{(b)} \ W_2 = W_1 \tag{7),(8}$$

METHOD OF SOLUTION

Considering the boundary conditions (2), we put the deflection W_1 as

$$W_1 = A\sin(\pi x/L) \tag{9}$$

where A is an unknown parameter. Considering equations (7) and (8), substituting equation (9) into equations (5) and (6) and applying the Fourier transform to equations (3), (5) and (6), we obtain

$$\phi_{,zz} - \xi^2\phi = 0 \tag{10}$$

$$\left.\begin{aligned} &\phi^+ - \phi^- = -Qw \quad \text{at } z = d,\ d+h \\ &\phi^+ - \phi^- = Qw \quad \text{at } z = -d-h,\ -d \\ &\phi^+_{,z} - (1+\chi)\phi^-_{,z} = 0 \quad \text{at } z = |d|,\ |d+h| \end{aligned}\right\} \text{for the case (a)} \tag{11}$$

$$\left.\begin{aligned} &\phi^+ - \phi^- = -Qw \\ &\phi^+_{,z} - (1+\chi)\phi^-_{,z} = 0 \end{aligned}\right\} \text{at } z = |d|,\ |d+h| \text{ for the case (b)} \tag{12}$$

where

$$\phi = \int_{-\infty}^{\infty} \Phi e^{-i\xi x} d\xi/\sqrt{2\pi}\ , \quad w = \sqrt{\pi/2}\cdot A(1 + e^{i\xi L})/[(\pi/L)^2 - \xi^2] \tag{13}$$

A solution to equation (10) vanishing at $|z| \to \infty$ is given by

$$\left.\begin{aligned} &\phi^+ = c_1 e^{-|\xi|z} \ \text{for } z < d+h\ , \quad \phi^+ = c_2 e^{|\xi|z} \ \text{for } z < -d-h \\ &\phi^+ = c_3\cosh|\xi|z + c_4\sinh|\xi|z \quad \text{for } |z| < d \\ &\phi^- = c_5\cosh|\xi|z + c_6\sinh|\xi|z \quad \text{for } d \leqq z \leqq d+h \\ &\phi^- = c_7\cosh|\xi|z + c_8\sinh|\xi|z \quad \text{for } -d-h < z < -d \end{aligned}\right\} \tag{14}$$

where c_j $(j = 1\text{-}8)$ are arbitrary functions of ξ and determined from the boundary conditions (11) and (12) as follows:

$$\left.\begin{aligned} &c_5 = Qw(\beta + 2\alpha_1\cosh|\xi|d)/(\alpha_2\beta - \alpha_1\gamma),\quad c_7 = -c_5,\ c_8 = c_6 \\ &c_6 = -Qw(\gamma + 2\alpha_2\cosh|\xi|d)/(\alpha_2\beta - \alpha_1\gamma),\quad c_3 = 0 \end{aligned}\right\} \text{for the case (a)} \tag{15}$$

$$\left.\begin{aligned} &c_5 = Qw(\beta + 2\alpha_1\sinh|\xi|d)/(\alpha_2\beta - \alpha_1\delta)\ ,\quad c_7 = c_5,\ c_8 = -c_6 \\ &c_6 = -Qw(\beta + 2\alpha_2\sinh|\xi|d)/(\alpha_2\beta - \alpha_1\delta),\quad c_4 = 0 \end{aligned}\right\} \text{for the case (b)} \tag{16}$$

where the expressions of c_1, c_2, and c_3 or c_4 are omitted, while α_1, α_2, β, γ and δ are given by

$$\left.\begin{aligned} &\alpha_1 = (1+\chi)\cosh|\xi|(d+h) + \sinh|\xi|(d+h) \\ &\alpha_2 = (1+\chi)\sinh|\xi|(d+h) + \cosh|\xi|(d+h),\quad \beta = \chi\sinh 2|\xi|d \\ &\gamma = \chi\cosh 2|\xi|d - \chi - 2,\quad \delta = \chi\cosh 2|\xi|d + \chi + 2 \end{aligned}\right\} \tag{17}$$

Substituting equations (15) and (16) into equations (14) applying the Fourier inversion transform to the resulting equations and then substituting these into equation (4), we obtain the magnetic body couple P_1 as follows:

$$P_1 = -\frac{\chi^2 A B_0^{\,2}}{2\mu_0(1+\chi)} \int_{-\infty}^{\infty} \frac{\xi^2(1 + e^{i\xi L})e^{-i\xi x}}{|\xi|[(\pi/L)^2 - \xi^2](\alpha_2\beta - \alpha_1\gamma)}$$

$$\times\ \{(\beta + 2\alpha_1\cosh|\xi|d)[\sinh|\xi|(d+h) - \sinh|\xi|d]$$

$$\times (\gamma + 2\alpha_2 \cosh|\xi|d)[\cosh|\xi|(d+h) - \cosh|\xi|d]\}d\xi \quad \text{for the case (a)} \tag{18}$$

$$P_1 = -\frac{\chi^2 A B_0{}^2}{2\mu_0(1+\chi)} \int_{-\infty}^{\infty} \frac{\xi^2(1+e^{i\xi L})e^{-i\xi x}}{|\xi|[(\pi/L)^2 - \xi^2](\alpha_2\beta - \alpha_1\delta)}$$

$$\times \{(\beta + 2\alpha_1 \sinh|\xi|d)[\sinh|\xi|(d+h) - \sinh|\xi|d]$$

$$- (\delta + 2\alpha_2 \sinh|\xi|d)[\cosh|\xi|(d+h) - \cosh|\xi|d]\}d\xi \quad \text{for the case (b)} \tag{19}$$

Substituting equations (9), (18) and (19) into equations (1) and applying the Galerkin method, we finally obtain the buckling magnetic field $\overline{B}_c$ as follows:

$$\overline{B}_c = 2(1+1/\chi)^2 / \int_0^{\infty} s(1+\cos s) f(s)/(\pi^2 - s^2)^2 ds \tag{20}$$

where the integral can be evaluated numerically,

$$f(s) = [\![\{\sinh 2\eta s + 2(1+1/\chi)\cosh(\eta+\zeta)s\cosh\eta s[1+\chi+\tanh(\eta+\zeta)s]/(1+\chi)$$

$$\times [\sinh(\eta+\zeta)s - \sinh\eta s] - \{\cosh 2\eta s - 1 - 2/\chi + 2(1+1/\chi)$$

$$\times \cosh(\eta+\zeta)s\cosh\eta s[1 + (1+\chi)\tanh(\eta+\zeta)s]/(1+\chi)\}[\cosh(\eta+\zeta)s$$

$$- \cosh\eta s]]\!] / \cosh(\eta+\zeta)s\{\sinh 2\eta s\{[1 + (1+\chi)\tanh(\eta+\zeta)s]/(1+\chi)$$

$$- (\cosh 2\eta s - 1 - 2/\chi)[1 + \chi + \tanh(\eta+\zeta)s]/(1+\chi)\} \quad \text{for the case (a)} \tag{21}$$

$$f(s) = [\![\{\cosh 2\eta s + 1 + 2/\chi + 2(1+1/\chi)\sinh\eta s\cosh(\eta+\zeta)s$$

$$\times [1 + \chi + \tanh(\eta+\zeta)s]/(1+\chi)\}[\sinh(\eta+\zeta)s - \sinh\eta s]$$

$$- \{\sinh 2\eta s + 2(1+1/\chi)\sinh\eta s\cosh(\eta+\zeta)s[1 + (1+\chi)\tanh(\eta+\zeta)s]/(1+\chi)\}$$

$$\times [\cosh(\eta+\zeta)s - \cosh\eta s]]\!] / \cosh(\eta+\zeta)s\{(\cosh 2\eta s + 1 + 2/\chi)$$

$$\times [1 + (1+\chi)\tanh(\eta+\zeta)s] - \sinh 2\eta s[1 + \chi + \tanh(\eta+\zeta)s]\}(1+\chi)$$

$$\text{for the case (b)} \tag{22}$$

and

$$\overline{B}_c = 8L^3 B_0^2/\pi^3\mu_0 D, \quad \eta = d/L, \quad \zeta = h/L \tag{23}$$

Considering that χ may be extremely large in forromagnetic materials, we will examine the limit of the distance between two plates. In the case with $\eta = 0$, where two plates are in contact,

$$\left.\begin{aligned} f(s) &= \chi\sinh\zeta s(2 - \operatorname{sech}\zeta s) && \text{for the case (a)} \\ &= 1/[1 + (1/\chi)\coth\zeta s] && \text{for the case (b)} \end{aligned}\right\} \tag{24}$$

Here it is to be noted that the buckling magnetic field becomes very small in the foregoing case (a). In the case with $\eta = \infty$, where two plates are separated by infinite distance,

$$f(s) = 2/[1 + (1/\chi)\coth(\zeta s/2)] \quad \text{for the cases (a) and (b)} \tag{25}$$

The latter agrees with the case of the single plate obtained by Moon [3].

NUMERICAL RESULTS

Using the value of 10^{-2} for the thickness ratio ζ, calculations are carried out to clarify the effect of the distance between two plates on the buckling magnetic field for the various values of the magnetic susceptibility χ. The results are illustrated in Figure 2. In this figure, the ordinate represents the ratio of the buckling magnetic field $\bar{B}_{c2}$ of two plates to that $\bar{B}_{c1}$ of single one and the abscissae the ratio $2\eta/\zeta(=2d/h)$ of the distance between two plates to the thickness of the plate. From this Figure it can be seen that the decrease in the distance between two plates reduces remarkably the buckling magnetic field and the buckling mode is symmetric with respect to the x axis.

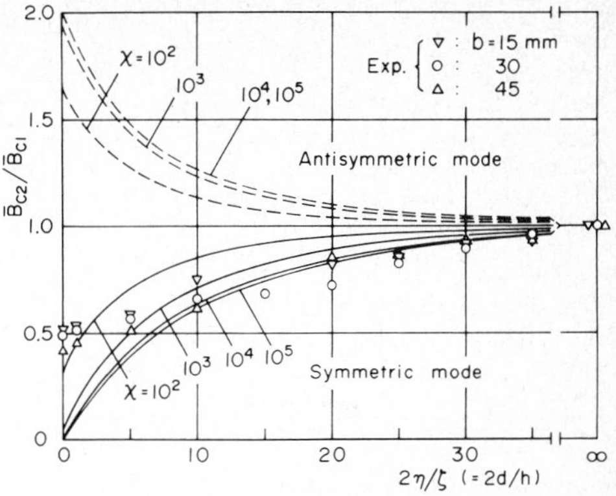

Figure 2
Buckling Magnetic Fields vs. Distance Between Two Plates

EXPERIMENT

For convenience, experiments were conducted for cantilever beam-plate specimens made of low carbon steel to check the results predicted by the preceding theoretical model. The specimens were the same as the ones used in the test of Miya, Hara and Someya [11]. Hence, Young's modulus $E = 2.15 \times 10^{11} Pa$ Poisson's ratio $\nu = 0.28$ and the permeability $\mu_r(=\chi+1) = 620$ in MKS units. The values of *1* and *150*mm were chosen for the thickness and length of the straight specimens, respectively, and *15*, *30*, and *45*mm for the width. The uniform magnetic field was generated by an electromagnet. Its intensity was controlled by a *d-c* current supply. The two specimens with the same configuration and property were clamped at one end in brass vise with a small distance so that these wide plane were parallel to the pole face of the magnet. This pole was circular and its diameter was *250*mm. A gap distance between two pole faces was *200*mm. The magnetic field between two pole faces was measured with a gaussmeter. The deflection was measured at a central point of one plate face with an apparatus exerting negligible small contact-force. The magnetic field and the deflection were recorded with a *X-Y* recorder.

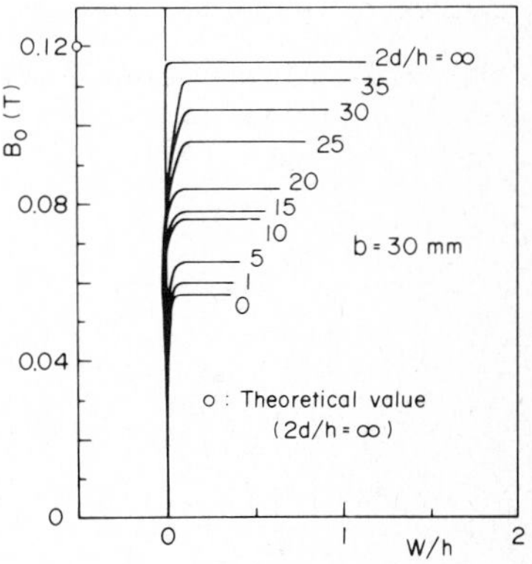

Figure 3
Magnetic Fields vs. Deflection

In each test, the magnetic field was continuously increased from zero until the cantilever snapped suddenly into a large deformed position. If no clear buckling magnetic field could be obtained, the whole assembly was adjusted slightly to remove the misalignment of the specimen and the vise. The magnetic field was increased again after the specimen had been demagnetized by a generator of *a-c* magnetic field. The buckling magnetic field was determined only when the clear buckling was reproducible.

The buckling magnetic field of single plate was first measured to check the validity of the present experiment, and then that of two nearby plates for the various values of their distance. The experimental results are shown in Figures 2 and 3. In this case, both $\overline{B}_{c1}$ and $\overline{B}_{c2}$ taken in Figure 2 were the experimental values. It can be seen from Figure 3 that the experimental buckling magnetic field of single plate agreed almost with the theoretical one, and that the small prebuckling deflection appeared for the case of two plates. It can be seen from Figure 2 that an decrease in the distance between two plates reduced the buckling field, but its quantity was less than the theoretical one, when the distance was very small. Further it was observed in the experiment that the snap-through buckling of two nearby plates occurred in the symmetric mode.

CONCLUSION

Limited experimental data supported the reduction of the buckling magnetic field due to the approach of two plates and the occurrence of the symmetric buckling mode. When two plates were very nearly located, however, the experimental results did not agree with the theoretical ones. The following points may be considered as the cause for this discrepancy: the nonuniformity of the magnetic field, the inevitable initial defect of the specimen and the notable difference in the boundary condition between the theoretical model and the experimental one.

ACKNOWLEDGMENT

The authors wish to express their appreciation to Associate Professor K. Miya, University of Tokyo, for preparing the specimens and valuable discussion.

REFERENCES

[1] Moon, F. C. and Pao, Y.-H., Magnetoelastic buckling of a thin plate, Jrnl. Appl. Mech. 35 (1968) 53-58.
[2] Moon, F. C. and Pao, Y.-H., Vibration and dynamic instability of a beam-plate in a transverse magnetic field, Jrnl. Appl. Mech. 36 (1969) 92-100.
[3] Moon, F. C., The mechanics of ferroelastic plates in uniform magnetic field, Jrnl. Appl. Mech. 37 (1970) 153-158.
[4] Popelar, C. H., Postbuckling analysis of a magnetoelastic beam, Jrnl. Appl. Mech. 39 (1972) 207-211.
[5] Wallerstein, D. V. and Peach, M. O., Magnetoelastic buckling of beams and thin plates of magnetically soft material, Jrnl. Appl. Mech. 39 (1972) 451-455.
[6] Popelar, C. H. and Bast, C. O., Anexperimental study of the magnetoelastic postbuckling behavior of a beam, Exp. Mech. 12 (1972) 537-542.
[7] Pao, Y.-H. and Yeh, C.-S., A linear theory for soft ferromagnetic elastic solids, Int. Jrnl. Engng. Sci. 11 (1973) 415-436.
[8] Dalrymple, J. M., Peach, M. O. and Viegelahn, G. L., Magnetoelastic buckling of thin magnetically soft plates in cylindrical mode, Jrnl. Appl. Mech. 41 (1974) 145-150.
[9] Dalrymple, J. M., Peach, M. O. and Viegelahn, G. L., Magnetoelastic buckling : theory vs. experiment, Exp. Mech. 16 (1976) 26-31.
[10] Dalrymple, J. M., Peach, M. O. and Viegelahn, G. L., Edge effect influence on magnetoelastic buckling of rectangular plates, Jrnl. Appl. Mech. 44 (1977) 305-310.
[11] Miya, K., Hara, K. and Someya, K., Experimental and theoretical study on magnetoelastic buckling of a ferromagnetic cantilevered beam-plate, Jrnl. Appl. Mech. 45 (1978) 355-360.
[12] Miya, K., Takagi, T. and Ando, Y., Finite-element analysis of magnetoelastic buckling of ferromagnetic beam plate, Jrnl. Appl. Mech. 47 (1980) 377-382.

The Mechanical Behavior of Electromagnetic Solid Continua
G.A. Maugin (editor)
Elsevier Science Publishers B.V. (North-Holland)

FINITE ELEMENT ANALYSIS OF DYNAMIC DEFORMATION OF FUSION REACTOR COMPONENTS DUE TO ELECTROMAGNETIC FORCES

Kenzo Miya

Nuclear Engineering Research Laboratory,
University of Tokyo,
Tokaimura, Ibaraki-Prefecture,
JAPAN

A structural design of fusion reactor components, as is well recognized recently, is very important due to severer conditions of high temperature, heavy irradiation by high energy neutrons and strong magnetic field. Higher magnetic stresses will be introduced in first wall and blanket structures due to transient eddy current interacting with an external field, and also generated in superconducting magnet structures due to an interaction between larger coil currents. Objective of the present paper is placed on a derivation of mathematical formulation convienient for a finite element analysis, and a demonstration of numerical results on the eddy current problems and magnetomechanical behavior of superconducting magnets. For a verification of the theoretical consideration, comparisons of the numerical results with experimental ones are stated with remarkable agreements between them.

1. Introduction

In these twenty years a new engineering field has been developed, which is related to an interaction between the elastic, thermal and electromagnetic fields. A motivation for its development has been a possibility of its applications to problems appeared in geophysics and acousties [1], [2].

Recently the electromagnetomechanical problems have been recognized important as to an engineering feasibility of fusion power reactor many of which components are to be used in strong magnetic field [3], [4]. These components will be loaded by Lorentz force which is a result of the interaction between magnetic field and an induced current [5], [6]. If a ferromagnetic material is used in the magnetic field, it would deform due to the magnetic stress resulting from an interaction between magnetic induction $\underset{\sim}{B}$ and induced magnetized moment $\underset{\sim}{M}$ [7], [8]. Moon [7] showed the experimental evidence of magnetoelastic buckling of a ferromagnetic plate inserted in a steady magnetic field.

Analytical solutions for the specific problems as to the eddy current were presented by some investigators. Dodd [9], [10] showed analytical solutions for axisymmetrical problems of the eddy current. One of these solutions is for a coil above a semi-infinite conducting material covered with an another plate. Geometric complications appearing in three dimensions are particularly common in the analysis of eddy current which is induced in metallic components of a magnetic fusion power reactor. Several studies associated with the eddy current calculation [1], [12], are presented.

On the other hand structural analysis of superconducting magnets is also essential because of huge electromagnetic forces acting on them. In the structural analysis the magnetic stiffness, which is a natural consequence of perturbed magnetic energy according to a mutual inductance change caused by its deformation, is a key concept determining a critical current for the magnet to buckle and a decrease of frequence with a current increase. Experimental evidence of the magnetomechanical instability of the toroidal coils was first demonstrated by Moon [13]. Miya and his co-workers have carried out the systematic studies on the toroidal coil, the solenoidal coil and the helical coil from a viewpoint of experimental verification of the theory of magnetomechanics [14-18].

In the present paper the theoretical background is first described for the eddy current problem and then the magnetomechanical behavior of superconducting magnets and second the numerical results are demonstrated together with the experimental results.

2. Theoretical consideration on magnetomechanics of fusion reactor components

2.1 Field equations for eddy current

One of underlying difficulties in obtaining general numerical solutions of electromagnetic field problems is a choice of variables with which the field is expressed. However, adoption of a current vector potential could simplify a mathematical treatment and an introduction of boundary conditions when obtaining numerical solutions for the eddy current. As described by Maxwell equations, the current and the magnetic flux are linked with each other as follows.

$$\mathrm{rot}\ \underset{\sim}{H} = \underset{\sim}{J} \qquad (1) \qquad\qquad \mathrm{rot}\ \underset{\sim}{E} = -\partial \underset{\sim}{B} / \partial t \qquad (2)$$

In the eddy current problems, interactions of the field intensity and the magnetic induction, the current flow and the electric field are controlled by the material properties of a conductor as expressed by

$$\underset{\sim}{B} = \mu \underset{\sim}{H} \qquad (3) \qquad \text{and} \qquad \underset{\sim}{J} = \sigma \underset{\sim}{E} \qquad (4)$$

where

$\underset{\sim}{H}$ = magnetic field intensity	$\underset{\sim}{B}$ = magnetic induction
$\underset{\sim}{J}$ = current density	$\underset{\sim}{E}$ = electrical field intensity
μ = magnetic permeability	σ = electrical conductivity
h = thickness of shell	

From the condition that both flow vectors of $\underset{\sim}{B}$ and $\underset{\sim}{J}$ are continuous, divergences of them are zero,

$$\mathrm{div}\ \underset{\sim}{B} = 0 \qquad (5) \qquad\qquad \mathrm{div}\ \underset{\sim}{J} = 0 \qquad (6)$$

A so-called magnetic vector potential $\underset{\sim}{A}$, defined by equation, $\underset{\sim}{B} = \mathrm{rot}\ \underset{\sim}{A}$, gives a simple and convenient method for description of the transverse magnetic problem.

An analogous current vector potential V , defined by eq. (7), also gives a simple and convenient method for the transverse electric problem:

$$\underset{\sim}{J} = \mathrm{rot}\ \underset{\sim}{V} \qquad \text{or} \qquad \underset{\sim}{I} = \mathrm{rot}\ \underset{\sim}{\Psi} \qquad (7)$$

where $\underset{\sim}{I} = h\underset{\sim}{J}$ and $\underset{\sim}{\Psi} = h\underset{\sim}{V}$

The current flow field problem is described by eq. (8), which is obtained from eq. (2), (4) and (7):

$$\mathrm{rot}\ (\mathrm{rot}\ V) = -\sigma \partial \underset{\sim}{B} / \partial t \qquad \text{or} \qquad \mathrm{rot} \cdot \mathrm{rot}\ \underset{\sim}{\Psi} = -\sigma h\, \partial \underset{\sim}{B} / \partial t \qquad (8)$$

2.2 Finite element formulation in orthogonal curvilinear co-ordinates

Orthogonal curvilinear co-ordinates (u,v,w) are introduced to solve the transient differential eq. (8) in a domain represented with a three-dimensionally extended curved shell. Unit vectors $\underset{\sim}{u},\underset{\sim}{v},\underset{\sim}{w}$ are along u, v, w-axes, respectively, as shown in Fig. 1. The unit vector $\underset{\sim}{w}$ is always set perpendicular to the curved shell surface. Since the eddy current density $\underset{\sim}{J}$ always flows in the thin shell, the current vector potential defined by eq. (7) is also perpendicular to the surface as is given by

$$\underset{\sim}{V} = \underset{\sim}{w}\, V \quad \text{or} \quad \underset{\sim}{\Psi} = \underset{\sim}{w}\, \Psi \tag{9}$$

namely $\underset{\sim}{V} = (0,0,V)$. Here it can be proved that $\operatorname{div} \underset{\sim}{V} = \underset{\sim}{0}$, because u- and v-components of $\underset{\sim}{V}$ are zero and V is a function of u and v only.

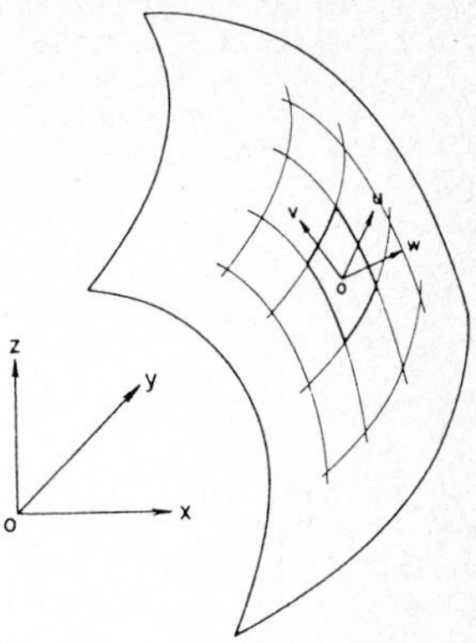

Fig. 1. Global and co-ordinates system

The magnetic induction in eq. (8) can be considered as a sum of magnetic inductions resulting from an external current source and the eddy current itself:

$$\underset{\sim}{B} = \underset{\sim}{B}_o + \underset{\sim}{B}_e \tag{10}$$

where $\underset{\sim}{B}_o$ is the magnetic induction produced by the external current, and $\underset{\sim}{B}_e$ the magnetic induction by the eddy current. The magnetic induction B_e can be calculated by the law of Biot-Savart as follows:

$$B_e = \frac{\mu}{4\pi h}\int \frac{\underset{\sim}{I}(r')\times\underset{\sim}{R}}{R^3}\, d\tau' \tag{11}$$

where $\underset{\sim}{R} = \underset{\sim}{r} - \underset{\sim}{r}'$, $R = |\underset{\sim}{R}|$

To integrate eq. (11), a derivative of 1/R in a source point is needed:

$$\nabla'(\frac{1}{R}) = \frac{\underset{\sim}{R}}{R^3} = (\frac{1}{h_1}\frac{\partial}{\partial u'}(\frac{1}{R}), \frac{1}{h_2 \partial}\frac{\partial}{v'}(\frac{1}{R}), \frac{\partial}{h_3\partial w'}(\frac{1}{R})) \tag{12}$$

where h_1, h_2 and h_3 are norms and $h_3 = 1$. And from eq. (7) and (9),

$$\underset{\sim}{I} = \frac{\underset{\sim}{u}}{h_2}\frac{\partial\Psi}{\partial u'} - \frac{\underset{\sim}{v}}{h_1}\frac{\partial\Psi}{\partial u'} \tag{13}$$

A vector product of $\underset{\sim}{J}(r')$ and $\underset{\sim}{R}/R^3$ is given by

$$I(r') \times \frac{\underset{\sim}{R}}{R^3} = \underset{\sim}{u}\,[-\frac{1}{h_1}\frac{\partial\Psi}{\partial u'}\frac{\partial}{\partial w'}(\frac{1}{R})] + \underset{\sim}{v}\,[-\frac{1}{h_2}\frac{\partial\Psi}{\partial v'}\frac{\partial}{\partial w'}(\frac{1}{R})]$$
$$+ \underset{\sim}{w}\,[\frac{1}{h_2^2}\frac{\partial\Psi}{\partial v'}\frac{\partial}{\partial v'}(\frac{1}{R}) + \frac{1}{h_1^2}\frac{\partial\Psi}{\partial u'}\frac{\partial}{\partial u'}(\frac{1}{R})] \tag{14}$$

Components in u and v directions do not give any contribution to the eddy current and thus the third term in equation (14) must be substituted in eq. (11). The normal component of $\underset{\sim}{B}_e$ is given by

$$B_w(\underset{\sim}{r}) = \frac{\mu}{4\pi h}\int [\frac{1}{h_1^2}\frac{\partial\Psi}{\partial u'}\frac{\partial}{\partial u'}(\frac{1}{R}) + \frac{1}{h_1^2}\frac{\partial\Psi}{\partial v'}\frac{\partial}{\partial v'}(\frac{1}{R})]d\tau'$$
$$= \frac{\mu}{4\pi h}\int \nabla'\ \Psi\cdot\nabla'(\frac{1}{R})d\tau' \tag{15}$$

Applying Gauss's divergence theorem and Dirac's δ-function to eq. (15) and considering that contribution from the ends of the shell to the integral in eq. (15) is very small, eq. (15) reduces to

$$B_w(\underset{\sim}{r}) = \frac{\mu\Psi}{h} + \frac{\mu}{4\pi h}\left[\iint_{top} \Psi(u',v')\frac{\partial}{\partial w'}\left(\frac{1}{R}\right)ds' - \iint_{bot} \Psi(u',v')\frac{\partial}{\partial w'}\left(\frac{1}{R}\right)ds'\right] \tag{16}$$

In general the second term must be included for the exact distribution of the eddy current. However, it should be noted that it may be neglected for the low and intermediate frequency range. Substituting eq. (16) in eq. (10) and then the result in eq. (8) we obtain,

$$\sigma\mu\frac{\partial\Psi}{\partial t} - \frac{1}{h_1h_2}\left\{\frac{\partial}{\partial u}\left(\frac{h_2}{h_1}\frac{\partial\Psi}{\partial u}\right) + \frac{\partial}{\partial v}\left(\frac{h_1}{h_2}\frac{\partial\Psi}{\partial v}\right)\right\}\frac{\partial}{\partial v}\left(\frac{h_1}{h_2}\frac{\partial\Psi}{\partial v}\right)\} + \sigma h\frac{\partial B_{on}}{\partial t} +$$
$$\frac{\sigma\mu}{4\pi}\left\{\iint_{top}\frac{\partial\Psi(u',v')}{\partial t}\frac{\partial}{\partial w'}\left(\frac{1}{R}\right)ds' - \iint_{bot}\frac{\partial\Psi(u',v')}{\partial t}\frac{\partial}{\partial w'}\left(\frac{1}{R}\right)ds'\right\} = 0 \tag{17}$$

where B_{on} is normal component of $\underset{\sim}{B}_o$. Considering that a variation of $\underset{\sim}{\Psi}$ could be a weighted residual when applying Galerkin's method to eq. (17) and integrating it by parts, we obtain,

$$\iint\left[\sigma\mu\, h_1h_2\frac{\partial\Psi}{\partial t}\delta\Psi + \frac{h_2}{h_1}\frac{\partial\Psi}{\partial u}\frac{\partial\Psi}{\partial u} + \frac{h_1}{h_2}\frac{\partial\Psi}{\partial v}\frac{\partial\Psi}{\partial v} + \frac{\sigma\mu h_1h_2}{4\pi}\left\{\iint_{top}\frac{\partial\Psi'}{\partial t}\frac{\partial}{\partial w'}\left(\frac{1}{R}\right)ds' -\right.\right.$$
$$\left.\left.\iint_{top}\frac{\partial\Psi'}{\partial t}\frac{\partial}{\partial w'}\left(\frac{1}{R}\right)ds'\right\} + \sigma\, h_1h_2\frac{\partial B_{on}}{\partial t}\delta\Psi\right]dudv = 0 \tag{18}$$

With a shape function $[N]$ we can express $\Psi(u,v)$ in terms of a nodal vector $\{\Psi^e\}$ like,

$$\Psi = [N]\{\Psi^e\} \qquad \text{and} \qquad \delta\Psi = [N]\{\Psi^e\} \tag{19}$$

Introducting eq. (19) in eq. (18) and then omitting the term $\delta\{\Psi^e\}$, we obtain

$$\{[P] + [\Omega]\}\left\{\frac{\partial\Psi}{\partial t}\right\} + [Q]\{\Psi\} = -\{R\} \tag{20}$$

where $[P] = \iint \sigma\mu\, h_1h_2\, [N]^T[N]dudv$, $\{R\} = \iint \sigma h\, h_1h_2\, [N]^T \frac{\partial B_{on}}{\partial t} dudv$

$$[Q] = \iint \left(\frac{h_2}{h_1}\frac{\partial[N]^T}{\partial u}\frac{\partial[N]}{\partial u} + \frac{h_1}{h_2}\frac{\partial[N]^T}{\partial v}\frac{\partial[N]}{\partial v}\right)dudv$$

$$[\Omega] = \iint \frac{\sigma\mu}{4\pi} h_1h_2\, [N]^T \left\{\iint [N'](L^+-L^-)h_1'h_2'\, du'dv\right\}dudv$$

and $L^+ = \frac{\partial}{\partial w'}\left(\frac{1}{R}\right)\Big|_{w=o,\ w'=\frac{h}{2}}$ $\qquad L^- = \frac{\partial}{\partial w'}\left(\frac{1}{R}\right)\Big|_{w=o,\ w'=\frac{h}{2}}$

An evaluation of teh matrix $[\Omega]$ is important but complicated and thus its evaluation is performed numerically. The stream function Ψ is obtained from subsequent solutions of eq. (20) and the eddy current is deduced from eq. (7).

2.3 Basic equations for magnetomechanics of superconducting magnets

Every type of electromagnetic force acting on the toroidal coils can be deduced from the magnetic energy stored in the coil system as stated in the theory of electricity and magnetism. To relate the magnetic energy to the force and thereby to derive the basic equations the energy method based on the Lagrangian is fully helpful. The Lagrangian L is constructed like

$$L = T - U + W \ , \tag{21}$$

where T is the kinetic energy, U is the elastic strain energy and W is the magnetic energy. They are expressed by

$$T = \frac{1}{2}\sum_i \iint \rho \left(\frac{\partial w}{\partial t}\right)^2 h dxdy \ , \quad U = \frac{1}{2}\sum_i \iint \{\varepsilon\}^T\{\sigma\}dxdy \ ,$$

$$W = \sum_{i \neq m}\sum \iint \underset{\sim}{A}\cdot\underset{\sim}{J}\ hdxdy \ , \tag{22}$$

where h = thickness of coil, ρ = equivalent density of coil,
w = lateral deflection of coil, $\{\varepsilon\}$ = generalized strain,
$\{\sigma\}$ = generalized stress, A = vector potential
$\underset{\sim}{J}$ = current density .

Here the coil is treated like a plane and if so, an application of the theory of plate bending is possible. Integrations and summations in eq. (22), therefore, are carried out in the coil plane and over all the toroidal coils. The double summation in eq. (22), whereas is not included the energy due to a self-inductance, corresponds to a mutual inductance calculation which is a sum of each magnetic energy stored in any two of the toroidal coils. The following Lagrange's equation must be satisfied for an electromechanical system in general,

$$\frac{d}{dt}\left(\frac{\partial L}{\partial \dot{w}}\right) - \frac{\partial L}{\partial w} = F \ , \tag{23}$$

where $\dot{w} = \partial w/\partial t$ and F is an external force acting on the coil. A substitution of eq. (22) in eq. (21) and then eq. (21) in eq. (23) yields

$$\sum_i \iint \rho\ddot{w}h\ dxdy + \sum_i \frac{\partial}{\partial w}\left[\frac{1}{2}\iint\{\varepsilon\}^T\{\sigma\}dxdy\right] - \sum_i \frac{\partial}{\partial w}\left[\sum_m \iint \underset{\sim}{A}\cdot\underset{\sim}{J}\ hdxdy\right] = \sum F_i \ . \tag{24}$$

This is the equation of motion of the toroidal field coils. The first and second terms in eq. (24) are independent of the relative displacements between coils while the third term is coupled with movements of the rest of the coils. A physical consideration of the magnetomechanical phenomena common in most case can allow us to classify the problems into two cases. The first case represents the typical problem such that the power supply provides an electrical energy to the coil system to satisfy a constant voltage condition. In this case the current J is maintained constant and eq. (24) can be treated under the constant current condition. The second case stands for a persistent mode of operation of the superconducting magnets where a voltage and an electrical resistance are zero. In this case, as is well-known, the magnetic flux is conserved for any motion and deformation of the coil. It should be noted here that only the first case of the constant current condition is discussed here, and furthermore that despite of the constant current condition a magnetic energy perturbation due to the directional change of the current density vector, $\underset{\sim}{J}$, must be taken into consideration.

2.4 Finite element implementation of magnetic stiffness with general expressions for the perturbation of magnetic energy

The perturbation of the magnetic energy is caused by a translational motion and deformation of the current carrying elastic body. For its evaluation, the following expression for the magnetic energy is used.

$$W = \iint \underset{\sim}{A}\cdot\underset{\sim}{I}\,dxdy\ , \qquad (25)$$

where A and I is the vector potential and the current density vector per unit width in the elastic body, respectively. The perturbation of the magnetic energy is generally given by

$$\Delta W = \iint \Delta\underset{\sim}{A}\cdot\underset{\sim}{I}\,dxdy + \iint \underset{\sim}{A}\cdot\Delta\underset{\sim}{I}\,dxdy\ . \qquad (26)$$

The directional change of the current density vector, $\Delta\underset{\sim}{I}$, is generally not zero even when its absolute is kept constant. The third term of eq. (24) is dependent on relative displacements between the conductors. The current-carrying elements are shown in Fig. 2 together with notations used in the following. Two Cartesian coordinates systems denoted by $x_i^{(1)}$ and $x_i^{(2)}$, (i=1,3), are fixed to the respective finite elements. Vectors $\underset{\sim}{R}_0$ and $\underset{\sim}{r}$ are defined by

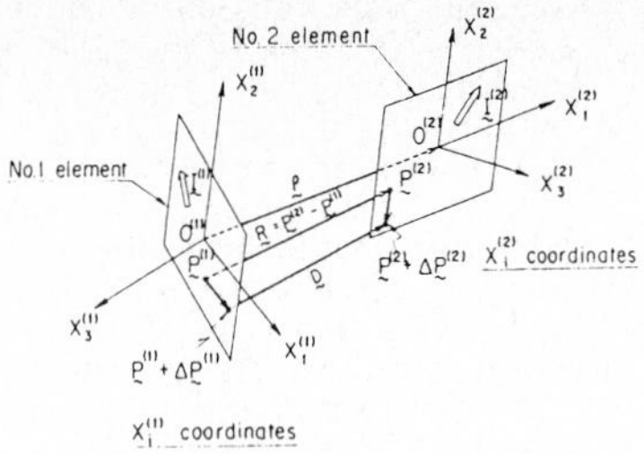

Fig. 2. Defintion of two Cartesian coordinates and current carrying finite elements

$$\underset{\sim}{R}_0 = (\bar{x}_i^{(1)} - x_i^{(1)}) \qquad , \qquad \underset{\sim}{r} = (\bar{u}_i^{(1)} - u_i^{(1)})\ . \qquad (27)$$

$a_i^{(k)}$ is defined like

$$a_i^{(1)} = x_i^{(1)} - T_{ij}\cdot x_j^{(2)} - \rho_i\ ,\quad a_i^{(2)} = x_i^{(2)} - T_{ji}\cdot x_j^{(1)} + T_{ji}\cdot\rho_j\ . \qquad (28)$$

where $\underset{\sim}{\rho}$ is a position vector stretching to the origin $0^{(2)}$ from $0^{(1)}$ as shown in Fig. 2. And a distance after deformation, D, is given by,

$$\frac{1}{D} = \frac{1}{R_0} - \frac{1}{2R_0^3}(2a_i^{(k)}\cdot u_i^{(k)} + u_i^{(k)}\cdot u_i^{(k)} - 2T_{ij}\cdot u_i^{(1)}\cdot u_j^{(2)}) + \frac{3}{2R_0^5}(a_i^{(k)}\cdot a_j^{(\ell)}\cdot u_i^{(k)}\cdot u_j^{(\ell)})\ . \qquad (29)$$

Let the magnetic energy generated by the two current carrying elements be denoted with W_{21}. And then it is given by

$$W_{21} = \frac{\mu_0}{4\pi}\iint \underset{\sim}{I}^{(1)}dS^{(1)}\iint \frac{\underset{\sim}{I}^{(2)}}{D}dS^{(2)}\ , \qquad (30)$$

where $dS^{(m)} = dx_1^{(m)}dx_2^{(m)}$, $\underset{\sim}{I}^{(m)} = I^{(m)}(\Psi_1^{(m)},\Psi_2^{(m)},\Psi_3^{(m)})$, (m=1,2)

The contracted notation is not applied.
This is an expression for the magnetic energy stored in the two finite elements in Fig. 2 after deformation. The first term $1/R_0$ in the above integrand corresponds to the magnetic energy before the deformation. The second term of eq. (26) can be evaluated with a calculation of the directional change of the current $\Delta\underset{\sim}{I}$. A system matrix equation deduced from eq. (23) is after all given like,

$$[M]\{\ddot{q}\} + \{[K_e] - [K_m]\}\{q\} = \{F\} \qquad (31)$$

where [M], $[K_e]$ and $[K_m]$ are the mass, elastic stiffness and magnetic stiffness matrices, respectively. As stated previously, $[K_m]$ is a potential source of a strong interaction between the deformation of current carrying elastic body and the magnetic field.

3. Discussions on numerical and experimental results

3.1 Numerical solutions for eddy current distribution

First wall and blanket structures of a fusion reactor are exposed to rigorous sources of the eddy current which are caused by changes of a great amount of plasma current, and of huge currents flowing in toroidal and poloidal field coils. There are two cases for the plasma current change, which are a rapid motion and a quick disruption. The quick disruption of the plasma current, called a major plasma disruption, induces an eddy current in the first wall and blanket structures. In Fig. 3 is shown a schematic distribution of eddy current in a one-sixth sector of a vacuum vessel with its development. The plasma disruption is assumed to be given by $I(t)=Io\ \exp(-t/\tau)$ where Io is 5.0×10^{6} Amps and τ is a decay constant. In Fig. 4 are shown changes of eddy current at a point " B " where the decay constant τ is a parameter. The eddy current becomes larger as the decay constant does shorter. In Fig. 5 is shown an experimental apparatus for a measurement of eddy current distribution. The eddy current is induced in a stainless steel circular plate by a pulsed current of a solenoidal coil from a condenser bank. For a detection of the current a pick-up coil is inserted into the plate. The experimental results are plotted in Fig. 6 with the numerical results. The numerical results represent magnetic inductions for a comparison because signals from the pick-up coil are converted into the magnetic inductions. An agreement between the experimental and numerical results is fairly good in a range of $r<25$ mm. In a region beyond the radius of 25 mm the agreement is not good. A reason for this discrepancy could originate in a misalignment of the pick-up coil to the circular plate.

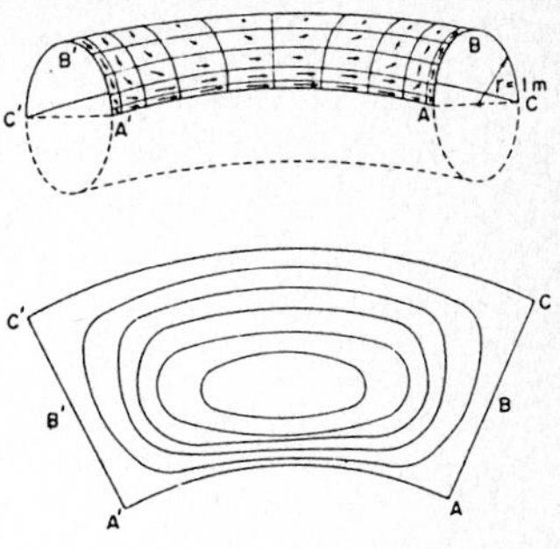

Fig. 3. Distribution of eddy current density at $t=20.0$ msec ($\Delta t=0.5$ msec, $\tau=10.0$ msec, $\eta=\frac{1}{3}$)

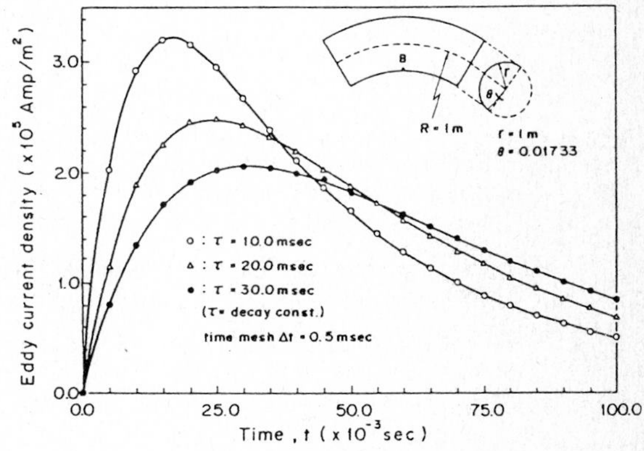

Fig. 4. Changes of eddy current density J at point $B(\eta=\frac{1}{3})$

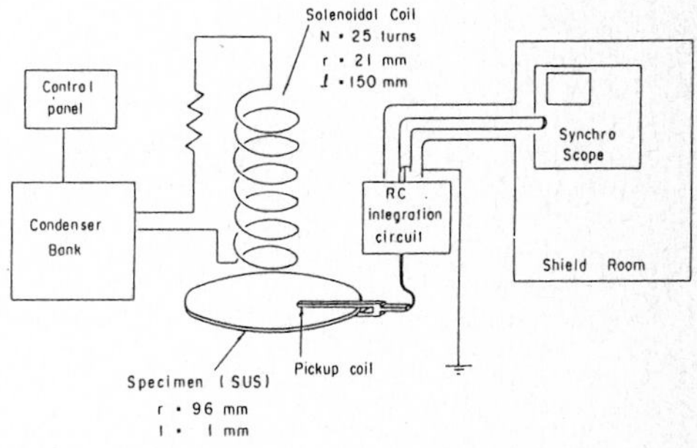

Fig. 5. Experimental apparatus

A bellows structure has a possibility to be applied to a fusion reactor to serve as a component of higher electrical resistance. There are many examples of its application to experimental devices for plasma confinement study. A bellows shown in Fig. 7 is attached to sectored thick vacuum vessels at both ends to get one-turn electrical resistance higher to a desirable value. Numerical results of the stress analysis is shown in Fig. 8. Elastic deformation of the bellows is caused by an electromagnetic force due to an interaction between the toroidal field and a saddle shaped eddy current. The saddle shaped current is generally induced when there exists a larger electrical resistance between the sectored vacuum vessels where only a small portion of the eddy current can pass through the bellows. The stress analysis was performed with use of the finite element code accounting for axisymmetric geometry and non-symmetric loading in combination with the eddy current code. As is shown in Fig. 8, a bending stress in a meridional direction σ_s^b is the biggest and all the membrane stresses are very small.

3.2 Numerical solutions for magnetomechanical behavior of superconducting magnets

The author and his co-workers have carried out many experiments on the magnetomechanical behavior of various types of magnets [18]. Here are shown numerical and experimental results of three coil partial torus. A test coil sits at a centre between two side coils and a vibration exciting coil is installed inside the test coil to cause it to vibrate. Those coils were made with use of superconducting wires because use of normal conductor can not afford enough high current density necessary for a generation of larger electromagnetic force. An example of the vibration of the coil in a magnetic field generated by the side coils is shown in Fig. 9 where the numerical results are compared with the experimental one. FEM-FLT is a name of the computer code [17] where an integration of eq. (31) in a time is performed with use of the Fast Laplace Transform algorithm [19]. Since the magnetic stiffness [Km] contains a squared coil current, then is expected a linear relationship between a squared frequency and the squared coil current. A validation of the conjecture is proved by a solid line and blank points from the experiment in Fig. 10. The current for a zero frequency corresponds to a buckling current and in this case it is 46.2 Amp/turn which is very close to the measured value of 47.8 Amp/turn from the buckling test. Solid circles in Fig. 10 were obtained from the plastically deformed coil and the frequency is smaller due to a decrease of elastic rigidity. A reason for the magnetomechanical buckling and the decrease of frequency with an increase of the coil current can be easily found for the toroidal coil array at least. It is an existence of the magnetic stiffness in a mathematical

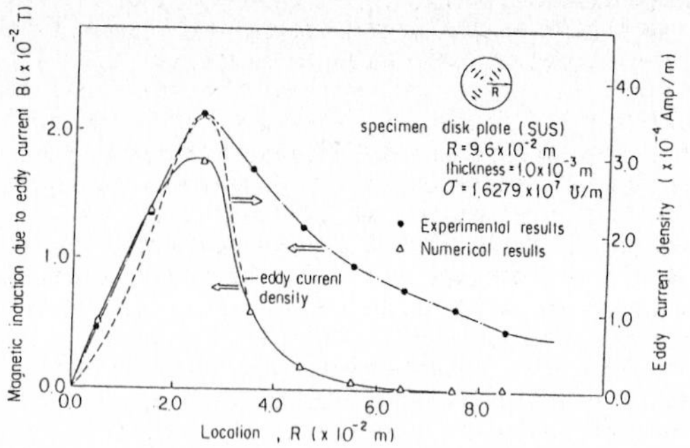

Fig. 6. Distribution of magnetic induction due to eddy current at $t=60\times10^{-6}$ sec

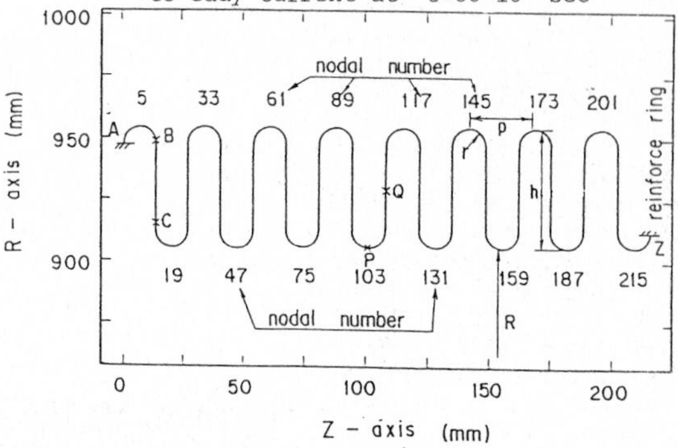

Fig. 7. Configuration of bellows

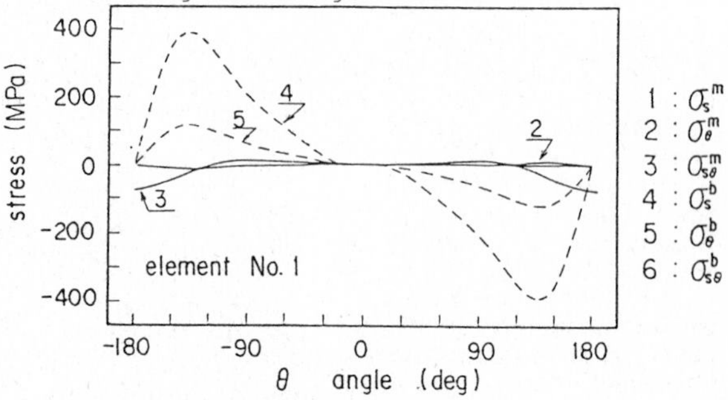

Fig. 8. Stress distributions due to saddle shaped force

sense or a balanced force acting on the coil in a physical sense. Then a question occurs as to a possibility of the magnetoelastic buckling for the helical coil. A conclusion is such that the instability of the double helical coil depends on its geometry, mainly on its pitch. An experimental apparatus of the double helical coil is shown in Photo. 1, which is remarked with apparatus B in Fig. 11. The apparatus A is almost the same one with B but transient electromagnetic force induced by vibration exciting coils is basically different. Blank circles from the experiment in Fig. 11 demonstrat that the linear decrease of the squared frequency with an increase of the squared current holds true in the double helical coils of 3.5 pitch. In some other cases such as the second mode of 1.0 pitch, the coil is stable, namely the frequency increase is assured for the increase of the current.

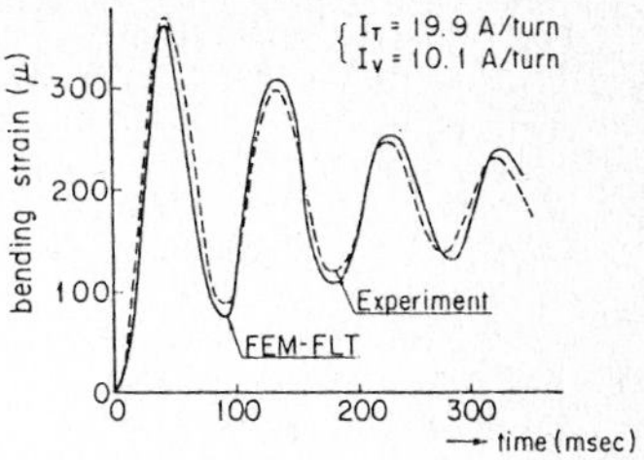

Fig. 9. Comparison of experimental results with numerical ones

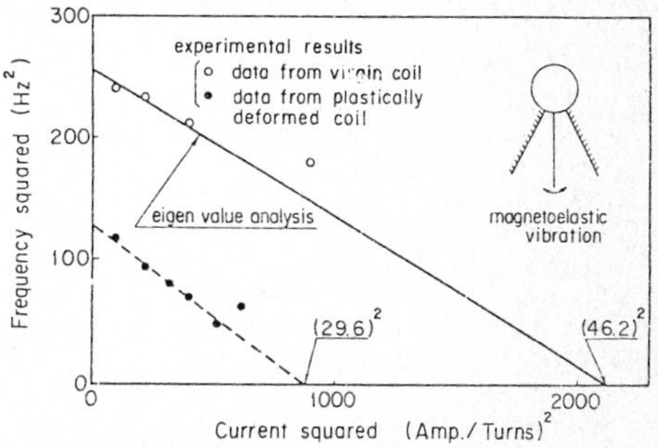

Fig. 10. Frequency-current dispersion curves for three coil partial torus

It is understood from eq. (31) that the magnetic stiffness could play an important role in a static deformation of the current carrying coil. The deformation is determined by the equation {[Ke]-[Km]}{q} = {F} from eq. (31) omitting the inertia term. A conventional magnet stress analysis is performed with an omission of the [Km] term Its result might not be accurate if a design current of the coil is more than half the buckling current. An experimental verification of the effect of the magnetic stiffness is clearly demonstrated in Fig. 12. If the magnetic stiffness does not effect the static deformation, the experimental results marked with blank circles should be on the line without the magnetic stiffness. Bending strains from the numerical analysis including the magnetic stiffness are shown with a dotted curve. This line approaches to an asymptote given by the theoretical value I_C=29.1 Amps. The buckling current from the experiment is 31.7 Amps and if the numerical values is corrected so as to give the buckling current, 31.7 Amps, the theoretical values agrees completely with the experimental values. This results clearly indicates that the stress analysis of the magnets must include the magnetic stiffness.

Acknowledgement

The author would like to express his sincere gratitude to Mr. K. Someya for his assistance in the experiments and to Mr. M. Uesaka, Mr. S. Hanai and

Mr. T. Takaghi for their cooperative work on experiments and numerical analyses.

References

[1] Paria, G., Proceedings of Vibration Problems, Warsaw, Vol.5, Vol.1 (1964)
[2] Kaliski, S. and Rogula, D., ibid, Vol.2, No.1 (1961)
[3] Cain, W.D., Proc. of 6-th Symp. on Engineering Problems of Fusion Research, San Diego (1975)
[4] Pardue, R.M. and Johnson, Jonson, N.E., ibid.
[5] Miya, K., An, S. and Ando, Ando, Y., ibid.
[6] Miya, K., Takagi, T. and Tabata, Y., Proc. of 7-th Symp. on Engineering Problems of Fusion Research, Knoxville (1977)
[7] Moon, F.C. and Pao, Y.H., ASME J. Appl. Mech., Vol.35, Series E, (1968)
[8] Popelar, C.H. and Bast, C.O., Experimental Mech., Vol.12, No.12 (1972)
[9] Dodd, C.V. and Deeds, W.E., J. Appl. Phys. Vol.39 (1968)
[10] Dodd, C.V., Cheng, C.C. and Deeds, W.E., J. Appl. Phys. Vol.45, 2 (1974)
[11] Yeh, H.T., Proc. 7-th Symp. on Engineering Problems of Fusion Research, Knoxville (1977)
[12] Kameari, A. and Suzuki, Y., ibid.
[13] Moon, F.C., J. Appl. Phys., Vol.47, No.3 (1976)
[14] Miya, K. and Takagi, T., ASME J. Pressure Vessel Tech., 100 (1978)
[15] Miya, K., Takagi, T. and Uesaka, M., ASME J. Mech. Des., 41 (1980)
[16] Miya, K., Takagi, T., Uesaka, M. and Someya, K., ASME J. Appl. Mech. 49 (1982)
[17] Miya, K., Uesaka, M. and Moon, F.C., ASME J. Appl. Mech., 49 (1982)
[18] Miya, K. and Uesaka, M., Nucl. Eng. Des., 72 (1982)
[19] Cooley, J.W. and Tukey, J.W., Math. Comp., Vol.19 (1965)

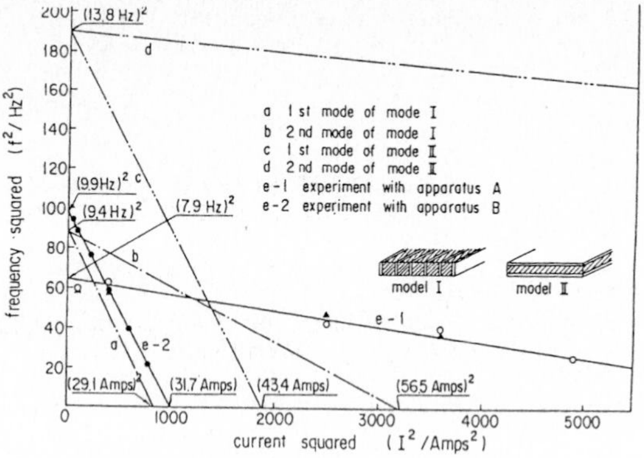

Fig. 11. Frequency-cuurent dispersion curves for 3.5 pitch straight type double helical coil

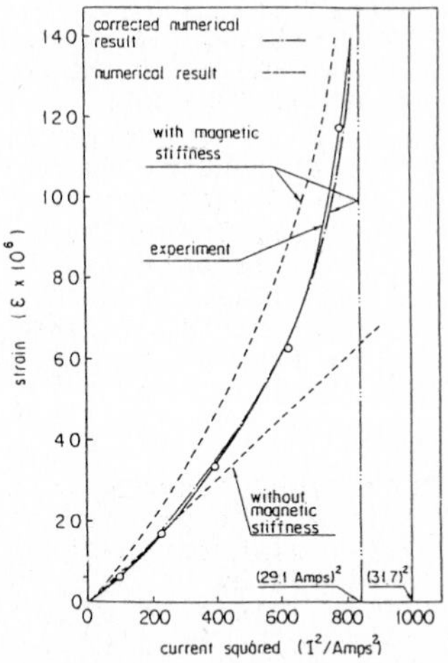

Fig. 12. Effect of magnetic stiffness on static deformation of the helical coil (apparatus B)

The Mechanical Behavior of Electromagnetic Solid Continua
G.A. Maugin (editor)
Elsevier Science Publishers B.V. (North-Holland)

ABOUT ELECTROMAGNETIC LOAD, STRESS STATE OF CURRENT-CARRYING TOROIDAL SHELL

R.N. Ovakimian

Chief of Department, Mechanics Institute of
Armenian Academy of Sciences
Yerevan, Armenia
USSR

The electromagnetic loads which arise at interaction of toroidal coil's current with its own magnetic field are approximately determined. The values of current density and magnetic field intensity at any point of tore cross-section are obtained. By means of shell's membrane theory the stress state of tore is determined. It is shown that due to the electromagnetic load in the shell both of tensile meridional and compressible toroidal strengths arise. This state differs from case p=const., when one has only tensible strength. The engineering formulas, to determine the strengths are given. The numerical results are given in the case of superconducting Tokamak-7.

It is known that Tokamak is used as fusion reactor [1]. In the first Tokamaks the problem of stabilization and plasma retention from contact with tore walls, has been investigated. This problem has been solved by using superposition of different magnetic fields. Among these fields the toroidal one is the most important. Toroidal field is created by current-carrying meridional coil. The superconducting coil is used for increasing intensity of the magnetic field. In the present time when the demonstrative Tokamak is being designed, the engineering problems have great importance. Here first of all, we should mention the strength calculations of the toroidal shell.

1. Let the constant electric J = const flow on toroidal winding of coil. The coil is covered by coaxial toroidal shell of large radius a and small radius R (Fig.1). In the coaxial clearance the cryogenic liquid may be flowing. The radii of curvature of shell middle surfaces with unit-vectors $\vec{e}_1, \vec{e}_2, \vec{e}_n$ are

$$R_1=\frac{a}{\sin\varphi}+R=a\,\frac{1+\alpha\sin\varphi}{\sin\varphi}, \quad R_2=R \tag{1}$$

where the coefficient $\alpha=R/a$ characterises the tore's geometry ($0<\alpha<1$). Since the shell is closed, $0\leqslant\theta<2\pi$, $0\leqslant\varphi<2\pi$.

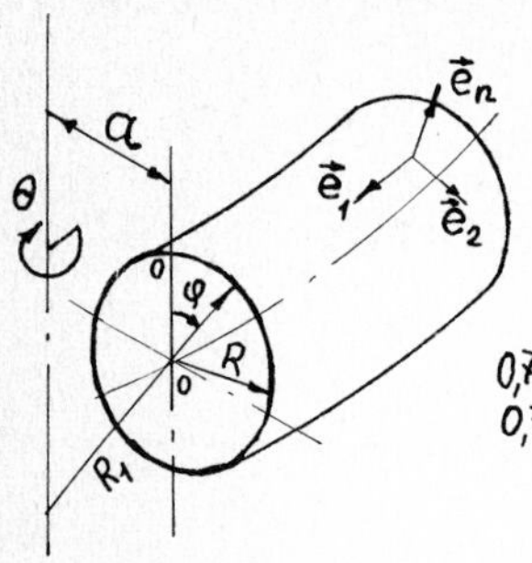

Fig. 1

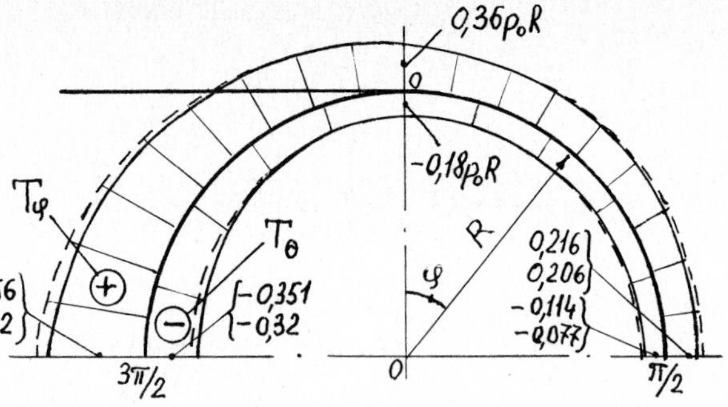

Fig. 2

Let the coil with small pitch wind uniformly on the tore. Then on an unit of lengths of tores arc in toroidal direction we have the number of winding

$$n = \frac{N}{2\pi a\,(1+\alpha \sin\varphi)} \qquad (2)$$

where N is the total number of coil windings. In the case of compact (without clearence) winding on tores inner circumference of radius $a-R$ we have relation $N\cdot\delta = 2\pi(a-R)$, where δ is the thickness of wire. Then taking into account (2) we have

$$n = \frac{1}{\delta}\,\frac{1-\alpha}{1+\alpha \sin\varphi} \qquad (3)$$

Henceforth we consider this case of compact winding. Approximately we consider that the current has only meridional direction (though, of course, the toroidal currents always take place). Assuming the cross-section of wire to be small in comparison with tore's cross-section, according to [2] the current is supposed to be surface one

$$\vec{i} = nJ\vec{e}_2 = \frac{J}{\delta}\,\frac{1-\alpha}{1+\alpha \sin\varphi}\,\vec{e}_2 \qquad (4)$$

In this case in [3] for tore's magnetic field the relation has been obtained

$$\vec{H} = -H_\theta \vec{e}_1 = -i(a)\,\frac{a}{r}\,\vec{e}_1 \qquad (5)$$

where r is the distance, between the axis of rotation and the points of tore's body, $i(a)$ - surface's density of current at $r=a$. On tore's surface $r_s = a(1+\alpha \sin\varphi)$. At the points $\varphi=0$, $\varphi=\pi$ $i(a) = J\frac{1-\alpha}{\delta}$. Then, according to (5)

$$\vec{H}_s = -\frac{J}{\delta}\,\frac{1-\alpha}{1+\alpha \sin\varphi}\,\vec{e}_1 \qquad (6)$$

Electromagnetic pressure on toroidal shell's surface is determined by relation

$$\vec{p} = \vec{i} \times \vec{B}_s \tag{7}$$

where in S.I.-units, $\vec{B}_s$ is magnetic induction, μ is relative magnetic permeability of the medium, $\mu_0 = 4\pi \cdot 10^{-7}\ N/A^2$ magnetic permeability of vacuum.
In the (7), using (4), (6) we shall obtain the expression for vector pressure

$$\vec{p} = p_0 \frac{(1-\alpha)^2}{(1+\alpha \sin\varphi)^2} \vec{e}_n \tag{8}$$

which has the direction of the normal $\vec{e}_n$ to the shell surface, where $p_0 = \mu\mu_0 (J/\delta)^2$.

According to (8), the electromagnetic pressure is distributed continuously along the shell surface. Actually in toroidal shell there are, of course, regions without any current. But our task is the qualitive investigation of current-carrying shell, so we suppose that pressure is "spread" along the shell surface [3]. On the inner circumference of the tore, where $\sin\varphi = -1$, pressure is maximal, at the distant circumference, where $\sin\varphi = 1$, pressure is minimal. When ratio α is decreased, the difference between pressure values is also decreased. In the limit when $\alpha \to 0$ $(\alpha \to \infty)$ $p \to p_0$.

We can mention that the load is symmetrical at upper and lower parts of circumference of radius R

2. For toroidal shell the exact solution of shell equilibrium's differential equations entails with great mathematical difficulties, even in the case of constant internal pressure. That is why there are few works concerning the theory of toroidal shells. The importance of investigations of Tokamak may change it. By now we can mention works of F. Moon [4,5] where the stability of superconducting toroidal rings used in Tokamak has been investigated.

Without prejudice to physical essence of problem the stress state of toroidal shell can be approximately described by means of membrane's theory of shell [6]. Such as the external load does not depend angle, the tangential strength $S=0$. Then equations of equilibrium of shell are

$$\left.\begin{aligned} \frac{1}{A_1 A_2}\left(\frac{d A_1 T_\varphi}{d\varphi} - T_\theta \frac{dA_1}{d\varphi}\right) &= 0 \\ \frac{T_\theta}{R_1} + \frac{T_\varphi}{R_2} &= p_n \end{aligned}\right\} \tag{9}$$

where A_1, A_2 - Lame's parameter, $A_1 = R_1 \sin\varphi = a(1+\alpha\sin\varphi)$, $A_2 = R_2 = R$; T_θ, T_φ - strength along $\vec{e}_1, \vec{e}_2$ respectively.

From system (9) we shall obtain the lineary differential equations

$$\frac{dT_\varphi}{d\varphi} + T_\varphi \frac{1+2\alpha \sin\varphi}{1+\alpha \sin\varphi} ctg\varphi = p_o R \frac{(1-\alpha)^2}{(1+\alpha \sin\varphi)^2} ctg\varphi \quad (10)$$

General solution of (10) has the form

$$T_\varphi = p_o R (1-\alpha)^2 \frac{\ln(1+\alpha \sin\varphi) + C}{\alpha \sin\varphi (1+\alpha \sin\varphi)} \quad (11)$$

Let us use the condition of finiteness of values T_φ. At $\sin\varphi \to 0$ in (11) $\lim\limits_{\sin\varphi \to 0} \frac{\ln(1+\alpha \sin\varphi)}{\alpha \sin\varphi} = 1$ and so $C = 0$.

Substituting (11) in the second equation of (9) we obtain

$$T_\theta = -p_o R \frac{(1-\alpha)^2}{\alpha \sin\varphi} \left[\frac{\ln(1+\alpha \sin\varphi)}{\alpha \sin\varphi} - \frac{1}{1+\alpha \sin\varphi} \right] \quad (12)$$

Values T_θ also satisfyes the condition of finiteness at $\sin\varphi \to 0$, which can be easibly shown by removing the uncertainty of the expression (12). From expression (12) follows the interesting and unexpected result. Under action of the load of type (8) in the shell arises the compresible toroidal strength, while at p =const., strength T_θ always is tensible.

In the existing Tokamak $\alpha \leqslant 0{,}4$. It gives us possibility to simplify formulas. Within the errors of 5%, in the formulas (11),(12) we may restrict by two members of expansion of function

$$\ln(1+\alpha \sin\varphi) = \alpha \sin\varphi - \frac{\alpha^2 \sin^2\varphi}{2} + \ldots$$

$$T_\varphi = p_o R \frac{(1-\alpha)^2}{2} \frac{2-\alpha \sin\varphi}{1+\alpha \sin\varphi}, \quad T_\theta = -p_o R \frac{(1-\alpha)^2}{2} \frac{1-\alpha \sin\varphi}{1+\alpha \sin\varphi} \quad (13)$$

which is convenient for the engineering calculations.

In the Fig. 2 the values of strength are presented, which are calculated in the case α = 0,4. The dotted line correspond to forlumas (11), (12), the continuous line - to approximate formulas (13). For comparison with the expressions (13), let us bring corresponding formulas, obtained in [7], by means of membrane's theory for the tore at $p = p_o$ = const

$$T_\varphi = \frac{p_o R}{2} \frac{2+\alpha \sin\varphi}{1+\alpha \sin\varphi}, \quad T_\theta = \frac{p_o R}{2} = const. \quad (14)$$

Correctness of (11), (12) or (13) and also (14), may be proved mathematically. Let us write down $p_n = p_o \frac{(1-\beta)^2}{(1+\beta \sin\varphi)^2}$, where instead of α in (8) we have the other constant β. Solving the differential equation (10) with this p_n we obtain

$$T_\varphi = p_o R \frac{(1-\beta)^2}{1+\alpha \sin\varphi} \left[\frac{1-\alpha/\beta}{1+\beta \sin\varphi} + \alpha \frac{\ln(1+\beta\sin\varphi)}{\beta^2 \sin\varphi} \right] \tag{15}$$

When $\beta=\alpha$ value T_φ coincides with (11). When $\beta \to 0$ $p_n \to p_o$ and the value T_φ coincides with (14). The corresponding formulas can be obtained also for values T_θ. These computations show the important part of the load kind in the toroidal shell.

3. Let us determine strength (stress) in the tore for a case of Tokamak-7 [8] with superconducting winding. Here we have $a=1{,}22$m, $R=0{,}5$m, $\alpha=0{,}41$. In "T-7" when $J=4{,}8$ kA there is the maximal induction $B = 4$ T (apparently at the point $\sin\varphi=-1$). Assuming $\mu=1$, due to formula (8) we ontain at $\sin\varphi=-1$ $p_{max}=p_o=12{,}4$ MPa (123 atm); at $\sin\varphi=1$ $p_{min}=2{,}12$ MPa (21 atm), which is almost 6 times smaller than p_{max}.

The presenting data show that there are large electromagnetic forces acting in some parts of Tokamak. In the point $\sin\varphi=-1$; where the stress is maximal, $T_{\varphi max}=44{,}1\cdot10^5$ N/m, $T_{\theta max}=-25{,}8\cdot10^5$ N/m according to (13). Assuming thickness of shell $h=10^{-2}$m we obtain meridional stress $\sigma_{\varphi max}=441$ MPa (≈ 4500 kG/sm^2) and toroidal (compressible) stress $\sigma_{\theta max}=-258$ MPa (≈ -2630 kG/sm^2). Taking into account that breaking strength for some type of chilled steels is $\sigma \approx$ ≈ 11.000 kG/sm^2, we can conclude that electromagnetic loads bring to large mechanical stresses in the Tokamak.

Finally, the author wishes to thank pr.L.Agalovian, especially dr.M.Belubekian for helpful suggestions and also dr. K.Kazarian for discussing this paper.

REFERENCES

[1] Atomies science and technics in USSR. (Atomizdat,M.,1977).
[2] TAMM I.E. "Foundations of electricity theory"(Science,M., 1976).
[3] DOJNIKOV N.I. "Determination of magnetic field of current flowing on tore's surface" Zh.T.Ph., V.34, (1964),1769-1780.
[4] MOON F.C. "Experimental on Magnetoelastic Buckling in a Superconducting Torns." Journal of Applied Mechanics, v.46, (1979), 145-150.
[5] MOON F.C. "Buckling of a Superconducting Ring in a Toroidal Magnetic Field." Journal of Applied Mechanics, v.46,(1979), 151-155.
[6] NOVOZHILOV V.V. "Theory of thin shells" (Sudpromizdat,L.,1962).
[7] PHEODOSIEV V.I. "Selected problems and tasks of strength of materials" (Science, M., 1973).
[8] IVANOV D.P. "T-7" - the first tokamak with superconducting windings", Priroda, 12 (1978), 121-123.

The Mechanical Behavior of Electromagnetic Solid Continua
G.A. Maugin (editor)
Elsevier Science Publishers B.V. (North-Holland)

ELECTRICALLY CONDUCTING ORTHOTROPIC CYLINDRICAL SHELL IN AXIAL AND RADIAL MAGNETIC FIELDS

Emanuel S. Bobrov

Francis Bitter National Magnet Laboratory
Massachusetts Institute of Technology
Cambridge, Massachusetts
U.S.A.

The existing analytical solutions for stresses and displacements in solenoids are based on consideration of either plane-stress or plane-strain models. These do not always provide the necessary information concerning the three-dimensional mechanical behavior of the winding and its interaction with the surrounding structure, such as the coil form and its flanges. This information includes radial and axial displacements near the ends of the winding as well as the presence of shear stress and its magnitude.

This paper develops techniques for calculation of radially and axially dependent stresses and strains in superconducting solenoids possessing anisotropic mechanical and thermal properties. The method is based on modelling a solenoid winding subject to mechanical, thermal, and electromagnetic loads as an orthotropic cylindrical shell with appropriate boundary conditions.

INTRODUCTION

A superconducting solenoid magnet represents a nonhomogeneous anisotropic mechanical system the behavior of which, under applied loads, determines the mechanical performance of the magnet. Strains and stresses in the magnet are caused by conductor tension during the winding process, anisotropic differential contraction at cooldown, and radial and axial Lorentz forces.

The mechanical behavior of wound impregnated solenoids is essentially three-dimensional. The radial, circumferential, axial, and transverse shear stresses vary along both the radial and axial coordinates. The variation of these quantities is caused not only by the axial variation of the magnetic field components, but also by the influence of the boundary conditions at the ends of the winding.

The problem of stress analysis of solenoid windings has been addressed by numerous authors in more than 30 publications. A detailed review of most of these works can be found in reference [1]. Without exception, the mechanical behavior of solenoids is analyzed in those papers on the basis of a two-dimensional plane-stress consideration.

All the plane-stress models of solenoid windings use the same basic simplification that the transverse shear stress does not contribute to the mechanical behavior of the system. As has been demonstrated by Arp [2] who compared the numerical results generated by his two-dimensional analytical model with a three-dimensional finite element analysis, such simplification can lead to noticeable inaccuracies in the calculated stress and displacement distributions.

In this paper we propose to model a solenoid magnet subject to mechanical, thermal, and electromagnetic loads, as a thick orthotropic cylindrical shell. The equations of the shell are derived by minimizing the strain energy of the cylinder, after assuming a given distribution of stresses across the build of the winding, satisfying the equilibrium equations and the boundary conditions on the cylindrical surfaces.

BASIC EQUATIONS OF A ROTATIONALLY SYMMETRIC CYLINDRICAL SHELL

In the coordinate system (z,θ,ζ) associated with the middle surface of the cylinder shown in Figure 1 the stress equilibrium equations can be written in the following form

$$\frac{d\sigma_z}{dz} + \frac{1}{r}\frac{d}{d\zeta}\left(r\tau_{rz}\right) + F_z = 0 \tag{1}$$

$$\frac{d\tau_{rz}}{dz} + \frac{1}{r}\frac{d}{d\zeta}\left(r\sigma_r\right) - \frac{\sigma_\theta}{r} + F_r = 0 \tag{2}$$

where σ_z, σ_r, and σ_θ are, respectively, the axial, radial, and circumferential stress; τ_{rz} is the shear stress; F_z and F_r are the axial and radial components of the body force. As shown in Figure 1

$$r = R + \zeta, \tag{3}$$

where R is the radius of curvature of the middle surface of the cylinder.

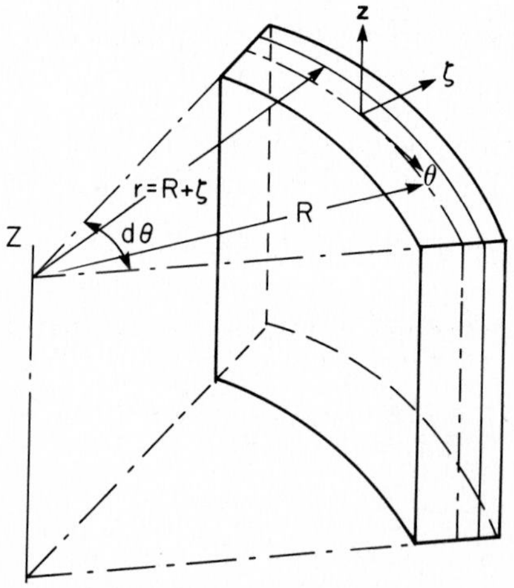

Figure 1

Coordinate system associated with the middle surface of a cylinder

We will assume a linear distribution of both the axial and radial components of the magnetic field, B_z and B_r across the build h of the winding. This yields

$$F_z = j\,B_r = a_z + b_z\zeta \tag{4}$$

$$F_r = j\,B_z = a_r + b_r\zeta \tag{5}$$

where j is the current density, and

$$a_z = j\,\frac{B_r^i + B_r^o}{2} \tag{6}$$

$$b_z = j\,\frac{B_r^o - B_r^i}{h} \tag{7}$$

$$a_r = j\,\frac{B_z^i + B_z^o}{2} \tag{8}$$

$$b_r = j\,\frac{B_z^o - B_z^i}{h} \tag{9}$$

The superscripts i and o denote, respectively, the inner and the outer surfaces of the cylinder.

On the surfaces of the cylinder

$$\zeta = \pm\frac{h}{2} \tag{10}$$

we will require the following boundary conditions to be satisfied

$$\sigma_r\left(\frac{h}{2}\right) = -p_o,\ \sigma_r\left(-\frac{h}{2}\right) = -p_i,\ \tau_{rz}\left(\pm\frac{h}{2}\right) = 0 \tag{11}$$

where p_i and p_o are the radial pressures applied respectively, to the inner and outer surfaces of the cylinder.

The following distribution of stresses σ_z, σ_θ, and τ_{rz} is assumed across the thickness of the cylinder

$$\sigma_z = \left(\frac{N_z}{h} + \frac{12\,M_z\zeta}{h^3}\right)\left(1 + \frac{\zeta}{R}\right)^{-1} \tag{12}$$

$$\sigma_\theta = \frac{N_\theta}{h} + \frac{12\,M_\theta\zeta}{h^3} \tag{13}$$

$$\tau_{rz} = \frac{3}{2h}\,Q\left(1 - \frac{4\zeta^2}{h^2}\right)\left(1 + \frac{\zeta}{R}\right)^{-1} \tag{14}$$

where N_z, N_θ, and Q are the stress resultants, and M_z and M_θ are the stress couples.

Substituting σ_z, σ_θ, and τ_{rz} from Eqs. (12) - (14) into the equilibrium equations (1) - (2) and integrating across the thickness, with the surface conditions (11), we obtain the following equilibrium equations

$$\frac{dN_z}{dz} + a_z h + \frac{b_z h^3}{12R} = 0 \tag{15}$$

$$\frac{dM_z}{dz} - Q + \frac{a_z}{R} + b_z\,\frac{h^3}{12} = 0 \tag{16}$$

$$\frac{dQ}{dz} - \frac{N_\theta}{R} + a_r h + \frac{b_r h^3}{12R} + \left(1 - \frac{h}{2R}\right)p_i - \left(1 + \frac{h}{2R}\right)p_o = 0 \tag{17}$$

We also arrive at the expression for the radial stress distribution as a function of ζ

$$\left(1 + \frac{\zeta}{R}\right)\sigma_r = -\frac{1}{2Rh}\left(1 - \frac{4\zeta^2}{h^2}\right)\left(N_\theta\zeta + 3M_\theta\right) +$$

$$+\frac{h}{4}\left(1 - \frac{4\zeta^2}{h^2}\right)\left[a_r\left(\frac{2\zeta}{h} + \frac{h}{2R}\right) + \frac{b_r h}{2}\left(1 + \frac{\zeta}{R}\right)\right] -$$

$$- \frac{1}{2} p_i \left(1 - \frac{h}{2R}\right)\left[1 - \frac{\zeta}{h}\left(3 - \frac{4\zeta^2}{h^2}\right)\right] -$$

$$- \frac{1}{2} p_o \left(1 + \frac{h}{2R}\right)\left[1 + \frac{\zeta}{h}\left(3 - \frac{4\zeta^2}{h^2}\right)\right] \tag{18}$$

Following the procedure given by Trefftz [3] and Reissner [4] the stress-strain relations are obtained by minimizing the strain energy of the cylinder expressed in terms of stress resultants and stress couples with equilibrium equations (15) -(17) as side conditions, using Lagrangean multipliers.

$$\varepsilon_{zm} = \frac{N_z - \nu_{\theta z} N_\theta}{E_z h} - \frac{M_z}{E_z hR} + \frac{\nu_{zr}}{E_r h}\left[\frac{M_\theta}{R} - \frac{b_r h^3}{12} + \frac{h}{2}\left(p_i + p_o\right)\right] - \delta_z , \tag{19}$$

$$\varepsilon_{\theta m} = \frac{N_\theta - \nu_{z\theta} N_z}{E_\theta h} + \frac{M_\theta}{E_\theta hR} + \frac{\nu_{zr} M_z}{5E_r hR} + \frac{\nu_{\theta r}}{E_r h}\left[\frac{6M_\theta}{5R} - \frac{b_r h^3}{12} + \frac{h}{2}\left(p_i + p_o\right)\right] - \delta_\theta , \tag{20}$$

$$\kappa_{zm} = \frac{12\left(M_z - \nu_{\theta z} M_\theta\right)}{E_z h^3} - \frac{N_z}{E_z hR} + \frac{\nu_{zr}}{E_r h}\left[\frac{1}{5}\,\frac{N_\theta}{R} - a_r h + \frac{6}{5}\left(p_i + p_o\right)\right], \tag{21}$$

$$\kappa_{\theta m} = \frac{12\left(M_\theta - \nu_{z\theta} M_z\right)}{E_\theta h^3} + \frac{N_\theta}{E_\theta hR} + \frac{\nu_{zr} N_z}{E_r hR} + \frac{\nu_{\theta r}}{E_r h}\left[\frac{6N_\theta}{5R} - \frac{a_r h}{5} + \frac{6}{5}\left(p_i + p_o\right)\right] -$$

$$- \frac{\delta_\theta - \delta_r}{R} , \tag{22}$$

$$\gamma_{rzm} = \frac{6Q}{5G_{rz} h} \tag{23}$$

where δ_z, δ_r, and δ_θ are the respective integrated thermal contractions, E's, ν's, and G are the appropriate moduli.

The effective midsurface strains and curvature changes are expressed in terms of the effective midsurface linear and angular displacements u, w, and β.

$$\varepsilon_{zm} = \frac{du}{dz} \tag{24}$$

$$\varepsilon_{\theta m} = \frac{w}{R} \tag{25}$$

$$\gamma_{rzm} = \beta + \frac{dw}{dz} \tag{26}$$

$$\kappa_{zm} = \frac{d\beta}{dz} \tag{27}$$

$$\kappa_{\theta m} = 0 \tag{28}$$

NUMERICAL IMPLEMENTATION AND RESULTS

The stress resultants and couples, and the effective midsurface displacements and rotation entering the basic system of equations derived above, vary only with the axial coordinate z. This implies that the system of equations (15) - (17) and (19) - (28) can be reduced to a single sixth order ordinary differential equation to which closed form complementary solutions can be found. Some difficulties might be expected in the derivation of the partial solution to that resolvent equation. This primarily depends on the possibility of an accurate analytical approximation of the magnetic field distributions along the inner and outer surfaces of the winding. We will discuss the analytical solutions to the basic systems of equations in a future publication [5]. The results presented in this paper have been obtained numerically.

A computer code to implement the finite difference solution of the equilibrium equations (15) - (17) and the constitutive equations (19) - (28), with appropriate static or geometric boundary conditions at the winding ends $z = \pm b$ (b is one half of the winding axial length), has been developed. The data input to run the code is minimal. It includes the inner and the outer radii of the solenoid, the winding material mechanical and thermal properties, a tabulated distribution of the axial and radial components of the magnetic field along the inner and outer surfaces, tabulated or computed (if interaction with other structures is involved) values of surface radial pressures, and six logical constants specifying the boundary conditions at the winding ends.

Compared to a finite element analysis the CPU time required for the code execution, is neglibibly small.

The case presented here as an example of a three-dimensional mechanical analysis of a superconducting solenoid, is one of the NbTi sections of the 14 T NMR magnet, currently under development at FBNML. The magnet consists of three Nb_3Sn high field inserts and five medium and low field NbTi coils. The coil under discussion is the main NbTi section, with the dimensions

$$r_{in} = 0.09 \text{ m}, \quad r_{out} = 0.11 \text{ m}, \quad b = 0.25 \text{ m}$$

The average current density in this tentative design is 98 MA/m^2. The distribution of the axial and radial components of the magnetic field along the cylindrical surfaces of the winding is given in Table 1.

TABLE 1

Field distribution along the cylindrical surfaces

	r	z/b				
	(m)	0.000	0.250	0.500	0.750	1.000
B_z (T)	0.09	8.046	8.217	8.382	7.874	5.588
	0.11	5.503	5.782	6.248	6.105	4.255
B_r (T)	0.09	0.000	-0.050	0.438	2.154	3.409
	0.11	0.000	-0.134	0.341	1.846	4.590

Two cases of the boundary conditions were considered at the winding ends. In the first case the winding ends were free of any constraints, in the second case the radial and axial movements of the ends were constrained.

The results of the three-dimensional mechanical analysis for both types of boundary conditions are presented in Figure 2. It shows the variation of the circumferential and radial stresses at the inner and outer radii of the winding, along the axis. In both types of boundary conditions substantial edge effects in both the circumferential and radial stress can be observed. As could have been expected the edge effect is much stronger in the case of the restricted displacements at the end of the magnet.

The coil under discussion is a relatively long solenoid. This explains the fact that along one half of the magnet length, on both sides of the midplane the stresses are not sensitive to the change of the boundary conditions at the ends.

In the regions close to the ends transverse shear stresses are present in both cases, and have approximately the same order of magnitude as the corresponding radial stresses. Again, in the case of restricted displacements a strong edge effect in the distribution of the shear stress is observed. Its range is about two times wider than in the case of unrestricted displacements.

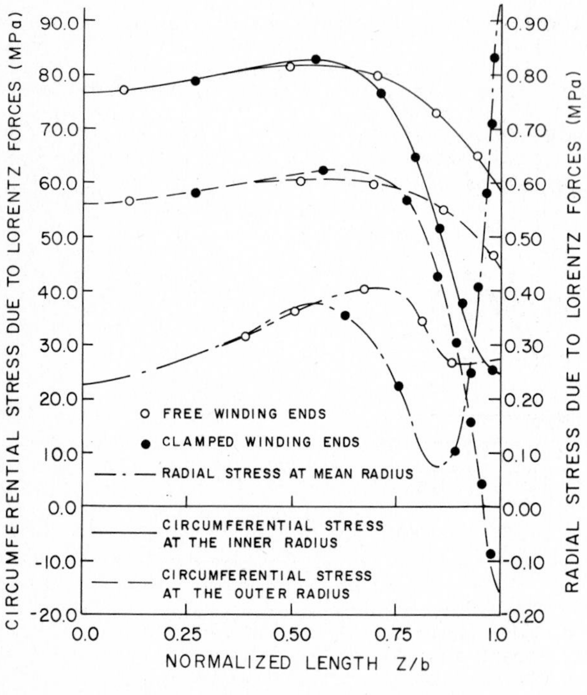

Figure 2

Circumferential and radial stresses generated by the radial Lorentz forces, for two types of boundary conditions at the coil ends.

In Figure 3, we compare the distributions of the circumferential and radial stresses across the build of the coil at the midplane, obtained by two different analyses. The curves shown in solid lines have been generated by the solution developed in this paper. The broken lines illustrate the results obtained from the exact two-dimensional solution [1]. This comparison indicates that in relatively long solenoids, far from the coil ends, the effect of transverse shear stress on the coil mechanical beharior is negligible, and the two-dimensional exact solutions are applicable.

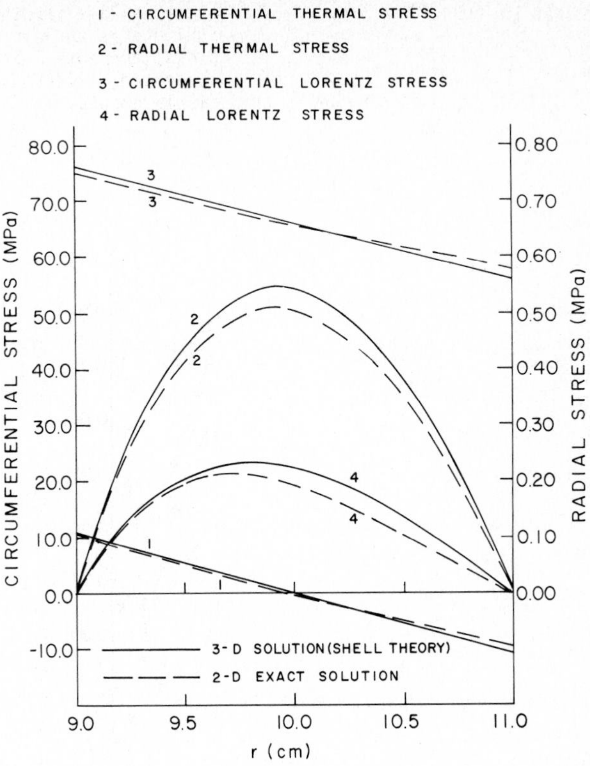

Figure 3

Comparison of thermal and electromagnetic stress distributions across the build of the coil, obtained by two analysis methods.

REFERENCES

[1] Bobrov, E.S., Williams, J.E.C., Stresses in superconducting solenoids, in: Moon, F.C. (ed.), Mechanics of superconducting structures (ASME, New York, 1980).

[2] Arp, V., Stresses in superconducting solenoids, J. Appl. Phys. 48 (1977) 2026 - 2036.

[3] Trefftz, E.G., Ableitung der Schalenbiegungsgleichungen mit dem Castigliano-schen Prinzip, Z. Angew. Math. Mech. 15 (1935) 101 - 108.

[4] Reissner, E., On some problems in shell theory, in: Goodier, J.N. and Hoff, N.J. (eds.), Structural Mechanics, Proc. First Symp. Naval Structural Mechanics, (Pergamon Press, New York, 1960).

[5] Bobrov, E.S., Analytical solutions to the problem of three-dimensional mechanical behavior of superconducting solenoids, to be published.

The Mechanical Behavior of Electromagnetic Solid Continua
G.A. Maugin (editor)
Elsevier Science Publishers B.V. (North-Holland)

MAGNETOELASTIC STABILITY OF SUPERCONDUCTING SHELLS IN A MAGNETIC FIELD

G.E. Bagdasarian

Chief of Magnetoelasticity Department, Institute of Mechanics, Armenian Academy of Sciences, Yerevan, Armenia
USSR

Based in special behaviour of supercunducting shells in a magnetic field both perturbance motion equations and corresponding boundary conditions from magnetoelasticity non-linear equations and boundary conditions are deduced. By means of formulated boundary problem for cylindrical and spherical shells both statical and dynamical stability problems are considered as for stationary and so non-stationary initial magnetic field.

1. GENERAL EQUATIONS AND BOUNDARY CONDITIONS

Let the isotropic thin shell with uniform thickness $2h$ be in the non-uniform magnetic field $\vec{H}_0$. It is supposed that the shell is made of elastic material and covered with thin layer of superconducting alloy. Elastic properties of the shell material are characterised by elasticity E, Poisson's ratio ν and density ρ. The shell is in the vacuum and is considered in the orthogonal system $\alpha_1, \alpha_2, \alpha_3$ so that the coordinate axes α_1, α_2 coincides with the lines of principal curvatures of middle surface.

It is known, that when the superconducting body placed in a magnetic field, on its thin surface layer there arise screening currents preventing the magnetic field penetration into the body. Pushing-out of magnetic field leads to a change of magnetic field intensity in the region outside the body. This change of intensity takes place due to the superposition of initial magnetic field $\vec{H}_0$ and the magnetic field $\vec{H}^\circ$ of screening currents. Due to it non-perturbance magnetic field $\vec{H} = \vec{H}_0 + \vec{H}^\circ$ is obtained from solution of the following magnetostatic problem in the outside region [1]

$$\mathrm{rot}\,\vec{H}^\circ = 0, \quad \mathrm{div}\,\vec{H}^\circ = 0; \quad \vec{n}_0 \cdot \vec{H}\,\big|_S = 0, \quad \lim_{r\to\infty} \vec{H}^\circ = 0 \tag{1.1}$$

where $\vec{n}_0$ is the unit vector of outside normal to non-deformed surface S, $\vec{r}$ is the radius-vector of the considered point.

Since the magnetic field does not penetrate into the shell, on shell surface Maxwell's tensor components have discontinuities. Due to these discontinuities there appears a magnetic pressure $\vec{P}_0$ which is determined by [1]

$$\vec{P}_0 = -(8\pi)^{-1} \vec{H}^2 \vec{n}_0 \tag{1.2}$$

Under the action of $\vec{P}_0$ the initial non-perturbance state arises in the shell which is characterized by displacement vector $\vec{u}_0$ and elastic stress tensor $\hat{\sigma}_0$. The above-mentioned values are determined from the following linear equations of equilibrium and boundary conditions

$$\operatorname{div}\hat{\sigma}_0 = 0\,; \quad \vec{n}_0\cdot\hat{\sigma}_0\big|_S = \vec{P}_0\big|_S \tag{1.3}$$

The characteristics of perturbance motion ($\vec{u}_0+\vec{u}$, $\hat{\sigma}_0+\hat{\sigma}$, $\vec{P}_0+\vec{P}$, $\vec{H}+\vec{h}$) have to satisfy the nonlinear equations and boundary conditions on the shell deformed surface. Assuming perturbancies to be small, the above-mentioned equations and boundary conditions are linearised according to [2,3]. As a result we have the following linear equations of perturbance motion [4]:
in the shell body

$$\operatorname{div}[\hat{\sigma}+\hat{\sigma}_0(\nabla\vec{u})^*] = \rho\ddot{\vec{u}}, \tag{1.4}$$

outside the shell body

$$\operatorname{rot}\vec{h} = 0, \quad \operatorname{div}\vec{h} = 0 \tag{1.5}$$

and linearised boundary conditions on the surface S

$$\vec{n}_0\cdot\hat{\sigma} = \vec{P}, \tag{1.6}$$

$$\vec{n}_0\left[\vec{h}+\vec{H}(\nabla\vec{u})^*\right] = 0 \tag{1.7}$$

Here

$$\hat{\sigma} = E(2+2\nu)^{-1}\left[2\nu(1-2\nu)^{-1}\hat{E}(\operatorname{div}\vec{u})+\nabla\vec{u}+(\nabla\vec{u})^*\right] \tag{1.8}$$

$$P = \vec{n}_0\cdot\hat{T}, \quad T_{ik} = (4\pi)^{-1}(H_i h_k + H_k h_i - \delta_{ik}\vec{H}\cdot\vec{h}) \tag{1.9}$$

where $\vec{E}$ is the unit tensor, ∇ is the Hamilton's operator, $(\nabla\vec{u})^*$ is the transposed tensor $\nabla\vec{u}$, $\hat{T}$ is Maxwell's tensor for perturbance state. Using of Kirchoff-Love's hypothesis and proceeding as in the classical shell theory from (1.4) and taking into account (1.6)-(1.9), we will obtain the system of differential equations of the shell stability in respect to displacements u, v, w of shell middle surface. In particular for the cylindrical shell we have [4]

$$\partial_1(\partial_1 u+\nu_1\partial_2 v)+\nu_2\partial_2(\partial_2 u)+\nu R^{-1}\partial_1 w - D_1\left[R^{-1}T_2^0\partial_1 w + \right.$$
$$\left. +N_1^0\partial_1(\partial_1 w)+N_2^0\partial_1(\partial_2 w)-(4\pi)^{-1}H_1^+(H_1^+\partial_1 w+H_2^+\partial_2 w)\right] = \rho_1\ddot{u},$$

$$\partial_2(\partial_2 v+\nu_1\partial_1 u)+\nu_2\partial_1(\partial_1 v)+R^{-1}\partial_2 w - h^2(3R)^{-1}\partial_2(\Delta w +$$
$$+R^{-2}w)+D_1\left[R^{-1}S^0\partial_1 w - N_1^0\partial_1(\partial_2 w)-N_2^0\partial_2(\partial_2 w)-R^{-2}N_2^0 w+\right.$$
$$\left.+(4\pi)^{-1}H_2^+(H_1^+\partial_1 w+H_2^+\partial_2 w)\right] = \rho_1\ddot{v},$$

$$D\left[\Delta^2 w+\nu R^{-2}\partial_1(\partial_1 w)+R^{-2}\partial_2(\partial_2 w)+3(Rh^2)^{-1}(\partial_2 v+\nu\partial_1 u+R^{-1}w)\right]+$$
$$+2\rho h\ddot{w}-T_1^\circ\partial_1(\partial_1 w)-2S^\circ\partial_1(\partial_2 w)-T_2^\circ\partial_2(\partial_2 w)-R^{-2}T_2^\circ w+(4\pi)^{-1}(H_1^+h_1^+ + \qquad (1.10)$$
$$+H_2^+h_2^+)+(4\pi)^{-1}h\{\partial_1[H_1^+(H_1^+\partial_1 w+H_2^+\partial_2 w)]-\partial_2[H_2^+(H_1^+\partial_1 w+H_2^+\partial_2 w)]\}=0\ ,$$

$$\nu_1=\frac{1+\nu}{2}\ ,\ \nu_2=\frac{1-\nu}{2}\ ,\ D=\frac{2Eh^3}{3(1-\nu^2)}\ ,\ D_1=\frac{1-\nu^2}{2Eh}\ ,\ \rho_1=\rho\frac{1-\nu^2}{E}\ .$$

Here R is the radius of shell's middle surface, Δ is the two-dimensional Laplace's operator; $T_1^\circ, T_2^\circ, S^\circ, N_1^\circ, N_2^\circ$ are stresses, which determine the initial non-perturbance state of the shell and are solutions of the problem (2.3). By means of index "+" the corresponding values on the shell surface $\alpha_3 = h$ are noted.

In the equations (1.10) there are unknown boundary values the tangential components h_1^+ and h_2^+ of induced magnetic field on the shell surface $\alpha_3=h$. These unknown values have to be determined from the equations (1.5), condition (1.7) and from the conditions of perturbances damping at infinity. According to Kirchhoff--Love's hypothesis the condition (1.7) have the form

$$h_3^+ = H_1^+\partial_1 w + H_2^+\partial_2 w \qquad (1.11)$$

By means of potential function φ

$$\vec{h} = \operatorname{grad}\varphi \qquad (1.12)$$

the determination of perturbance magnetic field $\vec{h}$, according to (1.5) and (1.12), is reduced to the solution of the following Neiman's external problem in the domain $\alpha_3>h$

$$\Delta_1\varphi = 0\ ;\qquad \left.\frac{\partial\varphi}{\partial\alpha_3}\right|_{\alpha_3=h} = h_3^+(\alpha_1,\alpha_2,t) \qquad (1.13)$$

Here Δ_1 is the three-dimentional Laplace's operator.

2. CYLINDRICAL SHELL STATIC STABILITY

Let us consider the stability problem for superconducting cylindrical shell of infinite length in the uniform magnetic field

$$\vec{H}_0 = H_0(\vec{e}_2\sin\theta - \vec{e}_3\cos\theta)\ ,\quad \theta R=\alpha_2 \qquad (2.1)$$

where $\vec{e}_2$ and $\vec{e}_3$ are unit vectors along the coordinate axes α_2 and α_3.

The additional magnetic field of surface currents $\vec{H}^\circ$ is determined from the solution of (1.1) problem. Using this solutions from (1.3) for non-perturbance state strength according to (2.1) we obtain

$$T_2^\circ = -RH_0^2(12\pi)^{-1}(3+\cos 2\theta)\ ,\ \mathcal{N}_2^\circ = RH_0^2(6\pi)^{-1}\sin 2\theta \qquad (2.2)$$

Henceforth we shall consider a particular case when perturbancies don't depend on α_1. Then the solution of (1.13) is expressed by Dini's integral. Using this integral from (1.12) we can determine $\overset{+}{h}_2$

Substituting (2.1), (2.2) and the obtained expression for $\overset{+}{h}_2$ into (1.10) the stability problem under consideration is reduced to solution of the system of integro-differential equations in respect to displacements v, w. Expressing the solution of this system in the form

$$v = \sum_{n=2}^{\infty} v_n \sin n\theta\ ,\quad w = \sum_{n=2}^{\infty} w_n \cos n\theta$$

and using the common method of orthogonolization we obtain the regular infinite system of uniform linear algebraic equations in respect to unknown coefficients v_n, w_n. From the condition of the existence of non-trivial solution of this system we obtain the following expression for the critical value of the external magnetic field

$$H_{0*} = 5hR^{-1}(2\pi Eh)^{1/2}\left[3R(1-\nu^2)\right]^{-1/2} \qquad (2.3)$$

3. DYNAMICAL STABILITY OF CYLINDRICAL SHELL

Let us consider an infinite cylindrical shell along which alternating electrical current with linear electrical density

$$\vec{I} = (I_0 + I_1\cos\omega t)\vec{e}_1 \qquad (3.1)$$

is flowed.

Interaction of current (3.1) with its own magnetic field originates the magnetic pressure $\vec{P}_0$. Under the action of $\vec{P}_0$ the initial membrane stress-state with strength T_2°

$$T_2^\circ = -2\pi RC^{-2}(I_0 + I_1\cos\omega t)^2\ ,\quad c = 3\cdot 10^{10}\ \mathrm{cm/sec} \qquad (3.2)$$

arises in the shell.

Expressing solution of (1.10) in the form

$$v = \varphi_n(t)\sin n\theta\ ,\quad w = f_n(t)\cos n\theta \qquad (3.3)$$

and taking into account (3.2) and the obtained values $\overset{+}{h}_2$, we get differential equations for determining the unknown functions $f_n(t)$

$$\ddot{f}_n(t) + \Omega_n^2(1 - 2\mu_{1n}\cos\omega t - 2\mu_{2n}\cos 2\omega t)f_n(t) = 0 \qquad (3.4)$$

$$\Omega_n^2 = \Omega_{on}^2 - \gamma_n^2(2c^2)^{-1}(I_1^2+2I_o^2)\,,\quad \gamma_n^2 = \pi(n-1)^2(\rho hR)^{-1}$$

$$\mu_{1n} = \gamma_n^2(c\Omega_n)^{-2}I_oI_1\,,\quad \mu_{2n} = \gamma_n^2(2c\Omega_n)^{-2}I_1^2$$

$$\Omega_{on}^2 = D(n^2-1)^2(2\rho hR^4)^{-1}$$

From (3.4), according to [5], the boundaries of principal parametric resonance are determined by following expressions:

for the region near to the frequency $2\Omega_n$

$$\omega_* = 2\Omega_{on}I_{o*}^{-1}(I_{o*}^2 - I_o^2 - 2^{-1}I_1^2 \pm I_oI_1)^{1/2} \tag{3.5}$$

for the region near to the frequency Ω_n

$$\omega_* = \Omega_{on}I_{o*}^{-1}(I_{o*}^2 - I_o^2 - 2^{-1}I_1^2 \pm 4^{-1}I_1^2)^{1/2} \tag{3.6}$$

In (3.5), (3.6) I_{o*} is the critical value of stationary current $I_o\vec{e}_1$ beyond which the shell is statically unstable [6]

$$I_{o*} = c\Omega_{on}\gamma_n^{-1} \tag{3.7}$$

4. STATICAL STABILITY OF SPHERICAL SHELL

Let us consider the stability problem of the superconducting spherical shell in non-uniform magnetic field $\vec{H}_o$. This field arises from two current-carrying rings between which the shell is placed. Such a system of currents may fix the shell in space without any contact.

Proceeding as in the paragraph 2 for the critical values of the ring direct currents J_i (i = 1,2) we obtain

$$J_{i*} = \frac{2c}{3a^2}\left(\frac{2Eh}{\pi R}\right)^{1/2}(a^2+z^2)^{3/2}\left[B_{nk} - (-1)^i\frac{3\gamma(a^2+z^2)}{8zE\,B_{nk}}\right] \tag{4.1}$$

where a is the radius of the rings, z is the distance from centre of the rings to sphere's centre; γ is the mass of the unit volume of shell material, and

$$B_{nk}^2 = (\lambda_n - 2)\left[1+\delta^2(\lambda_n - 1+\nu)A_{kn}\right]^{-1}$$

$$\delta^2 = h^2\left[3R^2(1-\nu^2)\right]^{-1},\quad \lambda_n = n(n+1)\,,$$

$$A_{kn} = \left[(2n-1)(2n+3)\right]^{-1}\left\{4(1+\nu)(\lambda_n-1+\nu)^{-1}\left[2n^4+4n^3-2n-k^2(2n^2+2n-3)\right]+\right.$$

$$\left.+\left[n(2n+1)\right]^{-1}\left[2n^6-11n^5-20n^4-7n^3-8n^2+8n+k^2(10n^4+31n^3+n^2-30n+12)\right]\right\}$$

$$(n=2,3,4,\dots;\ k \leq n)$$

One of the formulae of (4.1) is obtained from the stability condition, and the other one is obtained from the condition of the non-contact fixing of the shell.

REFERENCES:

[1] LANDAU, L. and LIFSHITZ, E., Electrodynamics of Continium Media (Gostekhizdat, Moscow, 1957).

[2] NOVOZHILOV, V.V., Foundations of Nonlinear Theory of Elasticity (Gostekhizdat, Moscow, 1948).

[3] AMBARTSUMIAN, S.A., BAGDASARIAN, G.E., BELUBEKIAN, M.V., The Magneto-elasticity of Thin Shells and Plates (Science, Moscow, 1977).

[4] BAGDASARIAN, G.E. and MKRTCHIAN, P.A., Stability of a superconducting cylindrical shell in magnetic field. Izv., AN Arm. SSR, Mekhanika, v.34, No. 6 (1981) 36-48.

[5] BOLOTIN, V.V., Dynamic Stability of Elastic Systems (Gostekhizdat, Moscow, 1956).

[6] OVAKIMIAN, R.N., About the stability of cylindrical current--carrying of the infinite conductance, Izv., AN Arm.SSR, Mekhanika, v.22, No.4 (1969), 59-67.

The Mechanical Behavior of Electromagnetic Solid Continua
G.A. Maugin (editor)
Elsevier Science Publishers B.V. (North-Holland)

THE INFLUENCE OF FINITE SPECIMEN DIMENSIONS ON THE MAGNETO-ELASTIC BUCKLING OF A CANTILEVER

A.A.F. van de Ven

Eindhoven University of Technology
Department of Mathematics and Computing Science
Eindhoven
The Netherlands

The magnetoelastic buckling of a soft ferromagnetic and elastic beam having a wide rectangular cross-section is discussed. The magnetic fields inside the rectangular cross-section are approximated by the corresponding ones for an elliptic cross-section. These fields are used in the derivation of a buckling criterion, which accounts for the finite width of the beam. The results of this approach are compared with experimental data as given by a.o. Moon and Pao, and good agreement is found. It is thus shown that the influence of the finite width of the beam on the buckling load is essential.

1. INTRODUCTION

The magnetoelastic buckling of a ferromagnetic beam or plate inserted into a magnetic field has been a subject of several studies, both theoretical and experimental, during the past decade. The first contribution was given by Moon and Pao, [1], in 1968, who treated the case of a cantilevered beam-plate placed in a uniform transverse magnetic field. They deduced a buckling criterion under the assumption that the plate was infinitely wide and long. Moon and Pao found a great discrepancy between their theoretical predictions and experimental results (carried out, of course, on finite specimens). Several investigators have tried to explain this discrepancy ([2]-[8]). In [4], [5] and [7] further experimental data are given, which all show about the same order of deviation between theoretical and experimental buckling values. All of these treatments inevitably lead to one definite conclusion (which will be justified by the results of this paper), viz. the above mentioned discrepancy must be due to the finite width (and, eventually, finite length) of the beam.

The influence on the buckling value of the field distortion due to the finite width of the specimen was already mentioned in [2] and [3], but this effect was for the first time investigated on a more fundamental basis (using finite-element analysis) by Miya, Hara and Someya, [7]. In a subsequent paper, [8], Miya, Takagi and Ando also took into account the finite length of the beam. They found a correction in the basic field strength B_0 occurring in Moon's buckling criterion. A shortcoming of this analysis is that Miya et al calculated only the magnetic fields for the beam in its unbuckled shape (i.e. the rigid-body fields). However, for a consistent buckling theory, also the perturbations in the magnetic fields due to the deflection of the beam are needed. The calculation of the latter fields for a finite rectangular cross-section is a tremendously difficult problem. To overcome these difficulties we introduce as an auxiliary problem the calculation of the magnetic fields (rigid-body and perturbations) for a beam having a finite elliptic cross-section (the length of the beam is still assumed infinite). The latter problem can be formulated and solved in a completely analytical way. These calculations will be published in a forthcoming paper [9]. The results of this study are used here as approximations for (the mean values of) the field strength in the corresponding (i.e. same width and thickness) rectangular cross-section. This method yields an estimate for the stresses at the beam-surface and, hence, for the magnetic load on the beam-plate.

In the next section this procedure is evaluated, resulting into a buckling criterion for the cantilevered beam having a rectangular cross-section, in which the influence of the finite width is taken into account. This evaluation is presented here only very briefly; for more detailed derivations we refer to [9], or to [10] and [11] in which analogous treatments are described.

In the final section the results of this approach are compared with experimental data available from literature (especially [1]). This comparison shows that our theoretically predicted buckling values are in very good (both qualitative and quantitative) agreement with experiments.

2. BUCKLING CRITERION

A slender soft ferromagnetic elastic beam having a wide rectangular cross-section (length ℓ, width $2a$, thickness $2b$; $\ell >> a >> b$) is placed in a uniform magnetic field $\underline{B}_0$. The beam is built in at one side (cantilever) and the field is normal to the wide face of the beam. A cartesian coordinate system is chosen with the z-axis along the central line of the beam and with the y-axis (the short axis) coinciding with the $\underline{B}_0$-direction.

When the external field strength B_0 reaches a critical value the beam will suddenly deflect (buckle). In its deflected state the beam is acted upon by surface stresses of magnetic origin, which are given by (cf. [10], where it is also shown that the magnetic body force may be neglected)

$$t_{yy} = \pm \frac{\mu^2}{4\pi} \frac{\partial \Phi^{(0)}}{\partial y} \frac{\partial \varphi}{\partial y}\bigg|_{y=\pm b} . \tag{2.1}$$

Here t_{yy} is the normal stress on the upper or lower surface of the beam and μ is the magnetic permeability ($\mu >> 1$ for soft ferromagnetic materials). Moreover, $\Phi^{(0)}$and φ are magnetic potentials (both satisfying the Laplace equation) referring to the rigid body state and to the perturbations on that state, respectively. These perturbations are due to the deflections of the beam. By integrating t_{yy} over the boundary of the cross-section we find the normal load per unit of length $q(z)$ acting on the beam, which serves as load parameter in the classical beam equation

$$EIV^{iv}(z) = q(z) , \qquad (I = \tfrac{4}{3}ab^3) , \tag{2.2}$$

where E is Young's modulus and $V(z)$ represents the deflection of the beam. When this beam equation, together with the boundary conditions for a cantilever, has a non-trivial (non-zero) solution (recall that $q(z)$ is proportional to $V(z)$) we say that B_0 has reached its critical, or buckling, value.

What now still remains to be done is to construct a solution for the magnetic potentials $\Phi^{(0)}$ and φ. As said before, for a finite rectangular cross-section this is an extremely difficult problem. Therefore, we have replaced the rectangular cross-section by the corresponding elliptic one. Moreover, we have assumed the beam infinitely long. The latter problem can completely be solved in a closed form. This is done in an other (forthcoming) paper [9], and we shall use here the results of that paper.

As concerns the rigid-body potential $\Phi^{(0)}$, it turns out that the influence of the finite width of the cross-section is small (i.e. $O(b/a)$) and in fact reduces to zero together with the thickness-to-width ratio. In [9], we have derived that

$$\Phi^{(0)} = -\frac{(1+\beta)}{\mu} B_0 y \,, \qquad (\beta = \frac{b}{a} << 1) \,. \tag{2.3}$$

On the other hand, the influence of the finite width on the perturbed potential is essential, and does remain so even for $\beta \to 0$.

Provided that $V(z)$ satisfies (apart from an irrelevant rigid-body translation)

$$V''(z) + \lambda^2 V(z) = 0 \,, \tag{2.4}$$

where for the cantilever ($V'(0) = V''(\ell) = 0$) the parameter λ is given by

$$\lambda = \pi/2\ell \,, \tag{2.5}$$

the Laplace equation for φ for the elliptic cross-section can be solved by separation of variables, according to

$$\varphi = \Phi(u,v)V(z) \,, \tag{2.6}$$

where u and v are elliptic coordinates.
For slender beams, that is for

$$0 < \varepsilon := \frac{\pi a}{4\ell} << 1 \,, \tag{2.7}$$

the following solution for $\Phi(u,v)$ is obtained in [9]

$$\Phi(u,v) = -\frac{B_0(1 + O(\varepsilon^2))}{1 + 2\beta\mu\varepsilon^2\kappa} \left[1 + \tfrac{1}{2}(1-\beta^2)\varepsilon^2 \cosh 2u\right] \,, \tag{2.8}$$

where

$$\kappa = -\gamma - \ln[(1+\beta)\varepsilon/2] \,, \qquad (\gamma = 0.5772) \,. \tag{2.9}$$

In (2.8) the term $(2\beta\mu\varepsilon^2\kappa)$ is maintained, because μ is so large that, even for $\varepsilon << 1$ and $\beta << 1$, this term is not negligible with respect to unity; moreover the $O(\varepsilon^2)$-coefficient of $\cosh 2u$ is maintained because we are especially interested in the normal derivative $(\partial\Phi/\partial u)$ rather than in Φ itself. From (2.8) we deduce that along the boundary (given by $u = u_1 = \tfrac{1}{2}\ln[(1+\beta)/(1-\beta)]$)

$$\mu\frac{\partial\Phi}{\partial u}\bigg|_{u_1} = -\frac{B_0}{\kappa\Lambda}(1 + O(\varepsilon^2)) \,, \tag{2.10}$$

where

$$\Lambda = 1 + (2\mu\beta\varepsilon^2\kappa)^{-1} \,. \tag{2.11}$$

Hence, the normal derivative of the perturbed potential is uniform (up to $O(\varepsilon^2)$) along the boundary of the elliptic cross-section.

We now proceed as follows. We use (2.10) as an approximation for the normal derivative of the perturbed potential for the rectangular cross-section. In doing so, we obtain from (2.1), (2.3) and (2.10) the following expression for q(z)

$$q(z) = 2a\, t_{yy}\Big|_{y=-b}^{b} = \frac{a\mu}{\pi}\frac{\partial\Phi^{(0)}}{\partial y} \cdot \frac{\mu}{a}\frac{\partial\varphi}{\partial u} V(z) = \frac{(1+\beta)B_0^2}{\pi\kappa\Lambda} V(z) . \tag{2.12}$$

Furthermore, as a consequence of (2.4) we have

$$V^{iv}(z) = -\lambda^2 V''(z) = \lambda^4 V(z) . \tag{2.13}$$

Substitution of (2.12) and (2.13), with λ according to (2.5), into the beam equation (2.2) yields

$$\frac{2\pi a}{3\ell}\left(\frac{\pi b}{2\ell}\right)^3 EV(z) = \frac{(1+\beta)B_0^2}{\pi\kappa\Lambda} V(z) . \tag{2.14}$$

Equation (2.14) implies that B_0 reaches its critical value when

$$\left(\frac{B_0}{\sqrt{4\pi E}}\right)_{cr} = \sqrt{\frac{2\varepsilon\kappa\Lambda}{1+\beta}}\ \sqrt{\frac{1}{3}\left(\frac{\pi b}{2\ell}\right)^3} . \tag{2.15}$$

The second square root in the right-hand side of (2.15) represents the buckling value as obtained by Moon and Pao, [1]. In the next section, the criterion (2.15) will be compared with the theoretical and experimental results of Moon and Pao, and others.

3. CONCLUSIONS

In order to get a first rough estimate of the behaviour of the solution (2.15) under varying dimension ratios (b/ℓ), ε and β we subsequently

neglect β with respect to unity,
assume μ to be so large that $\Lambda \approx 1$, and, finally,
discard the influence of ε and β on the value of κ .

With these approximations (2.15) reduces to

$$\left(\frac{B_0}{\sqrt{4\pi E}}\right)_{cr} = \sqrt{2\kappa\varepsilon}\ \sqrt{\frac{1}{3}\left(\frac{\pi b}{2\ell}\right)^3} . \tag{3.1}$$

Since only two out of the three dimension ratios (b/ℓ), ε and β are independent, we have to make a choice which parameter besides (b/ℓ) is taken as the second independent variable.

Firstly, we consider B_{0cr} as function of (b/ℓ) at fixed values of ε. In that case (3.1) shows that B_{0cr} is proportional to the 3/2 power of the thickness-to-length ratio, as was also found by Moon and Pao, [1]. However, unlike [1], the coefficient in (3.1) depends on the width-to-length ratio ε. Moreover, the numerical value of the coefficient in (3.1) is essentially different from the one according to [1] (i.e. $\sqrt{2\kappa\varepsilon} \neq 1$). This leads to a substantially improved agreement between the theoretical buckling values and the experimental data of [1] (see Table 1).

If, on the other hand, β is chosen as independent variable we have to replace in (3.1) ε by

$$(1/2\beta)(\pi b/2\ell) \ ,$$

yielding

$$\left(\frac{B_0}{\sqrt{4\pi E}}\right)_{cr} = \sqrt{\frac{\kappa}{\beta}}\ \sqrt{\frac{1}{3}\left(\frac{\pi b}{2\ell}\right)^4} = \sqrt{\frac{\kappa}{3\beta}}\left(\frac{\pi b}{2\ell}\right)^2 . \tag{3.2}$$

Hence, at fixed β values B_{0cr} is proportional to the second power (instead of the 3/2 one) of the thickness-to-length ratio. In this respect we refer to [1], Fig. 6, and especially to the experimental results for the HY-MU-80-samples (which have fixed β values and $\mu = 6 \times 10^4$). Indeed, the latter data can be fitted with more accuracy by a second power law than by a 3/2 power relation.

Some further conclusions that can be drawn from (2.15) are listed below.

i) At fixed thickness-to-length ratio(b/ℓ) the buckling value B_{0cr} increases with increasing width-to-length ratio ε. This statement receives experimental confirmation from Dalrymple, Peach and Viegelahn [4]. In [4], Fig. 4, the variation of the critical buckling field with the plate width (at constant thickness and length) is shown. This experimental curve corresponds, at least qualitatively, very well with our criterion (2.15) (or its simplified version (3.1)). A complete quantitative comparison is not possible, because not all of the needed numerical values for the material coefficients (e.g. E and μ) are given in [4]. However, when these values are chosen such that in one point the theoretical buckling field is equal to the corresponding experimental field of [4], the experimental data of [4] can be fitted with good accuracy by formula (2.15) over the complete range.

ii) At fixed thickness-to-length ratio (b/ℓ) the buckling value B_{0cr} decreases with increasing thickness-to-width ratio β. Again this statement receives experimental confirmation from literature; this time from Miya, Hara and Someya (cf. [7], Fig. 3). In addition we note that the results of the latter paper also support statement i).

iii) The influence of the permeability μ on the buckling value B_{0cr} decreases with increasing thickness-to-length ratio (b/ℓ). Calculations of B_{0cr} based on (2.15) for $\beta = 5.6 \times 10^{-2}$ show a variation of about 30 percent in the buckling value as μ runs from 10^4 to 10^6 at a thickness-to-length ratio of

$$b/\ell = 10^{-3} \ ,$$

while the corresponding variation is only 7 percent if

$$b/\ell = 3 \times 10^{-3}$$

We conclude by giving a table in which for fixed values of β and μ and varying (b/ℓ) our theoretical buckling values are compared with some of the theoretical and experimental ones of Moon and Pao (cf. [1], esp. Fig. 6).

TABLE 1. Theoretical and experimental buckling values.

$\frac{b}{\ell} \times 10^3$	$\left[\left(\frac{B_0}{\sqrt{4\pi E}}\right)_{cr} \times 10^4\right]_{(2.15)}$	$\left[\left(\frac{B_0}{\sqrt{4\pi E}}\right)_{cr} \times 10^4\right]_{[1],Th.}$	$\left[\left(\frac{B_0}{\sqrt{4\pi E}}\right)_{cr} \times 10^4\right]_{[1],Exp.}$
$\beta = 5.6 \times 10^{-2}$ and $\mu = 6 \times 10^4$.			
3	0.96	1.87	0.9 - 1.0
2.5	0.69	1.42	0.65 - 0.70
2	0.46	1.02	0.45 - 0.50
$\beta = 2.4 \times 10^{-2}$ and $\mu = 10^4$.			
1.8	0.55	0.87	0.51
1.5	0.40	0.66	0.30 - 0.33
1.15	0.26	0.44	0.24
0.9	0.18	0.31	0.17 - 0.20

These tables show a good agreement between the theoretical buckling values predicted by (2.15) and the experimental data of Moon and Pao. This agreement is so evident that we believe that this justifies the conclusion that the discrepancy between the theoretical and experimental results of Moon and Pao, et al is substantially due to the incorrectness of the infinite-size assumption. The results of this paper clearly show that the influence of the finite width of the beam on the buckling value is essential and remains so even for very small thickness-to-length ratios.

REFERENCES

[1] Moon, F.C. and Yih-Hsing Pao, Magnetoelastic Buckling of Thin Plates, J.A.M. 35 (1968) 53-58.
[2] Wallerstein, D.V. and Peach, M.O., Magnetoelastic Buckling of Beams and Thin Plates of Magnetically Soft Material, J.A.M. 39 (1972) 451-455.
[3] Dalrymple, J.M., Peach, M.O. and Viegelahn, G.L., Magnetoelastic Buckling of Thin Magnetically Soft Plates in Cylindrical Mode, J.A.M. 41 (1974) 145-149.
[4] Dalrymple, J.M., Peach, M.O. and Viegelahn, G.L., Magnetoelastic Buckling: Theory vs. Experiment, J. of Experimental Mechanics 17 (1976) 26-31.
[5] Popelar, C.H. and Bast, C.O., An Experimental Study of the Magnetoelastic Post-buckling Behavior of a Beam, J. of Experimental Mechanics 13 (1972) 537-542.
[6] Popelar, C.H., Postbuckling Analysis of a Magnetoelastic Beam, J.A.M. 13 (1972) 207-211.
[7] Miya, K., Hara, K. and Someya, K., Experimental and Theoretical Study on Magneto-elastic Buckling of a Ferromagnetic Cantilevered Beam-Plate, J.A.M. 45 (1978) 355-360.
[8] Miya, K., Takagi, T. and Ando, Y., Finite-Element Analysis of Magnetoelastic Buckling of Ferromagnetic Beam-Plate, J.A.M. 47(1980) 377-382.
[9] Ven, A.A.F. van de, Magnetoelastic Buckling of Beams having an Elliptic Cross-Section, (forthcoming).
[10] Ven, A.A.F. van de, Magnetoelastic Buckling of Thin Plates in a Uniform Trans-verse Magnetic Field, J. of Elasticity 8 (1978) 297-312.
[11] Ven, A.A.F. van de, Magnetoelastic Buckling of Magnetically Saturated Bodies, Acta Mechanica 47 (1983).

AUTHOR INDEX

TNA = text not available at the time of print